172
Structure and Bonding

Aims and Scope

The series *Structure and Bonding* publishes critical reviews on topics of research concerned with chemical structure and bonding. The scope of the series spans the entire Periodic Table and addresses structure and bonding issues associated with all of the elements. It also focuses attention on new and developing areas of modern structural and theoretical chemistry such as nanostructures, molecular electronics, designed molecular solids, surfaces, metal clusters and supramolecular structures. Physical and spectroscopic techniques used to determine, examine and model structures fall within the purview of *Structure and Bonding* to the extent that the focus is on the scientific results obtained and not on specialist information concerning the techniques themselves. Issues associatedwith the development of bonding models and generalizations that illuminate the reactivity pathways and rates of chemical processes are also relevant.

The individual volumes in the series are thematic. The goal of each volume is to give the reader, whether at a university or in industry, a comprehensive overviewof an area where new insights are emerging that are of interest to a larger scientific audience. Thus each review within the volume critically surveys one aspect of that topic and places it within the context of the volume as a whole. The most significant developments of the last 5 to 10 years should be presented using selected examples to illustrate the principles discussed. A description of the physical basis of the experimental techniques that have been used to provide the primary data may also be appropriate, if it has not been covered in detail elsewhere. The coverage need not be exhaustive in data, but should rather be conceptual, concentrating on the new principles being developed that will allow the reader, who is not a specialist in the area covered, to understand the data presented. Discussion of possible future research directions in the area is welcomed.

Review articles for the individual volumes are invited by the volume editors.

In references *Structure and Bonding* is abbreviated *Struct Bond* and is cited as a journal.

More information about this series at http://www.springer.com/series/430

D. Michael P. Mingos
Editor

50 Years of Structure and Bonding – The Anniversary Volume

With contributions by

Y. Bian · R.D. Chapman · R. Clérac · X. Duan · D.G. Evans · P.A. Gale · A. Hauser · J. Jiang · R.P. Kelly · T.M. Klapötke · P. Köhler · R. Liang · D.M.P. Mingos · K. Prassides · D. Rabinovich · C. Reber · D. Reinen · P.W. Roesky · Y. Takabayashi · R. Tian · M. Wei · R.E.P. Winpenny · H.-L. Wong · V.W.-W. Yam · M.C.-L. Yeung

Springer

Editor
D. Michael P. Mingos
University of Oxford
Oxford, United Kingdom

ISSN 0081-5993 ISSN 1616-8550 (electronic)
Structure and Bonding
ISBN 978-3-319-35136-0 ISBN 978-3-319-35138-4 (eBook)
DOI 10.1007/978-3-319-35138-4

Library of Congress Control Number: 2016954966

Printed on acid-free paper

This Springer imprint is published by Springer Nature
The registered company is Springer International Publishing AG
The registered company address is: Gewerbestrasse 11, 6330 Cham, Switzerland

Preface

Volume 1 of *Structure and Bonding* was published in 1966 and the Preface suggested that "a valuable service is performed by bringing together up-to-date authoritative reviews from the different fields of modern inorganic chemistry, chemical physics and biochemistry, where the general subject of chemical bonding involves (usually) a metal and a small number of associated atoms. These arrangements are of importance for symmetry considerations by crystallographers, for the application of group theory and molecular orbital theory to chromophores and because of their widespread occurrence in polyatomic molecules, in complex ions in solution, in vitreous materials, in minerals and, not least, in biological materials and organisms. We are especially interested in the role of "the complex metal-ligand" moiety and purposely avoiding the more classical organic chemistry and even organo-metallic chemistry – for which many review journals are already available – and wish to direct attention towards borderline subjects. We are convinced that these borderline areas receive less attention than they justify, academic studies tend too often to be compartmentalized whereas technological interest too often lack sufficient fundamental understanding. We hope that his series may help to bridge the gaps between some of these different fields and perhaps provide in the process some stimulation and scientific profit to the reader".

Fifty years on and 175 volumes later, *Structure and Bonding* still plays a very important role in presenting authoritative reviews on structural and bonding issues in chemistry. To celebrate this significant landmark, we have invited recent editors and authors, who have contributed highly cited and noteworthy volumes and chapters in recent volumes of the series, to write chapters which highlight important new developments in their chosen area and indicate significant future developments. This volume marking the 50th anniversary starts with a historical introduction by myself, which traces the early days of the series and its social and intellectual context, and a summary of the contributions of the members of the editorial board. This is followed by noteworthy chapters by Philip Gale on anion receptors based on organic frameworks, Richard Winpenny on single molecule magnets, Thomas Klotke on recent progress in the development of high energy density materials, David Evans and Xue Duan on layered double hydroxide materials. Peter Roesky has contributed a chapter on sigma bonded metathesis and polymerisation of 1,3-dienes by rare earth complexes, Yasuhiro Takabayashi and Kosmas Prassides on fullerene superconductivity and Dan Rabonovich on the role of scorpionates in synthetic bioinorganic chemistry. The important optical properties of transition metal complexes and their theoretical interpretation are covered in chapters by Jianzhuang Jiang and Yongzhong Bian on phthalocyanine based functional materials, and Vivian Yam has contributed a very detailed chapter on photofunctional molecular materials. Andreas Hauser and Christian Reber discuss the spectroscopic properties and chemical bonding in transition metal complexes. Professor Dirk Reinen was a member of the original editorial board of *Structure and Bonding* and we are very lucky that he has contributed with P Kohler a chapter on ligand field theory and discusses the fascinating colours of solid iron(III) oxides, an area which he first analysed in an early volume of *Structure and Bonding*.

I thank all of these authors very much for their timely contributions and hope that their efforts will stimulate the successful launching of future volumes which will mark the next 50 years in the series' distinguished history.

Oxford, UK D. Michael P. Mingos
December 2015

Contents

Struct Bond (2016) 172: 1–18
DOI: 10.1007/430_2015_196

Published online: 10 November 2015

Structure and Bonding: The Early Days

D. Michael P. Mingos

Abstract In order to understand the launching in 1966 of *Structure and Bonding*, it is necessary to appreciate the factors which contributed to the emergence of inorganic chemistry as an equal branch of chemistry. A variety of social and economic factors contributed to the transformation of inorganic chemistry from an essentially descriptive subject into an intellectual equal of organic and physical chemistry. The aims and distinctive features of *Structure and Bonding* are identified with reference to the initial preface and the composition of the editorial board. The research interests and characteristics of some of the founding editorial board members are introduced and used as a basis for highlighting the important topics which were covered in the initial 50 volumes. Subsequent changes in the character of the journal are reviewed and used to introduce the present anniversary volume.

Keywords Bioinorganic chemistry • Biomineralization • Bioorganometallics • Chelate and macrocyclic effect • Crystal field theory • Diagnostic reagents • Electronic spectral data • Hard and soft acids and bases • Inorganic drugs • Ligand field theory • Magnetism • Metalloenzymes • Metalloproteins • Stability constants

Contents

D.M.P. Mingos (✉)
Inorganic Chemistry Laboratory, Oxford University, South Parks Road, Oxford 1 3QR, UK
e-mail: Michael.mingos@seh.ox.ac.uk

1 Introduction

Jay Labinger in *Up from Generality: How Inorganic Chemistry Finally Became a Respectable Field* [1] has documented the dramatic upsurge in the number of inorganic chemists and their enhanced respectability in the US academic community during the 1950s. He attributed the growth of inorganic chemistry to the following factors:

1. The inadequate recognition of the merits and independence of the field resulting from its reputation as a descriptive sideshow.
2. Key resources were made available from federal sources and industrial laboratories which resulted in greater opportunities for employment, research funding and a greater range of publication outlets. This had its origins in the Manhattan Project started during World War II.
3. The Russian Sputnik project resulted in a large injection of additional federal funding in the 1950s, which resulted in greater research activity in university and industrial laboratories, where inorganic chemists were required for the development of rocket fuels and materials, suitable for space travel.
4. Inorganic chemistry underwent a significant intellectual shift which transformed it from a mainly descriptive and phenomenological branch of chemistry into a more mechanistic and theoretically driven scientific activity. This enhanced its intellectual appeal in the eyes of organic and physical chemists and attracted bright research students.

The resurgence was not limited to the USA, and European and Japanese inorganic chemists also contributed significantly to the growth of the subject. In 1956, Ronald Nyholm moved from Australia to the Chair of Inorganic Chemistry at University College London and proclaimed in his inaugural lecture the coming renaissance of inorganic chemistry [2, 3]. He identified its key drivers as "the growth of theoretical techniques of quantum mechanics to an extent permitting widespread chemical application" and "those new optical, electrical and magnetic techniques of physical measurement by which structure can be investigated". Therefore, it was not surprising that Nyholm was one of the key figures in the launching of *Structure and Bonding* and remained on the editorial board until his untimely death in a tragic car accident in 1971 [4]. Nyholm's vision was supported by Chatt, Emeleus, Schwarzenbach, Malatesta, Bjerrum, Orgel, Hieber, Fisher and Wilkinson (after his return from Harvard to Imperial College), who established creative and internationally recognised groups in Europe capable of attracting postdoctoral workers form the USA as part of the Marshall Plan and Fulbright fellowships. Although the majority of historians of chemistry have concurred with Nyholm's analysis and enthusiasm, there were those who have added that inorganic chemists' contributions to the Manhattan Project and the subsequent expansion of the periodic table after the war not only provided new elements for study, suitable radioisotopes for mechanistic studies and generous federal funding. Furthermore, the growth of molecular biology in the 1950s [5, 6 and references therein], which followed the elucidation using X-ray measurements of the structure of DNA, and the X-ray structural determinations of vitamin B_{12}, myoglobin, haemoglobin and metalloenzymes underscored the importance of metal ions in biological systems.

The most significant interdisciplinary area which emerged in the 1950s was organometallic chemistry, which was brought into focus by the outstanding success of the Zeigler–Natta catalysts for the production of polythene and polypropylene, and the beauty and novelty of ferrocene and its surprising aromatic properties. Bioinorganic chemistry emerged from the realisation that inorganic elements played a crucial role in biological systems. Solid-state chemistry, rebranded as materials chemistry, grew out of the need for new materials for the burgeoning computer industry and narrowed the gap between fundamental science and the commercial applications. Interdisciplinary topics have provided a fertile bed for expanding traditional subjects, but their success encourages a centrifugal force, which increasingly encourages the release of independent subjects. The intellectual glue which initially held these diverging subdisciplines of inorganic chemistry together in the 1960s was ligand field theory, which linked physics and chemistry and provided a theoretical framework for interpreting the spectral and magnetic properties of transition metal ions, the rates of their substitution and redox reactions and the tuning of the relative stabilities of oxidation states by ligand modification [7, 8].

The Preface to Volume 1 of *Structure and Bonding* published in 1966 noted that:

> a valuable service is performed by bringing together up-to-date authoritative reviews from the different fields of modern inorganic chemistry, chemical physics and biochemistry, where the general subject of chemical bonding involves (usually) a metal and a small number of associated atoms. These arrangements are of importance for symmetry considerations by crystallographers, for the application of group theory and molecular orbital theory to chromophores and because of their widespread occurrence in polyatomic molecules, in complex ions in solution, in vitreous materials, in minerals and, not least, in biological materials and organisms. We are especially interested in the role of "the complex metal-ligand" moiety and purposely avoiding the more classical organic chemistry and even organo-metallic chemistry – for which many review journals are already available – and wish to direct attention towards borderline subjects. We are convinced that these borderline areas receive less attention than they justify, academic studies tend too often to be compartmentalized whereas technological interest too often lack sufficient fundamental understanding. We hope that his series may help to bridge the gaps between some of these different fields and perhaps provide in the process some stimulation and scientific profit to the reader.

The founding editors and their primary research interests are summarised below:

C.K. Jørgensen Cyanamid, Geneva, Switzerland – ligand field theory
J.B. Neilands, University of California, Berkeley, USA – metal ions in biology

R.S. Nyholm, University College London, UK – magnetochemistry, co-ordination chemistry, metal–metal bonding
D. Reinen, University of Bonn, Germany – spectroscopic properties of metal ions and ligand field theory
R.J.P. Williams, University of Oxford, UK – stabilities and bonding preferences of complexes, metal ions in biology, protein structures using NMR, biomineralisation

Not surprisingly, the editors' research interests closely reflect the areas which have been associated with the renaissance of inorganic chemistry discussed above. The only surviving member of this distinguished editorial board is Professor Dirk Reinen, who kindly agreed to contribute a chapter to this special volume. It recounts some of his early impressions and also summarises some of his recent spectroscopic results. I was scheduled to meet Professor Bob Williams earlier this year to discuss the background to the setting up of the journal, but he was admitted to the hospital on the day of our planned meeting, and he died soon afterwards. He was a great loss to the subject because he had a unique ability to identify important unanswered questions at the interfaces between chemistry, biology and geology, analyse them with the precision of a physical scientist and articulate his conclusions in a forceful and clear manner. I had looked forward to our meeting because I knew that he would give me a very honest account of the origins of *Structure and Bonding* and the personalities involved.

The following chemists have subsequently served as members of the editorial board – their start dates precede groups of names:

1968: M.J. Clarke, Boston College, USA; J.B. Goodenough, Oxford University, UK; J.A. Ibers, Northwestern University, USA; R. Weiss, University of Strasbourg, France; 1984: D.M.P. Mingos, Oxford University, UK; G.A. Palmer, Rice University, USA; P.J. Sadler, Birkbeck College, UK; 1999: A.J. Bard, University of Texas, USA; P. Day, Royal Institution, London, UK; Prof. T. Kunitake, Kyushu University, Japan; J.-P. Sauvage, University of Strasbourg, France; F. Wudl, University of California, Los Angeles, USA; I.G. Dance, University of New South Wales, Australia; T.J. Meyer, University of North Carolina, USA; H.W. Roesky, University of Göttingen, Germany; A. Simon, Max-Planck – Institut für Festkörperforschung, Germany; 2007; X. Duan, Beijing University of Chemical Technology, P.R. China; G. Parkin, Columbia University, USA; 2008: L.H. Gade, University of Heidelberg, Germany; 2010: K.R. Poeppelmeier, Northwestern University, USA; 2011: F.A. Armstrong, Oxford University, UK; 2012: T.M. Takano, Kyoto University, Japan

2 Early Review Chapters

2.1 Ligand Field Theory

Bethe [7], a physicist, developed an ionic model (crystal field theory) in the 1920s to account for the characteristic spectra in the visible region of transition metal ions in terms of the splittings of d orbitals in the octahedral and tetrahedral ligand environments created in the solid state. In the isolated metal atom, the *d* orbitals have the same energy, i.e. they are degenerate. The presence of the nonspherical ligand field in the octahedral and tetrahedral environments found in the solid state and in complexes causes the degeneracy of these orbitals to be removed in a way that may be analysed using a combination of symmetry arguments and perturbation theory. Van Vleck [8] adapted crystal field theory to accommodate the partial covalency of the metal–ligand bond in the 1930s and used it to understand the paramagnetic properties of transition metal ions. In the early 1950s, Orgel and Griffiths recognised the broader chemical implications of the crystal field approach and established that the *ligand field theory* [9–11] could be used to rationalise a wide range of properties. Orgel's book was particularly influential in introducing the basic concepts of ligand field theory in a non-mathematical manner [10, 11] which could be extended to new applications. Irving and Williams used the *ligand field theory* to interpret their comparative stability constant measurements, and Williams recognised the importance of ligand field stabilisation energies in mineralogy and biomineralisation [12, 13]. The subsequent development theoretical models for transition metal chemistry are captured in the volumes of *Structure and Bonding* summarising the contributions and influence of Carl Ballhausen [14, 15].

Given the historical background to the founding of *Structure and Bonding*, it will come as no surprise that the early volumes contained reviews analysing the spectral properties of fluorides, chalcogenides, cyanides and organic Schiff base ligands [16–20]. The specific electronic spectral features of five-co-ordinate complexes were also discussed [21]. The effects of tetragonal and Jahn–Teller distortions were evaluated in references [22–24].

Professor Christian Kluxbüll Jørgensen, who was one of the founding members of the editorial board, was a true polymath, whose research interests spanned the whole of the periodic table and encompassed wide tracts of chemistry physics and mathematics. As a student, he amazed his teachers with his knowledge of astronomy, geography, philosophy as well as chemistry. For Volume 1 of the series, he contributed a review entitled "Recent Progress in Ligand Field Theory" [25], and he subsequently published 20 reviews between 1966 and 1989. In the 1950s working in Bjerrum's laboratory in Copenhagen (Carl Ballhausen was a fellow student), he was one of the first to recognise that the spectra of a wide range of d^n and f^n metal complexes could be interpreted using theoretical models originally developed initially by Bethe and Van Vleck and being reintroduced to chemists by Orgel and Griffiths. With Schäffer he was responsible for formulating the angular overlap model, and early volumes of *Structure and Bonding* developed the applications [26–28]. He also recognised the importance of spin-allowed electronic transitions and did much to measure and interpret the charge-transfer spectra of metal complexes and used the data to propose optical electronegativities. As Schönherr has noted in the two volumes of *Structure and Bonding* dedicated to Jørgensen [29, 30], "It was developed on the basis of Jørgensen's encyclopaedic knowledge of inorganic chemistry and atomic spectroscopy and of his remarkable function as an autodidact in the field of quantum mechanics even though he remained largely a sceptic with regard to computational efforts in chemistry". In addition, he will be remembered for his definition of the nephelauxetic effect, which results from the reduction in interelectron repulsion energies when a metal ion forms a complex and may be related to the covalency of the metal–ligand bond. He also drew attention to the symbiotic effects associated with the aggregation of hard and soft bases, and he was one of the first chemists to identify the differences between innocent and non-innocent ligands [31–34]. He also was willing to address more controversial and speculative topics such as the plausible chemistry of yet undiscovered elements, elementary particles such as quarks and the heavy elements synthesised in supernovae and A-type stars. He and Reisfeld published numerous papers and reviews on the optical and luminescent properties of rare earth metals [34–36]. Electronic spectral analyses were also extended to the rare earth and actinide elements [37–42]. The luminescent properties of the lanthanide ions have proved to be particularly important in the development of colour display devices. Jørgensen was also quick to recognise the usefulness of photoelectron spectroscopy for calibrating theoretical studies and encouraged others to publish novel applications of photoelectron spectroscopy to inorganic systems [43–49].

As the computer power available to theoreticians expanded exponentially, there was an increased tendency to calculate the spectroscopic data using ab initio techniques [50–52]. The emergence of organometallic chemistry also led to greater interest in complexes containing synergic ligands. Consequently, carbonyl [53, 54] and sulphur dioxide complexes [55] of the transition metals were discussed in detail, and the bonding in sandwich compounds was analysed in more detail than that provided initially by Orgel and Moffitt [56–58].

2.2 New Spectroscopic Methods

Nyholm in his inaugural lecture noted the importance of new spectroscopic and magnetic techniques in characterising inorganic compounds, and the early volumes of *Structure and Bonding* reported the applications of Mössbauer spectroscopy [58], nuclear quadrupole spectroscopy [59] and resonance Raman effects [60]. Gillard and Mitchell reviewed the application of optical rotary dispersion and circular dichroism for determining the absolute configurations of transition metal complexes [61, 62].

2.3 Stabilities of Complexes

Although Alfred Werner (Nobel Prize 1913) defined the basic principles of co-ordination chemistry at the turn of the twentieth century, it was not until 1941 that Jannik Bjerrum developed a general method for determining the stability constants of metal complexes [63]. This long delay occurred because of the difficulty of measuring the concentrations of the numerous species in solution and therefore required the development of the glass electrode and pH metre [64]. After the Second World War, the Bjerrum methodology was applied to a wide range of complexes, and the data were analysed using graphical methods, because computers were not widely available. The relevant mathematical formulae underlying the graphical techniques were summarised in the comprehensive textbook published by Rossotti and Rossotti in 1961 [65]. As *Structure and Bonding* was being launched, the first computer programmes were developed, and this made more complex systems accessible to analysis [66–68]. Consequently, early volumes of *Structure and Bonding* contained many contributions from the leaders in the field, who were now accumulating reliable data and beginning to draw generalisations. Reviews by Williams, Ahrland, Gutmann, Drago and Hudson featured in the early volumes of the series [69–75].

Irving, Williams and Calvin's systematic study of the equilibrium constants for first-row transition metal ions in the late 1940s and 1950s had concentrated on the relative stabilities of (M^{2+}) complexes with simple ligands, and this led to the definition of the Irving–Williams series in 1952. They proposed that the relative stabilities of complexes followed the order [76–79]:

$$\mathrm{Mn} < \mathrm{Fe} < \mathrm{Co} < \mathrm{Ni} < \mathrm{Cu} > \mathrm{Zn}$$

This order was interpreted using a combination of crystal field stabilisation energies and ionic size effects. In the 1950s, Griffiths and Orgel had introduced inorganic chemists to *ligand field theory*, which had been developed by Bethe and Van Vleck in the 1930s, and Irving and Williams used the concepts to rationalise this stability order. The general trend was related to the decrease in ionic radius from Mn^{2+} to Zn^{2+}, but superimposed on this is the ligand field stabilisation energy (LFSE) which increases from zero for high-spin d^5 manganese(II) to reach a maximum at nickel (II) and returns to zero for d^{10} zinc(II). Although copper(II) has a smaller LFSE than nickel(II), these complexes are subject to a Jahn–Teller effect, which results in distorted octahedral complexes and at times lower co-ordination numbers, which favour higher stability constants. A related example of increasing stability constants across a series of the periodic table resulting from a decrease in ionic radius was confirmed for the series of (M^{3+}) lanthanide ions (the lanthanide contraction) [80]. The 4f orbitals are more core like for the lanthanide ions, and consequently the ligand field effects do not contribute as large an effect as that noted above for the transition metals ions.

The measurement of stability constants of related complexes with monodentate and polydentate ligands led to a thermodynamic understanding of the chelate effect and the macrocyclic effect. In particular, the conversion of the equilibrium constants into free energies and its partitioning into enthalpic and entropic contributions showed that the chelate and macrocyclic effects had their origins in entropic effects [81–83]. Significant reviews which appeared in *Structure and Bonding* summarising the structural and chemical consequence included Truter's "Structures of Alkali Metal Complexes with Organic Ligands" and Winkler's "Kinetics and Mechanism of Alkali Metal Complex Formation in Solution" which built on Pedersen's important discovery of crown ethers at the DuPont laboratories, in Wilmington, USA [84–86]. Jean-Marie Lehn published a review on "Design of Organic Complexing Agents Strategies Towards Properties" in *Structure and Bonding* in 1973 [87]. His outstanding research on complexes of cryptand ligands and supramolecular chemistry resulted in him sharing the Nobel Prize in 1987 with Pedersen and Cram. The biological implications of these new classes of ligands which were capable of co-ordinating alkali and alkaline earth metal ions were discussed in reviews in *Structure and Bonding* by Simon, Morf and Meier and Siegel [88, 89].

From Werner's initial studies, it established some ligands with lone pairs on more than one atom exhibited ambidentate properties, i.e. they form complexes through alternative donor atoms. The most widely studied ambidentate ligands were thiocyanate SCN^-, which formed a co-ordinate bond either through S or N and NO_2^-, which bonded through either N or O. In the early 1960s, when crystallographic determinations were rather rare, the identification of the co-ordination mode of ambidentate ligands was greatly facilitated by the application of infrared studies. These studies highlighted that if the donor atoms belonged to the first and second long rows of the periodic table, their donor properties are sufficiently different for isomeric complexes to be isolated.

The early synthetic studies of Mann, Chatt and Jensen established that tertiary phosphines [90] were capable of forming very stable complexes with the platinum metals, and this led to the isolation of kinetically inert alkyl and hydride complexes of these metals [91]. Ahrland, Chatt and Davies [92] completed detailed stability constant determinations in 1958, which lead them to conclude that metal ions fell into two distinct groups: Class A metal ions formed stronger complexes with ligands whose donor atoms are N, O or F rather than those with P, S or Cl, and Class B metal ions exhibited the reverse relative stabilities. For example, Ni^{2+} forms stronger complexes with amines than with phosphines, but Pt^{2+} forms stronger complexes with phosphines than with amines. Subsequently, Pearson generalised the observation to organic and inorganic systems and re-expressed it as the *hard and soft acids and bases theory* (HSAB theory) [93, 94]. Class A metals are hard acids and Class B metals are soft acids. HSAB theory predicts that *hard acids* form stronger complexes with *hard bases* and *soft bases* form stronger complexes with *soft acids*. Pearson's analysis rationalised this behaviour by suggesting that hard–hard interactions are predominantly electrostatic in origin, whereas soft–soft interactions involve predominantly covalent interactions.

Examples of hard acids are H^+, light alkali ions (Li through K all have small ionic radius), Ti^{4+}, Cr^{3+}, Cr^{6+} and BF_3. Examples of hard bases are OH^-, F^-, Cl^-, NH_3, CH_3COO^- and CO_3^{2-}. Examples of soft acids are CH_3Hg^+, Pt^{2+}, Pd^{2+}, Ag^+, Au^+, Hg^{2+}, Hg_2^{2+}, Cd^{2+} and BH_3. Examples of soft bases are H^-, R_3P, SCN^- and I^-.

The early volumes of *Structure and Bonding* had many contributions which reviewed the implications of the HSAB theory and also tried to unpick the fundamental electronic origins of the generalisations. References [95, 96] provide specific examples of chapters published in the early volumes of *Structure and Bonding*. To this day, it remains a useful semi-quantitative generalisation and a useful pedagogical tool. In 1983, Pearson and Parr extended the qualitative HSAB theory which developed a quantitative quantum mechanically based definition of chemical hardness [97–100], which is proportional to the second derivative of the total energy of a chemical system with respect to changes in the number of electrons at a fixed nuclear environment. Pearson, Parr and other authors developed this theme in Volume 80 (1993) of *Structure and Bonding* and entitled "Chemical Hardness".

In 2011, Herbert Mayr et al. published a critical review of the HSAB theory. Consecutive analysis of various types of ambident organic system led them to propose that an older approach based on thermodynamic/kinetic control describes reactivity of organic compounds perfectly well. They argued that the HSAB principle actually fails and should be abandoned in the rationalization of the ambident reactivities of organic compounds [41].

2.4 Bioinorganic Chemistry

Syphilis first made its impact in Western Europe in the 1490s, which implicated (although not proved) that it was brought back from the New World by sailors, who participated in the Christopher Columbus' expeditions. The treatment favoured in Europe for syphilis was mercury in the form of "unguentum Saracenicum" (Saracen's ointment), which had been a staple of Arab medicine for the treatment of scabs, psoriasis, leprosy and other skin diseases. This opened up the possibility that other metal-containing compounds may have medical applications, although it took several hundred years before it was established that metal ions occurred at the active sites of many biological molecules. In Victorian times, gold compounds and colloids were also used to treat rheumatoid arthritis, and in the early 1900s, Paul Ehrlich demonstrated that organoarsenic ("arsenicals") were also effective in the treatment of syphilis. Barnett Rosenberg's serendipitous discovery of the anticancer properties of cisplatin (*cis*-$[PtCl_2(NH_3)_2]$) in 1965 coincided (serendipitously) with the launch of *Structure and Bonding* [101–105]. Therefore, it is not surprising that the early volumes contained reviews on platinum and gold drugs by Williams, Thomson and Sadler [106, 107].

The first protein structures completed in the late 1950s had underlined the importance of metal ions in enzymes, proteins and vitamins. Urease was shown

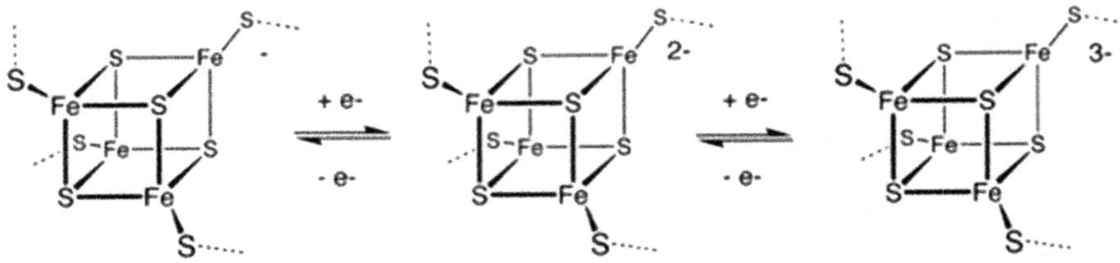

Fig. 1 4Fe–4S clusters serve as electron relays in proteins

to have nickel and carboxypeptidase zinc at the active sites. The structural studies of Kendrew and Perutz on myoglobin and haemoglobin provided a detailed understanding of the role of iron(II) in oxygen transport and the allosteric behaviour of the four subunits in the latter. The concerted model of allostery postulates that protein subunits are connected in such a way that a conformational change in one subunit is necessarily conferred to all other subunits. For haemoglobin, simple co-ordination chemistry concepts derived from ligand field theory were used to understand how conformational changes at one iron atom could be transmitted to the other iron–porphyrin centres. Dorothy Hodgkin's X-ray structural determination of vitamin B_{12} demonstrated that cobalt was located at the centre of a tetrapyrrole corrin macrocycle and also has a metal–carbon bond [108]. The emergence of this significant new research area at the interface of inorganic and biochemistry was recognised by some of the founding members of the editorial board of *Structure and Bonding*, notably Bob Williams and Joe Neilands. *Structure and Bonding* therefore played an important role in promoting bioinorganic chemistry and demonstrating how modern spectroscopic and structural techniques could be used to provide insights into the active sights of proteins and enzymes. The greater understanding of the electronic and chemical properties of metal ions afforded by *ligand field theory* also provided a theoretical framework for relating the multiple roles of the metal ions in biology to principles previously established for co-ordination chemistry. The following themes attracted particular attention in *Structure and Bonding* in the 1960s and 1970s.

2.4.1 Enzymology

In many biologically important reactions involving enzymes, metal ions and water molecules come together at the catalytically active sites of the enzymes, e.g. carbonic anhydrase, metallophosphatases and metalloproteinases. Reviews in *Structure and Bonding* sought to understand and define the function of these metalloproteins [109, 110]. Several reviews also tackled the difficult problem of understanding the role of chlorophyll in photosynthesis [111–113].

Metal-containing electron transfer proteins are also common and can be organised into three major classes: iron–sulphur proteins (e.g. rubredoxins, ferredoxins and Rieske proteins), blue copper proteins and cytochromes. The widely observed 4Fe–4S clusters serve as electron relays as shown in Fig. 1. These electron

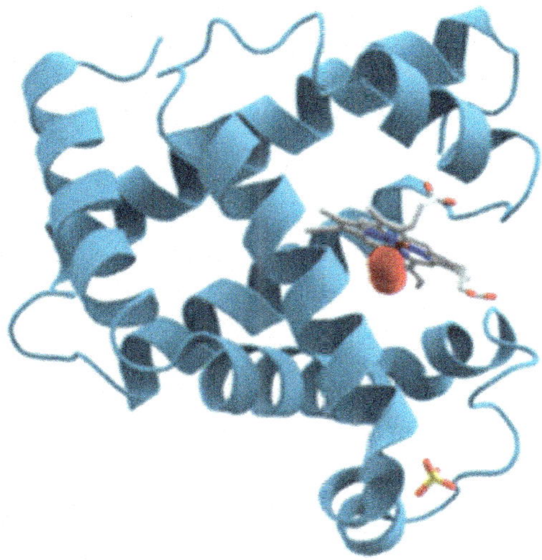

Fig. 2 Structure of myoglobin

transport proteins are complementary to the non-metal electron transporters nicotinamide adenine dinucleotide (NAD) and flavin adenine dinucleotide (FAD). The nitrogen cycle makes extensive use of metals for their redox interconversions. These topics were comprehensively reviewed in *Structure and Bonding* [114–120].

Biological processes make extensive use of metals such as iron, copper and manganese in order to utilise the dioxygen in the atmosphere [121–130]. Haem that is utilised by red blood cells in the form of the role of haemoglobin for oxygen transport by red blood cells is perhaps the most studied metal system in biology, but myoglobin (shown in Fig. 2), haemocyanin and hemerythrin represent examples of other oxygen transport systems. Oxidases and oxygenases are common throughout nature and use dioxygen for energy generation, e.g. cytochrome c oxidase, or small molecule oxidation, e.g. cytochrome P450 oxidases and methane monooxygenase. Some metalloproteins are designed to protect biological systems from the potentially harmful effects of dioxygen and hydrogen peroxide, e.g. peroxidases, catalases and superoxide dismutases. Dioxygen first appeared in our atmosphere from photosynthetic processes, which involved a metalloprotein as an oxygen-evolving complex.

More recently, it has been shown that iron haem proteins play an important function in the neurotransmitting functions of nitric oxide.

2.4.2 Bioorganometallic Chemistry

The surprising observation of a cobalt–carbon bond in vitamin B_{12} (methylcobalamin) was subsequently followed by other examples and led to their collective description as bioorganometallics. Bioorganometallic enzymes and proteins include the hydrogenases, FeMoco in nitrogenase (see Fig. 3). Unicellular

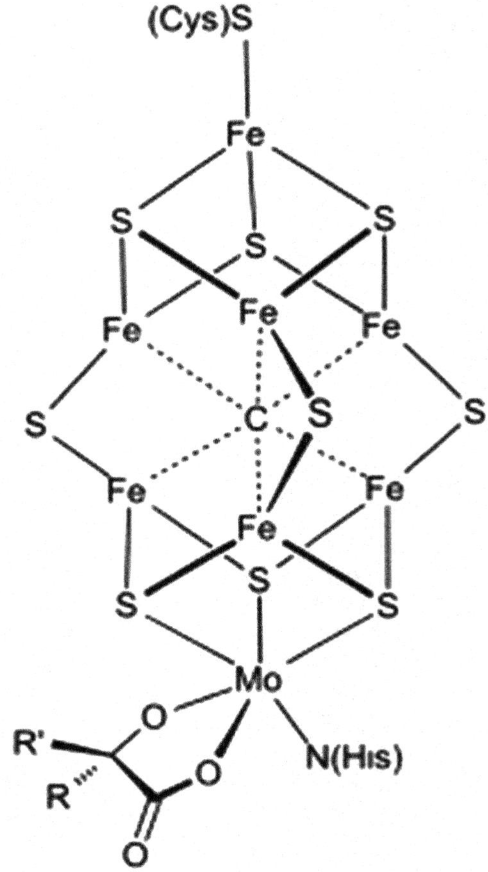

Fig. 3 Structure of FeMoco, the catalytic centre of nitrogenase

organisms are particularly able to utilise organometallic intermediates, and early reviews in *Structure and Bonding* assessed whether methylation of Hg^{2+} may result from *trans*-metallation of metal–carbon bonds. In the late 1960s, Minamata disease was related to the formation of $[HgMe]^+$ [108].

It was noted above that a number of drugs are based on metal complexes. It also became apparent that metal complexes could be used to improve the resolution of diagnostic techniques. Thus, MRI (magnetic resonance imaging) contrasting agents contain gadolinium ions, and technetium complexes are used as medical radio-tracers. Carbon monoxide-releasing molecule complexes have been developed to suppress inflammation by releasing small amounts of carbon monoxide. The cardiovascular and neuronal importance of nitric oxide has been examined and reviewed in recent volumes of *Structure and Bonding* [131].

2.4.3 Biomineralization

Biomineralization is the process by which living organisms produce minerals, which have the appropriate dimensions and mechanical properties to either harden or stiffen tissue materials, which are organically based. Examples include silicates in algae and diatoms, carbonates in invertebrates and calcium phosphates and carbonates in vertebrates. Other examples include copper, iron and gold deposits involving bacteria. Biologically formed minerals often have more specific uses such as magnetic sensors in magnetotactic bacteria (Fe_3O_4), gravity-sensing devices ($CaCO_3$, $CaSO_4$, $BaSO_4$) and iron storage and mobilisation ($Fe_2O_3{\cdot}H_2O$ in the protein ferritin). Since extracellular iron is strongly involved in inducing calcification, its control is essential in developing shells. The protein ferritin plays an important role in controlling the distribution of iron [132–136].

3 More Recent Developments

When the Volume 100 of *Structure and Bonding* was published in 2001, Springer decided that it was appropriate to review the series, and I was asked to act as the series editor. From my detailed analysis of the publication record of the series presented above, it was clear that *Structure and Bonding* together with *Progress in Inorganic Chemistry* and *Advances in Inorganic and Radiochemistry* (both published by the Wiley Group) had played a very important and distinguished role in the renaissance in inorganic chemistry articulated by Sir Ron Nyholm, but it was appropriate to undertake a review. It had already been decided not to publish contributions as they were submitted and move towards themed volumes. I was therefore encouraged to recruit an editorial board who would stimulate the commissioning of volumes which were scientifically sound and reflected the current concerns of chemists in the area of structure and bonding. In my forward, the new editorial board noted that the subject had moved forward in the 25 years since it was founded and redefined the aims of the series as follows:

> We expect the scope of *Structure and Bonding* series to span the entire periodic table and address structure and bonding issues wherever they may be relevant. Therefore, it is anticipated that there will be reviews dealing not only with the traditional areas of chemical bonding based on valence problems and dynamics, but also nanostructures, molecular electronics, supramolecular structures, surfaces and clusters. These represent new and emerging areas of chemistry.

These aims have been maintained, and we are marking the 50th anniversary by some excellent review chapters which illustrate and amplify these aims. The authors of the chapters were invited and chosen on the basis of an analysis of the most cited volumes in recent years. I thank the contributors for their excellent and accessible reviews which clearly define the power of modern structural and bonding techniques. We hope that the readers will find their contributions interesting and relevant. If you feel we have omitted an important and timely topic, then please do write to us with a short summary and a list of potential authors. We are currently making plans for the journal's next 50 years and thereby ensure that the strong tradition set by the original editorial board is maintained in the future.

Acknowledgements The editorial staff at the Springer offices in Heidelberg have made an enormous contribution to the success of the journal, and on behalf of all the editors past and present and contributors, I should like to thank them very much for their patience and professionalism.

References

1. Labinger J (2013) Up for generality – how inorganic chemistry finally became a respectable field, Springer, briefs in molecular science, history of chemistry. Springer, Heidelberg
2. Nyholm RS (1957) The renaissance of inorganic chemistry. J Chem Educ 34:166–169
3. Nyhom RS (1961) Tilden lecture: electronic configurations and structures of transition metal complexes. Proc Chem Soc 273–296
4. Williams RJP, Hale JD (1972) RS Nyholm – an appreciation. Struct Bond 15:1–2
5. Roberts K, Raff M, Alberts B, Walter P, Lewis J, Johnson A (2002) Molecular biology of the cell, 4th edn. Garland Science, New York
6. Astbury WT (1961) Molecular biology or ultrastructural biology? Nature 190:1124–1124
7. Bethe HA (1929) Splitting of terms in crystals. Ann Physik 3:133–209
8. Van Vleck JH (1932) Theory of the variations in paramagnetic anisotropy among different salts of the iron group. Phys Rev 41:208–232
9. Orgel LE, Griffith JH (1957) Ligand field theory. Quart Rev 11:381–393
10. Orgel LE (1952) The effects of crystal fields on the properties of transition metal ions. J Chem Soc 4756–4761
11. Orgel LE (1960) Introduction to transition metal chemistry. Methuen, London
12. Williams RJP (1959) Deposition of trace elements in a basic magna. Nature 184:44–44
13. Burns RG (1993) Mineralogical applications of crystal field theory, 2nd edn. Cambridge University Press, Cambridge
14. Mingos DMP, Day P, Dahl JP (2012) Molecular electronic structures of transition metal complexes I and II. Struct Bond 142:1–211
15. Mingos DMP, Day P, Dahl JP (2012) Molecular electronic structures of transition metal complexes I and II. Struct Bond 143:229
16. Allen GC, Warren KD (1971) The electronic spectra of hexafluoro complexes of the first transition series. Struct Bond 9:49–138
17. Allen GC, Warren KD (1974) The electronic spectra of hexafluoro complexes of the second and third transition series. Struct Bond 19:105–165
18. Moreau-Colin M (1972) Electronic spectra and structural properties of complex tetracyanides of platinum palladium, nickel. Struct Bond 10:167–190
19. Müller A, Baran EJ, Diemann E, Jørgensen CK (1976) Electronic spectra of tetrahedral oxo, thio, seleno complexes formed by elements of the beginning of the transition groups. Struct Bond 14:23–47
20. Daul C, Schläpfer CW, von Zelewsky A (1979) The electronic structure of cobalt (II) complexes with schiff bases and related ligands. Struct Bond 36:129–171
21. Ciampolini M (1969) Spectra of 3d five-coordinate complexes. Struct Bond 6:52–93
22. Oelkrug D (1971) Absorption spectra and ligand field parameters of tetragonal 3d transition metal fluorides. Struct Bond 9:1–26

23. Reinen D (1969) Ligand field spectroscopy and chemical bonding in Cr^{3+} containing oxide solids. Struct Bond 6:30–51
24. Reinen D, Friebel C (1971) Local and Cooperative jahn-teller interactions in model structures, spectroscopic and structural evidence. Struct Bond 37:1–60
25. Jørgensen CK (1965) Recent progress in ligand field theory. Struct Bond 1:3–31
26. Schäffer CE (1968) A perturbation representation of weak covalent bonding. Struct Bond 5:68–95
27. Schäffer CE (1973) Two symmetry parameterization of the angular overlap model of the ligand field. Relation to the crystal field model. Struct Bond 14:69–101
28. Smith DW (1978) Applications of the angular overlap model. Struct Bond 35:87–118
29. Schönherr T (2004) Optical spectra and chemical bonding in inorganic compounds. Struct Bond 106:1–255
30. Schönherr T (2004) Optical spectra and chemical bonding in inorganic compounds. Struct Bond 107:1–301
31. Duffy JA (1977) Optical electronegativity and nephelauxetic effect in oxide systems. Struct Bond 32:147–166
32. Jørgensen CK (1966) Electric polarizability, innocent ligands and spectroscopic oxidation states. Struct Bond 1:224–248
33. König E (1971) The nephelauxetic effect, calculation and accuracy of the interelectronic repulsion parameters I. Cubic high-spin d^2, d^3, d^7 and d^8 systems. Struct Bond 9:175–212
34. Jørgensen CK (1969) Valence shell expansion studied by ultraviolet spectroscopy. Struct Bond 6:94–115
35. Reisfeld R (1973) Spectra and energy transfer of rare earths in inorganic glasses. Struct Bond 13:53–98
36. Reisfeld R (1975) Radiative and non-radiative transitions of rare earths in inorganic glasses. Struct Bond 22:123–175
37. Peacock RD (1975) The intensities of lanthanide f-f transitions. Struct Bond 22:83–122
38. Sinha SP (1976) A systematic correlation of the properties of the f-f transition metal ions. Struct Bond 30:1–64
39. Wilson JA (1977) A generalized configuration-dependent band model for lanthanide compounds and conditions for interconfiguration fluctuations. Struct Bond 32:57–91
40. Blasse G (1976) The influence of charge transfer and rydberg states on the luminescence properties of lanthanides and actinides. Struct Bond 26:43–79
41. Bleijenberg KC (1980) Luminescence of uranite centres in solids. Struct Bond 42:97–128
42. Baker EC, Halstead GW, Raymond KN (1976) The structure and bonding of 4f and 5f series of organometallic compounds. Struct Bond 25:21–66
43. Furlani C, Cauletti C (1978) He(I) photoelectron spectra of *d*- metal compounds. Struct Bond 35:119–169
44. Green JC (1981) Gas phase photoelectron spectra of d and f-block organometallic compounds. Struct Bond 43:37–112
45. Jørgensen CK (1975) Photoelectron spectroscopy non-metallic solids and consequences for quantum chemistry. Struct Bond 24:1–58
46. Jørgensen CK (1976) Deep lying valence orbitals and problems of degeneracy and intensities in photoelectron spectra. Struct Bond 30:141–192
47. Bradshaw AW, Cederbaum LS, Domcke W (1975) Ultraviolet photoelectron spectroscopy gases absorbed on metal surfaces. Struct Bond 24:133–170
48. Watson RE, Perlman ML (1975) X-Ray photoelectron spectroscopy. Applications to metal and alloys. Struct Bond 24:83–132
49. Somorjai GA, van Hove MA (1979) Adsorbed monolayers on sold surfaces. Struct Bond 38:1–140
50. Cook DB (1978) The approximate calculation of molecular electronic structures as a theory of valence. Struct Bond 35:37–86

51. Gerloch M, Harding JH, Woolley G (1981) The context and application of ligand field theory. Struct Bond 46:1–46
52. Schutte CJH (1971) The ab initio calculation of molecular vibrational frequencies and force constants. Struct Bond 9:213–263
53. Braterman PS (1972) Spectra and bonding in metal carbonyls part A. Struct Bond 10:57–86
54. Braterman PS (1976) Spectra and bonding in metal carbonyls part B. Struct Bond 26:1–42
55. Ryan RR, Kubas GJ, Moody DC, Eller PG (1981) Structure and bonding of transition metal sulfur dioxide complexes. Struct Bond 46:47–100
56. Warren KD (1976) Ligand field theory of metal sandwich complexes. Struct Bond 27:45–159
57. Warren KD (1977) Ligand field theory of *f*- orbital metal sandwich complexes. Struct Bond 33:97–137
58. Bearden AJ, Dunham AW (1970) Iron electronic configurations in proteins: studies in mössbauer spectroscopy. Struct Bond 8:1–52
59. Van Bronswyk W (1970) The application of nuclear quadrupole spectroscopy to the study of transition metal compounds. Struct Bond 7:87–113
60. Clark RJH, Steart B (1979) The resonance Raman effect. The review of theory and of applications in inorganic chemistry. Struct Bond 36:1–80
61. Gillard RD, Mitchell PR (1970) The absolute configuration of transition metal complexes. Struct Bond 7:46–86
62. Blauer G (1974) Optical activity of conjugated proteins. Struct Bond 18:69–129
63. Bjerrum J (1941) Metal-ammine formation in aqueous solution. Haase, Copenhagen
64. Beck MT, Nagypál I (1990) Chemistry of complex equilibria. Horwood, Chichester
65. Rossotti FJC, Rossotti H (1961) The determination of stability constants. McGraw-Hill, New York
66. Dyrssen D, Ingri N, Sillen LG (1961) Pit-mapping - a general approach to computer refinement of stability constants. Acta Chem Scand 15:694–696
67. Ingri N, Sillen LG (1964) High-speed computers as a supplement to graphical methods, IV. An ALGOL version of LETAGROP-VRID. Arkiv Kemi 23:97–121
68. Sayce IG (1968) Computer calculations of equilibrium constants of species present in mixtures of metal ions and complexing reagents. Talanta 22(12):1397–1421
69. Ahrland S (1968) Thermodynamics of complex formation between hard and soft acceptors and donors. Struct Bond 5:118–149
70. Ahrland S (1973) Thermodynamics of the stepwise formation of metal-ion complexes in aqueous solution. Struct Bond 15:167–188
71. Williams RJP, Hale JD (1965) The classification of acceptors and donors in inorganic reactions. Struct Bond 6:249–281
72. Drago RS (1973) Quantitative evaluation and donor-acceptor interactions. Struct Bond 15:73–139
73. Gutmann V, Mayer U (1972) Thermochemistry of the chemical bond. Struct Bond 10:127–151
74. Guttmann V, Mayer U (1973) Redox properties: changes effected by co-ordination. Struct Bond 15:141–166
75. Hudson RF (1966) Displacement reactions and the concept of soft and hard acids and bases. Struct Bond 1:221–223
76. Irving HMN, Williams RJP (1953) The stability of transition-metal complexes. J Chem Soc 3192–3210
77. Calvin M, Melchior NC (1948) The stability of chelate complexes IV: the effect of the metal ion. J Am Chem Soc 70:3270–3273
78. Williams RJP (1958) The stability of transition metal complexes. Trans Faraday Soc 26:182–188
79. Irving HM, Williams RJP (1948) The order of stability constants. Nature 162:746–747
80. Morris DFC (1967) Ionic radii and enthalpies of hydration of ions. Struct Bond 4:63–82
81. Schwarzenbach G (1952) Der Chelateffekt. Helv Chim Acta 35:2344–2359

82. Cabinness DK, Margerum DW (1969) Macrocyclic effect on the stability of copper (II) tetramine complexes. J Am Chem Soc 91:6540–6541
83. Lindoy LF (1990) The chemistry of macrocyclic ligand complexes. Cambridge University Press, Cambridge
84. Pedersen CJ (1967) Cyclic polyethers and their complexes with metal salts. J Am Chem Soc 89:7017–7036
85. Truter MR (1973) Structures of organic complexes with alkali metal ions. Struct Bond 16:71–111
86. Winkler R (1972) Kinetics and mechanism of alkali metal complex formation in solution. Struct Bond 10:1–24
87. Lehn J-M (1973) Design of organic complexing agents strategies towards properties. Struct Bond 16:1–69
88. Simon W, Morf WE, Meier PCH (1973) Specificity for alkali and alkaline earth cations of synthetic and naturally organic complexing agents in membranes. Struct Bond 16:113–160
89. Siegel FL (1973) Calcium binding proteins. Struct Bond 17:221–268
90. Leigh GJ, Winterton N (2002) Modern co-ordination chemistry – the legacy of Joseph Chatt. Royal Society of Chemistry, Cambridge
91. Chatt J (1962) Tilden lecture: hydrido and related organic complexes of the transition metals. Proc Chem Soc 318–326
92. Ahrland S, Chatt J, Davies NR (1958) The relative affinities of ligand atoms for acceptor molecules and ions. Quart Rev 12:265–276
93. Pearson RG (1963) Hard and soft acids and bases. J Am Chem Soc 85:3533–3539
94. Pearson RG (1997) Chemical hardness: applications from molecules to solids. Springer, Heidelberg
95. Drago RS, Wong N, Bilgrien C, Vogel C (1987) E and C parameters from Hammett substituent constants and use of E and C to understand cobalt-carbon bond energies. Inorg Chem 26:9–14
96. Gutmann V (1978) The donor-acceptor approach to molecular interactions. Springer, Heidelberg
97. Parr RG, Pearson RG (1983) Absolute hardness: companion parameter to absolute electronegativity. J Am Chem Soc 105:7512–7516
98. Pearson RG (2005) Chemical hardness and density functional theory. J Chem Sci 117:369–377
99. Mayr H (2011) Farewell to the HSAB treatment of ambident reactivity. Angew Chem Int Ed 50:6470–6505
100. Frausto da Silva JJR, Williams RJP (1976) The uptake of elements by biological systems. Struct Bond 29:67–121
101. Lippard SJ, Berg JM (1994) Principles of bioinorganic chemistry. University Science, Sausilito
102. Kraatz H-B, Metzler-Nolte N (2006) Concepts and models in bioinorganic chemistry. Wiley, New York
103. Bertini I, Gray HB, Stiefel EI, Valentine JS (2007) Biological inorganic chemistry. University Science, Sausilito
104. Kaim W, Schwederski B (1994) Bioinorganic chemistry: inorganic elements in the chemistry of life. Wiley, New York
105. Que L Jr (ed) (2000) Physical methods in bioinorganic chemistry. University Science, Sausilito
106. Thomson AJ, Williams RJP, Reslova F (1972) The chemistry of complexes related to cis- Pt $(NH_3)_2Cl_2$: an antitumour drug. Struct Bond 11:1–46
107. Sadler PJ (1976) The biological chemistry of gold: a metallo drug and heavy atom label with variable valency. Struct Bond 29:171–214
108. Wood JM, Brown DG (1972) The chemistry of vitamin B_{12} enzymes. Struct Bond 11:47–105

109. Bertini I, Luchninat C, Scozzafava A (1981) Carbonic anhydrase: an insight into the zinc binding site and into the active site and into the active cavity through metal substitution. Struct Bond 48:45–91
110. Cheh AM, Neilands JP (1976) The δ-aminovulinte dehdratases: molecular and environmental properties. Struct Bond 29:123–169
111. Livorness J, Smith T (1982) The role of manganese in photosynthesis. Struct Bond 48:11–44
112. Maggiora GM, Ingraham LL (1967) Chlorophyll triplet states. Struct Bond 2:126–159
113. Rüdiger W (1980) Phytochrome, a light receptor of plant photomorphogenesis. Struct Bond 40:101–140
114. Bray RC, Swann JC (1972) Molybdenum containing enzyme. Struct Bond 11:107–144
114. Kimura T (1968) Biochemical aspects of iron-sulfur linkage in non-heme iron proteins with special reference to "andrenodoxin". Struct Bond 5:1–40
116. Fee JA (1975) Copper proteins – a system containing the "blue" copper centre. Struct Bond 23:1–60
117. Zumft WG (1976) The molecular basis of nitrogen fixation. Struct Bond 29:1–65
118. Xavier AV, Moura JJ, Moura I (1981) Novel structures in iron sulfur proteins. Struct Bond 43:187–213
119. Fuhrhop J-H (1974) The oxidation states and reversible redox reactions of metalloporphyrins. Struct Bond 18:1–67
120. Xavier AV, Moura JJ, Moura I (1981) Novel structures in iron sulfur proteins. Struct Bond 40:101–140
121. Hill HAO, Roder A, Williams RJP (1970) The chemical nature and reactivity of cytochrome P-450. Struct Bond 8:123–151
122. Neilands JB (1966) Naturally occurring non-porphyrin iron compounds. Struct Bond 1:59–1081
123. Neilands JB (1972) Evolution of biological iron binding centres. Struct Bond 11:145–170
124. Que L Jr (1980) Non-heme iron dioxygenases. Structure and mechanism. Struct Bond 40:39–72
125. Schneider W (1975) Kinetics and mechanism of metalloporphyrin formation. Struct Bond 23:123–166
126. Bayer E, Schretzmann P (1967) Reversible oxygenierung von metallkomplexen. Struct Bond 2:181–250
127. Buchanan BB (1966) The chemistry and function of ferredoxin. Struct Bond 1:109–148
128. Weissbluth M (1967) The physics of hemoglobin. Struct Bond 2:1–125
129. Smith DW, Williams RJP (1970) The spectra of ferric haemes and haemoproteins. Struct Bond 7:1–45
130. Trautwein A (1974) Mössbauer spectroscopy of heme proteins. Struct Bond 20:1–87
131. Clarke MJ, Fackler PH (1982) The chemistry of technetium: towards improved diagnostic agents. Struct Bond 50:57–78
132. Crichton RR (1973) Ferritin. Struct Bond 17:67–134
133. Frausto da Silva JJR, Williams RJP (1991) The biological chemistry of the elements – the inorganic chemistry of life. Oxford University Press, Oxford
134. Williams RJP (1970) The biochemistry of sodium, potassium, magnesium and calcium. Quart Rev 24:331–360
135. Adrian DC (1986) Trace metals in the terrestrial environment. Springer, Heidelberg
136. Lindskog S (1970) Cobalt(II) in metalloenzymes. A reporter of structure-function relationships. Struct Bond 8:153–196

Struct Bond (2016) 172: 19–34
DOI: 10.1007/430_2015_174

Published online: 7 August 2015

Anion Receptors Based on Organic Frameworks: Recent Advances

Philip A. Gale

Abstract This chapter looks at key advances in anion receptors based on organic frameworks since 2008 including the development of halogen-bonding systems, C–H hydrogen bond donors, new transmembrane anion transporters and the roles anions can play in self-assembly processes.

Keywords Anion transport · Anions · Halogen bonding · Hydrogen bonding · Self-assembly · Supramolecular chemistry

Contents

1 Introduction

The development of anion receptor chemistry has continued apace since the publication of the *Structure and Bonding* volume on the *Recognition of Anions* edited by Ramón Vilar [1]. Our contribution to that collection provided an introduction to anion receptors based on organic frameworks [2]. In the years since the initial publication of that volume, we have seen new functional groups used to

P.A. Gale (✉)
Chemistry, University of Southampton, Southampton SO17 1BJ, UK
e-mail: philip.gale@soton.ac.uk

complex anions including new classes of receptor containing C–H hydrogen bond donor groups in addition to compounds that employ halogen bonds or combinations of halogen and hydrogen bonds to bind anionic guests. However, perhaps the most significant advances in the area have been in the application of anion receptors to new challenges including the use of anions to template the assembly of interlocked structures and the use of anion receptors to mediate the transport of anions across lipid bilayer membranes – with the potential for these compounds to be used as future treatments for diseases such as cystic fibrosis and cancer. This update will therefore highlight some of the new approaches to anion complexation that have been developed since the publication of *Structure and Bonding* and provide an overview of progress in the areas of lipid bilayer and anion transport and anions in self-assembly.

2 New Approaches to Anion Complexation

A variety of receptors containing C–H hydrogen bond donor groups have been reported in recent years. In 2008, Li and Flood reported the synthesis of macrocycles containing 1,2,3-triazole heterocycles as C–H hydrogen bond donors for the complexation of anions [3–5]. The initial report of the synthesis of compound **1** (Scheme 1) [4] included an UV/vis titration study of this compound with tetrabutylammonium chloride in dichloromethane which demonstrated that the macrocycle binds chloride with a stability constant of $130{,}000 \pm 30{,}000\ M^{-1}$. ^1H NMR spectroscopic studies showed that the triazole C–H proton shifted downfield by 0.73 ppm in the presence of 10 equivalents of chloride (added as the tetrabutylammonium salt in CD_2Cl_2). The authors proposed that the C–H hydrogen bonding is aided by the large dipole present in the triazole ring, the positive end of which is aligned with the C–H hydrogen bond donor.

More recently Flood and co-workers have developed another new macrocycle containing C–H hydrogen bond donors named cyanostar (compound **7** Scheme 2) [6]. Again this system relies on the presence of a dipole to aid hydrogen bonding from a cyanostilbene C–H group (Scheme 2a). The macrocycle was shown to form strong complexes with normally weakly coordinating anions such as PF_6^-, ClO_4^- and BF_4^- via the formation of 2:1 receptor/anion sandwich complexes in 40% $MeOH/CH_2Cl_2$ as determined by UV/vis titration techniques.

Sindelar and co-workers have recently reported that carboxylate-functionalised bambusuril macrocycle **9** can form remarkably strong complexes with anions in buffered water solution with unusual selectivity (Fig. 1) [7]. The water-soluble macrocycle was prepared with pendant carboxylate groups according to the synthetic procedure shown in Scheme 3. The anion complexation properties of this system were studied by ^1H NMR titration techniques in D_2O in the presence of K_2DPO_4 at a pD of 7.1 Anions with lower association constants (F^-, Cl^-, CN^-, IO_4^- and ReO_4^-) (Table 1) formed complexes with **9** with fast-exchange kinetics on the NMR timescale allowing the binding isotherms to be used to calculate

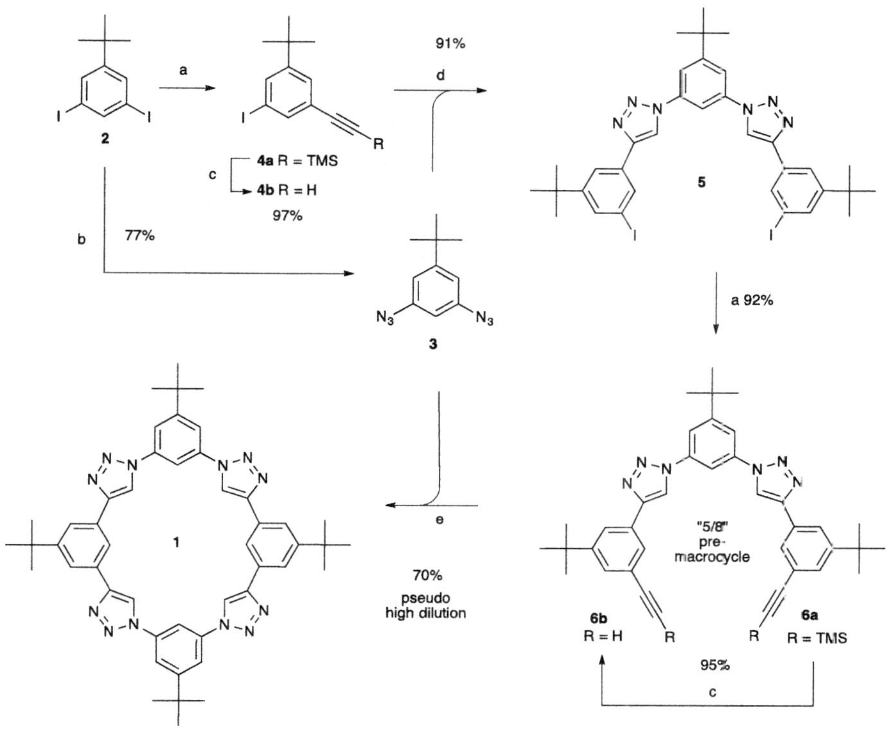

Scheme 1 The synthesis of triazole containing macrocycle **1**: (*a*) TMSCCH, $[PdCl_2(PPh_3)_2]$, CuI, iPr_2NH, THF, Ar, 8 h; (*b*) NaN_3, CuI, DMEA, sodium ascorbate, $EtOH/H_2O/PhMe$ (7:3:1), reflux, Ar, 1 h; (*c*) KF, MeOH, THF, 8 h; (*d*) $CuSO_4$, sodium ascorbate, $EtOH/H_2O/PhMe$ (7:3:1), Ar; (*e*) dropwise addition of 3 and 6b over 10 h to CuI, DBU, PhMe, Ar, stirring, 4 h. *TMS* trimethylsilyl, *DMEA N,N′*-dimethylethylenediamine, *DBU* 1,8-diazabicyclo[5.4.0]undec-7-ene

association constants. More strongly bound anions (Br^-, NO_3^-, PF_6^-, BF_4^-, I^- ClO_4^-) bound with slow-exchange kinetics, and therefore their association constants were calculated using competition experiments with another anion (normally chloride). The compound forms exceptionally strong complexes with anions in water and most interestingly with anions that are normally weakly coordinating such as PF_6^-, BF_4^- and ClO_4^-. The authors attribute this behaviour to the anion binding in the hydrophobic pocket provided by the macrocycle with the anion stabilised by multiple weak C–H···anion hydrogen-bonding interactions.

Receptors that use halogen bonds [8] to bind anionic guests have also been explored by a number of research groups. Beer and co-workers synthesised a bidentate halogen-bonding bromoimidazoliophane receptor $\mathbf{12}^{2+}.2PF_6^-$ for anion recognition in partially aqueous media [9]. During the synthesis (Scheme 4), both *syn* and *anti* conformers are formed (due to the steric bulk of the bromine substituents resulting in a preorganised anion binding site) which were separated by repeated recrystallisation of the bromide salts in methanol. The anion complexation

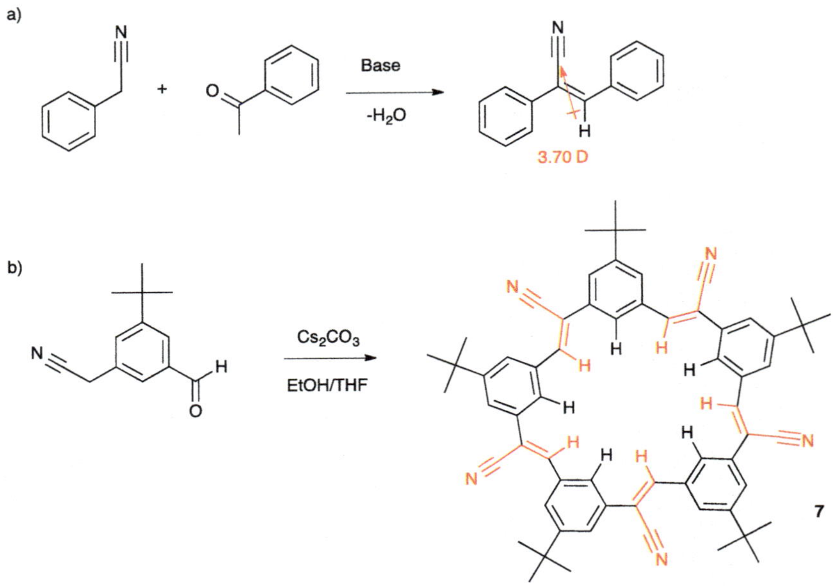

Scheme 2 (**a**) The general preparation of cyanostilbene via a Knoevenagel condensation with calculated dipole moment shown and (**b**) the one pot synthesis of the cyanostar molecule **7**

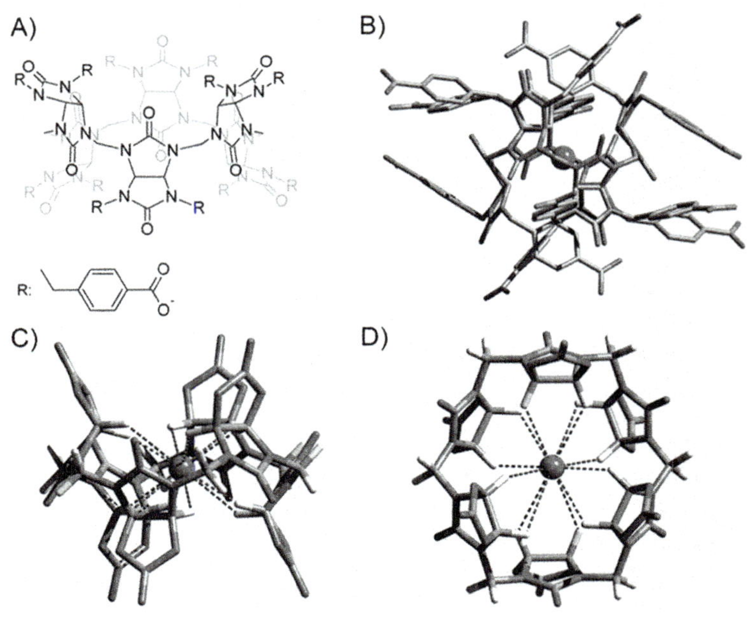

Fig. 1 (**a**) The structure of macrocycle **9**. (**b**) A representation of the chloride complex formed by compound **9** with *side* (**c**) and *top* (**d**) *views* of the complex with hydrogen atoms and benzyl substituents omitted for clarity. Weak C–H···Cl⁻ interactions are shown as *dashed lines*. Reproduced with permission from Yawer et al. [7]. Copyright 2015 Wiley-VCH

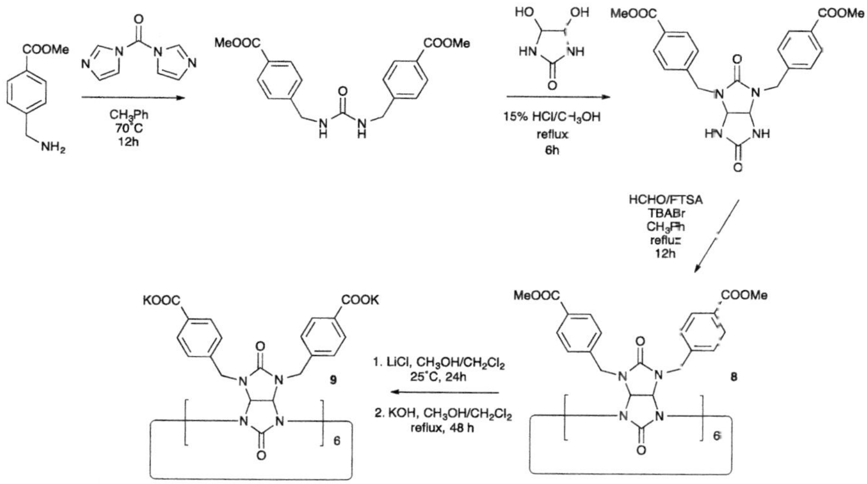

Scheme 3 The synthesis of water-soluble bambusuril macrocycle **9**

Table 1 Association constants K_a[a] for 1:1 complexes of bambusuril macrocycle **9** with various anions determined by ^{1}H NMR spectroscopy (D_2O, 20 mM K_2DPO_4, pD 7.1, 303.15 K)

Anion	K_a (M^{-1})
F^-	1.1×10^2
Cl^-	9.1×10^2
CN^- [b]	1.1×10^3
IO_4^-	6.5×10^3
ReO_4^-	3.0×10^4
Br^-	1.4×10^5
NO_3^-	4.8×10^5
PF_6^-	2.2×10^6
BF_4^-	4.3×10^6
I^-	1.0×10^7
ClO_4^-	5.5×10^7

[a]Standard deviations calculated from two independent measurements are lower than 10%
[b]CN^- is largely protonated under experimental conditions

properties of $\mathbf{12}^{2+}.2PF_6^-$ were measured by ^{1}H NMR titration techniques in 9:1 CD_3OD/D_2O solution and compared to the protic analogue $\mathbf{13}^{2+}.2PF_6^-$. The results show that the halogen bond-based system $\mathbf{12}^{2+}.2PF_6^-$ forms significantly stronger complexes with bromide and iodide in this competitive solvent mixture than analogue $\mathbf{13}^{2+}.2PF_6^-$ (Table 2). An X-ray crystal structure of the bromide complex of $\mathbf{12}^+$ is shown in Fig. 2 with Br···Br^- bond lengths of 3.217(1) Å.

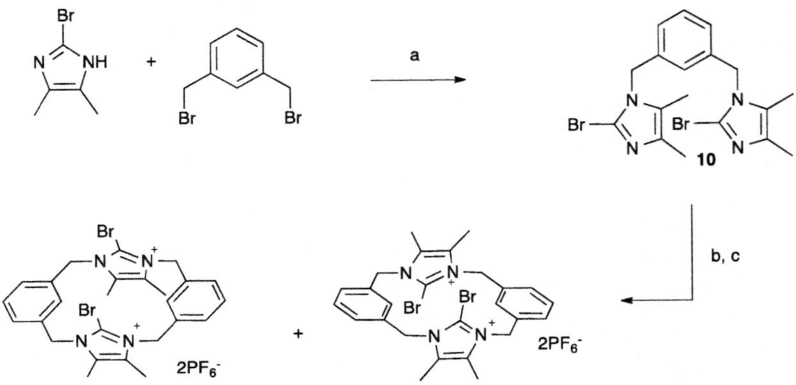

$12^{2+}.2PF_6^-$ (*syn*) $11^{2+}.2PF_6^-$ (*anti*)

Scheme 4 Synthesis of receptors $\mathbf{11}^{2+}.2PF_6^-$ and $\mathbf{12}^{2+}.2PF_6^-$. Reagents and conditions: (*a*) NaOH (1 M in water), acetonitrile, reflux, yield 80%; (*b*) 1,3-bis(bromomethyl)benzene, acetonitrile, 82°C, yield: 82%; (*c*) wash with saturated NH_4PF_6(aq)

Table 2 Association constants K_a for $\mathbf{12}^{2+}.2PF_6^-$ (*syn*) and $\mathbf{13}^{2+}.2PF_6^-$ with halide ions[a]

Receptor	Anion[b]	K_a (M^{-1})[c]
$\mathbf{12}^{2+}.2PF_6^-$	F^-	–[d]
	Cl^-	<10
	Br^-	889 (37)
	I^-	184 (15)
$\mathbf{13}^{2+}.2PF_6^-$	F^-	–[d]
	Cl^-	133 (12)
	Br^-	130 (10)
	I^-	102 (7)

[a]In 9:1 CD_3OD/D_2O at 295 K
[b]Anions have been used as a tetrabutylammonium salt
[c]Obtained from monitoring the internal phenyl proton
[d]No detectable evidence of binding observed; error in parentheses

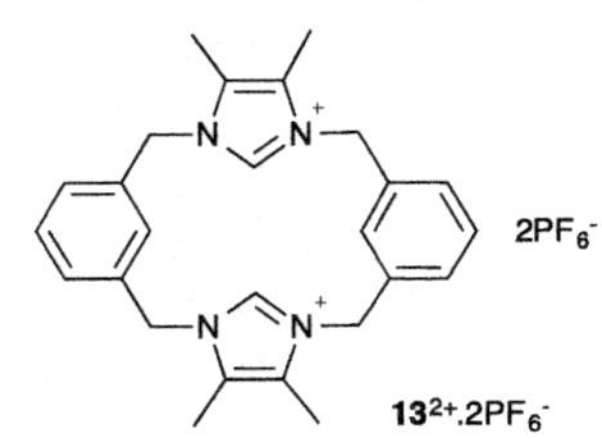

$13^{2+}.2PF_6^-$

Taylor and co-workers have combined hydrogen and halogen-bonding groups in a series of receptors including compound **14a** [10]. The anion complexation properties of this receptor were compared to analogue **14b** which does not contain an iodine substituent and therefore cannot form halogen-bonding interactions to anionic guests. The stability constants for a range of anionic guests were elucidated

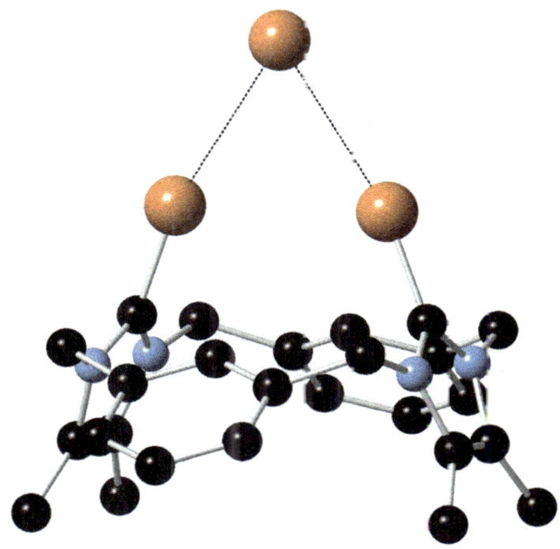

Fig. 2 The X-ray crystal structure of the bromide complex of macrocycle 12^{2+}. Hydrogen atoms, methanol and lattice bromide anions are omitted for clarity

Table 3 Association constants (K_a) of receptors **14a** and **14b** with halide anions and free energy contributions of the halogen bond interaction ($\Delta\Delta G_{XB}$)

Receptor	Anion	K_a $(M^{-1})^a$	$\Delta\Delta G_{XB}$ (kcal/mol)
14a	BzO^-	3.3×10^3	-0.7 ± 0.1
14b	BzO^-	1.0×10^3	
14a	Cl^-	2.4×10^3	-2.0 ± 0.1
14b	Cl^-	86	
14a	$H_2PO_4^-$	8.3×10^2	-0.8 ± 0.1
14b	$H_2PO_4^-$	2.0×10^2	
14a	Br^-	7.0×10^2	-2.0 ± 0.1
14b	Br^-	24	
14a	I^-	1.4×10^2	-1.7 ± 0.1
14b	I^-	8.5	
14a	TsO^-	54^b	-0.2 ± 0.1
14b	TsO^-	38^b	
14a	HSO_4^-	30^b	-0.3 ± 0.1
14b	HSO_4^-	19^b	
14a	NO_3^-	21	-0.2 ± 0.1
14b	NO_3^-	14	

[a]Determined by fitting changes in ^1H NMR and ^{19}F-NMR chemical shift as a function of anion concentration to a 1:1 binding isotherm (tetrabutylammonium cation, CD_3CN solvent, carried out in duplicate: uncertainty in K_a values estimated to be 20%). The reported values are the averages of K_a determinations using changes in chemical shift for one urea N_H proton and one fluorine substituent

[b]Association constants for TsO^- and HSO_4^- were determined by ^{19}F-NMR because of significant broadening of the signals corresponding to the NH protons in the ^1H NMR spectrum

by ^1H and ^{19}F NMR titration techniques and the difference in binding energy between the two compounds estimated. The results shown in Table 3 reveal the preference for halogen-bonding interactions to halides over oxo-anionic guests.

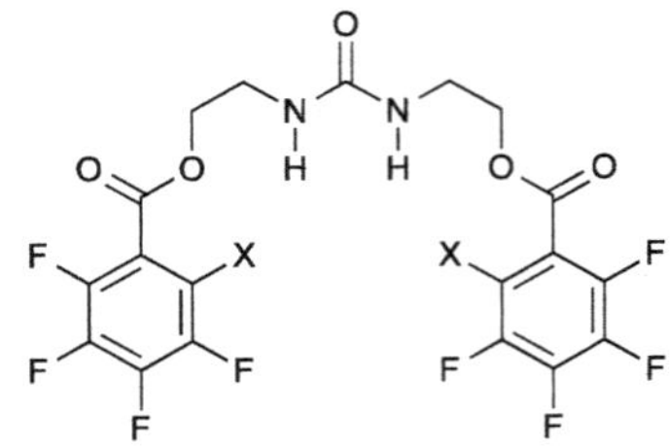

14a: X = I
14b: X = F

Anion–π interactions have come under the spotlight over the last 7 years. In 2008 Hay and Bryantsev published an important article that summarised the experimental and theoretical evidence for the nature of the adducts formed between anions and arenes [11]. The interactions of anions with the π-systems were analysed and three binding modes that were distinguished by their geometry and degree of covalency identified. The authors found that addition of electron-withdrawing substituents to aromatic rings can result in strong hydrogen bonds from the C–H groups of the aromatic systems in excess of $-100\ \mathrm{kJ\ mol^{-1}}$. Nucleophilic anions such as fluoride or cyanide were found to form strong σ-complexes with electron-deficient arenes, whilst anion–π interactions were found to be most likely with large, charge-diffuse anions such as hexafluorophosphate, tetrafluoroborate and perchlorate.

Gamez and co-workers published a review of anion–π interactions in 2011 which in addition to surveying synthetic supramolecular examples of this type of complex also included a search of the Protein Data Bank [12]. This revealed short anion–π distances in some protein structures including glutathione S-transferase from *Xylella fastidiosa*. In this structure, a chloride is found located over the indole ring of a tryptophan residue with $d_{A-\pi} = 3.72$ Å. Interestingly a recent report on ketosteroid isomerase provides compelling evidence that anion–aromatic interactions between the anionic Asp38 general base and two phenylalanine residues Phe54 and Phe116 may position the general base in the active site of the protein in lieu of a hydrogen-bonding network in this case via C–H anion interactions [13].

3 Anion Transport Across Lipid Bilayers

The development of small-molecule anion transporters to facilitate the transport of anions across lipid bilayer membranes has progressed significantly since 2008. Up until this point, most studies had focused on chloride transport; however, since then studies have expanded to other anions including bicarbonate [14, 15], the transport of which, like chloride, is impaired through the cystic fibrosis transmembrane regulator protein in the epithelial cell membranes of cystic fibrosis patients. There has also been a move from more complex supramolecular transporters to simpler more "drug-like" molecules [16]. For example, in 2013 Gale and co-workers conducted a quantitative structure–activity relationship (QSAR) study using

Hansch analysis [17] on a series of simple thiourea-based anion transporters **15–36** in order to determine the molecular parameters that should be optimised to produce the most effective transmembrane transporters. In this series of compounds, it was found that transport was predominantly determined by the lipophilicity (log P) of the transporter but that the Hammett constant of the substituent in the *para*-position of the aromatic ring (which was shown to correlate with the affinity of the transporters for anions) and the size of the transporters played significant albeit lesser roles [18].

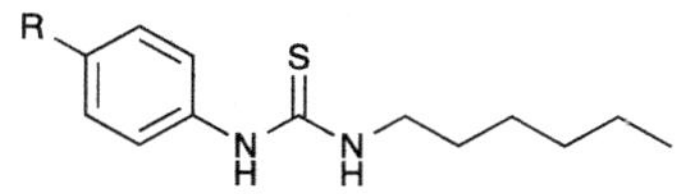

15	R = Br	**26**	R = O(CO)Me
16	R = CF_3	**27**	R = OCF_3
17	R = Cl	**28**	R = OEt
18	R = CN	**29**	R = OMe
19	R = $COCF_3$	**30**	R = SMe
20	R = COMe	**31**	R = SO_2Me
21	R = COOMe	**32**	R = CH_3
22	R = F	**33**	R = CH_2CH_3
23	R = H	**34**	R = $(CH_2)_2CH_3$
24	R = I	**35**	R = $(CH_2)_3CH_3$
25	R = NO_2	**36**	R = $(CH_2)_4CH_3$

Other very active yet structurally simple transporters that have been developed by Gale and co-workers include squaramide-based anion transporters, e.g. **37** [19], and bis-ureas based on an *ortho*-phenylenediamine scaffold, e.g. **38** [20, 21].

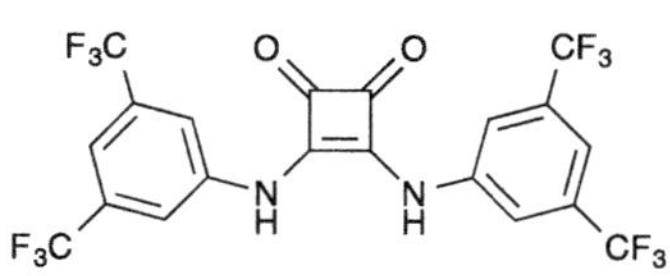

37

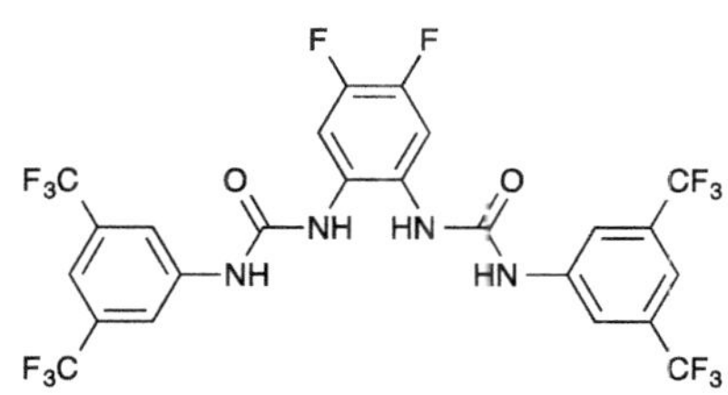

38

Davis and co-workers have employed hydrogen bond donor groups in combination with lipophilic scaffolds to produce highly effective transmembrane chloride transporters such as compound **39** [22]. This compound was used in a study by Kros, Davis and co-workers using giant unilamellar vesicles (GUVs) containing encapsulated lucigenin – a fluorescent indicator for chloride that is quenched by the presence of the anion [23]. This allowed for visualisation of individual vesicles by fluorescence microscopy and the direct observation of fluorescence quenching of internal lucigenin due to influx of chloride into the GUVs mediated by compound **39**.

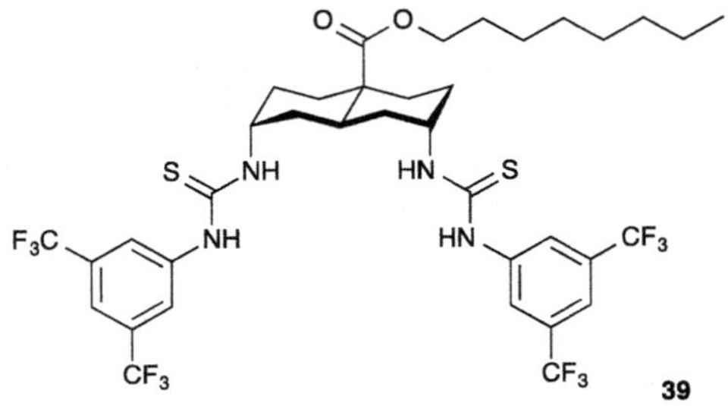

39

The next step in the development of synthetic transmembrane anion transporters is to study transport processes in cells and the effects of transport on cell viability. This work has begun recently with the effects of anion transporters on a range of cancer cell lines studied upon addition of tren-based tris-ureas and tris-thioureas [24], indole-functionalised ureas and thioureas [16] and tambjamines [25] with all these systems depolarising acidic compartments within the cells and triggering apoptosis. Recently Gale, Sessler, Shin and co-workers have shown that pyridine-dicarboxamide strapped calixpyrroles **40** and **41** can mediate the transport of chloride in Fischer rat thyroid (FRT) cells that express yellow fluorescent protein (the protein fluorescence is partially quenched upon influx of chloride into the cells). The transporters were also shown to concomitantly trigger the influx of sodium into a range of cancer lines through endogenous sodium channels present in the cell membranes. Further studies demonstrated that this transporter-channel coupled influx of sodium chloride resulted in an increase in reactive oxygen species within the cells and subsequent apoptosis [26].

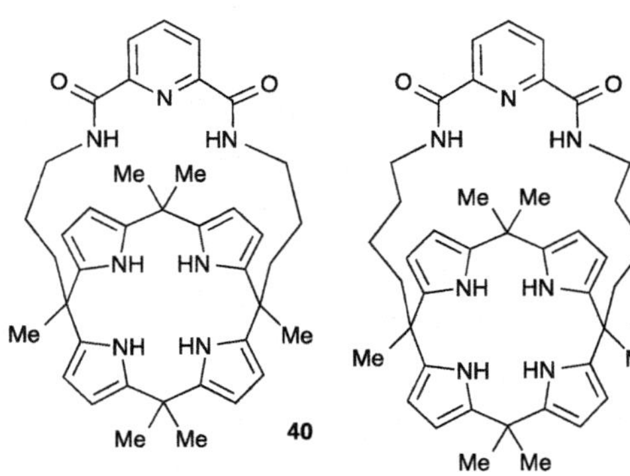

40 41

Matile and co-workers have developed a number of channel- and carrier-type membrane transport systems employing halogen-bonding interactions. These include functionalised calixarenes that function as ditopic ion transport systems [27] and transmembrane halogen-bonding cascades, e.g. **42** [28]. In 2012 Matile and co-workers reported that, remarkably, very simple organic molecules such as trifluoroiodomethane (which is gaseous under ambient conditions) are capable of mediating anion transport offering the possibility of halogen-bond-based transporters that may be delivered by inhalation [29].

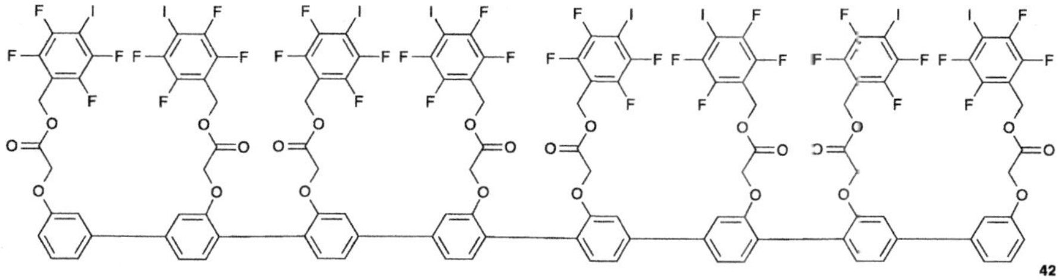

42

Returning to the hydrogen-bonding arena, the use of lipids themselves as transporters has begun to be explored. J. T. Davis and co-workers have shown that the sphingolipid ceramide **43**, which forms transmembrane pores at high concentrations, can act as a discrete molecular transporter at lower transporter to lipid ratios [30]. The compound forms complexes with anions through the formation of OH and NH hydrogen bonds and is capable of transporting chloride and bicarbonate through EYPC lipid bilayers.

Smith and co-workers have functionalised lipids with anion binding urea groups (**44** and **45**) and also prepared a control compound that has a very low affinity for chloride (**46**) [31]. Compounds **44** and **45** act as chloride transporters via a relay mechanism wherein the anion is complexed by the urea of a lipid in one leaflet of the membrane and transported to the centre of the bilayer where it is handed to a functionalised lipid in the other leaflet. Consequently transport only occurs when urea-functionalised lipids are present in both the inner and outer leaflets of the lipid bilayer membrane.

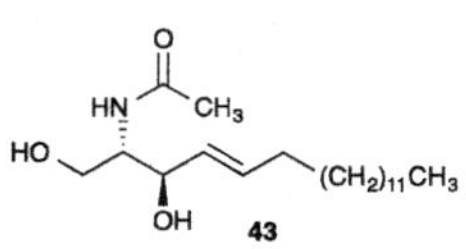

43

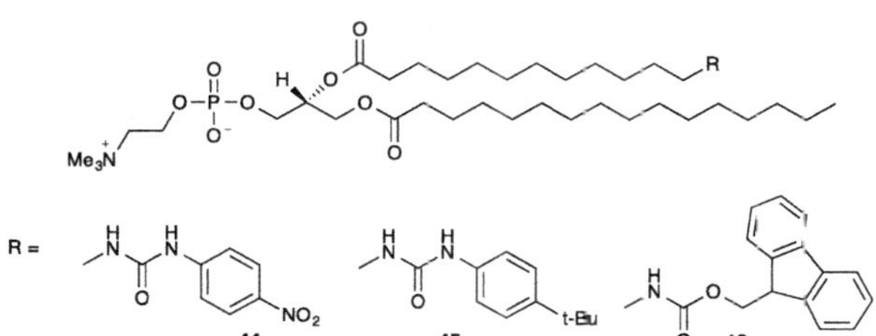

44 45 46

4 Assembly and Interlocked Structures

The use of anions to template self-assembled structures has also significantly progressed since 2008. One of the first examples of an anion-templated self-assembled structure was Lehn's chloride-templated pentameric circular double

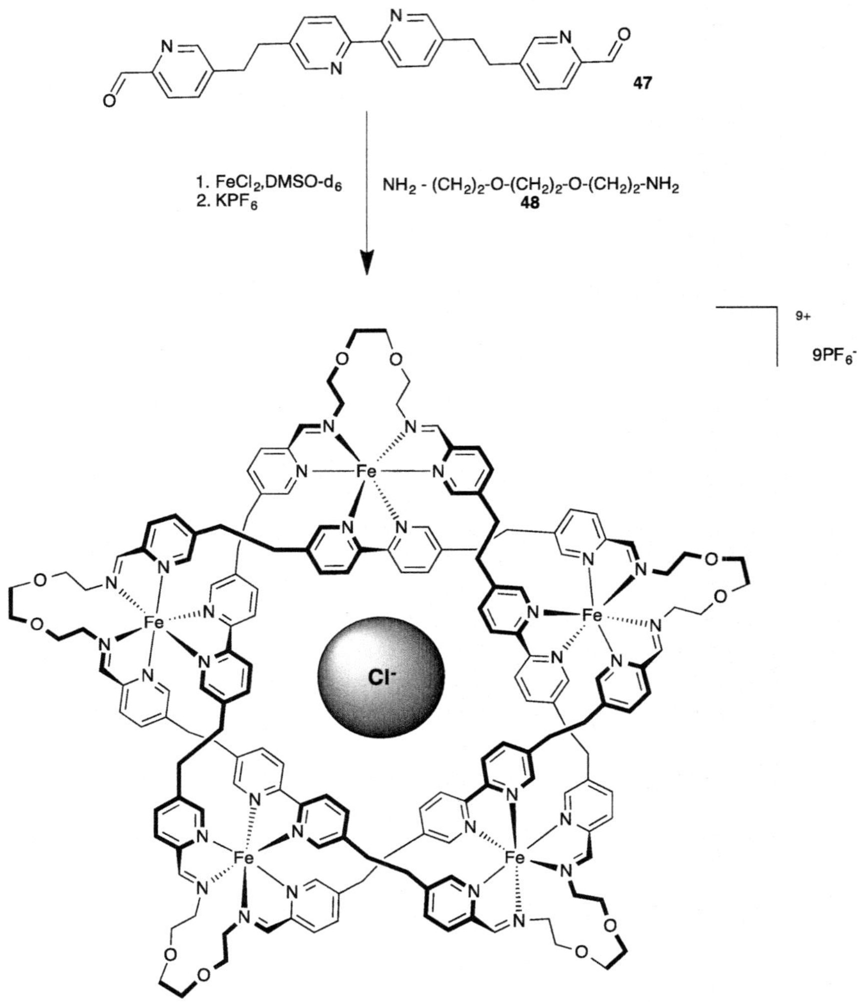

Scheme 5 Synthesis of the molecular pentafoil knot

helicate [32]. This structure was stabilised via C–H···Cl^- interactions in the centre of the complex. This metal containing organic framework demonstrated very clearly the role that anions can play in self-assembly processes. Leigh and co-workers have revisited this paradigmatic anion-templated structure, and by employing iron-coordinating pyridine ligands that are terminated by aldehyde groups (**47**) when the circular helicate is formed in the presence of diamine **48** (Scheme 5), adjacent terminal ends of the pyridine ligands are linked via imine bonds resulting in a continuous pentafoil knot structure around the templating chloride anion (Fig. 3) [33].

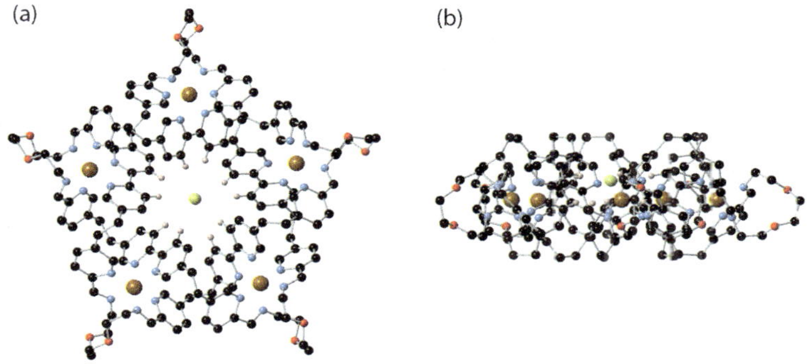

Fig. 3 (**a**) *Top* and (**b**) *side views* of the X-ray crystal structure of Leigh's chloride-templated pentafoil knot. Hydrogen atoms (except those binding the templating chloride anion) have been omitted for clarity

Table 4 Anion association constants K_a (M^{-1}) in water

K_a (M^{-1})					
		Cl^-	Br^-	I^-	SO_4^{2-}
Triazole receptors					
HB	Acyclic **49**	–[a]	–[a]	5	–[a]
	Rotaxane **51**	–[a]	10	20	–[a]
XB	Acyclic **50**	–[a]	15	40	–[a]
HB/XB	Rotaxane **52**	55	290	2,200	30
Amide receptors					
HB	Acyclic **53**	–[a]	–[a]	–[a]	–[a]
	Rotaxane **54**	15	35	50	–[a]

[a]No binding. Errors were estimated to be <10%

Beer and co-workers have continued their work on the synthesis of anion-templated rotaxanes and catenanes [34, 35]. Recently this group have used rotaxanes that employ a combination of halogen and hydrogen bonding to form strong complexes with halides in water [36]. Acyclic and rotaxane receptors **49–54** were synthesised that contain cyclodextrin stoppers. The compounds either contain solely halogen bond donors (acyclic receptor **50**), hydrogen bond donors (acyclic receptor **49**, rotaxane **51**, acyclic receptor **53** and rotaxane **54**) or a mixture of both (rotaxane **52**). Association constants were measured by ^{1}H NMR spectroscopy anion binding titrations in D_2O at 298 K. Anions were added as the sodium salts and association constants calculated using WinEQNMR2 with the chemical shift data of the internal protons of the binding site. Rotaxane receptor **52** which employs a combination of hydrogen and halogen bonds formed a particularly strong complex with iodide (2,200 M^{-1}) in water as compared to the hydrogen bond only analogue rotaxane **51** (20 M^{-1}) (Table 4).

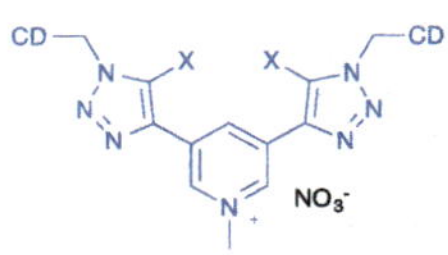

49 X = H
50 X = I

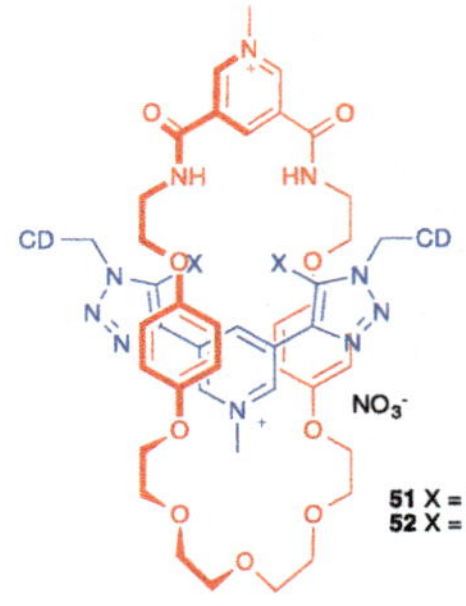

51 X = H
52 X = I

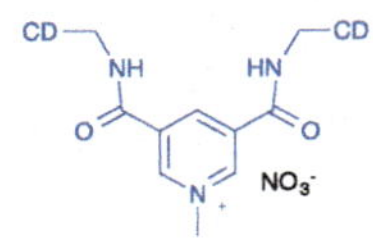

53

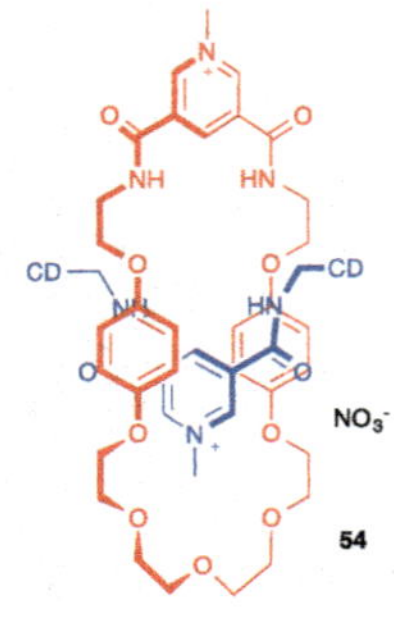

54

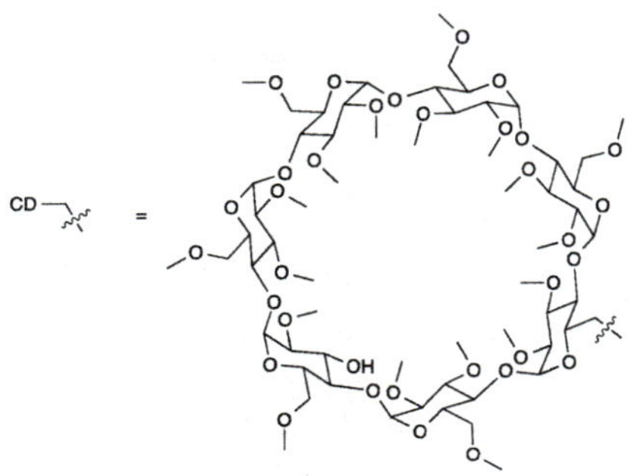

5 Conclusions

This short update has highlighted only a few of the advances in the supramolecular chemistry of anionic species since 2008. There have been many other significant steps forward over the last 7 years in the areas of extraction of anions from mixtures, the development of new sensors and sensing arrays for anionic species and in anion complexation in organocatalysis [37]. Anion receptor chemistry continues to provide new fundamental insights into supramolecular interactions, but we are also seeing exciting new applications for this important area of chemistry.

Acknowledgements PAG thanks the Royal Society and the Wolfson Foundation for a Research Merit Award.

References

1. Recognition of Anions (2008) Structure and bonding, vol 129. Springer, Berlin
2. Bates GW, Gale PA (2008) An introduction to anion receptors based on organic frameworks. Struct Bond 129:1–44

3. Li Y, Flood AH (2008) Strong, size-selective, and electronically tunable C–H··· halide binding with steric control over aggregation from synthetically modular, shape-persistent [34] triazolophanes. J Am Chem Soc 130:12111–12122
4. Li Y, Flood AH (2008) Pure C-H hydrogen bonding to chloride ions: a preorganised and rigid macrocyclic receptor. Angew Chem Int Ed 120:2689–2692
5. McDonald K, Hua Y, Flood A (2010) 1,2,3-Triazoles and the expanding utility of charge neutral CH··· anion interactions. In: Gale PA, Dehaen W (eds) Anion recognition in supramolecular chemistry, vol 24, Topics in heterocyclic chemistry. Springer, Heidelberg, pp 341–366
6. Lee S, Chen C-H, Flood AH (2013) A pentagonal cyanostar macrocycle with cyanostilbene CH donors binds anions and forms dialkylphosphate [3]rotaxanes. Nat Chem 5:704–710. doi:10.1038/nchem.1668
7. Yawer MA, Havel V, Sindelar V (2015) A bambusuril macrocycle that binds anions in water with high affinity and selectivity. Angew Chem Int Ed 54:276–279
8. Metrangolo P, Meyer F, Pilati T, Resnati G, Terraneo G (2008) Halogen bonding. Angew Chem Int Ed 47:6114–6127
9. Caballero A, White NG, Beer PD (2011) A bidentate halogen-bonding bromoimidazoliciphane receptor for bromide ion recognition in aqueous media. Angew Chem Int Ed 50:1845–1848
10. Chudzinski MG, McClary CA, Taylor MS (2011) Anion receptors composed of hydrogen- and halogen-bond donor groups: modulating selectivity with combinations of distinct noncovalent interactions. J Am Chem Soc 133:10559–10567
11. Hay BP, Bryantsev VS (2008) Anion–arene adducts: C–H hydrogen bonding, anion–pi interaction, and carbon bonding motifs. Chem Commun 2417–2428
12. Robertazzi A, Krull F, Knapp E-W, Gamez P (2011) Recent advances in anion-pi interactions. CrystEngComm 13:3293–3300
13. Schwans JP, Sunden F, Lassila JK, Gonzalez A, Tsai Y, Herschlag D (2013) Use of anion-aromatic interactions to position the general base in the ketosteroid isomerase active site. Proc Natl Acad Sci U S A 110:11308–11313
14. Davis JT, Gale PA, Okunola OA, Prados P, Iglesias-Sánchez JC, Torroba T, Quesada R (2009) Using small molecules to facilitate exchange of bicarbonate and chloride anions across liposomal membranes. Nat Chem 1:138–144
15. Casey JR (2006) Why bicarbonate? Biochem Cell Biol 84:930–939
16. Moore SJ, Wenzel M, Light ME, Morley R, Bradberry SJ, Gomez-Iglesias P, Soto-Cerrato V, Perez-Tomas R, Gale PA (2012) Towards "drug-like" indole-based transmembrane anion transporters. Chem Sci 3:2501–2509
17. Hansch C, Fujita T (1964) J Am Chem Soc 86:1616–1626
18. Busschaert N, Bradberry SJ, Wenzel M, Haynes CJE, Hiscock JR, Kirby IL, Karagiannidis LE, Moore SJ, Wells NJ, Herniman J, Langley GJ, Horton PN, Light ME, Marques I, Costa PJ, Felix V, Frey JG, Gale PA (2013) Towards predictable transmembrane transport: QSAR analysis of anion binding and transport. Chem Sci 4:3036–3045
19. Busschaert N, Kirby IL, Young S, Coles SJ, Horton PN, Light ME, Gale PA (2012) Squaramides as potent transmembrane anion transporters. Angew Chem Int Ed 51:4426–4430
20. Moore SJ, Haynes CJE, Gonzalez J, Sutton JL, Brooks SJ, Light ME, Herniman J, Langley GJ, Soto-Cerrato V, Perez-Tomas R, Marques I, Costa PJ, Felix V, Gale PA (2013) Chloride, carboxylate and carbonate transport by ortho-phenylenediamine-based bisureas. Chem Sci 4:103–117
21. Karagiannidis LE, Haynes CJE, Holder KJ, Kirby IL, Moore SJ, Wells NJ, Gale PA (2014) Highly effective yet simple transmembrane anion transporters based upon ortho-phenylenediamine bis-ureas. Chem Commun 50:12050–12053
22. Valkenier H, Judd LW, Li H, Hussain S, Sheppard DN, Davis AP (2014) Preorganized bis-thioureas as powerful anion carriers: chloride transport by single molecules in large unilamellar vesicles. J Am Chem Soc 136:12507–12512
23. Valkenier H, López Mora N, Kros A, Davis AP (2015) Visualisation and quantification of transmembrane ion transport into giant unilamellar vesicles. Angew Chem Int Ed 54:2137–2141

24. Busschaert N, Wenzel M, Light ME, Iglesias-Hernández P, Pérez-Tomás R, Gale PA (2011) Structure–activity relationships in tripodal transmembrane anion transporters: the effect of fluorination. J Am Chem Soc 133:14136–14148
25. Hernando E, Soto-Cerrato V, Cortes-Arroyo S, Perez-Tomas R, Quesada R (2014) Transmembrane anion transport and cytotoxicity of synthetic tambjamine analogs. Org Biomol Chem 12:1771–1778
26. Ko S-K, Kim SK, Share A, Lynch VM, Park J, Namkung W, Van Rossom W, Busschaert N, Gale PA, Sessler JL, Shin I (2014) Synthetic ion transporters can induce apoptosis by facilitating chloride anion transport into cells. Nat Chem 6:885–892
27. Vargas Jentzsch A, Emery D, Mareda J, Metrangolo P, Resnati G, Matile S (2011) Ditopic ion transport systems: anion-pi interactions and halogen bonds at work. Angew Chem Int Ed 50:11675–11678
28. Vargas Jentzsch A, Matile S (2013) Transmembrane halogen-bonding cascades. J Am Chem Soc 135:5302–5303
29. Jentzsch AV, Emery D, Mareda J, Nayak SK, Metrangolo P, Resnati G, Sakai N, Matile S (2012) Transmembrane anion transport mediated by halogen-bond donors. Nat Commun 3:905
30. Harrell WA Jr, Bergmeyer ML, Zavalij PY, Davis JT (2010) Ceramide-mediated transport of chloride and bicarbonate across phospholipid membranes. Chem Commun 46:3950–3952
31. McNally BA, O'Neil EJ, Nguyen A, Smith BD (2008) Membrane transporters for anions that use a relay mechanism. J Am Chem Soc 130:17274–17275
32. Hasenknopf B, Lehn J-M, Kneisel BO, Baum G, Fenske D (1996) Self-assembly of a circular double helicate. Angew Chem Int Ed 35:1838–1840
33. Ayme J-F, Beves JE, Leigh DA, McBurney RT, Rissanen K, Schultz D (2012) A synthetic molecular pentafoil knot. Nat Chem 4:15–20
34. Spence GT, Beer PD (2013) Expanding the scope of the anion templated synthesis of interlocked structures. Acc Chem Res 46:571–586
35. Langton MJ, Beer PD (2014) Rotaxane and catenane host structures for sensing charged guest species. Acc Chem Res 47:1935–1949
36. Langton MJ, Robinson SW, Marques I, Félix V, Beer PD (2014) Halogen bonding in water results in enhanced anion recognition in acyclic and rotaxane hosts. Nat Chem 6:1039–1043
37. Busschaert N, Caltagirone C, Van Rossom W, Gale PA (2015) Chem Rev. doi:10.1021/acs.chemrev.5b00099

Struct Bond (2016) 172: 35–48
DOI: 10.1007/430_2015_198

Published online: 20 January 2016

Single-Molecule Magnets and Related Phenomena

Rodolphe Clérac and Richard E.P. Winpenny

Abstract A personal perspective is given of progress in single-molecule magnets (SMMs) and related phenomena over the last decade. Progress is discussed under seven headings: "Lanthanide Single-Molecule Magnets", "Organometallic and Low Coordination Number SMMs", "Single-Chain Magnets", "SMMs on Surfaces and on the Route to Spintronic Devices", "Molecular Nanomagnets for Magnetic Cooling", "Molecular Nanomagnets for Quantum Information Processing" and "Photomagnets". In all areas, substantial progress has been made, but the proposition is made that prototype devices are now needed for the field to make further progress.

Keywords Magnetic cooling • Magnetic relaxation • Photomagnetism • Single-chain magnets • Single-molecule magnets • Transition metal clusters

Contents

R. Clérac
CNRS, CRPP, UPR 8641, 33600 Pessac, France

University of Bordeaux, CRPP, UPR 8641, 33600 Pessac, France

R.E.P. Winpenny (✉)
The School of Chemistry, The University of Manchester, Oxford Road, Manchester M13 9PL, UK
e-mail: richard.winpenny@manchester.ac.uk

1 Structure and Bonding

When one of us was a postgraduate at Imperial College, the physical appearance of the *Structure and Bonding* series was part of the reason I started to read it. They were thin, approachable volumes with the title and volume number in very large font at the top of the friendly green front cover. The reason I read more than one volume was that the content surveyed modern inorganic chemistry quickly and authoritatively, saving me a great deal of time. I found much that influenced my eventual research from these thin volumes, for example, reviews on polymetallic compounds [1, 2] and their magnetic properties [3] and reviews on the seemingly incomprehensible magnetic properties of high-spin cobalt(II) [4]. And then when working in Texas on gold clusters, I spent a great deal of time pouring over a lovely article by Schmid on how to build larger clusters [5] and an article by the editor of this volume on theoretical treatments of the bonding in such clusters [6]. The volume containing that second article also contained a fascinating review by the late Olivier Kahn concerning how to design magnetic molecules with predictable properties [7]; this review hugely influenced what I chose to do once I had an independent position. The juxtaposition of the review I needed to read preceding a review that pointed in a new direction was always part of the charm of S&B. It is one of the losses of the computer age that now we always find exactly the article we want quickly, and it is much more difficult to become productively distracted.

It was therefore with a great deal of pleasure that I accepted Peter Day's invitation to edit a volume of *Structure and Bonding*, on *Single-Molecule Magnets and Related Phenomena* [8]. The authors of the various chapters wrote excellent reviews, two are officially "highly cited", and the volume has received over 1,200 citations in total. It was also a great pleasure when the volume was updated in 2015 under the excellent editorship of Prof. Song Gao [9].

This short perspective on molecular nanomagnets (and related phenomena) has been co-written with Dr. Rodolphe Clérac – who contributed superb reviews to both of the previous volumes of *Structure and Bonding*.

2 Importance and Significance of Single-Molecule Magnets and Related Phenomena

The discovery in the early 1990s that a single molecule could retain magnetisation in the absence of a magnetic field was extremely exciting [10, 11]. At the time it was hoped that such molecules could revolutionise information storage; most magnetic information is stored in arrays of magnetic nanostructures, and, in 1993, a molecule was around 10^5 times smaller than the smallest feature in use. Sadly, in the first decade after the initial discovery, progress towards any technological applications was very slow.

The major challenge was the temperature at which these novel materials – rapidly named "single-molecule magnets" (SMMs) – operated. In the best examples, an extensive family of $\{Mn_{12}\}$ cages [12], the energy barrier to magnetic relaxation (U_{eff}) is around 70 K. This means that the magnetic information could only be stored at temperatures that require liquid helium for cooling. Commonly, such a temperature is considered to be too low for any real application, but it is important to understand that if extraordinary behaviour or performance of SMM-based devices could be discovered, then low temperatures would not be a problem; for example, liquid He is used to cool the superconductor coils in modern NMR spectrometers and magnetic resonance imagers.

The energy barrier arises from a combination of a high-spin ground state ($S = 10$ for $\{Mn_{12}\}$) and significant axial anisotropy of that ground state, normally reported as the axial zero-field splitting parameter D. For integer spins $U_{eff} = |D|S^2$, if D is negative, which conventionally is the sign chosen for easy-axis anisotropy (with $H = DS_z^2$). Until 2003, this largest energy barrier observed did not change very much. By contrast, the number of techniques available routinely for studying the magnetic behaviour increased dramatically. In the 1990s, characterisation of SMMs involved low-temperature direct current (d.c.) susceptibility measurements, and, as the temperatures involved were so low, there were few centres in the world that could discover SMMs [13, 14]. At the beginning of this century, the use of alternating current (a.c.) susceptibility measurements became common which became an alternative method to discover SMMs; the availability of a.c. susceptibility measurements led to a huge increase in the number of reported SMMs. High-frequency and high-field EPR spectroscopy became more common and even the use of other spectroscopic techniques such as magnetic circular dichroism (MCD) spectroscopy [15]. The use of MCD on a solution of a $\{Mn_{12}\}$ cage proved that the slow relaxation was due to individual molecules rather than any intermolecular interaction.

Two new groups of compounds were discovered in the early 2000s, which both offered higher energy barriers for magnetic relaxation. Firstly, one-dimensional (1D) systems were discovered in 2001 to exhibit slow dynamics of their magnetisation [16] as predicted in 1963 by R. J. Glauber in the frame of the Ising model [17]. These compounds were named single-chain magnets (SCMs) [18] by analogy of their properties to SMMs.

In contrast to SMMs for which the slow relaxation of the magnetisation is the signature of isolated anisotropic complexes (vide supra), the magnetisation dynamics in SCMs arises from the magnetic interactions between anisotropic repeating units along a single chain [19, 20]. As a consequence, the energy barriers found were immediately higher than for the SMMs as both exchange and anisotropy energies contribute to them. Because intrachain magnetic interactions are easier to control experimentally than the intrinsic magnetic anisotropy in SMMs, SCMs appear to be a promising alternative for applications. Nevertheless, the physics of the SCMs is still much less understood than for SMMs; it is particularly intriguing [17, 20] as it often falls between physics that can be assigned to single molecules and physics due to extended lattices.

The second group was based on monometallic complexes of the lanthanides, and there have been huge developments here since 2006; these are covered in the next section.

While technological applications were not forthcoming for SMMs, the field required chemists and physicists to work closely together to understand the physics of existing systems and to then synthesise new compounds to test the new theories developed. SMMs lie at the boundary of quantum systems, and it became possible to perform novel studies of quantum phase interference and parity effects [21] and other novel quantum physics [22]. The close collaborations also created an influx of new ideas around the physics of low-dimensional magnetic materials, including proposals to use molecule-based nanomagnets in quantum computing [23] and in magnetic cooling [24]. These ideas have largely inspired much of the work performed since 2006. The strong interactions between chemists and physicists are probably the most important outcome of the first two decades of studies of molecule-based nanomagnets.

3 Major Developments from 2006 to 2015

It is somewhat dangerous to pick out the major developments over the last decade or so; we restrict ourselves to seven headings, recognising we will have missed something. We apologise for any offence caused.

3.1 Lanthanide Single-Molecule Magnets

Beyond argument, the major change between 2006 and 2015 is the change in the elements studied; to 2006, the vast majority of SMMs involved polymetallic cage complexes of Mn(III), with an occasional report of an SMM involving a cage complex of another 3d-metal [15].

A report from 2003 by Ishikawa and co-workers changed the focus by reporting that the U_{eff} value for a terbium-phthalocyanine (Pc) sandwich complex was an order of magnitude higher than that for $\{Mn_{12}\}$ [25]. This has led to a huge amount of work studying the dynamic magnetism of lanthanide complexes, especially complexes of Dy(III) [26, 27]. The work coincided with the increase use of a.c. susceptibility as a method to study SMMs, and this has perhaps polluted the area a little; almost any Dy(III) complex can be made to show some slow relaxation of magnetisation if an expert is involved in designing the experiment.

The slow relaxation of the lanthanides arises because of their electronic structure. The strong spin-orbit coupling leads to ions with very high total angular momentum, J, and large magnetic moments, especially for the heavier lanthanides Dy(III), Ho(III), Tb(III) and Er(III). In the correct coordination geometries, and hence crystal fields, this can stabilise ground-state doublets with very high m_J values. For Dy(III), the most studied 4f-ion in this context, the ground-state doublet is often $m_J = \pm 15/2$ with the other doublets at higher energy. This is a very similar energy spectrum to that observed for 3d-SMMs, except that we are discussing m_J levels, not m_S levels, and the splitting of the doublets can, in principle, be controlled by the symmetry of the crystal fields.

The highest U_{eff} values now reported are 652 cm^{-1} for a derivative of $\{TbPc_2\}$ [28] and 585 cm^{-1} for Dy(III) doped into an yttrium alkoxide cage [29]. However, higher energy barriers have not led to the same increase in the "blocking temperature" T_B, i.e. the temperature at which the magnetic properties of the system show a dependence on history. The highest value published is around 14 K, for a radical-bridged Tb(III) dimer [30].

The reason for the disparity is that the relaxation mechanisms involved in lanthanide SMMs are far more complex than for 3d-SMMs. This is probably caused by the very large thermal barrier; in 3d-SMMs, the U_{eff} value is normally 60 cm^{-1} at the best, and hence other routes to relaxation do not really need to be considered to explain the dynamics. For lanthanides, other processes are clearly important, and a feature in the next few years will be the development of methods to measure, model and understand the relaxation processes in these new compounds.

The close relationship between the crystal field and the splitting of doublets in lanthanide complexes also allows challenges to be laid down to synthetic chemists in this area. For Dy(III) or Tb(III), the ideal crystal field is strongly axial, and the most axial crystal field that could be achieved is $D_{\infty h}$. It is difficult to imagine feasible lanthanide complexes that could have such high symmetry, but the extraordinary developments of metal-organic chemistry have allowed low coordination number 4f-complexes to be made. Unfortunately, the only such complex to approach linearity, a two-coordinate Sm(II) complex, contains an ion which is diamagnetic at low temperature [31]. However, calculations suggest that if a two-coordinate Dy(III) complex could be made, the U_{eff} values found would be double any thus far reported [31].

3.2 Organometallic and Low Coordination Number SMMs

Most lanthanide complexes are found in the +3 oxidation state, and they can exist in this oxidation state almost regardless of the ligands present. This allows ligands to be used that would create diamagnetic, or at least very low spin, complexes if used with d-block metals. Thus, it is possible to make SMMs involving cyclopentadienyl ligands [32] or bridging hydrides [33], or even bridging N_2^{3-} ligands [30], if the paramagnetic centre involved is a 4f-ion. This opens up many more ligands, and perhaps more importantly has brought molecular magnetism to the attention of extraordinarily talented synthetic chemists. Magnetic studies are now being reported of lanthanide complexes in the +2 oxidation states due to the synthetic skills of the Evans group [34].

This has led in turn to studies of low-coordinate 3d-metal complexes as SMMs [35]. Perhaps, the most interesting is a two-coordinate Fe(I) complex which has a huge U_{eff} value of 226 cm^{-1} [36], which is a little higher than the energy barrier in the equivalent two-coordinate Fe(II) complex. It is striking that this massive barrier has been seen in a linear compound, which has more recently been predicted to be the ideal geometry for 4f-SMMs [31].

The involvement of chemists capable of handling ragingly air-sensitive materials has also led to actinide SMMs, chiefly of uranium as both U(III) and U(V) [37] but also neptunium [38] and plutonium [39]. There are still too few examples known for it to be clear how far this work can extend. Understanding the magnetic properties of the 5f-ions could make a major contribution to understanding the electronic structure of these fascinating elements.

3.3 Single-Chain Magnets and Related Systems

While before 2006, the number of papers dedicated to SCMs or closely related systems was less than 30 [40], the last decade has seen the publication of more than 500 papers on this emerging field of research [20, 41–44]. Most of these reports are dedicated to the preparation of new SCMs for which serendipity plays a key role that contrasts with the designed SCM materials reported at the infancy of the field [16, 40]. Among this massive amount of results since 2006, it is worth emphasising the SCMs assembled from paramagnetic radical [16, 45–49] that promote strong magnetic interaction along the chain and thus often lead to the high energy barriers and blocking temperatures. This approach, trying to increase the intrachain interactions as much as possible, is definitely an interesting synthetic strategy to explore further in order to make SCMs at high temperatures, while attempting to organise one-dimensional coordination network of lanthanide ions or more generally weakly interacting spin carriers seems to be a dead end.

Unfortunately, this evolution towards the production of more and more SCM systems does not help to understand the tricky physics of these 1D systems that is much less developed than the chemistry and needs simple model compounds. While SCMs falling in the Ising limit (for which the magnetic anisotropy energy dominates the exchange energy) are now relatively well understood from a theoretical point of view [20], this is still not the case for more complicated materials in terms of spin and interaction topologies. In most of the reports, the characterisation of SCM properties relies mainly on the observation of the dynamics of the magnetisation by a.c. susceptibility, while detailed static (thermodynamic) and dynamic magnetic measurements are both required to prove unambiguously the SCM properties. The future developments of this field of research will necessary need the strong involvement of physicists working closer to chemists to understand the fascinating physics behind these materials.

Another interesting development in this field was to study the influence of the interchain magnetic interactions on the magnetisation dynamics of these systems. As suggested by Sessoli [50] commenting the work of Ishida and co-workers on a radical Co^{II} chain system [51], the intrinsic magnetisation dynamics of SCM units might induce magnetisation hysteresis loop with very large coercivity in 3D magnetically ordered materials. Indeed, Coulon, Miyasaka and co-workers demonstrated that the magnetisation dynamics of SCM is preserved for materials showing a 3D magnetic order, at least when the interchain couplings are weaker than the intrachain interactions [52, 53]. This conclusion was drawn for antiferromagnets and later for canted antiferromagnets [54], which also exhibit large M vs. H hysteresis loops (like classical magnets) due to the presence of SCMs in antiferromagnetic interactions in the crystal packing. These reports point towards the possibility to design new materials with hard magnet properties at high temperature incorporating SCMs but also any units possessing an intrinsic slow magnetisation dynamics like SMMs [52]. Detailed magnetic characterisation (static and dynamic) involving the application of a d.c. magnetic field are now essential to fully understand if a system is an SMM or SCM or possesses a magnetically ordered ground state.

3.4 SMMs on Surfaces and on the Route to Spintronic Devices

In 2006, the first experiments on deposition of SMMs on surfaces were being reported, and this preliminary work looked interesting but clearly had problems [55]. The chief problem was that $\{Mn_{12}\}$, the prototype SMM, decomposed when deposited on any surface.

In the following decade, two SMMs have been studied which are stable on surfaces which led to the idea that SMMs could become components of future molecular spintronic devices [56]. The two most studied SMMs in this context are derivatives of an $\{Fe_4\}$ stars [57], where beautiful X-ray MCD experiments show that the SMM properties are retained when the molecules are bound to surfaces

[58]. Unfortunately, $\{Fe_4\}$ does not have a very high U_{eff}, but clearly this work shows that SMMs could be placed on surfaces without destroying their properties.

The other molecules studied are the derivatives of $\{TbPc_2\}$ where some astonishing results have been reported [59]. These reports show a real promise for esoteric spintronic devices into the future. The most extraordinary results have come from the team of Ruben and Wernsdorfer and have included reports of molecular spin valves [60] and single-molecule transistors [61]. Perhaps, the most extraordinary work involves driving Rabi oscillations in the nucleus of a single Tb(III) ion [62]. This is revolutionary work. It is interesting that such results mirror activities in IBM laboratories, where studies are being pursued on the magnetism of individual metal atoms deposited by scanning tunnelling microscopes [63]. These results point towards devices that do not presently exist, i.e. they are not just a scaled-down version of existing technology. It is in such disruptive new devices that molecular nanomagnets could have impact and could justify liquid helium cooling.

3.5 Molecular Nanomagnets for Magnetic Cooling

The research in the 1990s on 3d-metal cages as SMMs produced a very large number of molecules with high-spin ground states. Unfortunately, not all of these molecules had significant anisotropy of the spin in the ground state, and hence quite a number of the highest-spin molecules were not SMMs. One such molecule is an $\{Fe_{14}\}$ cage made by the McInnes group [64]. This molecule has an $S = 25$ ground state and vanishingly small anisotropy. These parameters are ideal for use as molecular magnetic coolants, using the magnetocaloric effect (MCE).

The MCE is intrinsic to any paramagnet and occurs because the entropy of the system changes on magnetisation or demagnetisation [65]. On magnetisation, the system becomes more ordered; when an external magnetic field is switched off, this additional order is lost and hence the magnetic entropy goes up. If the system is ordered, then that additional entropy is gained by absorbing lattice phonons. This cools the system. The MCE is greatest for isotropic systems and is not so great for SMMs [24].

The observation of significant MCE for molecular nanomagnets has spawned a huge number of papers [66]. As with a.c. susceptibility measurements on Dy(III) complexes, it is quite straightforward, albeit expensive in time and liquid helium, to measure the MCE on isotropic polymetallic cage complexes. Unfortunately, the end result of all these studies is probably that the most important molecular parameter is the percentage, by weight of the isotropic metal present in the material studied; as a result, one of the highest MCE values reported is for $[Gd(OH)(CO_3)]_n$, which is not very far from the gadolinium oxides already used for this application [67]. However, some beautiful new physics has been seen, including signatures of quantum critical points in direct studies of the magnetic cooling by a $\{Gd_7\}$ disc [68].

3.6 Molecular Nanomagnets for Quantum Information Processing

The original idea by Leuenberger and Loss to use crystals of {Mn_{12}} to perform quantum computation [23] has been heavily cited, but comparatively little progress has been made to implement the experiments proposed. This may be because they would be technically extremely demanding.

A later proposal by the Loss group to use molecular nanomagnets with $S = 1/2$ ground states for quantum information processing has led to much more new chemistry [69]. The idea is that an $S = 1/2$ molecule is a two-level system ($m_S = +1/2$ and $m_S = -1/2$) and could function as a quantum bit, invariably abbreviated as qubit. There are many molecules that could so function; one family that has been extensively studied to this end is {Cr_7Ni} rings [70]. One question about using molecular nanomagnets in this context is the phase memory time T_m (also called the coherence time and, confusingly, the decoherence time), which is the time the phase of the information is retained before it decays. Pulsed electron paramagnetic resonance (EPR) spectroscopic studies were reported in 2007 [71] that T_m is around 3 μS if ligands are deuterated but is far shorter if the ligands have natural abundance ^1H presence. This suggested that a major cause of decoherence was electron-nucleus hyperfine interactions, and since 2007 several groups have worked to maximise T_m by minimising the number of H atoms present in molecules and by using simple complexes with rigid ligands. Such approaches have led to T_m values approaching 70 μS, for example, in copper dithiolate complexes [72]. This is extremely promising.

Other implementation strategies have been examined. Probably the most advanced involves simple organic radicals and has been pursued by the Takui group, who has reported a C-NOT gate using an organic di-radical [73]. The Aromí group has also proposed using the Kramer's doublets of hetero-bimetallic lanthanide complexes for two-qubit gates [74]. These two approaches have very different possible implementations but both involve g-engineering. In the organic di-radicals, the very high resolution possible for EPR spectroscopy of organic systems allows the different radicals to be addressed even though the g-values are quite close. In the dilanthanide case, the g-values are vastly different. Very recently, g-engineering has also been used for d-block spin centres, combining the {Cr_7Ni} rings with Cu(II) complexes [75]. Implementation of another universal quantum gate, the √SWAP gate, was proposed using di-radicals within a polyoxometalate [76].

These proposals and experiments show a great deal of promise for this approach, but it is now becoming essential for the area to retain credibility for a simple algorithm to be performed using molecular electron spins. It is also noticeable that all the studies being reported involve simple paramagnets, and not SMMs. However, it was the collaborations established while SMMs were being explored that has led to this somewhat different science of paramagnetic compounds.

3.7 Photomagnets

In the past decade, important research activity has been devoted to photoactive magnetic systems and in particular to magnets that can be controlled or induced by light irradiation, i.e. photomagnets. Three different kinds of photomagnets should be highlighted here: (1) photo-SMMs, (2) photo-SCMs and (3) photomagnets for which a 3D magnetic order is photoinduced. Interestingly, all these systems rely on two different photoactive mechanisms, the well-known spin-crossover properties of the Fe^{II} metal ion and the metal-metal electron transfer processes in cyanido-bridged metal ion assemblies (Prussian blue analogues). In 2012, Oshio and co-workers reported the first photoinduced SMM in an hexanuclear Fe/Co Prussian blue analogue [77], and shortly after in 2013, Long, Smith and co-workers discovered the same photomagnetic properties in mononuclear Fe(II) spin-crossover complexes [78, 79]. Photoinduced SCM properties were first demonstrated by Sato, Oshio and co-workers in one-dimensional Fe/Co and Fe/Fe Prussian blue analogues thanks to the intrachain electron transfer between Co and Fe metal ions through cyanido bridges [80–82] or to the spin-crossover properties of the Fe (II) site [83]. Related to these breakthrough results, it should be mentioned that Bogani and co-workers suggested that the magnetisation dynamics of SCM systems could be photo-controlled by domain-wall kickoff mechanism [84]. Even if this result requires further confirmation on different SCMs, it opens a completely new field of investigations with the ultimate goal to improve the optical switching methods for application.

In molecule-based materials, a photoinduced magnetic order was first discovered in 1996 by Hashimoto, Fujishima and co-workers in 3D Fe/Co Prussian blue analogues [85]; this system has subsequently been studied in great detail by Bleuzen, Verdaguer and co-workers [86]. Since then, only a few examples of photomagnets have been reported, most of them relying on the metal-metal electron transfer mechanism in Prussian blue analogues or related cyanido-based systems [86, 87]. In 2011, the discovery of the first photoinduced magnetic order based on a spin-crossover process is probably one of the most exciting results in the field of molecular magnetism for the past decade. By a judicious choice of the blocking ligand, Ohkoshi, Tokoro and co-workers were able to tune the ligand field of Fe^{II} metal ions in order to stabilise thermally and photoinduced spin-crossover processes at this site and at the same time incorporate this Fe^{II} unit into coordination networks that can promote magnetic interactions between spin carriers (in these materials, between Fe^{II} and Nb^{IV} through cyanido bridges) [88]. This inspiring result led to the design of new photomagnets modulating the Fe^{II} nitrogen-based ligands as reported by Sieklucka and co-workers [89] and Ohkoshi's group that was able to implement chirality to a cyanido-based Fe^{II}/Nb^{IV} photomagnet inducing remarkable magneto-optical switching properties [90].

4 A Prediction of Future Developments

While reviewing progress over the previous 10 years is difficult, making predictions concerning future developments is largely impossible. There is one development that is needed, if the area of molecular nanomagnetism is to continue to thrive.

The field needs some form of device. The area arose from observations of magnetic information storage in individual molecules at very low temperatures [10, 11]. Recent developments have suggested that the temperature at which information could be stored in molecules could rise to temperatures achievable with liquid nitrogen cooling [30, 31], if we can understand relaxation processes in lanthanide SMMs and learn to control them. However, given the maturity of the information storage industry how technologically applicable this will be is debateable. The use of molecular electron spins in quantum information processing might be a more productive area, simply because it would not be competing with such mature technology. Again, however, there is a need to perform some simple algorithms simply to maintain credibility.

The MCE bubble may be coming to an end, simply because studies tend to lead back towards very simple compounds such as [Gd(OH)(CO_3)]. However, possible applications of molecular MCE materials are much easier to see; for example, coating materials with magnetic coolants should be easier from molecular precursors than from inorganic solids. Again, the discovery of a useful device would make a massive difference to work in the area.

The extraordinary experiments reported by Ruben and Wernsdorfer and their co-workers [60–62] possibly show the most promise for exploitation, despite the fact that they are also the most technologically challenging. The justification for such an opinion is that such experiments cannot be performed by other means – the molecular nanomagnet is a key component, and there is no existing technology that would allow similar experiments at any temperature. The drawback of this approach is its lack of scalability; the method for making the devices is stochastic, and a major challenge must be to devise a method to make multiple equivalent single-molecule devices in a controlled and reproducible manner. Such an achievement could lead to implementation of molecular spintronic devices.

References

1. Spiro TG, Saltman P (1969) Struct Bond 6:116
2. Ludi A, Güdel HU (1973) Struct Bond 14:1
3. Griffith JS (1972) Struct Bond 10:87
4. Banci L, Bencini A, Benelli C, Gatteschi D, Zanchini C (1982) Struct Bond 52:37
5. Schmid G (1985) Struct Bond 62:51
6. Mingos DMP, Johnston RL (1987) Struct Bond 68:29
7. Kahn O (1987) Struct Bond 68:89
8. Winpenny REP (ed) (2006) Single-molecule magnets and related phenomena. Struct Bond 122
9. Gao S (ed) (2015) Molecular nanomagnets and related phenomena. Struct Bond 164

10. Caneschi A, Gatteschi D, Sessoli R, Barra AL, Brunel LC, Guillot M (1991) J Am Chem Soc 113:5873
11. Sessoli R, Tsai HL, Schake AK, Wang SY, Vincent JB, Folting K, Gatteschi D, Christou G, Hendrickson DN (1993) J Am Chem Soc 115:1804
12. Bagai R, Christou G (2009) Chem Soc Rev 38:1011
13. Barra AL, Debrunner P, Gatteschi D, Schulz CE, Sessoli R (1996) Europhys Lett 35:133
14. Aromí G, Brechin EK (2006) Struct Bond 122:1
15. McInnes EJL, Pidcock E, Oganesyan VS, Cheesman MR, Powell AK, Thomson AJ (2002) J Am Chem Soc 124:9219
16. Caneschi A, Gatteschi D, Lalioti N, Sangregorio C, Sessoli R, Venturi G, Vindigni A, Rettori A, Pini MG, Novak MA (2001) Angew Chem Int Ed 40:1760
17. Glauber J (1963) J Math Phys 4:294
18. Clérac R, Miyasaka H, Yamashita M, Coulon C (2002) J Am Chem Soc 124:12837
19. Ferbinteanu M, Miyasaka H, Wernsdorfer W, Nakata K, Sugiura K, Yamashita M, Coulon C, Clérac R (2005) J Am Chem Soc 127:3090
20. Coulon C, Pianet V, Urdampilleta M, Clérac R (2015) Struct Bond 164:143
21. Wernsdorfer W, Sessoli R (1999) Science 284:133
22. Sessoli R, Gatteschi D (2003) Angew Chem Int Ed 42:268
23. Leuenberger MN, Loss D (2001) Nature 410:789
24. Torres F, Hernández JM, Bohigas X, Tejada J (2000) Appl Phys Lett 77:3248
25. Ishikawa N, Sugita M, Ishikawa T, Koshihara S, Kaizu Y (2003) J Am Chem Soc 125:8694
26. Woodruff DN, Winpenny REP, Layfield RA (2013) Chem Rev 113:5110
27. Jiang S-D, Wang B-W, Gao S (2015) Struct Bond 164:111
28. Ganivet CR, Ballesteros B, de la Torre G, Clemente-Juan JM, Coronado E, Torres T (2013) Chem Eur J 19:1457
29. Blagg RJ, Ungur L, Tuna F, Speak J, Comar P, Collison D, Wernsdorfer W, McInnes EJL, Chibotaru LF, Winpenny REP (2013) Nat Chem 5:673
30. Rinehart JD, Fang M, Evans WJ, Long JR (2011) J Am Chem Soc 133:14236
31. Chilton NF, Goodwin CAP, Mills DP, Winpenny REP (2015) Chem Commun 51:101
32. Layfield RA, McDouall JJW, Sulway SA, Tuna F, Winpenny REP (2010) Chem Eur J 16:4442
33. Venugopal A, Tuna F, Spaniol TP, Ungur L, Chibotaru LF, Okuda J, Layfield RA (2013) Chem Commun 49:901
34. Meihaus KR, Fieser ME, Corbey JF, Evans WJ, Long JR (2015) J Am Chem Soc 137:9855
35. Craig GA, Murrie M (2015) Chem Soc Rev 44:2135
36. Zadrozny JM, Xiao DJ, Atanasov M, Long GJ, Grandjean F, Neese F, Long JR (2013) Nat Chem 5:577
37. Liddle ST (2015) Angew Chem Int Ed 54:8604
38. Magnani N, Colineau E, Eloirdi R, Griveau JC, Caciuffo R, Cornet SM, May I, Sharrad CA, Collison D, Winpenny REP (2010) Phys Rev Lett 104:197202
39. Magnani N, Colineau E, Griveau JC, Apostolidis C, Walter O, Caciuffo R (2014) Chem Commun 5:8171
40. Coulon C, Miyasaka H, Clérac R (2006) Struct Bond 122:1
41. Miyasaka H, Julve M, Yamashita M, Clérac R (2009) Inorg Chem 48:3420
42. Sun HL, Wang ZM, Gao S (2010) Coord Chem Rev 254:1081
43. Zhang W-X, Ishikawa R, Breedlove B, Yamashita M (2013) RSC Adv 3:3772
44. Gatteschi D, Vindigni A (2014) In: Bartolomé J et al (eds) Molecular magnets, Nanoscience and technology. Springer, Berlin, 191 p
45. Miyasaka H, Madanbashi T, Sugimoto K, Nakazawa Y, Wernsdorfer W, Sugiura K-I, Yamashita M, Coulon C, Clérac R (2006) Chem Eur J 12:7028
46. Bernot K, Bogani L, Caneschi A, Gatteschi D, Sessoli R (2006) J Am Chem Soc 128:7974
47. Tomkowicz Z, Rams M, Balanda M, Foro S, Nojiri H, Krupskaya Y, Kataev V, Bücher B, Nayak SK, Yakhmi JV, Haase W (2012) Inorg Chem 51:9983
48. Ishikawa R, Katoh K, Breedlove BK, Yamashita M (2012) Inorg Chem 51:9123

49. Vaz MGF, Allao Cassaro RA, Akpinar H, Schlueter JA, Lahti PM, Novak MA (2014) Chem Eur J 20:5460
50. Sessoli R (2008) Angew Chem Int Ed 47:5508
51. Ishii N, Okamura Y, Chiba S, Nogami T, Ishida T (2008) J Am Chem Soc 130:24
52. Coulon C, Clérac R, Wernsdorfer W, Colin T, Miyasaka H (2009) Phys Rev Lett 102:167204
53. Miyasaka H, Takayama K, Saitoh A, Furukawa S, Yamashita M, Clérac R (2010) Chem Eur J 16:3656
54. Bhowmick I, Hillard EA, Dechambenoit P, Coulon C, Harris TD, Clérac R (2012) Chem Commun 48:9717
55. Cornia A, Costantino AF, Zobbi L, Caneschi A, Gatteschi D, Mannini M, Sessoli R (2006) Struct Bond 122:133
56. Bogani L, Wernsdorfer W (2008) Nat Mater 7:179
57. Cornia A, Mannini M (2015) Struct Bond 164:293
58. Mannini M, Pineider F, Daniele C, Totti F, Sorace L, Sainctavit P, Arrio MA, Otero E, Joly L, Cezar JC, Cornia A, Sessoli R (2010) Nature 468:417
59. Katoh K, Isshiki H, Komeda T, Yamashita M (2012) Chem Asian J 7:1154
60. Urdampilleta M, Cleuziou J-P, Klyatskaya S, Ruben M, Wernsdorfer W (2011) Nat Mater 10:502
61. Vincent R, Klyatskaya S, Ruben M, Wernsdorfer W, Balestro F (2012) Nature 488:357
62. Thiele S, Balestro F, Balou R, Klyatskaya S, Ruben M, Wernsdorfer W (2014) Science 344:1135
63. Loth S, Baumann S, Lutz CP, Eigler DM, Heinrich AJ (2012) Science 335 196
64. Shaw R, Laye RH, Jones LF, Low DM, Talbot-Eeckelaers C, Wei Q, Milios CJ, Teat SJ, Helliwell M, Raftery J, Evangelisti M, Affronte M, Collison D, Brechin EK, McInnes EJL (2007) Inorg Chem 46:4968
65. Evangelisti M, Luis F, de Jongh LJ, Affronte M (2006) J Mater Chem 16:2534
66. Zheng Y-Z, Zhou G-J, Zheng Z, Winpenny REP (2014) Chem Soc Rev 43:1462
67. Chen YC, Qin L, Meng ZS, Yang DF, Wu C, Fu ZD, Zheng YZ, Liu JL, Tarasenko R, Orendac M, Prokleska J, Sechovsky V, Tong ML (2014) J Mater Chem A 2:9851
68. Sharples JW, Collison D, McInnes EJL, Schnack J, Palacios E, Evangelisti M (2014) Nat Commun 5:5321
69. Meier F, Levy J, Loss D (2003) Phys Rev Lett 90:047901
70. McInnes EJL, Timco GA, Whitehead GFS, Winpenny REP (2015) Angew Chem Int Ed 54:14244
71. Ardavan A, Rival O, Morton JJL, Blundell SJ, Tyryshkin AM, Timco GA, Winpenny REP (2007) Phys Rev Lett 98:057201
72. Bader K, Dengler D, Lenz S, Endeward B, Jiang S-D, Neugebauer P, van Slageren J (2014) Nat Commun 5:5304
73. Nakazawa S, Nishida S, Ise T, Yoshino T, Mori NR, Rahimi D, Sato K, Morita Y, Toyota K, Shiomi D, Kitagawa M, Hara H, Carl P, Höfer P, Takui T (2012) Angew Chem Int Ed 51:9860
74. Aguilà D, Barrios LA, Velasco V, Roubeau O, Repollés A, Alonso PJ, Sesé J, Teat SJ, Luis F, Aromí G (2014) J Am Chem Soc 136:14215
75. Fernandez A, Pineda EM, Muryn CA, Sproules S, Moro F, Timco GA, McInnes EJL, Winpenny REP (2015) Angew Chem Int Ed. doi:10.1002/anie.201504487
76. Lehmann J, Gaita-Ariño A, Coronado E, Loss D (2007) Nat Nanotechnol 2:312
77. Nihei M, Okamoto Y, Sekine Y, Hoshino N, Shiga T, Liu IPC, Oshio H (2012) Angew Chem Int Ed 51:6361
78. Feng X, Mathonière C, Jeon IR, Rouzières M, Ozarowski A, Aubrey ML, Gonzalez MI, Clérac R, Long J (2013) J Am Chem Soc 135:15880
79. Mathonière C, Lin HJ, Siretanu D, Clérac R, Smith JM (2013) J Am Chem Soc 135:19083
80. Liu T, Zhang YJ, Sato O (2010) J Am Chem Soc 132:8250
81. Dong DP, Liu T, Kanegawa S, Kang S, Sato O, He C, Duan CY (2012) Angew Chem Int Ed 51:5119

82. Hoshino N, Iijima F, Newton GN, Yoshida N, Shiga T, Nojiri H, Nakao A, Kumai R, Murakami Y, Oshio H (2012) Nat Chem 4:921
83. Liu T, Zheng H, Shiota Y, Hayami S, Mito M, Sato O, Yoshizawa K, Kanegawa S, Duan C (2013) Nat Commun 4:2826
84. Heintze E, El Hallak F, Clauss C, Rettori A, Pini MG, Totti F, Dressel M, Bogani L (2012) Nat Mater 12:202
85. Sato O, Iyoda T, Fujishima A, Hashimoto K (1996) Science 272:704
86. Bleuzen A, Marvaud V, Mathonière C, Sieklucka B, Verdaguer M (2009) Inorg Chem 48:3453
87. Ohkoshi SI, Tokoro H (2012) Acc Chem Rev 45:1749
88. Ohkoshi SI, Imoto K, Tsunobuchi Y, Takano S, Tokoro H (2011) Nat Chem 3:564
89. Pinkowicz D, Rams M, Misěk M, Kamenev KV, Tomkowiak H, Katrusiak A, Sieklucka B (2015) Inorg Chem 137:8795
90. Ohkoshi SI, Takano S, Imoto K, Yoshikiyo M, Namai A, Tokoro H (2013) Nat Photonics 8:65

Struct Bond (2016) 172: 49–64
DOI: 10.1007/430_2015_190

Published online: 7 November 2015

Progress in the Area of High Energy Density Materials

Thomas M. Klapötke and Robert D. Chapman

Abstract Great strides have been made in increasing performance and decreasing sensitivity in energetic materials since the first commercialization of nitroglycerine (NG) in the form of dynamite in 1867 by Alfred Nobel. However, the high energy manufacturers continue to rely on traditional chemicals to meet their needs. New energetic materials must be developed to extend their capabilities and handling capabilities. The new materials which have been prepared recently have led to new possibilities. Important advances have been made especially in the area of high-nitrogen compounds, organic difluoramine derivatives. Computational simulations have also led not only to a greater insight into the basic thermodynamics and kinetics of these materials but also their practical behavior in the field. This chapter summarizes new developments that have been achieved since Volume 126 of Structure and Bonding, which was published in 2007 and gave a comprehensive review of the field.

Since the book "High Energy Density Materials" was published in 2007 [Structure and Bonding, Vol. 125/2007: High Energy Density Compounds, T. M. Klapötke (vol. editor), D. M. P. Mingos (series editor), Springer, Berlin/Heidelberg, **2007**], significant advances have been made especially in the area of high-nitrogen compounds, organic difluoramine derivatives, and computational simulations.

T.M. Klapötke (✉)
Department of Chemistry, Ludwig Maximilian University of Munich, Energetic Materials Research, Munich 81377, Germany
e-mail: tmk@cup.uni-muenchen.de

R.D. Chapman
Chemistry Branch (Code 4F0000D), Research Division, Research and Intelligence Department, Naval Air Warfare Center Weapons Division, Naval Air Systems Command, China Lake, CA 93555-6106, USA
e-mail: rchapman@aol.com

Keywords Difluoramine compounds · Explosives · Fluorine explosives · High-nitrogen compounds · N-oxides

Contents

Abbreviations

BDNPA/F	1:1 Mixture of bis(2,2-dinitropropyl) acetal and bis(2,2-dinitropropyl) formal
CBNT	Carbonic dihydrazidinium bis[3-(5-nitroimino-1,2,4-triazolate)] DNAN dinitroanisole
FOX-7	1,1-Diamino-2,2-dinitroethene
HMX	Octogen
HNFX	3,3,7,7-Tetrakis(difluoramino)octahydro-1,5-dinitro-1,5-diazocine
LLM-105	2,6-Diamino-3,5-dinitropyrazine 1-oxide
NG	Nitroglycerine
NTO	Nitrotriazolinone
PrNQ	Propyl nitroguanidine
RDX	Hexogen
TKX-50	Bis(hydroxylammonium) 5,5′-bitetrazolate 1,1′-dioxide
TNT	Trinitrotoluene

1 FOX-7 and LLM-105

As part of a more extensive theoretical study the properties of 3,5-dinitro-2,6-diaminopyrazine (**I**), 3,5-dinitro-2,6-diaminopyrazine 1-oxide (**II**, LLM-105) and 3,5-dinitro-2,6-diaminopyrazine 1,4-dioxide (**III**) were studied [1] (Fig. 1).

The most important findings of this computational study can be summarized as follows:

1. The enthalpies of formation for 2,6-diamino-3,5-dinitropyrazine (**I**), 2,6-diamino-3,5-dinitropyrazine 1-oxide (**II**), and 2,6-diamino-3,5-dinitropyrazine 1,4-dioxide (**III**) were calculated using the complete basis set (CBS-4 M) method in order to obtain very accurate energies and compared with experimental values that were available for **I** and **II**.

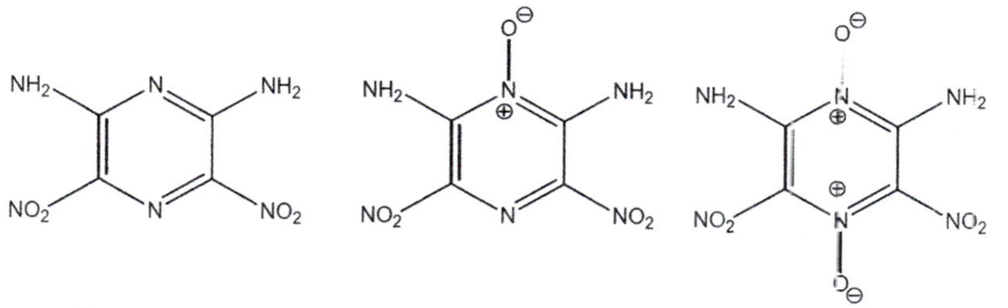

I II III

Fig. 1 Structures of 3,5-dinitro-2,6-diaminopyrazine (**I**), 3,5-dinitro-2,6-diaminopyrazine 1-oxide (**II**, LLM-105) and 3,5-dinitro-2,6-diaminopyrazine 1,4-dioxide (**III**). Reproduced from Gökçınar et al. [1], with kind permission of Elsevier

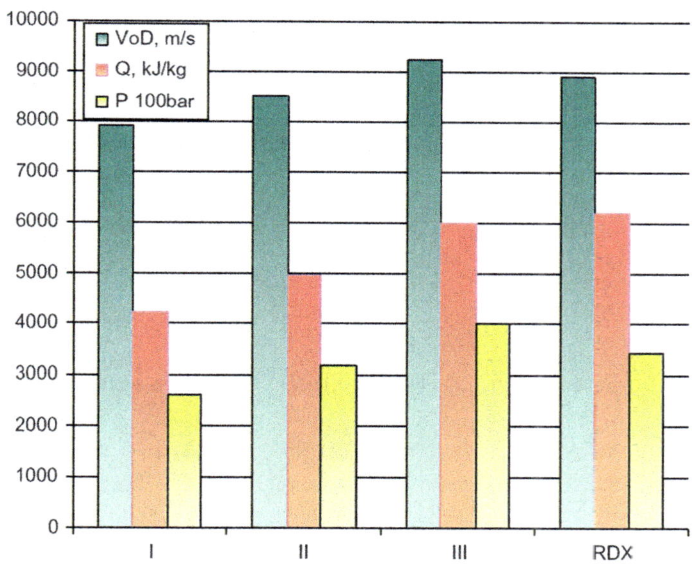

Fig. 2 Computed detonation parameters for **I**, **II**, **III** and RDX. Reproduced from Gökçınar et al. [1], with kind permission of Elsevier

2. The detonation parameters of **I**, **II**, and **III** were calculated and shown to increase from **I** via **II** to **III** (Fig. 2).
3. The dioxide **III** is expected to be a good candidate with calculated detonation velocity and detonation pressure even slightly higher than those of RDX (Fig. 2).
4. The computed electrostatic potentials of compounds **I**, **II**, and **III** revealed that attaching N-oxide substituents to the aromatic ring (compounds **II** and **III**) gives rise to a considerable imbalance between stronger positive regions and weaker negative ones which can be related to the impact sensitivities (Fig. 3). Although compound **III** would be expected to show higher sensitivity than compound **II** (LLM-105), since it exceeds the explosive power of **II** and its performance is

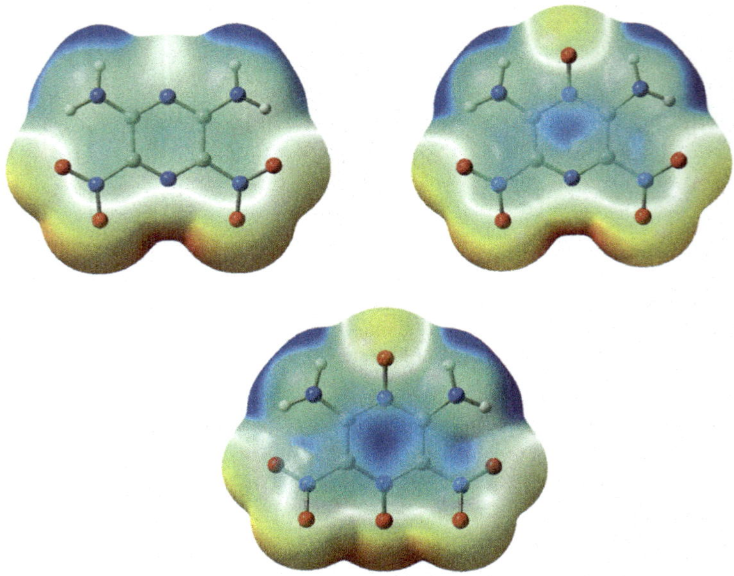

Fig. 3 Electrostatic potentials of compounds **I** (*left*), **II** (*right*) and **III** (*bottom*). (B3LYP/6-31G (d), 0.001e bohr^{-3} isosurface, energy values −0.06 H to +0.06 H); color coding: *red* (very negative), *orange* (negative), *yellow* (slightly negative), *green* (neutral), *turquoise* (slightly positive), *light blue* (positive), *dark blue* (very positive). Reproduced from Gökçınar et al. [1], with kind permission of Elsevier

Table 1 Computed and experimentally observed relative ^{15}N NMR chemical shifts d (ppm) of compound **II** (LLM-105)

	Abs. isotropic shielding	Rel. chem. shift, δ (ppm)	Exptl. chem. shift, δ (ppm)
H_3CNO_2	−116.8	0.0	0.0
N (amino)	+177.4	−294.2	−292.1
N-oxide	+17.0	−133.8	−139.8
N (second ring)	−25.5	−91.3	−106.6
N (nitro)	−102.4	−14.4	−16.1

comparable or even slightly higher than that of RDX, it is still very interesting and might be a suitable replacement explosive for RDX.

5. The computed ^{15}N NMR chemical shifts are in good agreement with the experimentally observed values for compound **II** (Table 1) which gives credence to the reliability of the computed chemical shifts for compounds **I** and **III**.

The results obtained in this study should encourage synthetic work in order to prepare 2,6-diamino-3,5-dinitropyrazine 1,4-dioxide (**III**) on a laboratory scale and to experimentally evaluate its properties, first and foremost its thermal stability.

A study investigating the reactivity of NTO and FOX-7 toward alkyl isocyanates may be of interest to chemists working with PBXs containing FOX-7, where curing with isocyanates might be involved [2].

The reactivity of the less-sensitive explosives NTO (5-nitro-1,2,4-triazolin-3-one) and FOX-7 (1,1-diamino-2,2-dinitroethene) toward simple alkyl isocyanates was investigated experimentally. New carboxamidic derivatives were synthesized by reacting the explosives with 1-isocyanatohexane (NTO and FOX-7) and 1,6-diisocyanatohexane (NTO only) under relatively mild conditions. The new products were characterized by a combination of ^{1}H and ^{13}C NMR and FTIR spectroscopy, single crystal X-ray diffraction, and differential scanning calorimetry.

2 High-Nitrogen-Containing Compounds

TKX-50 (bis(hydroxylammonium) 5,5′-bitetrazolate 1,1′-dioxide) is one of the most promising ionic salts being developed as a possible replacement for RDX [3–6]. It can be prepared on a >100 g-scale scale by the reaction of 1,1′-dihydroxy-5,5′-bitetrazole with dimethyl amine which forms the bis(dimethylammonium) 5,5′-bitetrazolate 1,1′-dioxide salt which is then isolated, purified, and subsequently reacted in boiling water with two equiv. of hydroxylammonium chloride to form TKX-50, dimethylammonium chloride, and HCl (Fig. 4).

TKX-50 shows high performance with respect to its detonation velocity and C-J pressure, but a desirable low sensitivity toward impact and friction (Fig. 5).

The thermal behavior of TKX-50 (bis(hydroxylammonium) 5,5′-bitetrazolate 1,1′-dioxide) and the kinetics of its thermal decomposition were studied using differential scanning calorimetry (DSC) and thermogravimetric analysis (TGA). The thermal decomposition of TKX-50 starts to occur within the range 210–250°C depending on the heating rate used.

By applying multiple heating rate DSC measurements and Ozawa's iso-conversional model free method, an activation energy of 34.2 kcal mol^{-1} and pre-exponential factor of 1.99×10^{12} s^{-1} were calculated from the DSC peak maximum temperature–heating rate relationship.

By applying non-isothermal TGA experiments and the Flynn–Wall iso-conversional model free method, it was found that the activation energy changes with conversion and lies between 34.7 and 43.3 kcal mol^{-1}, while the pre-exponential factor ranges from 9.81×10^{11} to 1.79×10^{16} s^{-1}.

The enthalpy of formation of TKX-50 was calculated as $\Delta H_f°$(TKX-50) = +109 kcal mol^{-1}. The experimentally determined value based on bomb calorimetry measurements is $\Delta H_f°$(TKX-50) = +113 ± 2 kcal mol^{-1}.

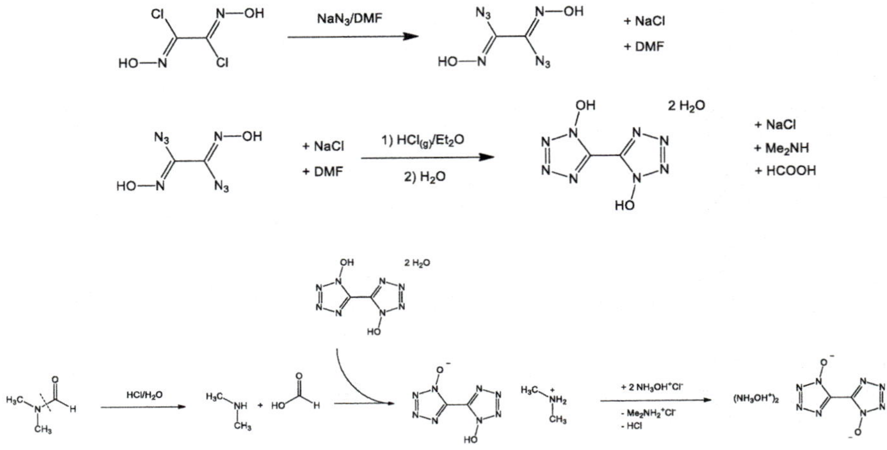

Fig. 4 Synthesis of TKX-50

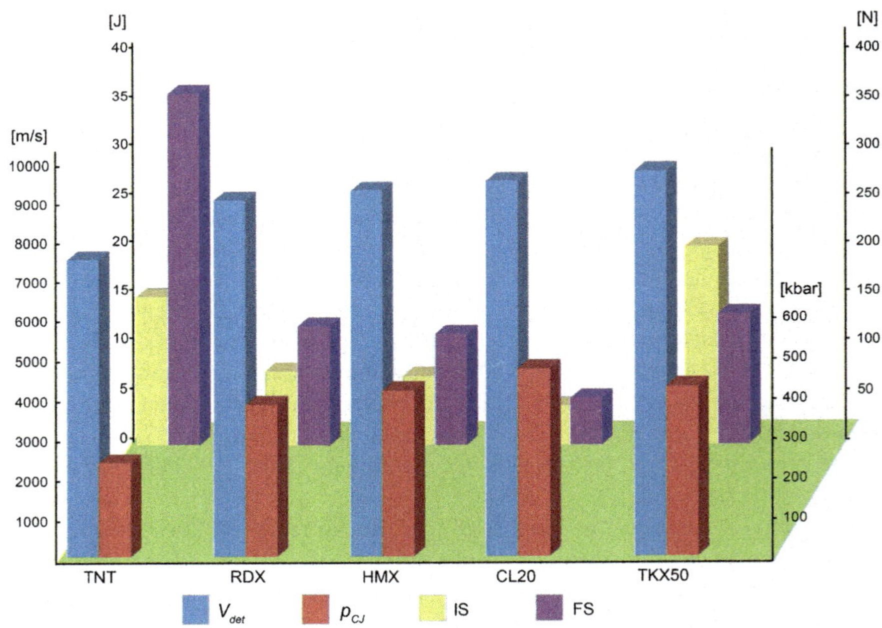

Fig. 5 Performance (detonation velocity and C-J pressure) and sensitivity data (impact and friction) of TKX-50 in comparison with other high explosives. Reproduced from T. M. Klapötke [6], with kind permission of de Gruyter]

The calculated detonation parameters as well as the equations of state for the detonation products (EOS DP) of the explosive materials TKX-50 and MAD-X1 (and also for several of their derivatives) were obtained using the computer program

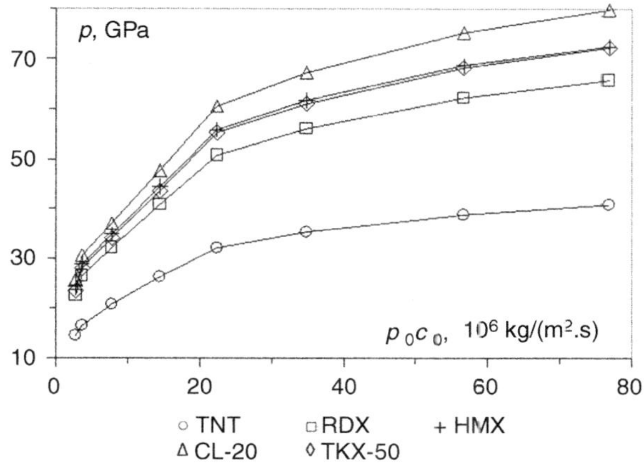

Fig. 6 The initial pressure on the interfaces of the explosive barrier from different materials (from polystyrene to tungsten). Reproduced from T. M. Klapötke [6], with kind permission of de Gruyter

EXPLO5 V.6.01. These values were also calculated for standard explosive materials which are commonly used such as TNT, PETN, RDX, and HMX, as well as for the more powerful explosive material CL-20 for comparison. The determination of the detonation parameters and EOS DP was conducted both for explosive materials having the maximum crystalline density and for porous materials of up to 50% in volume. The influence of the content of the plastic binder which was used (polyisobutylene up to 20% in volume) on all of the investigated properties was also examined.

Calculated results on shock wave loading of different inert barriers in a wide range of their dynamic properties under explosion on their surfaces of concrete size charges of different explosive materials in various initial states were obtained with the use of the one-dimensional computer hydrocode EP. Barriers due to materials such as polystyrene, textolite, magnesium, aluminum, zinc, copper, tantalum, or tungsten were examined (Fig. 6). Initial values of pressure and other parameters of loading on the interface explosive barrier were determined in the process of conducted calculations. The phenomena of propagation and attenuation of shock waves in barrier materials were also considered for all possible situations.

From these calculations, an essentially complete overview of the explosive properties and characteristics of shock wave action on barriers were obtained for several new and also for several standard explosive materials for comparison (Figs. 7, 8, and 9; Table 2). The results which were obtained suggest that in a wide range of their initial states (porosity, inert binders), the new explosive materials TKX-50 and MAD-X1 possess better explosive properties and shock wave action on practically every compact barrier which was considered, in comparison with standard explosive materials including the military explosive RDX. The large number of the calculated results which were obtained can be used to study the influence of different factors on the explosion and shock wave action of new explosive materials. Furthermore, some of these results can then be used to plan experiments to confirm the predicted properties.

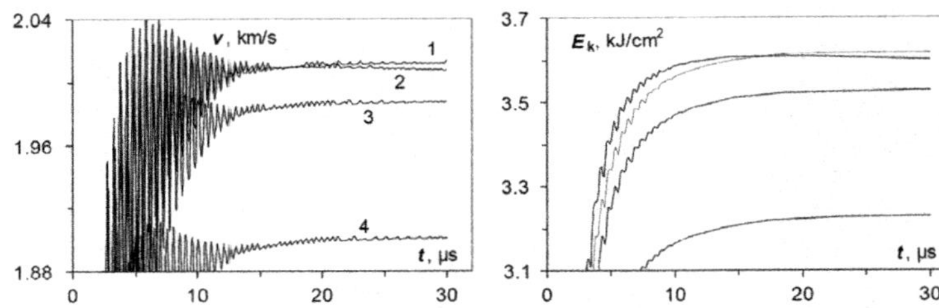

Fig. 7 Acceleration of copper plates by free charges of HMX (1), TKX-50 (2), MAD-X1 (3), and RDX (4). Reproduced from T. M. Klapötke [6], with kind permission of de Gruyter

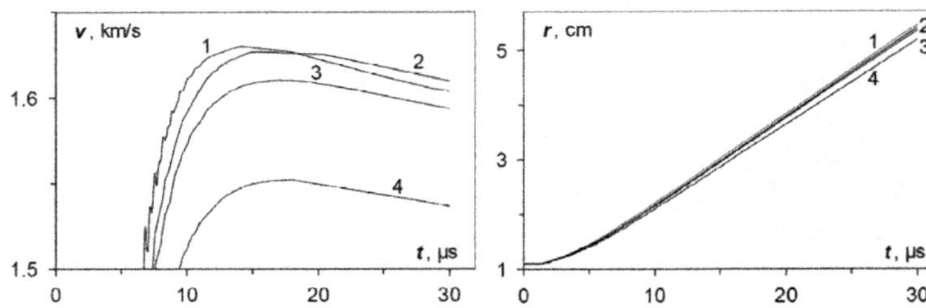

Fig. 8 Acceleration of tantalum cylindrical layers by charges of HMX (1), TKX-50 (2), MAD-X1 (3), and RDX (4). Reproduced from T. M. Klapötke [6], with kind permission of de Gruyter

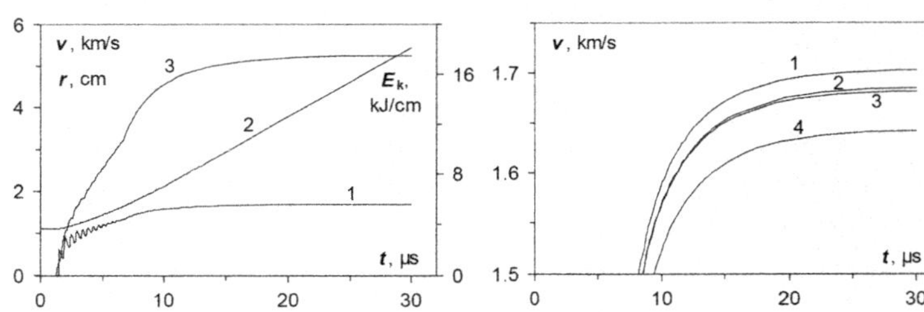

Fig. 9 Acceleration of a cylinder ($1 = v$; $2 = r$; $3 = E_k$) by a porous charge of TKX-50 (*left*) and by porous charges of HMX (1), TKX-50 (2), MAD-X1, (3) and RDX (4). Reproduced from T. M. Klapötke [6], with kind permission of de Gruyter

Table 2 Calculated cylinder energies for TKX-50 and RDX

Compound V/V_0	E_C/kJ cm^{-3}	% of standard			
		TATB	PETN	HMX	Cl-20
TKX-50					
2.2	−8.16	168	128	109	90
RDX					
2.2	−6.94	143	109	93	77

Table 3 Results of a small-scale shock reactivity test (SSRT)

Explosive	Mass (mg)	Dent (mg SiO_2)
RDX	504	589
CL-20	550	947
TKX-50	509	857

Table 4 Compatibility of TKX-50 performed according to a vacuum stability test (STANAG 4556 ED.1), 40 h, 100°C (compatible 0.0–3.0 mL)

Material	Reactivity of mix (mL)	Result
TKX-50	0.067	Pass
TKX-50 and CuO	0.62	Pass
TKX-50 and steel shavings	Negligible	Pass
TKX-50 and Al powder	0.13	Pass
TKX-50 and R8002	2.4	Pass
TKX-50 and BDNPA/F	0.94	Pass
TKX-50 and Chlorez wax	Negligible	Pass
TKX-50 and Viton	Negligible	Pass
TKX-50 and DNAN	Negligible	Pass
TKX-50 and TNT	Negligible	Pass
TKX-50 and NTO	Negligible	Pass
TKX-50 and HMX	Negligible	Pass
TKX-50 and PrNQ	1.2	Pass

An experimental small-scale shock reactivity test (SSRT) (Table 3) was in good agreement with the calculated cylinder energies (Table 2).

The detonation velocity was measured by the electrical method using an electronic timer and fiber-optic probes. Apparently TKX-50 has a large critical diameter (>3 cm) but shows a high detonation velocity of 9,460 m s^{-1} in a cylindrical 40 mm tube.

While TKX-50 shows good compatibility with other energetic fillers, binders, and plasticizers as well as with most metals and metal oxides (Table 4), its purity is crucial in order for it to exhibit good compatibility with all of the above for use in formulations.

Another very promising ionic energetic material is carbonic dihydrazidinium bis[3-(5-nitroimino-1,2,4-triazolate)] (CBNT) which was first synthesized in 2010 by Shreeve et al. [7] (Fig. 10). The salt is thermally stable up to 220°C and has a high density of 1.95 g cm^{-3} and very good calculated detonation parameters of VoD = 9,399 m s^{-1} and $p_{C\text{-}J}$ = 360 kbar.

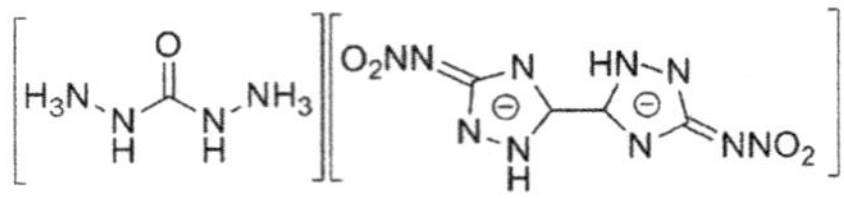

CBNT

Fig. 10 Structure of carbonic dihydrazidinium bis[3-(5-nitroimino-1,2,4-triazolate)] (CBNT). Reproduced from T. M. Klapötke [6], with kind permission of de Gruyter

3 Organic Difluoramine Derivatives

This review will emphasize recent developments in the synthesis and characterization of organic difluoramine derivatives (NF_2 being the difluoramino group), including theoretical treatments, since the previous review in this series [8], which discussed technical reasons for interest in this class of compounds and reviewed the literature on this class from the late 1980s to about 2007.

Relatively little literature on experimental aspects of organic difluoramine chemistry has appeared since 2007.

The Shreeve group prepared and characterized nitrogenous-cation salts of 5-(difluoramino)difluoromethyl]tetrazole, including ammonium, 1,5-diamino-4-methyltetrazolium, and 4-amino-1-methyl-1,2,4-triazolium [9].

A series of US patents [10–12] has issued related to a scalable one-pot process to prepare 3,3,7,7-tetrakis(difluoramino)octahydro-1,5-dinitro-1,5-diazocine (HNFX, **1**) from suitably protected 1,5-diazocine-3,7(2*H*,6*H*)-dione intermediates, especially the 1,5-bis(4-nitrobenzenesulfonyl) derivative, via the process of protolytic denosylation of the 3,3,7,7-tetrakis(difluoramino) intermediate – prepared by difluoramination of the dione reactant, as previously reported [13] – conducted in superacids such as trifluoromethanesulfonic. The resulting denosylated, protonated diazocinium salt readily underwent N-nitration in situ to HNFX (Scheme 1).

Octafluoropentaerythrityltetramine (octafluoro-PETA), which might be considered a difluoramino analog of PETN, was prepared by direct fluorination (with F_2/N_2) of tetraethyl pentaerythrityltetracarbamate, $C(CH_2NHCOOC_2H_5)_4$, in acetonitrile at subambient temperatures [14]. However, octafluoro-PETA's physical properties, especially its low melting point (40–42°C) and high volatility, were not conducive to practical applications in explosive formulations [15].

Conditions for iodometric determination of aqueous *N,N*-difluorourea (DFU) content – the precursor typically used for generation of difluoramine (HNF_2) via acidic hydrolysis – have been optimized by Kang et al. [16]. Although iodometric titration of aqueous DFU has long been prescribed, even in the very first report of difluorourea prepared and isolated by this method [17], the procedure had not been reported as optimized for accurate assessment of DFU concentration. The rather specific and non-ambient conditions reported for the optimized process may explain earlier discrepancies in DFU content measured after following a specified procedure for its preparation [18]: 1.21 M measured by a submitter vs. 0.99 M measured by a checker.

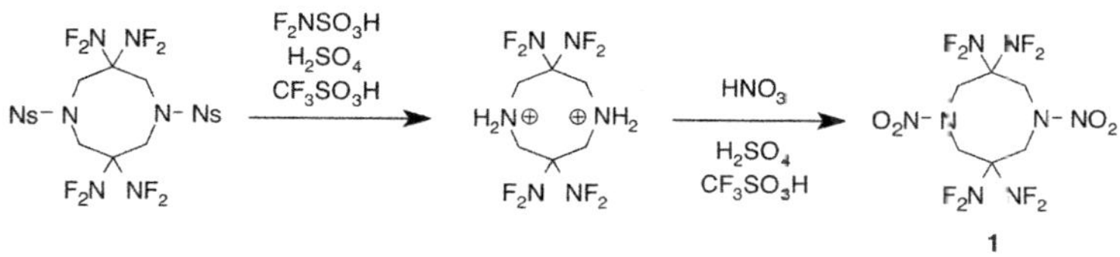

Scheme 1 A scalable preparation of HNFX (1): protolytic N-denosylation plus simple nitration

Belter (Y-Not Chemical Consulting) has prepared a variety of organic difluoramine derivatives (alkanes, arenes, ethers) by reactions of corresponding substrates with nitrogen trifluoride (NF_3) at 400°C in a vapor phase reactor [19]. *N,N*-Difluoraminoalkanes were also prepared by reactions of NF_3 with various alkylmagnesium (Grignard) reagents in diethyl ether solvent at subambient temperatures [20, 21], and their ^{19}F NMR spectra were analyzed in detail.

Zhang et al. [22] recently reported the preparation and additional characterizations of 3,3-bis(difluoraminomethyl)oxetane, which was previously described in patents by Archibald et al. [23, 24].

McPake et al. have described preparations of various *N,N*-difluoroanilines by direct fluorination (10% F_2/N_2 in acetonitrile solvent) of certain substituted polyfluorinated anilines using newly developed continuous flow methodology for this transformation [25].

A detailed study of infrared and Raman spectra of HNFX has been published by Weck et al. (University of Nevada at Las Vegas) [26]. Della Védova and coworkers measured vibrational properties of difluoraminocarbonyl fluoride, $F_2NC(O)F$, and difluoraminocarbonyl isocyanate, $F_2NC(O)NCO$ from their FTIR spectra, as well as the Raman spectrum of the latter and studied conformational spaces of both derivatives using the B3LYP and MP2 level of theory [27].

HNFX, upon detonation, has been demonstrated to be an efficacious sporicide against *Bacillus thuringiensis*, a surrogate for anthrax spores: a detonation of 2.90 g of HNFX in a sealed chamber of 8.5 L volume achieved a kill rate of >9 $\log_{10}$ orders of magnitude following a 5.0-second exposure after detonation [15, 28]. By comparison to a similar test with HMX – resulting in an apparent kill rate of only ~0.2 $\log_{10}$ reduction after 0.4 h of exposure following a detonation of 2.67 g of HMX – the high killing efficiency achieved by HNFX detonations was assessed to be a genuine chemical effect, caused by very reactive transient chemical species (such as atomic fluorine or formyl fluoride) formed during detonation and not due to extreme temperature or pressure conditions that occurred in those experiments.

Pan (Nanjing University of Science and Technology) and coworkers have used thermogravimetry and differential scanning calorimetry (TG/DSC) techniques to study thermal decomposition mechanisms of certain difluoramine monomers and polymers: poly[3-(difluoraminomethyl)-3-methyloxetane] (PDFAMO), including in binary mixtures with various other explosives [29, 30], and copolymers poly[3,3-bis (difluoraminomethyl)oxetane–tetrahydrofuran], poly[3-(difluoraminomethyl)-3-methyloxetane–tetrahydrofuran), and poly[3,3-bis(difluoraminomethyl)oxetane–3-

(difluoraminomethyl)-3-methyloxetane–tetrahydrofuran] [31]. Mass spectral fragmentation pathways of 3,3-bis(difluoraminomethyl)oxetane and 3-(difluoraminomethyl)-3-methyloxetane were studied by electron ionization mass spectrometry [32].

The large majority of publications on organic difluoramine derivatives, which have appeared since the last review, have dealt with computational aspects. The difluoramines are often subsets of collections of a range of substituted structures whose properties were calculated. The following structural systems have been studied in these theoretical treatments.

Fan and Ju (Nanjing University of Science and Technology) studied geometric and electronic structures, as well as heats of formation, of four-membered rings (cyclobutane, azetidines, oxetanes) with various energetic substituents, including NF_2 [33]. Ju et al. also studied, by density functional theory (DFT), the stability of the theoretical structure 1,3,5-triamino-2,4,6-tris(difluoramino)benzene in comparison to that of TATB [34]. Fan et al. reported the strain energies of cubane derivatives with different substituent groups, including NF_2, derived from total energies of the optimized geometric structures at the DFT-B3LYP/6-31G* level [35]. Other cubane derivatives have been studied by Campanelli et al. for electronic substituent effects induced in 4-substituted 1-phenylcubane derivatives, Ph-C_8H_6-X, including X = NF_2 [36]; the same group reported a similar study of electronic substituent effects across the 1,12-dicarba-*closo*-dodecaborane cage [37]. Various derivatives of "2-diazo-4,6-dinitrophenol" (DDNP) with additional energetic substituents, including NF_2, have been studied [38]. Ueda and Takahashi calculated ionization potentials and relaxation energies of a variety of small molecules, including NF_2CXO (X = H or F) [39]. A theoretical study of the vibrational spectra and thermodynamic properties of certain derivatives (including difluoramino) of 2,2′,4,4′,6,6′-hexanitrostilbene was reported by Wang et al. [40].

Thermodynamic properties of theoretical difluoramino derivatives of a wide variety of heterocyclic systems have been studied computationally: piperidines and diazocines, including HNFX [41]; 1,2,4,5-tetrazines [42]; pyrazoles [43]; a series of substituted 2,2′-bi-1*H*-imidazoles [44]; 3,3′-azo-1,2,4,5-tetrazine derivatives [45]; monocyclic furazans [46]; tetrazolo[1,5-*b*][1,2,4,5] tetrazine and 1,2,4-triazolo[4,3-*b*] [1,2,4,5]tetrazine derivatives [47]; various *C*- and *N*-(difluoramino)tetrazolium salts [48]; bifurazans and various bridged difurazanyl systems [49]; substituted 2,2′,4,4′,6,6′-hexanitroazobenzene derivatives [50]; 1,1′- and 5,5′-bitetrazoles and various bridged ditetrazolyl systems [51]; azo- and hydrazo-1,3,5-triazines [52]; guanidinium salts of various 4,6-substituted 2-nitramino-1,3,5-triazines [53, 54]; 2,2′-bi-1,3,5-triazines and various 2,2′-bridged ditriazinyl systems [55]; furazano[3,4-*b*]pyrazines [56]; 2,4,6-trinitropyridine 1-oxide derivatives [57]; various derivatives of 5-amino-1-(3,5-dinitro-2-pyridinyl)-3-nitro-1*H*-1,2,4-triazole [58]; various *C*- and *N*-difluoramino- and (difluoraminomethyl)-substituted *C*- and *N*-(trinitromethyl)tetrazoles and 1,2,4,5-tetrazines [59]; energetic nitrogen-rich salts of tetrazole-based anions with substituents such as NF_2, CH_2NF_2, CF_2NF_2, or $C(NO_2)_2NF_2$ [60]; tetrazole- and tetrazine-based high-density compounds with oxygen balance equal to zero [61];

poly(difluoramino)purines [62]; 2,2′-bipyridines and various 2,2′-bridged dipyridyl systems as well as their N-oxides [63]; variously substituted 1,2,4-triazoles, including the anionic forms of NF_2-substituted derivatives [64]; various pyridines [65]; and C-substituted 4,4′-azobi-1,2,4-triazoles [66].

4 Summary

Future research will lead to even more powerful, high-nitrogen high-oxygen explosives with enhanced and superior detonation parameters to fulfill the lethality requirements of all forces. New energetics will provide vastly increased energy content over RDX, up to several times the energetic performance. In addition, these materials will have a high energy density with a high activation energy.

With these capabilities, DoD will be able to develop new applications for energetic materials and vastly improve performance in current munitions. The increased performance of the next generation of energetics will enable DoD to meet the same strategic goals with fewer munitions and less energetic material. The increase in energy output can be harnessed in munitions that require a fraction of the amount of energetics to deliver the same payload or thrust profile. This will allow engineers to put energetics in munitions that have never had them before.

Acknowledgments Financial support of this work by the Ludwig Maximilian University of Munich (LMU), the US Army Research Laboratory (ARL), the Armament Research, Development and Engineering Center (ARDEC), the Office of Naval Research (ONR) under grant no. ONR.N00014-12-1-0538, and the Bundeswehr–Wehrtechnische Dienststelle für Waffen und Munition (WTD 91) under grant no. E/E91S/FC015/CF049 is gratefully acknowledged. The authors acknowledge collaborations with Dr. Muhamed Suceska (Brodarski Institut, Croatia) in the development of new computational codes to predict the detonation and propulsion parameters of novel explosives. We are indebted to and thank Drs. Betsy M. Rice and Ed Byrd (ARL, Aberdeen Proving Ground, MD) and Dr. Anthony Bellamy for many inspired discussions and their help preparing this chapter.

References

1. Gökçınar E, Klapötke TM, Bellamy AJ (2010) J Mol Struct (TheoChem) 953:18–23
2. Bellamy AJ, Contini, AE, Andrews MRG (2009) New trends in research of energetic materials. In: Proceedings of the seminar, 12th, Pardubice, Czech Republic, April 1–3 (Pt. 2) pp 473–480
3. Fischer N, Fischer D, Klapötke TM, Piercey DG, Stierstorfer J (2012) J Mater Chem 22(38): 20418–20422
4. Fischer N, Klapötke TM, Matecic MS, Stierstorfer J, Suceska M (2013) *TKX-50*, new trends in research of energetic materials, Part II, Czech Republic, pp 574–585
5. Golubev V, Klapötke TM, Stierstorfer J (2014) TKX-50 and MAD-X1 – a progress report, 40th international pyrotechnics seminar, Colorado Springs, CO, July 13–18
6. Klapötke TM (2015) Chemistry of high-energy materials, 3rd edn. de Gruyter, Berlin

7. Wang R, Xu H, Guo Y, Sa R, Shreeve JM (2010) J Am Chem Soc 32:11904
8. Chapman RD (2007) Struct Bond 125:123
9. Ye C, Gao H, Shreeve JM (2007) J Fluor Chem 128:1410
10. Chapman RD, Groshens TJ (2009) US Patent 7,563,889
11. Chapman RD, Groshens TJ (2009) US Patent 7,632,943
12. Chapman RD, Groshens TJ (2013) US Patent 8,444,783
13. Chapman RD, Welker MF, Kreutzberger CB (1998) J Org Chem 63:1566
14. Chapman RD, Hollins RA (2011) US Patent 8,008,527
15. Chapman RD, Hollins RA, Groshens TJ, Thompson D, Schilling TJ, Wooldridge D, Cash PN, Jones TS, Ooi GT (2014) *N,N*-Dihaloamine explosives as harmful agent defeat materials. Technical report DTRA-TR-14-26. Defense Threat Reduction Agency, Fort Belvoir, VA. http://www.dtic.mil/cgi-bin/GetTRDoc?Location=U2&doc=GetTRDoc.pdf&AD=ADA602478. Accessed 7 June 2015
16. Kang L, Liu J, Zhang M, Liu H (2011) Huaxue Tuijinji Yu Gaofenzi Cailiao 9(5):93
17. Lawton EA, Cain EFC, Shefhan DF, Warner M (1961) J Inorg Nucl Chem 17:188
18. Parker CO, Freeman JP (1970) Inorg Synth 12:307
19. Belter RK (2011) J Fluor Chem 132:961
20. Belter RK (2012) J Fluor Chem 137:73
21. Belter RK (2015) J Fluor Chem 175:110
22. Zhang M, Liu H, Gao B, Zhang L, Kang L, Zhang K (2012) Hanneng Cailiao 20:314
23. Archibald TG, Manser GE, Immoos JE (1993) US Patent 5,272,249
24. Archibald TG, Manser GE, Immoos JE (1995) US Patent 5,420,311
25. McPake CB, Murray CB, Sandford G (2013) Aust J Chem 66:145
26. Weck PF, Gobin C, Kim E, Pravica MG (2009) J Raman Spectrosc 40:964
27. Erben MF, Padro JM, Willner H, Della Védova CO (2009) J Phys Chem A 113:13029
28. Chapman RD (2012) US Patent 8,221,566
29. Li H, Wang W, Zhang L, Pan R (2013) Int Annu Conf ICT 44:40/1
30. Li H, Pan R, Wang W, Zhang L (2014) Propellants Explos Pyrotech 39:819
31. Li H, Pan R, Wang W, Zhang L (2014) J Therm Anal Calorim 118:189
32. Wang W, Li H, Zhang L, Pan R (2013) Int Annu Conf ICT 44:41/1
33. Fan XW, Ju XH (2007) J Comput Chem 29:505
34. Ju XH, Wang ZY, Xiao HM (2007) J Chin Chem Soc 54:313 (Taipei, Taiwan)
35. Fan XW, Ju XH, Xia QY, Xiao HM (2008) J Hazard Mater 151:255
36. Campanelli AR, Domenicano A, Ramondo F (2011) Struct Chem 22:449
37. Campanelli AR, Domenicano A, Hnyk D (2015) J Phys Chem A 119:205
38. Liu Y, Wang L, Wang G, Du H, Gong X (2012) J Mol Model 18:1561
39. Ueda K, Takahashi O (2012) J Electron Spectrosc Relat Phenom 185:301
40. Wang G, Gong X, Xiao H (2013) Adv Mater Res 742:202 (Durnten-Zurich, Switz)
41. Fan XW, Ju XH, Xiao HM (2008) J Hazard Mater 156:342
42. Wei T, Zhu W, Zhang X, Li YF, Xiao H (2009) J Phys Chem A 113:9404
43. Li YF, Fan XW, Wang ZY, Ju XH (2009) J Mol Struct (TheoChem) 896:96
44. Li YF, Wang ZY, Ju XH, Fan XW (2009) J Mol Struct (TheoChem) 907:29
45. Zhang JJ, Gao HW, Wei T, Wang CJ (2010) Wuli Huaxue Xuebao 26:3337
46. Zhang X, Zhu W, Xiao H (2010) Int J Quantum Chem 110:1549
47. Wei T, Zhu W, Zhang J, Xiao H (2010) J Hazard Mater 179:581
48. Zhang X, Zhu W, Wei T, Zhang C, Xiao H (2010) J Phys Chem C 114:13142
49. Zhang X, Zhu W, Xiao H (2010) J Phys Chem A 114:603
50. Liu Y, Gong X, Wang L, Wang G, Xiao H (2011) J Phys Chem A 115:1754
51. Zhu W, Zhang C, Wei T, Xiao H (2011) J Comput Chem 32:2298
52. Wang F, Du H, Zhang J, Gong X (2011) J Phys Chem A 115:11852
53. Wang F, Du H, Liu H, Gong X (2012) Chem Asian J 7:2577
54. Wang F, Zhang Q, Gong X, Li H, Zhao Z (2014) Struct Chem 25:1785
55. Pan Y, Zhu W, Xiao H (2012) J Mol Model 18:3125

56. Pan Y, Li J, Cheng B, Zhu W, Xiao H (2012) Comput Theor Chem 992:110
57. Liu H, Wang F, Wang G, Gong X (2013) Mol Simul 39:123
58. Liu H, Wang F, Wang GX, Gong XD (2013) J Phys Org Chem 26:30
59. Liu Z, Wu Q, Zhu W, Xiao H (2013) J Phys Org Chem 26:939
60. Shao Y, Zhu W, Xiao H (2013) J Mol Graph Model 40:54
61. Wu Q, Zhu W, Xiao H (2013) J Chem Eng Data 58:2748
62. Yan T, Sun G, Chi W, Li L, Li B, Wu H (2013) C R Chim 16:765
63. Liu H, Gong XD (2013) Struct Chem 24:471
64. Shao Y, Pan Y, Wu Q, Zhu W, Li J, Cheng B, Xiao H (2013) Struct Chem 24:1429
65. Lian P, Lai WP, Wang BZ, Wang XJ, Luo YF (2014) Asian J Chem 26 2357
66. Yang J, Yan H, Wang G, Zhang X, Wang T, Gong X (2014) J Mol Model 20:1

Struct Bond (2016) 172: 65–84
DOI: 10.1007/430_2015_205

Published online: 14 February 2016

Applications of Layered Double Hydroxide Materials: Recent Advances and Perspective

Rui Tian, Ruizheng Liang, Min Wei, David G. Evans, and Xue Duan

Abstract Layered double hydroxides (LDHs) are a class of layered materials with highly tunable composition and controllable morphology. The layered structure of LDHs enables fabrication of a large variety of intercalated, assembled, and hybrid materials which have a wide range of applications as photofunctional materials and bio-related materials, in catalysis and electrochemistry, and as multifunctional materials. The efficacy of LDHs in many applications can be attributed to the confined microenvironment of the guest species and the host–guest interactions. In this review, we will first summarize the synthetic strategies used to produce LDHs and LDH-based materials and then discuss their host–guest interactions and finally their applications.

Keywords Bio-application • Catalysis • Electrochemistry • Layered double hydroxides • Photofunctional

Contents

R. Tian, R. Liang, M. Wei (✉), D.G. Evans, and X. Duan
State Key Laboratory of Chemical Resource Engineering, Beijing University of Chemical Technology, Beijing 100029, P. R. China
e-mail: weimin@mail.buct.edu.cn

1 Preamble

In 2005 two of us (X.D. and D.G.E.) were invited by Mike Mingos to edit a volume of *Structure and Bonding* on layered double hydroxides (LDHs), also known as hydrotalcite-like materials. Since D.G.E. was once a DPhil student with Mike Mingos in Oxford, this "invitation" fell into the category of "can't be refused" (!), but in any case we were delighted to accept because we felt that the rapid growth in publications on LDHs (around 550 listed in *Web of Science* in 2005) meant that such a volume was timely. Together with our collaborators in Oxford, Guildford, and Clermont-Ferrand, we put together Volume 119 containing five reviews [1–5] on the structure, preparation, and applications of LDHs, LDH–polymer composites, and mechanistic and kinetic studies of intercalation in LDHs. These reviews have been cited over 750 times and three of them are in the top six most highly cited papers in *Structure and Bonding*. Since 2005, there has been an explosive growth in LDH-related publications with around 1,500 papers already in 2015 (January to end of October), and many new groups around the world have entered the area. We hope that our volume of *Structure and Bonding* may have helped some of them to become acquainted with the fascinating world of LDHs, and indeed on visiting some of these groups, we have been pleased on occasion to see (sometimes well-thumbed!) copies in the labs. Subsequently X.D. was invited to join the editorial board of *Structure and Bonding* and has been delighted to have the opportunity to raise the profile of the journal in China and invite some of the country's leading scientists to edit volumes on *Functional Phthalocyanine Molecular Materials* (Volume 135), *Structure–Property Relationships in Non-Linear Optical Crystals* (two volumes, 144 and 145), *Structures and Interactions of Ionic Liquids* (Volume 151), and *Structure and Modeling of Complex Petroleum Mixtures* (in press). In addition, one of us (M.W.) together with a former student and now independent researcher, Prof. Dongpeng Yan, has recently edited a volume on *Photofunctional*

Layered Materials (Volume 166) in which LDHs feature strongly, alongside a range of other layered materials.

In this review we summarize some of the most exciting recent developments in LDH chemistry which have occurred since our earlier edited volume of *Structure and Bonding* and take this opportunity to wish the journal a successful next 50 years!

2 Introduction

2.1 Definition of Layered Double Hydroxides (LDHs)

Layered double hydroxides (LDHs) are a large class of anionic layered materials which are based on the brucite structure with edge-sharing $M(OH)_6$ octahedra [1, 2]. LDHs can be expressed as $[M^{2+}{}_{1-x}M^{3+}{}_x(OH)_2]\ (A^{n-})_{x/n}{\cdot}mH_2O$ where M^{2+} and M^{3+} are divalent (such as Mg^{2+}, Zn^{2+}, Cu^{2+}, or Co^{2+}) and trivalent metal cations (such as Al^{3+}, Fe^{3+}, or Cr^{3+}) [3, 4]. The positive charges in the host layer can be balanced by anions within the hydrated interlayer galleries. The wide tunability in the host layer and ion-exchangeable properties of the interlayer guest ions make LDHs a versatile class of materials [5–7]. In addition, various efforts have been made to optimize the structure and functionality of LDHs and LDH-based materials, which have further expanded the applications of LDHs.

2.2 Preparation of LDHs

The traditional method to prepare LDHs involves coprecipitation followed by hydrothermal treatment [8]. By controlling the concentration of reactants, temperature, and aging time, LDHs with different particle sizes, morphologies, and crystallization degrees can be fabricated. Other methods, including in situ growth, electrosynthesis, and calcination–reconstruction, have also been developed in order to construct LDHs with controllable morphology and properties [9]:

1. In situ growth is a new method to fabricate LDHs grafted on various substrates in order to obtain a nanodevice with enhanced performance; various substrates such as aluminum sheets, conductive fabrics, and stretchable carbon nanotubes or fibers have been used (Scheme 1a) [10].
2. Electrosynthesis is a controllable approach to synthesis of LDHs as nanoarrays which exhibit ordered structure and uniform morphology. This method has been widely used in electrochemistry and photoelectrocatalysis (Scheme 1b) [11].
3. Calcination–reconstruction has been developed to obtain highly dispersed metal nanoparticles (NPs) with tunable chemical composition, size, shape, and crystallographic orientation (Scheme 1c) [12, 13]. Upon calcination, the LDHs

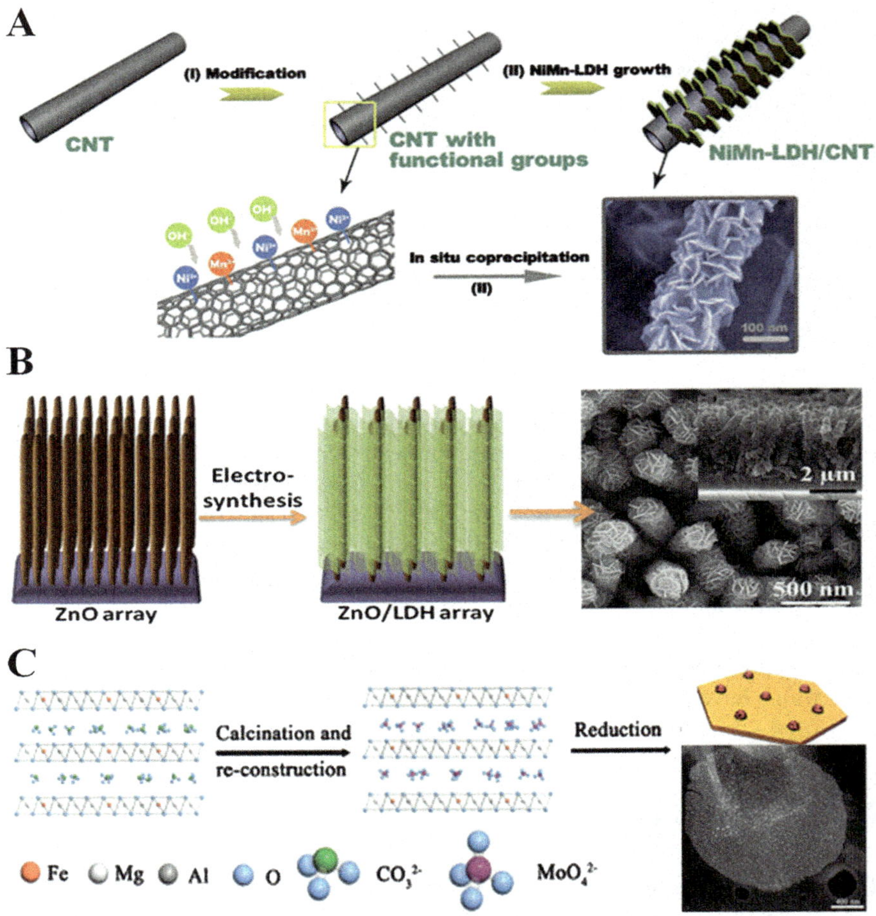

Scheme 1 Fabrication of LDHs by (**a**) in situ growth, (**b**) electrosynthesis, and (**c**) calcination–reconstruction methods (reprinted with permission from [10], [11] and [13], respectively)

decompose into mixed metal oxides (MMO), while rehydration can turn the MMO topotactically back to layered LDHs.

2.3 *Fabrication of LDH-Based Multifunctional Materials*

In order to fabricate LDH-based multifunctional materials, guest species with various functionalities have been intercalated into the interlayer galleries of LDHs, and LDHs with controllable structure or morphology have been designed [14]. The commonly employed fabrication methods are as follows:

1. Layer-by-layer (LbL) assembly
   Layer-by-layer assembly takes advantage of the electrostatic interactions between guest species and the LDH host. The basic LbL method relies on alternate dipping of a substrate into a suspension of positively charged LDH nanosheets and a solution of negatively charged guest anions. When positively charged ions or neutral molecules are desired as guest species, a new strategy involving co-assembly via a polyanion carrier has been proposed [15]. Moreover, assisted by an external field, such as a magnetic or electric field, ultrathin films (UTFs) with high guest loading, preferred orientation, and anisotropic properties can be obtained [16, 17]. Thus, a solid device with multifunctionality can be achieved based on assembly of guest anions and an LDH host. In addition, when the positively charged LDH nanosheets and negatively charged guest anions coexist in an aqueous medium, exfoliation–reassembly methods can be employed to assemble LDHs with guests and obtain nanohybrids with layered, core–shell, or porous structures.
2. Intercalation assembly
   In order to obtain powered or colloidal LDH-based multifunctional materials, intercalation assembly is an attractive choice [18, 19]. Direct intercalation can be achieved by the coprecipitation method, followed by hydrothermal treatment. When more than one type of guest species is to be intercalated or the guests cannot be directly accommodated in the LDH gallery, co-intercalation is employed [20, 21]. If the guest species are unstable in aqueous solutions containing the divalent or trivalent metal cations, it is possible to use ion-exchange methods to fabricate LDH–guest composites [22].

2.4 *Applications of LDHs*

LDHs and LDH-based composites have been widely used as photofunctional materials, as catalysts, and in electrochemistry and bio-related applications (Scheme 2) due to their tunable composition, layered structure, and confined microenvironment. In this review, we will summarize recent applications of LDHs and LDH-based materials and investigate the host–guest interaction and structure–function correlations [23–27].

3 Photofunctional Properties

3.1 *Introduction*

LDHs have been widely employed as photofunctional materials due to their layered structure, tunable composition, and outstanding thermal and chemical stability [28]. Luminescent LDHs with emission in the range of 350–550 nm can be obtained

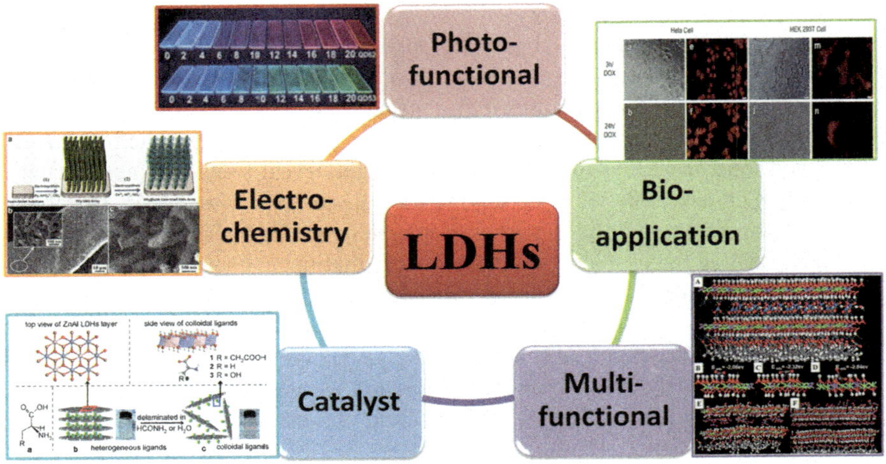

Scheme 2 Application of LDHs and LDH-based materials (reprinted with permission from [23], [24], [25], [26] and [78], respectively)

by virtue of surface defects which trap the excited electrons/holes and afford enhanced recombination [29, 30].

Assembly and intercalation of fluorescent guests into the interlayer galleries of LDHs is another procedure widely used in photofunctional applications. Various guest species, such as organic fluorescent polymers [31, 32], metal complexes [33–35], small anions [36, 37], quantum dots [38–40], and polyoxometalates [41–43], have been intercalated into LDHs. Taking account of the wide variety of possible guest species, various methods—including LbL assembly, co-assembly, direct intercalation (coprecipitation), ion exchange, and co-intercalation—have been employed [44–46]. Thus, based on the interaction between the LDH layers and guest species, the arrangement and distribution of interlayered guests can be tuned [47]. By tailoring the layer composition and charge density, the quantity of guest species can be adjusted, which results in different properties. Moreover, the confined environment of the LDH interlayer galleries can be expected to endow the guest species with novel performance and applications.

3.2 Host–Guest Interaction Between LDH Layers and Guest Species

The assembly and intercalation of photoactive guest species into the LDH interlayer galleries can endow the resulting LDH-based photofunctional materials with enhanced fluorescence performance due to the host–guest interactions and confinement effect induced by LDHs. By means of co-intercalation, many guest species, such as sulforhodamine B (SRB) [48], zinc phthalocyanines (ZnPc) [27],

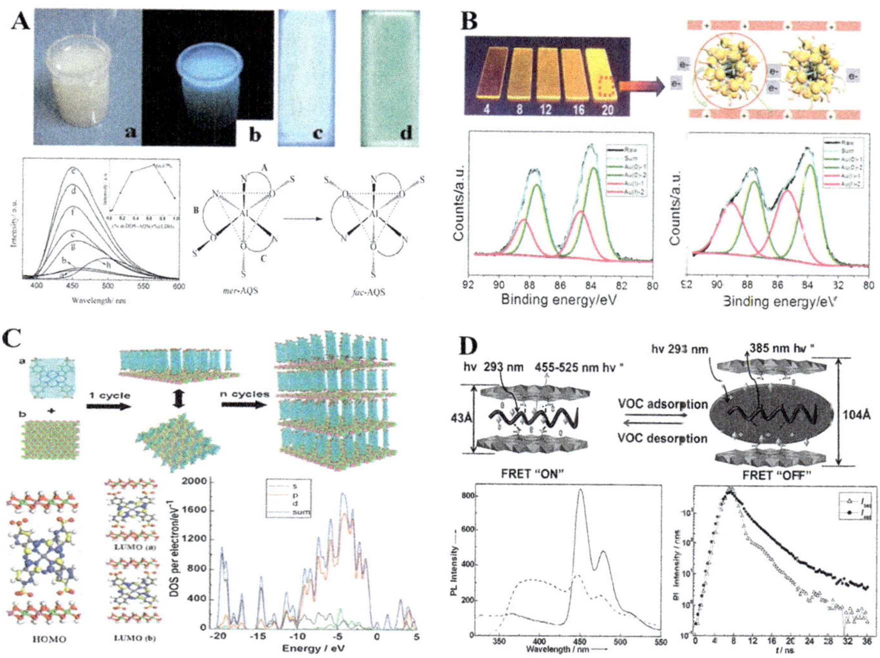

Fig. 1 Representations of host–guest interactions for (**a**) DDS–AQS/LDH samples, (**b**) (Au NCs/LDH)$_n$, (**c**) (ZnTSPc/LDH)$_n$, and (**d**) (perylene@PVK/LDH)$_n$ UTFs (reprinted with permission from [52], [53], [54] and [55], respectively)

fluorescein (FLU) [49], ammonium 1-anilinonaphthalene-8-sulfonate (ANS) [50], and benzocarbazole anions (BCZC) [51], have been intercalated with the aid of surfactants. By tuning the layer composition and the intercalation ratio, the distribution and arrangement of guest species can be precisely adjusted, which results in different emission wavelengths and intensities. Li et al. have co-intercalated bis (8-hydroxyquinolate-5-sulfonate)aluminum anion (AQS^{2-}) and dodecyl sulfonate (DDS) and tailored the fluorescence properties through adjustment of the intercalation content and host layer composition [52]. The host–guest interaction directly leads to structural changes in the intercalated AQS, and the observed luminescence transition can be attributed to the isomerization of meridional to facial AQS (Fig. 1a).

Layer-by-layer assembly provides an opportunity to immobilize the guest species and obtain UTFs with multifunctionality. An example involves the assembly of LDHs with Au nanoclusters (NCs), where the resulting UTF showed a dense and ordered localization of Au NCs on the LDH nanosheets [53]. The electrophilic effect of the LDH nanosheets effectively localizes the Au NCs and polarizes the charge distribution of the Au NCs (Fig. 1b). The confinement effect imposed by the LDH nanosheets effectively inhibits the non-radiative energy loss of Au NCs, leading to increased quantum yields (QYs) (from 2.69% to 14.11%) and prolonged fluorescence lifetimes (from 1.84 to 14.67 μs).

Frontier orbital analysis shows that multiple quantum-well structures can be built based on the HOMO and LUMO for the ZnTSPc/LDH system (Fig. 1c) [54]. The LDH monolayer functions as an energy well which hinders the interlayer gallery interactions of the ZnTSPc and contributes to the outstanding fluorescence behavior of the ZnTSPc chromophore. In addition, LDHs provide an isolated environment for interlayer Förster resonance energy transfer (FRET). The use of ultrathin LDH nanosheets results in immobilization of the guest species and thus guarantees the effective distance between the donor and acceptor in the FRET process (Fig. 1d). Moreover, the FRET state (on/off) and its efficiency can be tuned by changes in the interlayer distance induced by the adsorption and desorption of volatile organic compounds [55–57].

3.3 Applications in Display and Sensing

Through controlled variation of the guest species, various LDH-based photofunctional materials can be obtained. Introduction of LDHs can effectively enhance the fluorescence performance and stability of guest species. Some novel intercalated LDH composites with a specifically tailored arrangement in the LDH gallery have been used in displays [58, 59], polarized emission [60, 61], and sensing (Fig. 2). Materials with multicolor luminescence have been obtained by effective utilization of a diverse variety of chromophores (organic dyes, polymers, or quantum dots) [62]. Chemosensors for heavy metal ions (HMI) [63], biomolecules [64], and environmental pollutants [65] have been designed, and multifunctional

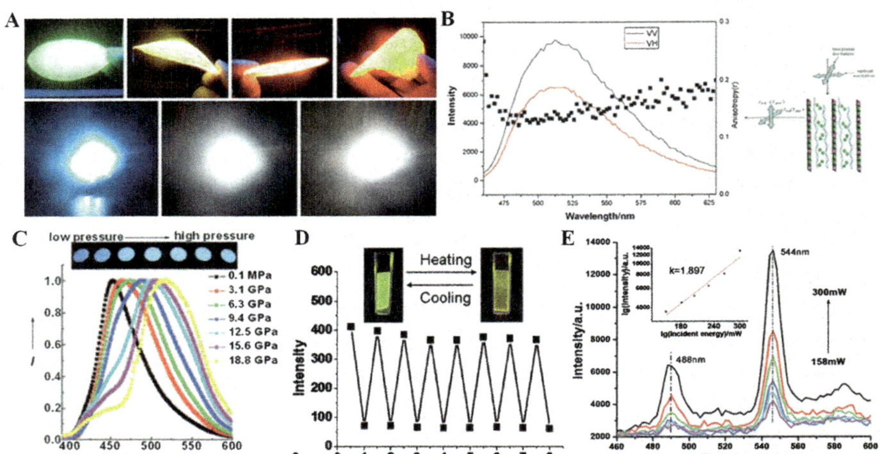

Fig. 2 Application of LDH-based photofunctional materials for (**a**) displays, (**b**) polarized fluorescence, (**c**) piezochromic materials, (**d**) temperature sensing, and (**e**) two-photon emission (reprinted with permission from [59], [36], [68], [67] and [61], respectively)

materials which respond to changes in pH [66], temperature [67], pressure [68], and light [69] have been constructed and investigated.

4 Bio-applications In Vitro and In Vivo

4.1 Advantages of LDH-Based Materials in Bio-applications

Recently, great attention has been focused on medical applications of LDHs because their 2D structure and intercalation properties allow novel bio-inorganic nanohybrids to be fabricated. Due to the high anionic exchange capacity of LDHs, various biomolecules such as drugs, genes, and nutrient ingredients can readily be immobilized in their interlayer galleries [70, 71]. By employing various synthetic routes such as coprecipitation, ion exchange, calcination–reconstruction, and exfoliation–reassembly, intercalated nanohybrids can be fabricated and employed in bio-LDH delivery hybrid systems for the storage, transport, and subsequent controlled release of bioactive species [72].

LDH materials possess numerous advantages as versatile carriers in drug/gene delivery and biomedicine. Firstly, their low cytotoxicity and excellent biocompatibility make LDHs an ideal drug nanocarrier. It has been reported that the cytotoxicity of LDHs is relatively low toward normal cells, and no cytotoxic effect is observed even at concentrations up to 1 mg/mL [73, 74]. Secondly, the dispersion and stability of guests can be significantly enhanced after intercalation into the LDH interlamellar galleries. Thirdly, intercalation or absorption of the drug, gene, or other target materials (e.g., proteins, nucleic acids, peptides, antibodies, or aptamers) by LDHs can avoid complicated ligand exchange reactions and surface modification. In addition, a controlled release of the interlamellar drug/gene, which is often necessary for pharmaceutical efficacy, can be realized by employment of LDHs.

4.2 Applications in Imaging, Medicine Delivery, and Therapeutics

LDHs can be combined with fluorescent, optical, magnetic, or radioactive materials so as to realize molecular imaging, disease diagnostics, and tumor visualization in a biological system. Yoon et al. have grafted fluorescein onto exfoliated Gd-doped LDHs, and the obtained materials simultaneously possess fluorescence and magnetic properties (Fig. 3a) [75]. Therefore, the resulting materials may find great application in fluorescence and magnetic resonance imaging (MRI) for the systematic study of the diseases in deep lesions with high resolution.

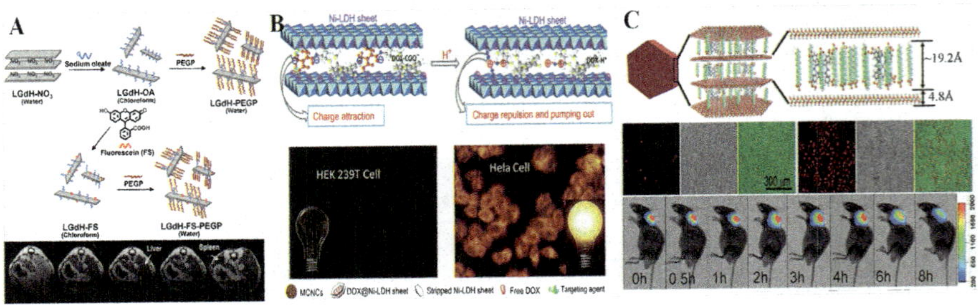

Fig. 3 Bio-related applications of LDH-based materials in (**a**) imaging, (**b**) drug delivery, and (**c**) therapy (reprinted with permission from [75], [78] and [27], respectively)

Moreover, anticancer drugs (such as methotrexate, doxorubicin (DOX), and 5-fluorouracil) and genes can be intercalated into LDHs and form bio-LDH nanohybrids. The tumor bioaccumulation ratio for such bio-LDH nanohybrids can be increased severalfold compared with the free drugs or genes at the tumor site, which shows the passive targeting ability of the materials. Such bio-LDH nanohybrids containing drugs or genes can also be effectively transferred into cells through clathrin-mediated endocytosis [76]. Furthermore, active targeting can be achieved through the incorporation of target moieties for specific interaction with, and recognition of, targeted cells [77]. A wide variety of receptor-specific targeting moieties including nucleic acids, peptides, aptamers, and proteins (antibodies) can be covalently grafted to the surface of the bio-LDH nanohybrids and result in molecular level targeting.

Current research also pays close attention to the development of theranostic bio-LDH composites so as to incorporate therapeutic agents with imaging probes. Li et al. developed hierarchically structured core/shell nanospheres by combination of DOX, fluorescein isothiocyanate (FITC), magnetic nanoparticles (MSPs), and folate (DOX@MSPs/Ni-LDH-folate) [78]. The amide bond in DOX-COOH intercalated in the interlayer galleries of an LDH can be cleaved in the acidic cytoplasm, and the protonated DOX drug is quickly released from the LDH into the cytoplasm (Fig. 3b). Shi's group also successfully developed Gd-doped LDH nanoparticles with Au nanoparticles deposited on the surface (LDH-Gd/Au) and explored diagnosis and simultaneous drug delivery monitored by noninvasive visualization in vitro and in vivo [79]. As both a drug vehicle and a diagnostic agent, the LDH-Gd/Au can be employed as a drug delivery system to transport DOX into the cancer cells via endocytosis and release DOX into the acidic cytoplasm which induces death of the cancer cells.

In addition, sustained release capability of LDHs is also an important merit for some drugs. A low-molecular-weight heparin (LMWH) (with a molecular mass (4–6 kDa)) is frequently used as an anticoagulant but suffers from a short half-life of 2–4 h [80]. When LMWH was intercalated into LDH interlayer galleries through the anion-exchange method, the resulting composite exhibited a prolonged half-life and enhanced stability in blood plasma. Furthermore, it has been found that the intercalation of guest species into LDH interlamellar galleries can effectively depress their aggregation and result in enhanced uniformity and stability. For example, a supramolecular photosensitizer was fabricated by intercalation of a zinc phthalocyanine (ZnPc) into LDH galleries, and the product showed an excellent photodynamic therapy (PDT) anticancer behavior (Fig. 3c). A high dispersion of ZnPc in the monomeric state in the interlayer region of the LDH matrix can be achieved on account of the host–guest and guest–guest interactions, which contributes to high singlet oxygen production efficiency and significantly enhanced anticancer behavior [27].

5 Catalysis

Due to their tunable composition, particle size, structure, and particle morphology, LDHs and LDH-based nanohybrids have been widely used in:

Heterogeneous catalysis, photocatalysis, and photoelectrocatalysis [81–83]. Diverse catalysts with high catalytic performance, good stability, and recyclability can be achieved using LDHs and LDH-based nanohybrids [13].

5.1 LDHs Themselves as Catalysts

Because of the hydroxyl groups on the LDH layers, there are abundant basic sites in LDHs accounting for their efficacy as heterogeneous solid base catalysts. Moreover, catalytically active transition metals can be anchored in the tunable host layer to obtain LDHs with high catalytic activity and selectivity [84, 85]. Various catalytically active transition metal species (e.g., Fe, Co, Ni, Mn, Cr, Ce, Ti, and Cu) can be incorporated into the host layer of LDHs, and these elements give LDHs good catalytic activity in electrocatalysis, photocatalysis, and photoelectrocatalysis [13, 86]. Hu et al. have studied the oxygen evolution reaction (OER) activity of NiFe and NiCo LDHs before and after exfoliation (Fig. 4a) and found that single-layer LDH nanosheets exhibit significantly enhanced catalytic activity compared with the unexfoliated precursor due to the increase in active site density and conductivity [87]. Similar results were also obtained for CoMn-LDHs [88].

5.2 LDHs as Precursors to Metal or Metal Oxide Catalysts

LDHs containing catalytically active elements (e.g., Fe, Co, Ni, or Pd) can be employed as precursors to fabricate metal or metal oxide catalysts based on the topological transformation of LDHs through a calcination–reconstruction process

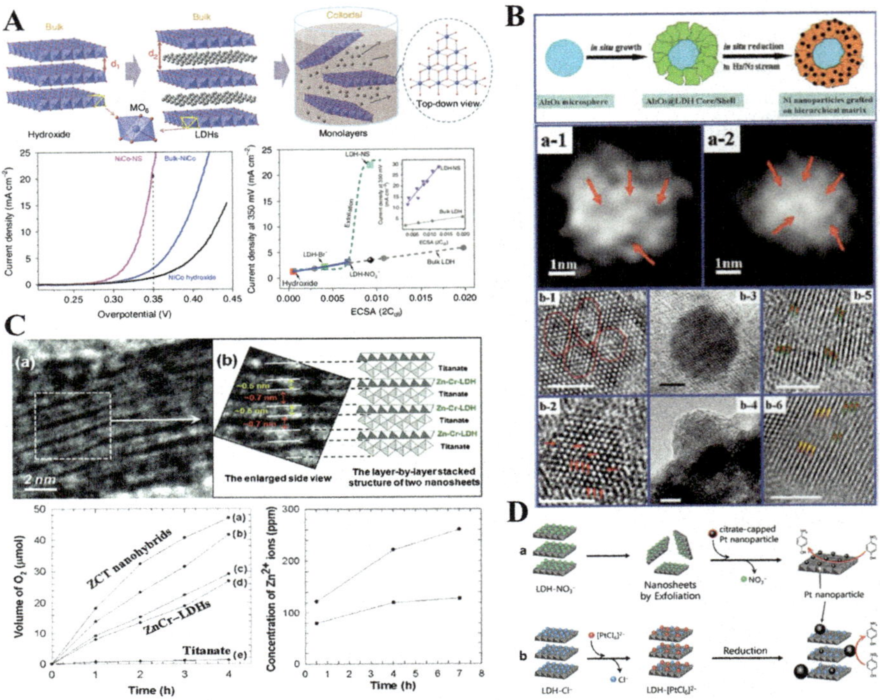

Fig. 4 LDHs employed in catalysis as (**a**) actual catalysts, (**b**) precursors for metal or metal oxide catalysts, (**c**) a host for intercalated catalysts, and (**d**) catalyst supports (reprinted with permission from [87], [93], [84] and [107], respectively)

[89–92]. The first such topological transformation of LDHs was reported by Wei et al., and well-dispersed and high-density Fe, Co, and Ni metal nanoparticles with good thermal stability were obtained [90]. He and coworkers have obtained Ni nanoparticles by in situ reduction of a NiAl-LDH precursor (Fig. 4b) [93]. The employment of LDHs effectively prevents the aggregation of Ni, and the strong anchoring effect leads to good distribution of Ni nanoparticles.

5.3 *LDH-Intercalated Hybrids as Catalysts*

Catalytically active species can be intercalated into the interlayer galleries of LDHs, and the confined environment of LDHs can lead to precise control over the arrangement and dispersion of the active species, which leads to enhanced reaction rates and excellent catalytic stability. Moreover, by tuning the charge density and composition of the host layer, the geometric and electronic states of the active sites can be finely tuned. Organic acids/bases, inorganic anion/ nanoparticles, metal complexes, biomolecules, and noble metal nanoparticles

[94–103] can be intercalated into LDHs giving materials with high catalytic activity. Hwang et al. have constructed ordered structures by LbL assembly of a ZnCr–LDH and layered titanium oxide (Fig. 4c), and the resulting heterolayer materials exhibit excellent catalytic activity for visible light-induced O_2 generation as well as high chemical stability [84].

5.4 LDHs as Catalyst Supports

In order to improve the recyclability of catalysts, LDHs have been used to effectively stabilize the active sites, and a high dispersion of the immobilized catalyst particles can be obtained [104, 105]. Au, Pd, and Pt can be anchored onto LDHs, and a synergetic interaction results in significantly enhanced catalytic activity and selectivity (Fig. 4d) [106, 107].

6 Electrochemistry

6.1 Introduction

Recently, energy storage and conversion in electrochemistry have attracted great attention as an alternative to traditional energy resources [108, 109]. For energy storage devices such as supercapacitors, a high-power density, super-high-cycle life span, and fast dynamic response rate are required if they are to have wide application [110–112].

LDH-based nanohybrids offer significant advantages as electrochemical materials. Firstly, taking advantage of the tunable composition of the LDH host layer, electrically active transition metals, such as Co, Mn, Fe, and Ni, can be introduced [113]. Secondly, diverse strategies can be exploited to fabricate LDHs with large surface area, suitable mesopores, or ordered arrays, which facilitates electron transport and improves power density [114, 115]. Thirdly, combination of LDHs and various electrically active substances affords hybrid materials with long-term stability and fast redox reactions [99].

6.2 Structure and Composition of LDH-Based Electro-materials

In order to obtain electrically active materials, a variety of structures (core–shell structure, hollow microspheres, layered structures or arrays) have been designed and fabricated (Fig. 5). A core–shell structure with excellent electrochemical

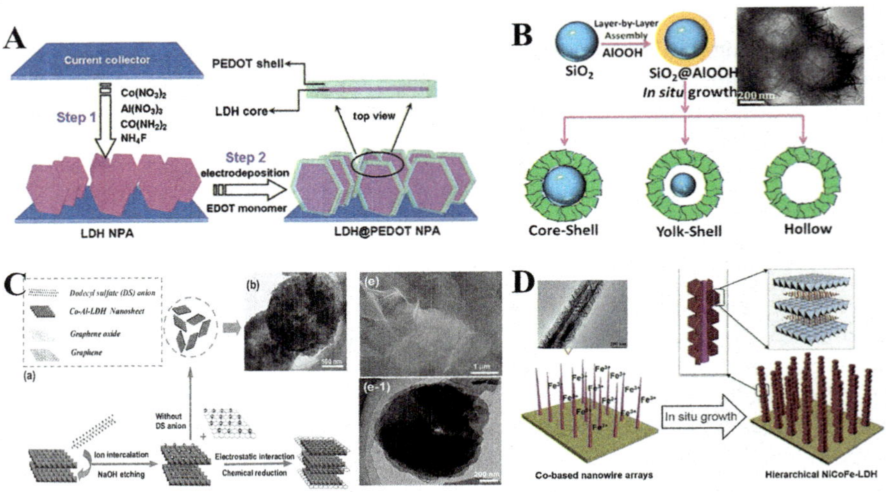

Fig. 5 Electrically active materials with different structures: (**a**) core–shell structure, (**b**) hollow microspheres, (**c**) layered structure, and (**d**) arrays (reprinted with permission from [116], [117], [120] and [121], respectively)

behavior can be constructed by electrodepositing poly (3,4-ethylenedioxythiophene) (PEDOT) on an LDH nanoplatelet core (Fig. 5a) [116]. The LDH core provides high energy storage capacity while the PEDOT shell contributes to fast electron transport. Moreover, NiAl-LDH microspheres with tunable interior architecture from core–shell to hollow structures have been designed via LbL deposition followed by an in situ growth technique (Fig. 5b) [117]. Due to the high surface area and mesopore distribution, the resulting hollow NiAl-LDH microspheres exhibit excellent capacitance much larger than that of core–shell and yolk–shell microspheres. A similar structure was constructed by employing a carbon hollow sphere as an interior shell and NiAl-LDH as an exterior shell, and large specific surface area with uniform mesoporous structure can be achieved [118]. The porous structure of LDHs is beneficial in terms of electro-activity [119], and the combination of porous LDHs and conductive graphene nanosheets endows the composites with significantly improved capacitance and ultrahigh rate performance capability (Fig. 5c) [120]. Figure 5d shows a novel hierarchical nanoarray of vertically aligned NiCoFe-LDHs on a Co-based nanowire array [121]; this material exhibits large anodic current density and excellent electrochemical durability.

Other LDHs with different host layer composition (e.g., NiMn, ZnCo, $Co^{II}Co^{III}$Al-LDHs) have also been investigated as electrically active materials [122–128]. Moreover, various materials (such as carbon materials, conductive polymers, metal oxides, and biomolecules) have been combined with LDHs to fabricate electrically active materials [129, 130]. Carbon materials have been widely investigated as a promising electrode material by virtue of their high electrical activity and low cost [131, 132]. Zhao et al. have fabricated a series of hierarchical nanostructures composed of LDH grafted on a carbon material backbone by in situ growth [133, 134], which exhibits excellent activity and high capacitance. Graphene oxide or graphene have also been combined with LDHs to obtain hybrid materials with high catalytic activity or charge conductivity [135–137]. In addition, Song et al. have fabricated a high-rate asymmetric supercapacitor by combining partially exfoliated graphite, CoNi-LDH, and polypyrrole (PPy), and fast electron/ion transportation and high specific capacitance were obtained [138].

7 LDH-Based Multifunctional Materials

The layered structure and rigid host layer of LDHs allow the fabrication of a wide variety of hybrid materials with diverse applications, and self-supporting multifunctional films as well as sols with magnetic properties have attracted particular attention [139–141].

7.1 Multifunctional Films

Gas barrier materials have been constructed by combining cellulose acetate (CA) or polyacrylic acid (PAA) with LDHs [25, 142]. LDHs can behave simultaneously as a physical blocking material and chemical absorption agent, accounting for the excellent gas barrier performance when compared with commonly used materials (Fig. 6a). Antireflection materials have been prepared through the calcination–rehydration procedure (Fig. 6b), and different transmittance and reflection effects can be achieved by switching between nonporous LDHs and calcined porous MMO materials [143]. Superhydrophobic materials can be constructed by preparing oriented NiAl- or ZnAl-LDH films on a porous aluminum oxide (PAO)/Al substrate (Fig. 6c) [144, 145]. By virtue of the microscale hemispherical protrusions on the surface, both micro- and nanoscale hierarchical structures can be obtained, accounting for large superhydrophobic surface areas. Moreover, assisted by an external AC electric field, LDH-polyphosphazene-based ionomer hybrids can be obtained and employed as anion-exchange membranes (Fig. 6d) [146]. Owing to the oriented LDH nanoplatelets, the ion-exchange capacity, mechanical properties, and ionic conductivity of the anion-exchange membranes can be significantly enhanced.

7.2 Multifunctional Powders or Colloids

LDH-based magnetic materials have been developed for medicine/protein separation and controlled release [147, 148]. There are three approaches to introduce

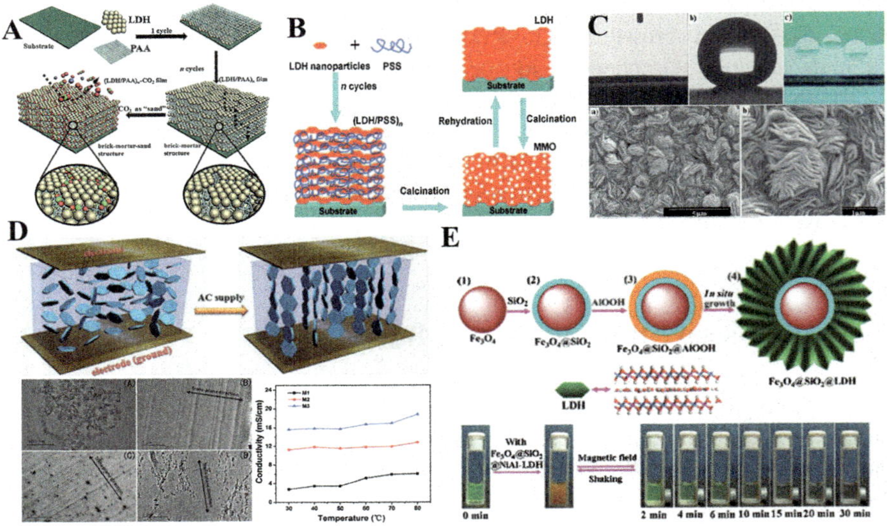

Fig. 6 Schematic illustrations of (**a**) a gas barrier material, (**b**) an antireflection material, (**c**) a superhydrophobic material, (**d**) an ion-conductive electrolyte, and (**e**) an absorption and separation material (reprinted with permission from [25], [143], [144], [146] and [114], respectively)

magnetic species: (1) incorporation of the magnetic element into the LDH host layers, such as $Ni^{II}Cr^{III}$-LDHs [149]; (2) intercalation of magnetic guests into the interlayer galleries of LDHs [150]; and (3) employing magnesium ferrite particles as a magnetic core to fabricate LDH-based core–shell structures [151]. Moreover, assisted by an external magnetic field, the recycling and reuse of these magnetic composites can be achieved (Fig. 6e) [114].

Fabrication of multifunctional LDH-based composite materials is still a fascinating area, and many efforts are ongoing [152–154].

8 Conclusion and Outlook

LDH-based compounds have become promising multifunctional materials which can be used as photofunctional materials in bio-applications, in catalysis, and in electrochemistry among other areas. Due to their tunable composition, ion-exchangeable properties, tailorable morphology, and inherent high stability, LDHs can serve as an effective building block or support for the fabrication of a wide variety of hybrid materials. Moreover, by virtue of the host–guest interaction and confinement effect induced by LDHs, the resulting materials usually exhibit superior performances when compared with the pristine guest species.

In the future, greater efforts are needed to explore the detailed structure–property correlations in LDH-based composites in order to develop their applications in new interdisciplinary areas. Firstly, how the confinement effect and host–guest interaction of LDHs functions remains a key scientific issue. Secondly, the connections between the microscopic structure and macroscopic performance need to be better understood. With persistent efforts and deepened understanding, we firmly believe that the large-scale preparation of novel multifunctional LDH materials can be realized, which may lead to many exciting industrial applications.

Acknowledgment We would like to thank all the coworkers cited in the references for their invaluable contributions to the work described here. This work was supported by the 973 Program (Grant No. 2014CB932103) and the National Natural Science Foundation of China (NSFC).

References

1. Evans DG, Slade RCT (2006) Struct Bond 119:1
2. He J, Wei M, Li B, Kang Y, Evans DG, Duan X (2006) Struct Bond 119:89
3. Taviot-Gueho C, Leroux F (2006) Struct Bond 119:121
4. Williams GR, O'Hare D (2006) Struct Bond 119:161
5. Li F, Duan X (2006) Struct Bond 119:193
6. Ruiz-Hitzky E, Aranda P, Darder M, Rytwo G (2010) J Mater Chem 20:9306
7. Liu J, Zhang G (2014) Phys Chem Chem Phys 16:8178
8. Shu X, Zhang W, He J, Gao F, Zhu Y (2006) Solid State Sci 8:634
9. Liu XL, Wei M, Wang ZL, Evans DG, Duan X (2008) J Phys Chem C 112:17517
10. Zhao J, Chen J, Xu S, Shao M, Zhang Q, Wei F, Ma J, Wei M, Evans DG, Duan X (2014) Adv Funct Mater 20:2938
11. Shao M, Ning F, Wei M, Evans DG, Duan X (2014) Adv Funct Mater 24:580
12. Darder M, López-Blanco M, Aranda P, Leroux F, Ruiz-Hitzky E (2005) Chem Mater 17:1969
13. Zhao MQ, Zhang Q, Zhang W, Huang JQ, Zhang Y, Su DS, Wei F (2010) J Am Chem Soc 132:14739
14. Khan AI, O'Hare D (2002) J Mater Chem 12:3191
15. Li S, Lu J, Ma H, Yan D, Li Z, Qin S, Evans DG, Duan X (2012) J Phys Chem C 116:12836
16. Shao M, Wei M, Evans DG, Duan X (2011) Chem Commun 47:3171
17. Shao M, Xu X, Han J, Zhao J, Shi W, Kong X, Wei M, Evans DG, Duan X (2011) Langmuir 27:8233
18. Costa AL, Gomes AC, Pillinger M, Gonçalves IS, de Seixas Melo JS (2015) Langmuir 31:4769
19. An Z, Zhang W, Shi H, He J (2006) J Catal 241:319
20. Shi W, Wei M, Evans DG, Duan X (2010) J Mater Chem 20:3901
21. Shi W, Sun Z, Wei M, Evans DG, Duan X (2010) J Phys Chem C 114:21070
22. Shi W, Wei M, Lu J, Evans DG, Duan X (2009) J Phys Chem C 113:12888
23. Wang J, Zhao L, Shi H, He J (2011) Angew Chem Int Ed 50:9171
24. Tian R, Liang R, Yan D, Shi W, Yu X, Wei M, Li LS, Evans DG, Duan X (2013) J Mater Chem C 1:5654
25. Dou Y, Pan T, Xu S, Yan H, Han J, Wei M, Evans DG, Duan X (2015) Angew Chem Int Ed 54:9673
26. Shao M, Li Z, Zhang R, Ning F, Wei M, Evans DG, Duan X (2015) Small 29:3530
27. Liang RZ, Tian R, Ma L, Zhang LL, Hu YL, Wang J, Wei M, Yan D, Evans DG, Duan X (2014) Adv Funct Mater 24:3144
28. Tian R, Yan D, Wei M (2015) Struct Bond 166:1
29. Xu K, Zhang Z, Chen G, Shen J (2014) RSC Adv 4:19218

30. Zhang Z, Chen G, Xu K (2013) Ind Eng Chem Res 52:11045
31. Yan D, Lu J, Wei M, Han J, Ma J, Li F, Evans DG, Duan X (2009) Angew Chem Int Ed 48:3073
32. Yan D, Lu J, Ma J, Wei M, Wang X, Evans DG, Duan X (2010) Langmuir 26:7007
33. Yan D, Lu J, Wei M, Ma J, Evans DG, Duan X (2009) Chem Commun 6358
34. Li S, Lu J, Xu J, Dang S, Evans DG, Duan X (2010) J Mater Chem 20:9718
35. Li S, Lu J, Ma H, Xu J, Yan D, Wei M, Evans DG, Duan X (2011) Langmuir 27:11501
36. Yan D, Lu J, Chen L, Qin S, Ma J, Wei M, Evans DG, Duan X (2010) Chem Commun 46:5912
37. Zheng SF, Lu J, Li W, Qin YM, Yan DP, Evans DG, Duan X (2014) J Mater Chem C 2:5161
38. Bendall JS, Paderi M, Ghigliotti F, Li Pira N, Lambertini V, Lesnyak V, Gaponik N, Visimberga G, Eychmüller A, Torres CMS, Welland ME, Gieck C, Marchese L (2010) Adv Funct Mater 20:3298
39. Cho S, Kwag J, Jeong S, Baek Y, Kim S (2013) Chem Mater 25:1071
40. Wu G, Wang L, Evans DG, Duan X (2006) Eur J Inorg Chem 2006:3185
41. Omwoma S, Chen W, Tsunashima R, Song YF (2014) Coord Chem Rev 58:258
42. Han Z, Guo Y, Tsunashima R, Song YF (2013) Eur J Inorg Chem 2013:1475
43. Sousa FL, Pillinger M, Sá Ferreira RA, Granadeiro CM, Cavaleiro AMV, Rocha J, Carlos LD, Trindade T, Nogueira HIS (2006) Eur J Inorg Chem 2006:726
44. Tian R, Li M, Teng H, Luo H, Yan D, Wei M (2015) J Mater Chem C 3:5167
45. Fogg AM, Freij AJ, Parkinson GM (2002) Chem Mater 14:232
46. Leroux F, Taviot-Guého C (2005) J Mater Chem 15:3628
47. Desigaux L, Belkacem MB, Richard P, Cellier J, Léone P, Cario L, Leroux F, Taviot-Guého C, Pitard B (2006) Nano Lett 6:199
48. Yan D, Lu J, Wei M, Evans DG, Duan X (2009) J Phys Chem B 113:1381
49. Shi W, He S, Wei M, Evans DG, Duan X (2010) Adv Funct Mater 20:3856
50. Sun Z, Jin L, Shi W, Wei M, Evans DG, Duan X (2011) Langmuir 27:7113
51. Yan D, Lu J, Ma J, Wei M, Evans DG, Duan X (2010) Phys Chem Chem Phys 12:15085
52. Li S, Lu J, Wei M, Evans DG, Duan X (2010) Adv Funct Mater 20:2848
53. Tian R, Zhang S, Li M, Zhou Y, Lu B, Yan D, Wei M, Evans DG, Duan X (2015) Adv Funct Mater 25:5006
54. Yan D, Qin S, Chen L, Lu J, Ma J, Wei M, Evans DG, Duan X (2010) Chem Commun 46:8654
55. Li Z, Lu J, Li S, Qin S, Qin Y (2012) Adv Mater 24:6053
56. Qin Y, Lu J, Li S, Li Z, Zheng S (2014) J Phys Chem C 118:20538
57. Gunawan P, Xu R (2009) J Phys Chem C 113:17206
58. Liang R, Xu S, Yan D, Shi W, Tian R, Yan H, Wei M, Evans DG, Duan X (2012) Adv Funct Mater 22:4940
59. Liang R, Yan D, Tian R, Yu X, Shi W, Li C, Wei M, Evans DG, Duan X (2014) Chem Mater 26:2595
60. Yan D, Lu J, Wei M, Evans DG, Duan X (2011) J Mater Chem 21:13128
61. Gao R, Zhao M, Guan Y, Fang X, Li X, Yan D (2014) J Mater Chem C 45:9579
62. Cho S, Jung S, Jeong S, Bang J, Park J, Park Y, Kim S (2013) Langmuir 29:441
63. Jin L, Guo Z, Sun Z, Li A, Jin Q, Wei M (2012) Sensor Actuat B-Chem 161:714
64. Jin L, Guo Z, Wang T, Wei M (2013) Sensor Actuat B-Chem 177:145
65. Sun Z, Jin L, Zhang S, Shi W, Pu M, Wei M, Evans DG, Duan X (2011) Anal Chim Acta 702:95
66. Shi W, Ji X, Wei M, Evans DG, Duan X (2012) Langmuir 28:7119
67. Liang R, Tian R, Shi W, Liu Z, Yan D, Wei M, Evans DG, Duan X (2013) Chem Commun 49:969
68. Yan D, Lu J, Ma J, Qin S, Wei M, Evans DG, Duan X (2011) Angew Chem Int Ed 50:7037
69. Wang XR, Lu J, Yan D, Wei M, Evans DG, Duan X (2010) Chem Phys Lett 493:333
70. Oh JM, Park DH, Choi SJ, Choy JH (2012) Recent Pat Nanotechnol 6:200

71. Choi SJ, Choy JH (2011) Nanomedicine 6:803
72. Park DH, Hwang SJ, Oh JM, Yang JH, Choy JH (2013) Prog Polym Sci 38:1442
73. Kriven WM, Kwak SY, Wallig MA, Choy JH (2004) MRS Bull 29:33
74. Xu ZP, Walker TL, Liu KL, Cooper HM, Lu GQ, Bartlett PF (2007) Int J Nanomedicine 2:163
75. Yoon YS, Lee KS, Im GH, Lee BI, Byeon SH, Lee JH, Lee IS (2009) Adv Funct Mater 19:3375
76. Khan AI, Lei L, Norquist AJ, O'Hare D (2001) Chem Commun 22:2342
77. Oh JM, Choi SJ, Lee GE, Han SH, Choy JH (2009) Adv Funct Mater 19:1617
78. Li D, Zhang YT, Yu M, Guo J, Chaudhary D, Wang CC (2013) Biomaterials 32:7913
79. Wang L, Xing H, Zhang S, Ren Q, Pan L, Zhang K, Bu W, Zheng X, Zhou L, Peng W, Hua Y, Shi J (2013) Biomaterials 34:3390
80. Gu Z, Thomas AC, Xu ZP, Campbell JH, Lu GQ (2008) Chem Mater 20:3715
81. Li C, Wei M, Evans DG, Duan X (2014) Small 10:4469
82. Fan G, Li F, Evans DG, Duan X (2014) Chem Soc Rev 43:7040
83. Li C, Wei M, Evans DG, Duan X (2015) Catal Today 247:163
84. Gunjakar JL, Kim TW, Kim HN, Kim IY, Hwang SJ (2011) J Am Chem Soc 133:14998
85. Kim SJ, Lee Y, Lee DK, Lee JW, Kang JK (2014) J Mater Chem A 2:4136
86. Morikawa M, Ahmed N, Yoshida Y, Izumi Y (2014) Appl Catal B-Environ 144:561
87. Song F, Hu X (2014) Nat Commun 5:1
88. Song F, Hu X (2014) J Am Chem Soc 136:16481
89. Koike M, Li D, Nakagawa Y, Tomishige K (2012) ChemSusChem 5:2312
90. Tian GL, Zhao MQ, Zhang B, Zhang Q, Zhang W, Huang JQ, Chen TC, Qian WZ, Su DS, Wei F (2014) J Mater Chem A 2:1686
91. Gao W, Li C, Chen H, Wu M, He S, Wei M, Evans DG, Duan X (2014) Green Chem 16:1560
92. He Y, Liang L, Liu Y, Feng J, Ma C, Li D (2014) J Catal 309:166
93. He S, Li C, Chen H, Su D, Zhang B, Cao X, Wang B, Wei M, Evans DG, Duan X (2013) Chem Mater 25:1040
94. Zhao S, Xu J, Wei M, Song YF (2011) Green Chem 13:384
95. Liu P, Wang C, Li C (2009) J Catal 262:159
96. Liu P, Wang H, Feng Z, Ying P, Li C (2008) J Catal 256:345
97. Reinholdt MX, Kirkpatrick RJ (2006) Chem Mater 18:2567
98. Dhakshinamoorthy A, Sharmila A, Pitchumani K (2010) Chem Eur J 16:1128
99. Zhao J, Kong X, Shi W, Shao M, Han J, Wei M, Evans DG, Duan X (2011) J Mater Chem 21:13926
100. Parida KM, Sahoo M, Singha S (2010) J Catal 276:161
101. Liu H, Zhao L, Wang J, He J (2013) J Catal 298:70
102. Bhattacharjee S, Dines TJ, Anderson JA (2008) J Phys Chem C 112:14124
103. Shi H, Yu C, He J (2010) J Catal 271:79
104. Yang XF, Wang A, Qiao B, Li J, Liu J, Zhang T (2013) Acc Chem Res 46:1740
105. Yang Z, Tjiu WW, Fan W, Liu T (2013) Electrochim Acta 90:400
106. Zhang F, Zhao X, Feng C, Li B, Chen T, Lu W, Lei X, Xu S (2011) ACS Catal 1:232
107. Lee JH, Kim H, Lee YS, Jung DY (2014) ChemCatChem 6:113
108. Peng Y, Chen Z, Le Z, Xu Q, Li H, Lu Y (2015) Chem Commun 51:12056
109. Zhao G, Li J, Jiang L, Dong H, Wang X, Hu W (2012) Chem Sci 3:433
110. Simon P, Gogotsi Y (2008) Nat Mater 7:845
111. Yang P, Xiao X, Li Y, Ding Y, Qiang P, Tan X, Mai W, Lin Z, Wu W, Li T, Jin H, Liu P, Zhou J, Wong CP, Wang ZL (2013) ACS Nano 7:2617
112. Yuan L, Lu XH, Xiao X, Zhai T, Dai J, Zhang F, Hu B, Wang X, Gong L, Chen J, Hu C, Tong Y, Zhou J, Wang ZL (2012) ACS Nano 6:656
113. Shao M, Wei M, Evans DG, Duan X (2013) Chem Eur J 19:4100
114. Shao M, Ning F, Zhao J, Wei M, Evans DG, Duan X (2012) J Am Chem Soc 134:1071
115. Chen H, Hu L, Yan Y, Che R, Chen M, Wu L (2013) Adv Energy Mater 3:1636

116. Han J, Dou Y, Zhao J, Wei M, Evans DG, Duan X (2013) Small 9:98
117. Shao M, Ning F, Zhao Y, Zhao J, Wei M, Evans DG, Duan X (2012) Chem Mater 24:1192
118. Xu J, He F, Gai S, Zhang S, Li L, Yang P (2014) Nanoscale 6:10887
119. Zhang L, Wang J, Zhu J, Zhang X, San Hui K, Hui KN (2013) J Mater Chem A 1:9046
120. Wu X, Jiang L, Long C, Wei T, Fan Z (2015) Adv Funct Mater 25:1648
121. Yang Q, Li T, Lu Z, Sun X, Liu J (2014) Nanoscale 6:11789
122. Yuan P, Zhang N, Zhang D, Liu T, Chen L, Liu X, Ma R, Qiu G (2014) Chem Commun 50:11188
123. Zou X, Goswami A, Asefa T (2013) J Am Chem Soc 135:17242
124. Sim H, Jo C, Yu T, Lim E, Yoon S, Lee JH, Yoo J, Lee J, Lim B (2014) Chem Eur J 20:14880
125. Liu X, Ma R, Bando Y, Sasaki T (2012) Adv Mater 24:2148
126. Shao M, Ning F, Zhao J, Wei M, Evans DG, Duan X (2013) Adv Funct Mater 28:3513
127. Zhao J, Xu S, Tschulik K, Compton RG, Wei M, O'Hare D, Evans DG, Duan X (2015) Adv Funct Mater 18:2745
128. Doménech A, Coronado E, Lardiés N, Gastaldo CM, Doménech-Carbó MT, Ribera A (2008) J Electroanal Chem 624:275
129. Liu Y, Zhang X, Tu X, Wei C, Lv L (2014) Anal Methods 6:4061
130. Ning F, Shao M, Zhang C, Xu S, Wei M, Duan X (2014) Nano Energy 7:134
131. Wang X, Yan C, Sumboja A, Yan J, Lee PS (2014) Adv Energy Mater 4:1301240
132. Liu X, Wang C, Dou Y, Zhou A, Pan T, Han J, Wei M (2014) J Mater Chem A 2:1682
133. Zhao J, Chen J, Xu S, Shao M, Yan D, Wei M, Evans DG, Duan X (2013) J Mater Chem A 1:8836
134. Zhao J, Lu Z, Shao M, Yan D, Wei M, Evans DG, Duan X (2013) RSC Adv 3:1045
135. Li Z, Shao M, Zhou L, Zhang R, Zhang C, Wei M, Evans DG, Duan X (2015) Adv Mater. doi:10.1002/adma.201505086
136. Long X, Li J, Xiao S, Yan K, Wang Z, Chen H, Yang S (2014) Angew Chem Int Ed 3:7584
137. Dong Y, Liu Y, Yin J, Zhao X (2014) J Mater Chem C 2:10386
138. Song Y, Cai X, Xu X, Liu XX (2015) J Mater Chem A 3:14712
139. Han J, Dou Y, Yan D, Ma J, Wei M, Evans DG, Duan X (2011) Chem Commun 47:5274
140. Tadanaga K, Furukawa Y, Hayashi A, Tatsumisago M (2010) Adv Mater 22:4401
141. Lee JH, Chang J, Cha JH, Jung DY, Kim SS, Kim JM (2010) Chem Eur J 16:8296
142. Dou Y, Xu S, Liu X, Han J, Yan H, Wei M, Evans DG, Duan X (2014) Adv Funct Mater 24:514
143. Han J, Dou Y, Wei M, Evans DG, Duan X (2010) Angew Chem Int Ed 49:2171
144. Zhang F, Zhao L, Chen H, Xu S, Evans DG, Duan X (2008) Angew Chem Int Ed 47:2466
145. Chen H, Zhang F, Fu S, Duan X (2006) Adv Mater 18:3089
146. Fan J, Zhu H, Li R, Chen N, Han K (2014) J Mater Chem A 2:8376
147. Pan D, Zhang H, Fan T, Chen J, Duan X (2011) Chem Commun 47:908
148. Zhang H, Zhang G, Bi X, Chen X (2013) J Mater Chem A 1:5934
149. Almansa JJ, Coronado E, Martí-Gastaldo C, Ribera A (2008) Eur J Inorg Chem 2008:5642
150. Coronado E, Martí-Gastaldo C, Navarro-Moratalla E, Ribera A, Galán- Mascarós JR (2013) J Mater Chem 9476
151. Zhang H, Pan D, Zou K, He J, Duan X (2009) J Mater Chem 19:3069
152. Wang Q, Tay HH, Zhong Z, Luo J, Borgna A (2012) Energy Environ Sci 5:7526
153. Wang Q, Gao Y, Luo J, Zhong Z, Borgna A, Guo Z, O'Hare D (2013) RSC Adv 3:3414
154. Lv X, Chen Z, Wang Y, Huang F, Lin Z (2013) ACS Appl Mater Interfaces 5:11271

Struct Bond (2016) 172: 85–118
DOI: 10.1007/430_2015_193

Published online: 25 November 2015

Catalytic Sigma-Bond Metathesis and the Polymerization of 1,3-Dienes by Rare-Earth Metal Complexes

Rory P. Kelly and Peter W. Roesky

Abstract This review is a selection of research highlights since 2010 in two main areas of rare-earth catalysis, but it is not meant to be an exhaustive treatment. Part 1 (Sect. 2) deals with advances in intra- and intermolecular hydroamination, hydrophosphination, and hydrosilylation. Part 2 (Sect. 3) covers progress in the polymerization of 1,3-dienes, and it is split into three main subsections: *cis*-1,4-selective and *trans*-1,4-selective polymerization, the 3,4-selective polymerization of isoprene, and copolymerization reactions of 1,3-dienes with alkenes.

Keywords Catalysis · Hydroamination · Hydrophosphination · Hydrosilylation · Lanthanides · Polymerization (polymerisation) of 1,3-dienes · Rare earths

Contents

R.P. Kelly and P.W. Roesky (✉)
Karlsruhe Institute of Technology, Karlsruhe, Baden-Württemberg, Germany
e-mail: roesky@kit.edu

Abbreviations

Ad	Adamantyl
Bn	Benzyl
BOPA	Bis(oxazolinylphenyl)amide
BTSA	Bis(trimethylsilyl)amide
cat.	Catalyst
Cp	Cyclopentadienyl
Cp*	Pentamethylcyclopentadienyl
Cy	Cyclohexyl
dipp	2,6-Diisopropylphenyl
dmba	α-Deprotonated dimethylaminobenzyl
dme	1,2-Dimethoxyethane
d_p	Pore diameter
DRIFT	Diffuse reflectance infrared Fourier transform
ee	Enantiomeric excess
Et	Ethyl
*i*Bu	Isobutyl
*i*Pr	Isopropyl
Ln	Rare-earth metal
Me	Methyl
Me_3TACD	1,4,7-Trimethyl-1,4,7,10-tetraazacyclododecane(1−)
*n*Bu	*n*-Butyl
Ph	Phenyl
PhMe	Toluene
PMS	Periodic mesoporous silica
r.t.	Room temperature
*t*Bu	*t*-Butyl
T_g	Glass transition temperature
thf	Tetrahydrofuran
T_m	Melting point (polymer)
TOF	Turnover frequency
σ-BM	Sigma-bond metathesis

1 Introduction

The formation of new C–X (X=C, H, N, etc.) bonds is one of the fundamental goals in synthetic chemistry. Given economic and environmental concerns, it is desirable to effect these transformations catalytically, and reactive metal-organic species have yielded spectacular results. The transition metals are typically capable of exhibiting multiple oxidation states, which allows them to undergo oxidative addition and reductive elimination reactions. Such reactions underpin many catalytic processes mediated by these metals. On the other hand, the rare-earth

(Ln) metals have an overwhelming tendency to adopt the +3 oxidation state. Divalent complexes are routinely available only for Eu, Yb, and Sm, while Ce is the only Ln metal that affords tetravalent coordination complexes. Consequently, Ln metal complexes are incapable of engaging in oxidative addition and reductive elimination reactions. Instead, σ-bond metathesis (σ-BM) and insertion reactions are the dominant transformations that Ln complexes participate in, and their pronounced reactivity is largely a result of the polarized nature of Ln–X bonds.

A prototypical example of σ-BM mediated by a Ln complex is the reaction between $[Lu(Cp^*)_2(CH_3)]$ and $^{13}CH_4$ (1) [1], and this highlights the impressive reactivity of Ln metal alkyl complexes.

$$\left[\mathrm{Lu}\left(\mathrm{Cp}^*\right)_2(\mathrm{CH}_3)\right] + {}^{13}\mathrm{CH}_4 \overset{70^\circ\mathrm{C}}{\rightleftharpoons} \left[\mathrm{Lu}\left(\mathrm{Cp}^*\right)_2\left({}^{13}\mathrm{CH}_3\right)\right] + \mathrm{CH}_4 \qquad (1)$$

Such σ-BM reactions are considered to be concerted processes with relatively nonpolar and highly organized four-membered transition states (2), and this hypothesis is supported by numerous experimental and theoretical studies [2–5].

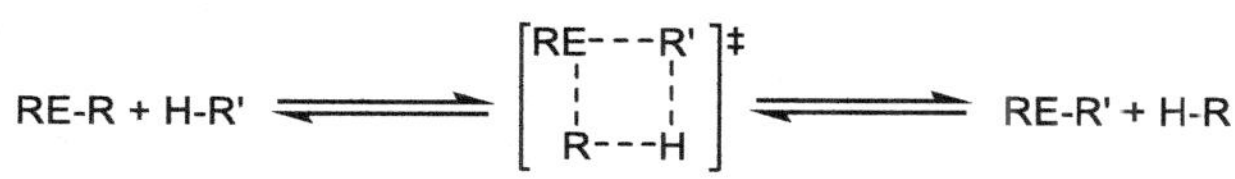

(2)

Insertion reactions are another hallmark of Ln complexes. Insertion reactions of relevance to this review relate to the polymerization of conjugated 1,3-dienes, but small molecules (e.g., CO [6], CO_2 [7] and SO_2 [8, 9]) can also insert into reactive Ln–X bonds. The insertion of 1,3-butadiene into the Sm–H bond of $[Sm(Cp^*)_2H]$ is shown below (Scheme 1) [10, 11]. Subsequent insertion reactions give rise to polymerization processes. It is important to note that σ-BM can compete with insertion, and this can be a major concern in polymerization reactions.

The reaction in (2) describes the initiation step of many important catalytic reactions involving Ln complexes. Such σ-BM reactions include hydroamination, hydrophosphination, and hydrosilylation.

The focus of the first part of this review is progress in these areas since the publication of *Molecular Catalysis of Rare-Earth Elements*, which appeared in *Structure and Bonding* in 2010 [12]. The second part of this review deals with advances in the polymerization of conjugated 1,3-dienes by Ln complexes. The overarching purpose of this review is to provide *research highlights* in these areas since 2010 instead of a comprehensive treatment. Moreover, an in-depth

Scheme 1 The insertion of 1,3-butadiene into the Sm–H bond of $[Sm(Cp^*)_2H]$ [10, 11]

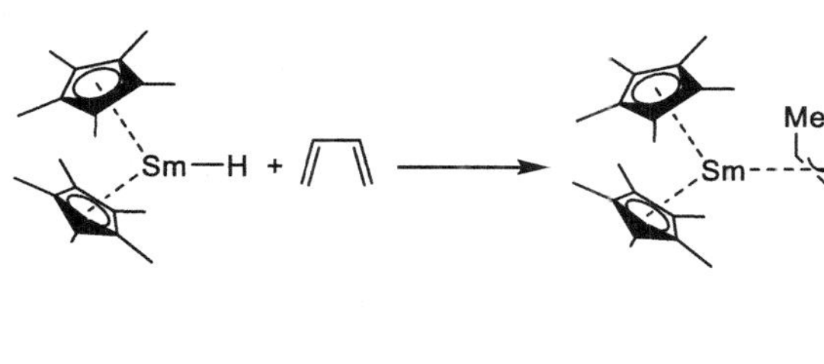

background to these topics is beyond the scope of this review, so the interested reader can consult previous reviews and references therein [12–18].

2 Part 1: Catalytic Sigma-Bond Metathesis

2.1 Hydroamination

Hydroamination is the addition of an N–H bond to an unsaturated bond, and it is a process that is used to produce amines for use in fine chemical, pharmaceutical, and agricultural industries. Ln catalysts have been at the forefront of investigations into hydroamination reactions, and interest in these catalysts remains strong.

As a result of the stronger donor ability of amines versus either alkenes or alkynes, Ln catalysts have been much more successful in intramolecular hydroamination than in intermolecular hydroamination, and this is where current research efforts mainly lie, with so-called "post-metallocene" catalysts continuing to attract much attention. In addition, the emergence of chiral catalysts has been one of the most exciting developments in recent years, and they have been used to achieve greater control over stereoselectivity.

2.1.1 Intermolecular Hydroamination

Chiral Catalysts

Although intermolecular hydroamination has typically proven problematic for Ln catalysts, in 2010, Hultzsch and coworkers reported the asymmetric hydroamination of unactivated alkenes with simple amines [19]. They used substituted binaphtholate Ln complexes (Fig. 1) to catalyze the intermolecular hydroamination of 1-alkenes with primary amines. Lu and Y catalysts were used and they exhibited high Markovnikov selectivity and conversion without the formation of by-products. Importantly, the catalysts are configurationally stable even at high temperatures (150–170°C), and this offers advantages over traditional lanthanocene complexes.

Related Y and Lu aminodiolate complexes have also been used in intra- and intermolecular hydroamination/cyclization (Scheme 2) [20]. Enantiomeric excesses (ees) for the hydroamination/cyclization reactions were as high as 92% using a Lu catalyst (**4**), while the ees for the intermolecular hydroamination of 1-heptene and 4-phenyl-1-butene using an Y catalyst (**3**) were 36% and 40%, respectively. In-situ-generated La catalysts did not show any improvement in performance. Overall, these catalysts perform worse than the aforementioned

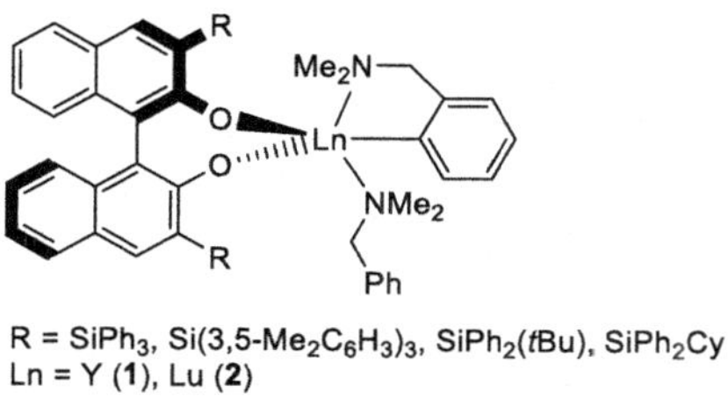

Fig. 1 Binaphtholate RE complexes used to catalyze the intermolecular hydroamination of 1-alkenes with primary amines [19]

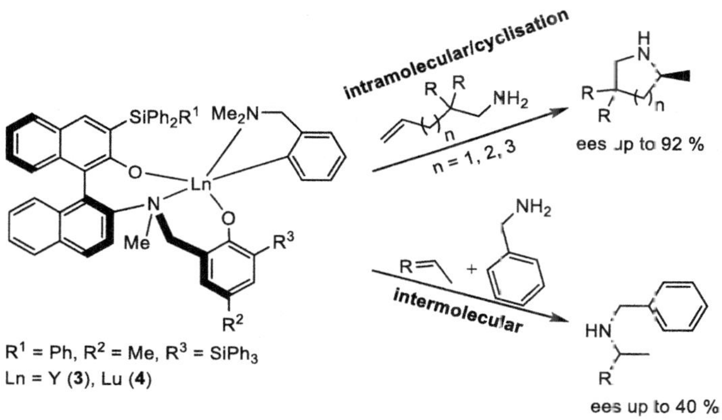

Scheme 2 Ln aminodiolate complexes used in intra- and intermolecular hydroamination/cyclization reactions [20]

binaphtholate complexes, and this was attributed to the presence of the additional amine functionality in the aminodiolate complexes.

Achiral Catalysts

Most Ln catalysts feature trivalent metal centers, but Trifonov and Carpentier recently reported divalent Yb amide complexes (Scheme 3) that catalyze the intermolecular hydroamination of styrene with pyrrolidine to give the anti-Markovnikov addition product exclusively [21]. Catalyst loadings of 2 mol% gave widely differing results. Complex **5** gave a conversion of just 12%, while **6** and **7** performed much better at 85% and 96%, respectively. The well-known complex, $[Yb\{N(SiMe_3)_2\}_2(thf)_2]$, was also tested and it gave a conversion of 37%. The superiority of **7** in this particular hydroamination reaction can possibly be attributed to the lower degree of coordinative saturation compared with the other complexes, although other ligand effects cannot be ruled out.

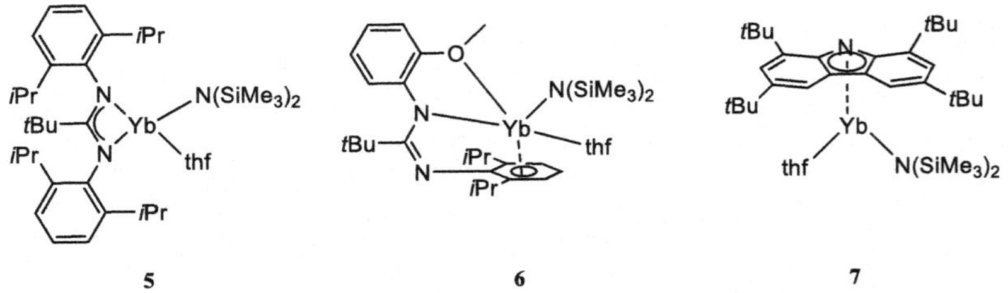

Scheme 3 Structures of divalent Yb catalysts used in the intermolecular hydroamination of styrene with pyrrolidine [21]

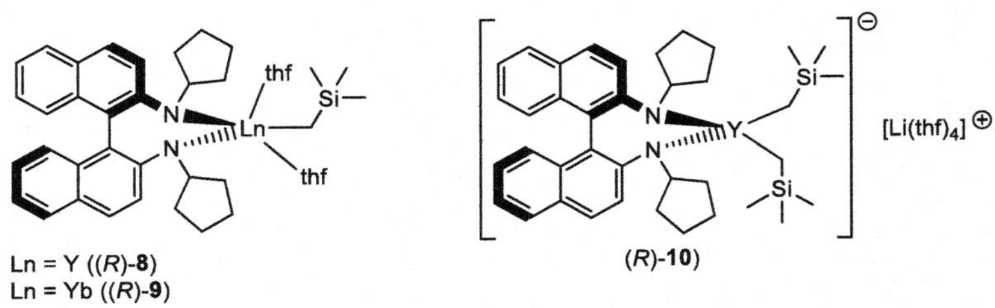

Fig. 2 Y and Yb *N*-substituted binaphthylamido neosilyl complexes [24]

2.1.2 Intramolecular Hydroamination

Chiral Catalysts

As documented in "Chiral Catalysts" of Section 2.1.1, substituted bis(naphthyl) complexes have become popular spectator ligands. Collin and coworkers continued this trend with their use of *N*-substituted binaphthylamido complexes to effect the intramolecular hydroamination/cyclization of aminoalkenes [22], and Livinghouse et al. also used this same ligand framework to prepare neosilyl complexes for use in intramolecular hydroamination [23]. Hannedouche and coworkers also used these same ligands to make neutral and "ate" neosilyl complexes of yttrium and ytterbium (**8–10**, Fig. 2) [24]. When compared with the neutral complexes, the "ate" complexes were less active and less regioselective (see Table 1). The neutral complexes were very active and proceeded with ees of up to 83%, which at the time was highest reported value for a rare-earth-catalyzed hydroamination of a secondary amine.

The same *N*-substituted binaphthylamido ligands were used to prepare alkyl "ate" complexes, which were prepared in situ for hydroamination reactions (**11–13**, Scheme 4) [25]. The lanthanum and samarium catalysts were generally more active and more enantioselective than the yttrium catalysts, and this is thought to be a function of the larger ionic radii of the former two metals. However, catalytic systems without any lanthanide present were by far the most active, although there was no improvement in the ees.

Table 1 Examples of Y and Yb *N*-substituted binaphthylamido neosilyl complexes in hydroamination reactions (catalyst loading = 6 mol%) [24]

Catalyst	Aminoalkene	Product	Temp. (°C)	Time (h)	Conversion (%)	ee (%)
(*S*)-**8**	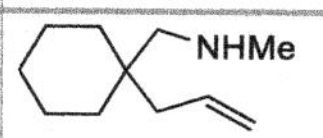	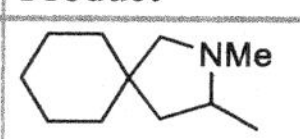	r.t.	0.17	95	83
(*S*)-**9**			r.t.	0.17	95	81
(*R*)-**10**			r.t.	0.17	95	−72[a]
(*S*)-**10**			r.t.	0.17	95	72
(*S*)-**8**	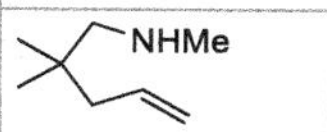	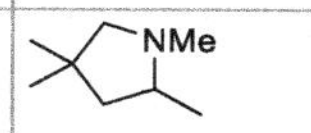	r.t.	0.17	95	80
(*S*)-**9**			r.t.	0.17	95	80
(*S*)-**10**			r.t.	84	95	60
(*S*)-**10**			50	16	85	66

[a]Change in the absolute configuration of the product

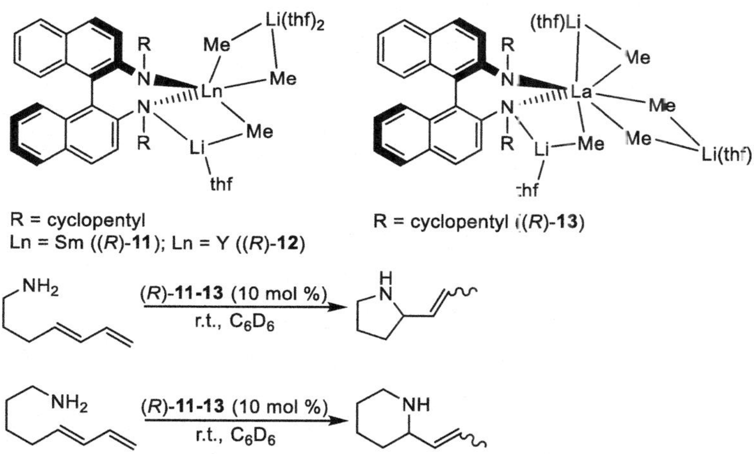

Scheme 4 Chiral *N*-substituted binaphthylamido methyl "ate" complexes for hydroamination/cyclization [25]

In follow-up work, Hannedouche and coworkers prepared (*R*)-**8**·LiCl, along with other related catalysts, and they found that the LiCl-containing catalysts of the cyclopentyl-substituted binaphthylamido ligands were more active and gave better ees than the complexes without LiCl [26]. This is evidence of a possible cooperative heterobimetallic effect, but the exact role of the LiCl in these reactions has not yet been determined.

In a shift away from binaphthylamido ligands, Roesky and coworkers prepared the first chiral rare-earth amidinate complex (Fig. 3) [27]. Complex **14** was tested in the hydroamination of various alkenes and alkynes, and high yields and ees up to 75% were obtained. In a follow-up paper, the yttrium (**15**) analogue was also prepared (Fig. 3) [28]. Complex **15** was generally more active than complex **14**; however, the enantioselectivity was lower, presumably a consequence of more space around the metal center.

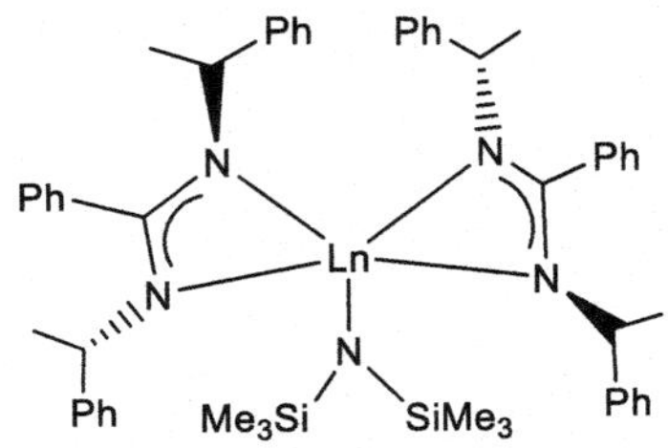

Fig. 3 The first chiral rare-earth amidinate complexes [27, 28]

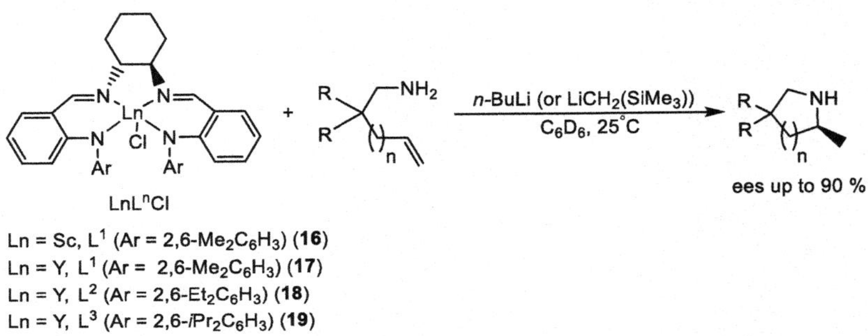

Scheme 5 Chiral tetraazane complexes used in hydroamination/cyclization reactions [29]

Mu and coworkers used substituted chiral tetraazane ligands to prepare a variety of rare-earth complexes for use in hydroamination reactions (Scheme 5) [29]. When used in tandem with an equimolar amount of an alkyllithium reagent, complexes **16–19** were used to effect the intramolecular hydroamination of various aminoalkenes. Good conversion and spectacular ees of up to 90% were recorded at room temperature. Complexes **17** and **19** were treated with $Li(NEt_2)$ to yield yttrium amide complexes, which could be used directly in hydroamination reactions without the use of an alkyllithium cocatalyst. However, the activity of the amide complexes was typically lower than that of the rare-earth/alkyllithium systems. Nevertheless, in one transformation, $[YL^3(NEt_2)Li(thf)_2]$ afforded an ee of 90%.

Ward and coworkers used the bis(oxazolinylphenyl)amide (*R*-BOPA) ligand to prepare a variety of lanthanide bis(alkyl) and bis(BTSA) (BTSA = bis (trimethylsilyl)amide) complexes (Fig. 4) [30]. Complexes **20a–c** are noteworthy for their thermal stability compared with CH_2SiMe_3 complexes, and this was attributed to replacing one methyl group with a phenyl group. Indeed, Zhang and Li reported the ring opening of oxazoline complexes when CH_2SiMe_3 co-ligands were present [31]. Complexes **20–25** were used in the hydroamination/cyclization of two related aminoalkenes (Scheme 6), and poor to modest ees were achieved. The yttrium amide complexes **21a–c** did not cyclize either aminoalkene, even under

Ln = Y, X = CH_2SiMe_2Ph, R = H, R' = *i*Pr **20a**, Bn **20b**
Ln = Y, X = CH_2SiMe_2Ph, R = Ph, R' = H **20c**
Ln = Y **21**, La **22**, Pr **23**, Nd **24**, Sm **25**; X = BTSA
R = H, R' = *i*Pr **a**, Bn **b**
R = Ph, R' = H **c**

X = CH_2SiMe_2Ph, R = H, R' = *i*Pr **20d**, Bn **20e**
X = CH_2SiMe_2Ph, R = Ph, R' = H **20f**

Fig. 4 Chiral rare-earth *R*-BOPA complexes [30]

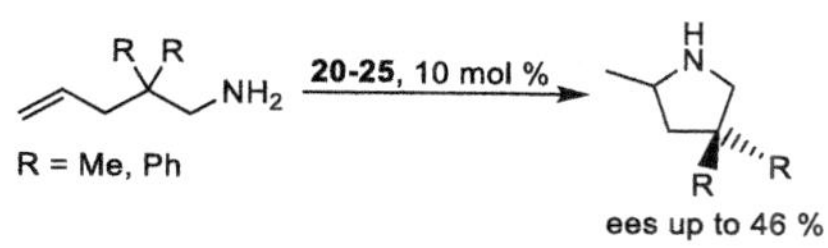

Scheme 6 Intramolecular hydroamination/cyclization catalyzed by chiral rare-earth *R*-BOPA complexes [30]

forcing conditions, but the alkyl complexes **20a–f** reacted swiftly with the geminal diphenylaminoalkene and afforded ees up to 44% (inversion of absolute configuration in some cases). The cationic complexes **20d–f** showed a slight reduction in selectivity. In contrast to the yttrium complexes, the complexes of the larger lanthanides were able to catalyze the transformation of the geminal dimethylaminoalkene, although the reactions proceeded sluggishly and with poor ees. Interestingly, with the phenyl-substituted BOPA complexes **22c–25c**, the cyclization of the geminal diphenylaminoalkene proceeded with significantly improved ees when the reactions were performed in toluene instead of benzene. Although beyond the scope of this review, the complexes exhibited luminescent properties and they were also investigated in the ring-opening polymerization of *rac*-lactide.

Although research into cyclopentadienyl (Cp) catalysts has waned, Sadow and coworkers published some impressive results with a chiral (cyclopentadienyl)bis(oxazolinyl)borato yttrium alkyl complex (**26**; Scheme 7) [32]. Complex **26** was incredibly efficient at promoting the hydroamination/cyclization of 3,3-diphenylpent-4-en-1-amine (Scheme 7). The reaction proceeded with complete conversion in 10 min with an ee of 94% (*S* isomer). A bis(dimethylamino)zirconium complex bearing the same supporting ligand also showed excellent activity (ee = 93%), but it afforded the *R* isomer instead. They also prepared other achiral and chiral catalysts with similar ligands but they did not afford as good results as **26**.

Other chiral catalysts have also been prepared, including amidate [33] and linked β-diketiminate [34] complexes, highlighting the multitude of chiral ligands that are available. The number of chiral Ln complexes used in hydroamination looks set to increase greatly, and the future holds great promise.

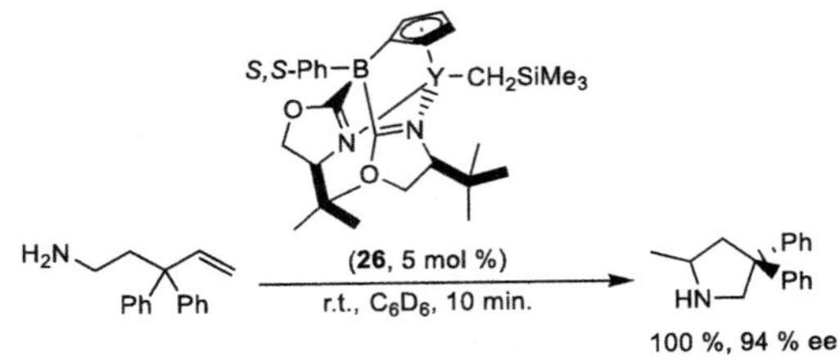

Scheme 7 A chiral (cyclopentadienyl)bis(oxazolinyl)borate yttrium alkyl complex used to catalyze the hydroamination/cyclization of 3,3-diphenyl-pent-4-en-1-amine [32]

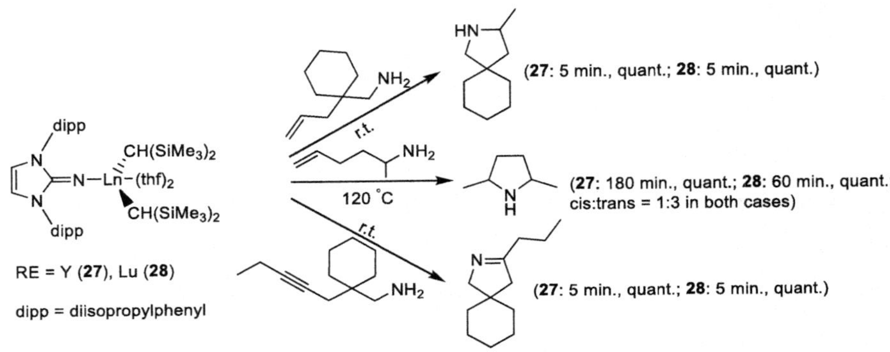

Scheme 8 Examples of hydroamination/cyclization by rare-earth imidazolin-2-iminato bis (neosilyl) complexes (catalyst loading = 5 mol%) [38]

Achiral Catalysts

Although research into hydroamination has seen a general shift away from metallocene (and related) catalysts, research is still being conducted in this area, e.g., the use of cyclopentadienylphosphazene bis(alkyl) complexes [35], half-sandwich indenyl bis(alkyl) complexes [36], and silicon-linked amido/indenyl BTSA complexes [37].

Highlighting the trend toward the use of non-metallocene catalysts, Tamm and Roesky and coworkers successfully employed yttrium and lutetium imidazolin-2-iminato bis(neosilyl) complexes to catalyze various hydroamination/cyclization reactions of aminoalkenes and aminoalkynes (see Scheme 8 for examples) [38]. The reactions proceeded regiospecifically and gave full conversion in most cases. Unlike metallocene catalysts, the activity of the yttrium and lutetium catalysts was generally not related to the ionic radius of the respective metals, but they were not as active as their metallocene counterparts. The catalytic applications of Ln imidazolin-2-iminato complexes were covered by Tamm and coworkers several years ago [39].

Anwander and coworkers found a novel way to exploit new reactivity from otherwise standard [Ln(BTSA)$_3$] complexes. Grafting of [Ln(BTSA)$_3$] (Ln = Y,

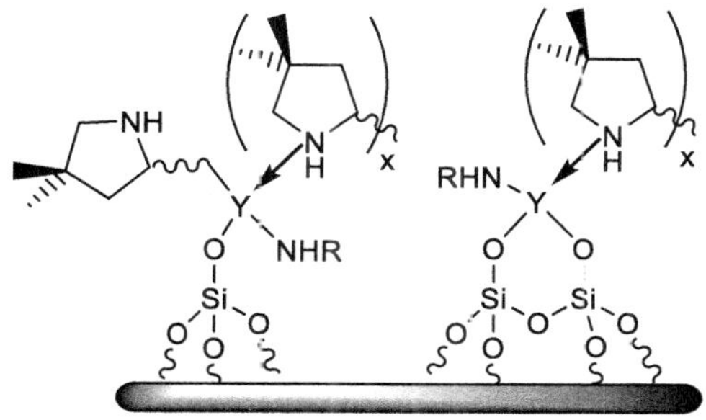

Fig. 5 Proposed surface species of recycled catalyst [Y(BTSA)$_3$]@SBA-15LP$_{-500}$ (Adapted from [40])

La, Nd) onto partially dehydroxylated periodic mesoporous silica (PMS) gave active, recyclable catalysts for the intramolecular hydroamination/cyclization of 2,2-dimethyl-4-penten-1-amine [40]. The activity of the grafted complexes was much higher than that of [Y(BTSA)$_3$], but it was dependent on the metal size, pore size, and particle morphology. SBA-15$_{-500}$ ($d_p = 7.9$ nm), SBA-15LP$_{-500}$ ($d_p = 16.6$ nm), and MCM-41$_{-500}$ ($d_p = 4.1$ nm) were the supports that were used, with the extra-large pore size of SBA-15LP$_{-500}$ proving the most successful at minimizing product inhibition and substrate diffusion effects. The activity of [Y(BTSA)$_3$]@SBA-15LP$_{-500}$ (TOF = 420 h^{-1} at 60°C) was much higher than that of [Y(BTSA)$_3$] (TOF = 54 h^{-1} at 50°C), and it could be collected and reused twice at much different temperatures, albeit with greatly reduced activity for each recycling step. The surface species of the recycled catalyst was investigated by DRIFT spectroscopy (the proposed surface structure is shown in Fig. 5).

Trifonov and coworkers prepared a series of thiazole-substituted amidopyridinate yttrium alkyl complexes (**29–31**, Fig. 6) and tested them in various hydroamination/cyclization reactions (Table 2) [41]. They also tested the complexes in the presence of [Ph$_3$C][B(C$_6$F$_5$)$_4$] as an alkyl group abstractor in order to generate cationic complexes in situ, and these were far more active than the neutral complexes. Cationic complexes derived from **30** and **31** were far superior to those of complex **29**, with the cationic complex produced from **31** giving the best results in all cases. The superiority of the cationic complexes can be attributed to a lack of steric saturation. Moreover, the coordinating thf molecule in **29** blocks access to a coordination site and lowers the Lewis acidity of the yttrium ion. Interestingly, complex **30** undergoes metal-to-ligand alkyl transfer with concomitant ring opening of the thiazole ring to yield **32** (Fig. 6), which afforded poor conversions in hydroamination reactions (Table 2).

2.2 Hydrophosphination

Hydrophosphination is the addition of a P–H bond to an unsaturated bond, and it is used to generate useful phosphines. Although conceptually very similar to

Fig. 6 Thiazole-substituted amidopyridinate yttrium alkyl complexes (**29–31**) and **32** [41]

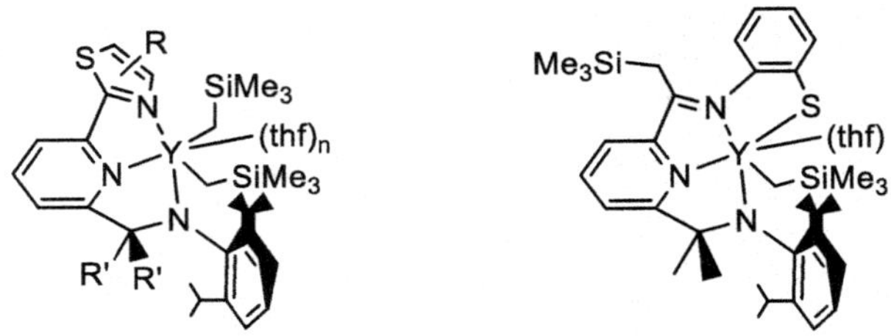

R = 5-Me, R' = H, n = 1 (**29**)
R = 5-Me, R' = Me, n = 0 (**30**)
R = benzo[d], R' = Me, n = 0 (**31**)

32

Table 2 Thiazole-substituted amidopyridinate alkyl yttrium complexes used in hydroamination/cyclization reactions in toluene (catalyst loading = 5 mol%) [41]

Catalyst	Aminoalkene	Product	Temp. (°C)	Time (h)	Conversion (%)
29[a]	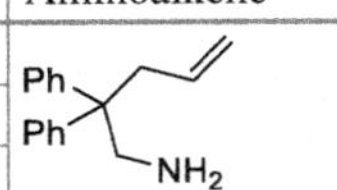	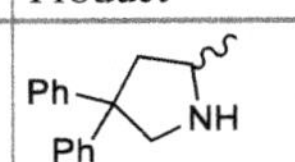	80	2	10
30			40	1.5	11
30[a]			80	0.75	96
31			80	7	30
31[a]			80	0.75	>99
32			80	7	23
32[a]			80	24	–
29[a]	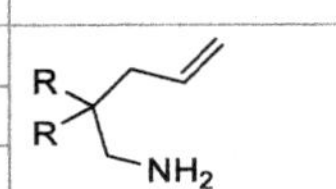	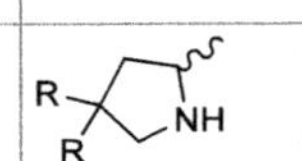	80	2	5
30[a]			40	2	81
31[a]	(R = $-(CH_2)_5-$)		80	0.75	>99

[a]$[Ph_3C][B(C_6F_5)_4]$ present as an activator

hydroamination, there are important differences in regard to metal size effects and rate-determining steps [42].

Progress in hydrophosphination mediated by rare-earth metals has long lagged behind research into hydroamination, and this continues to be the case. Nonetheless, some exciting results have been published in the last several years.

Schmidt and Behrle investigated the addition of various phosphines to a variety of heterocumulenes [43]. These reactions were catalyzed by $[La(dmba)_3]$ (**33**, dmba = α-deprotonated dimethylaminobenzyl) (Scheme 9). The yields of the reactions were modest to excellent. The acidity of the disubstituted phosphines that were used also had an effect, with diarylphosphines adding to the heterocumulenes at room temperature, whereas the less acidic di-*tert*-butylphosphine did not react at all, even at 80°C.

In a particularly comprehensive study, Carpentier and coworkers prepared divalent ytterbium and europium iminoanilide BTSA complexes (Fig. 7) and tested them in the hydrophosphination of styrene with diphenylphosphine (Scheme 10), as well as in both hydroamination/cyclization and intermolecular hydroamination

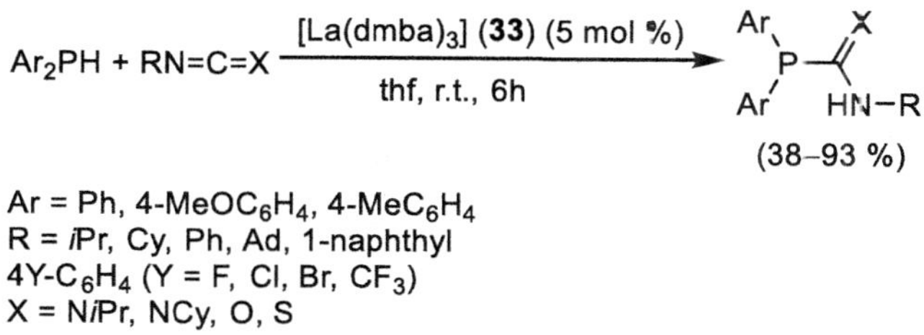

Scheme 9 Intermolecular hydrophosphination reactions catalyzed by $[La(dmba)_3]$ [43]

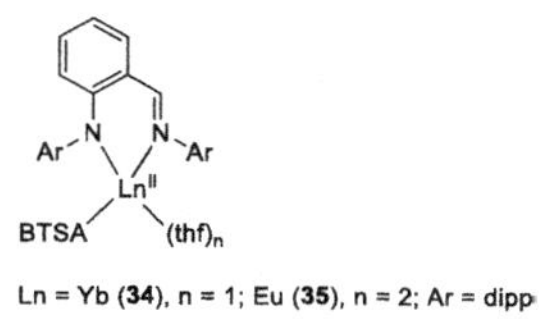

Fig. 7 Ytterbium and europium iminoanilide BTSA complexes used in hydrophosphination, hydroamination/cyclization, and intermolecular hydroamination reactions [44]

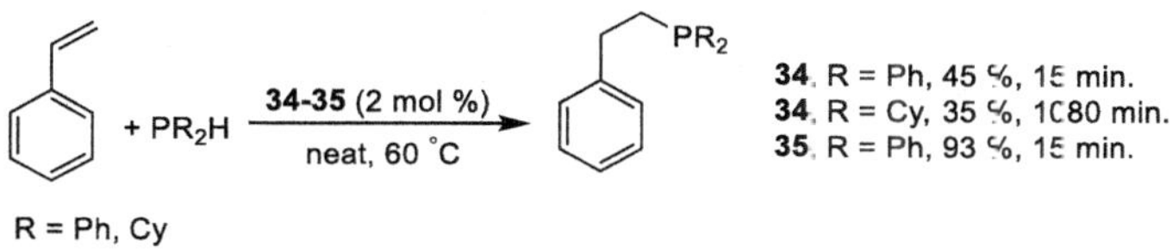

Scheme 10 Hydrophosphination reactions catalyzed by ytterbium and europium iminoanilide BTSA complexes **34** and **35** [44]

reactions [44]. They also prepared group 2 iminoanilide complexes but the results of these catalysts are not discussed here. As expected, the hydrophosphination of styrene with PPh_2H catalyzed by **34** proceeded much faster than the analogous reaction with PCy_2H, a consequence of the much lower acidity of PCy_2H. The anti-Markovnikov products were obtained in all reactions, with the larger europium catalyst **35** affording much greater conversion than **34**.

As outlined in "Achiral Catalysts" of Section 2.1.1, Trifonov and Carpentier recently reported divalent Yb complexes (**5–7**, Scheme 3) that were used to catalyze intermolecular hydroamination reactions [21]. They also tested the same complexes in various intermolecular hydrophosphination reactions of styrene, all of which yielded the anti-Markovnikov product (Scheme 11; results are summarized in Table 3). As in the hydroamination reactions, complex **7** provided the best results. Using PPh_2H, an impressive conversion rate of 92% was achieved after 4 h with a catalyst loading of 1 mol%, although poor conversion of 15% was achieved after 8 h with a catalyst loading of 0.5 mol%, indicating decay of the catalyst. This decay

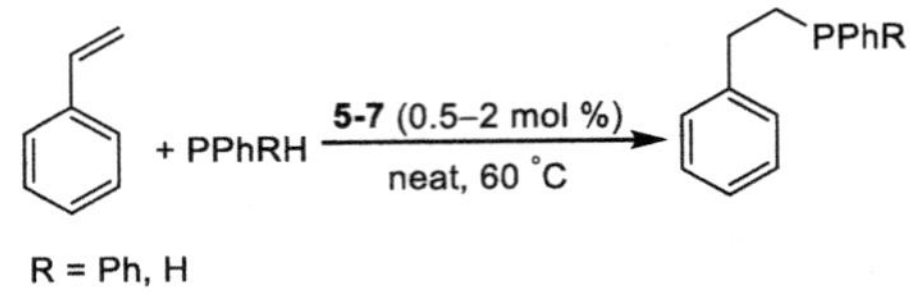

Scheme 11 Hydrophosphination reactions of styrene with various aryl phosphines, catalyzed by the divalent Yb complexes **5–7** [21]

Table 3 Results of the hydrophosphination of styrene with various aryl phosphines, catalyzed by the divalent Yb complexes **5–7** [21]

Catalyst (mol%)	Phosphine	Time (h)	Conversion (%)
5 (2.0)	Ph_2PH	0.25	27
5 (2.0)	Ph_2PH	2	55
6 (2.0)	Ph_2PH	0.25	17
6 (2.0)	Ph_2PH	2	27
7 (2.0)	Ph_2PH	0.25	31
7 (2.0)	Ph_2PH	2	71
7 (1.0)	Ph_2PH	4	92
7 (0.5)	Ph_2PH	8	15
7 (2.0)	$PhPH_2$	10	30
7 (2.0)	$PhPH_2$	60	61
$[Yb(BTSA)_2(thf)_2]$ (2.0)	Ph_2PH	2	24

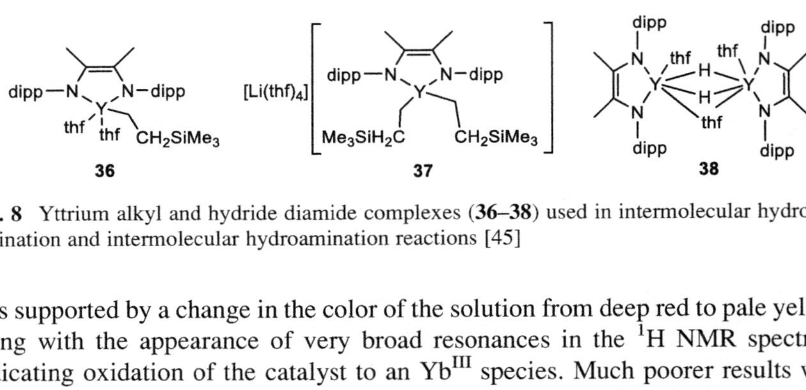

Fig. 8 Yttrium alkyl and hydride diamide complexes (**36–38**) used in intermolecular hydrophosphination and intermolecular hydroamination reactions [45]

was supported by a change in the color of the solution from deep red to pale yellow, along with the appearance of very broad resonances in the ^1H NMR spectrum, indicating oxidation of the catalyst to an Yb^{III} species. Much poorer results were achieved using $PPhH_2$, and only the primary hydrophosphination product was observed. Kinetic experiments with **7** showed that the reaction is zero order with respect to [phosphine] and first order with respect to [styrene] (partial order of 0.72 was found but the authors believed that this is due to experimental error). The reaction was found to be partial order (0.5) with respect to [catalyst]. This was suggested to relate to dimerization in solution, but further experiments did not confirm this hypothesis.

Trifonov and coworkers prepared a series of yttrium alkyl and hydride diamide complexes that they used in intermolecular hydrophosphination and intermolecular hydroamination (Fig. 8) [45]. All catalysts gave 100% conversion in the hydrophosphination reactions (2 mol% cat., 72 h, 70°C) of styrene and 4-vinylpyridine with PPh_2H. However, in the analogous hydrophosphination reaction with 1-nonene, only catalyst **37** afforded any of the product, albeit in low yield (17%). In the hydrophosphination of diphenylacetylene with $PPhH_2$, all complexes gave 100% conversion, but when PPh_2H was used, catalyst **37** was the only species that afforded full conversion. All complexes also catalyzed the double addition of $PPhH_2$ to styrene, with complex **37** proving to be the most active once again.

2.3 *Hydrosilylation*

Hydrosilylation is the addition of Si–H bonds across unsaturated bonds, and it is very useful in the preparation of organosilicon compounds. However, like hydrophosphination, rare-earth-catalyzed hydrosilylation has not been as extensively researched as hydroamination. Nonetheless, a handful of promising results have been reported in the past several years.

As seen in "Achiral Catalysts" of Section 2.1.2, Roesky and Tamm reported the imidazolin-2-iminato bis(neosilyl) rare-earth complexes **27** and **28** (Scheme 8) and explored their use in hydroamination reactions [38]. They also explored their activity in the hydrosilylation of 1-hexene and 1-octene with phenylsilane. Both catalysts showed remarkable activity, with all reactions going to completion in less than 5 min at room temperature. The anti-Markovnikov product was overwhelmingly favored in every case, with anti-Markovnikov/Markovnikov (AM:M) ratios of 99:1 being observed in all cases (Table 4).

In some particularly impressive work, Piers and coworkers reported the anilidobipyridyl complex **39**·$B(C_6F_5)_3$ (Fig. 9), which was formed by CO_2 insertion into an anilidobipyridyl organoscandium precursor followed by treatment with $B(C_6F_5)_3$ [46]. Complex **39**·$B(C_6F_5)_3$ proved to be a potent catalyst for the hydrosilylation of CO_2 with Et_3SiH to form the bis(silylacetal), $Et_3SiOCH_2OSiEt_3$. The proposed catalytic cycle is complex, though it seems that dissociation of small amounts of $B(C_6F_5)_3$ initiates the catalytic cycle by activating Et_3SiH. Further reduction to methane does not occur since the equilibrium between dissociation of **39**·$B(C_6F_5)_3$ into $B(C_6F_5)_3$ and **39** lies much in favor of **39**·$B(C_6F_5)_3$. A TON of 3,400 was found using low catalyst loadings (0.02%) and prolonged reaction times.

Table 4 Hydrosilylation of 1-hexene and 1-octene with phenylsilane, catalyzed by the imidazolin-2-iminato bis(neosilyl) rare-earth complexes **27** and **28** [38]

Catalyst	Alkene	Product	Time (min)	Yield (%) (AM:M)
27	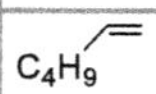	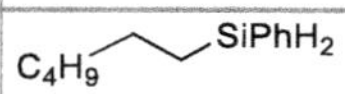	<5	99:1
28			<5	99:1
27	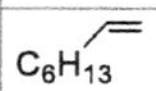	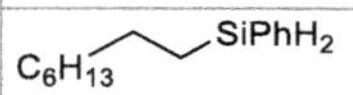	<5	99:1
28			<5	99:1

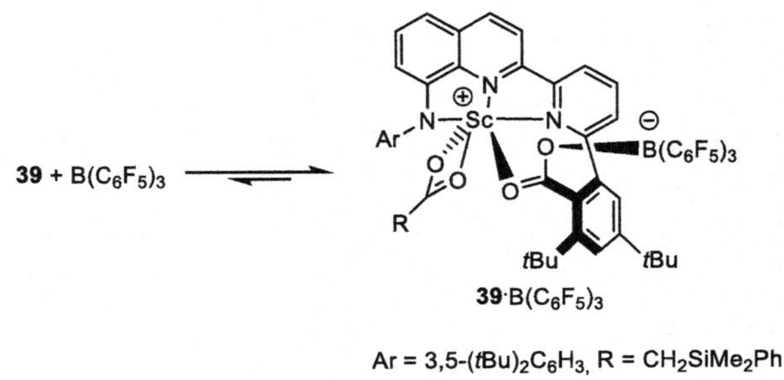

Fig. 9 An anilidobipyridyl complex **39**·B$(C_6F_5)_3$ used in the hydrosilylation of CO_2 with Et_3SiH to form $Et_3SiOCH_2OSiEt_3$ [46]

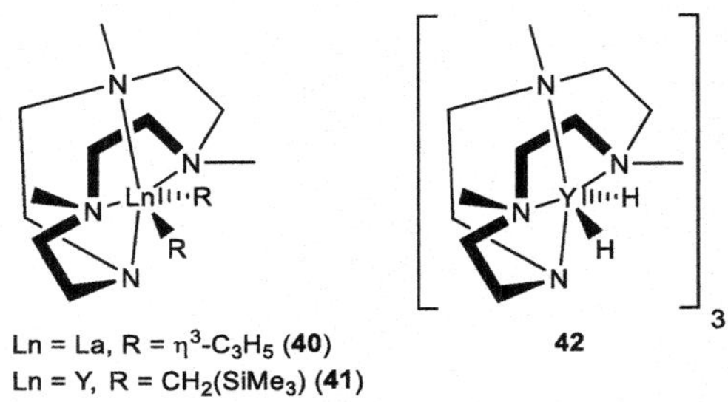

Fig. 10 Rare-earth metal Me_3TACD complexes used in hydrosilylation reactions [47]

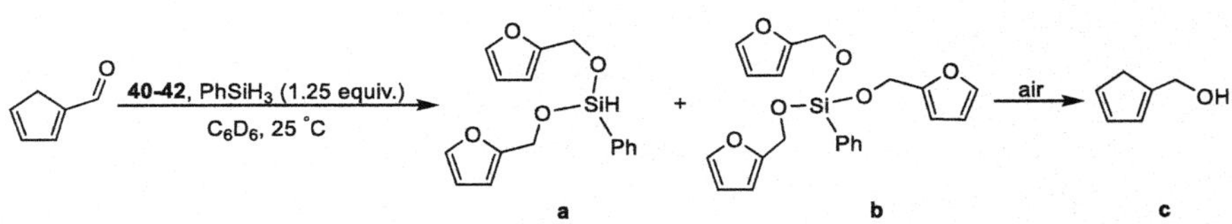

Scheme 12 Hydrosilylation of furfural with $PhSiH_3$ catalyzed by the rare earth Me_3TACD complexes **40–42** (**a**=bis(furfuryl)phenylsilane, **b**=tris(furfuryl)phenylsilane, **c**=2-furfuryl alcohol) [47]

Continuing the theme of multidentate ligands, Okuda and coworkers prepared a series of macrocyclic complexes bearing the 1,4,7-trimethyl-1,4,7,10-tetraazacyclododecane(1−) ligand (Me_3TACD), along with alkyl, allyl, or hydride co-ligands (Fig. 10) [47]. Although complexes other than the ones shown in Fig. 9 were also prepared, complexes **40–42** were the only ones used in hydrosilylation reactions (Scheme 12). The hydrosilylation of furfural with $PhSiH_3$ proceeded swiftly, with the yttrium complexes **41** and **42** proving much more reactive than the lanthanum complex **40** (Table 5).

Trifonov and coworkers utilized the bis(guanidinate) rare-earth hydride complexes, $[Ln\{(Me_3Si)_2NC(NR_2)\}_2(\mu\text{-}H)]_2$ (Ln=Y, R=*i*Pr (**43**); Ln=Lu, R=*i*Pr (**44**); Ln=Y, R=Cy (**45**); Ln=Lu, R=Cy (**46**)), to effect the hydrosilylation of non-1-ene with phenylsilane to afford the anti-Markovnikov product, $Ph(n\text{-}C_9H_{19})_2SiH_2$, in

Table 5 Results from the hydrosilylation of furfural with $PhSiH_3$, catalyzed by rare-earth Me_3TACD complexes **40–42**

Catalyst	Catalyst/furfural	Time	Conversion (%)	Product ratio	
				a	**b**
40	1:40	3 days	>98	31	69
41	1:40	<7 min	>98	81	19
42	1:40	<7 min	>98	18	82

good yields [48]. The reaction using **43** was studied in detail, and NMR experiments showed that the reaction is zero order with respect to $[PhSiH_3]$ and first order with respect to $[C_7H_{15}CH{=}CH_2]$. During studies with excess non-1-ene, it was found the silane that is formed, Ph(n-$C_9H_{19})_2SiH_2$, undergoes another reaction to afford the tertiary silane, Ph(n-$C_9H_{19})_2SiH$. This was the first time that a rare-earth-catalyzed reaction allowed for the double addition of an alkene to phenylsilane, underscoring both the utility of rare-earth complexes in hydrosilylation reactions and the dearth of research in this important field.

2.4 *Outlook*

Research into hydroamination promoted by rare-earth catalysts should remain strong, and it is likely that efforts will be increasingly directed to the development of a greater range of chiral complexes. Further research into less common divalent complexes is warranted by some promising results, and it is also likely that the number and effectiveness of rare-earth catalysts for intermolecular hydroamination will grow. The relatively new field of hydrophosphination should attract more attention, and many catalysts that have proven effective in hydroamination could be tested. Although interest into hydrosilylation is currently very low, there are a considerable number of potentially suitable complexes that could be trialed, and the area is primed for more intensive research.

3 Part 2: Polymerization of 1,3-Dienes

3.1 *Introduction*

Polymers derived from 1,3-dienes are among the most industrially important products made today, and they are mainly used in the production of synthetic rubbers. The properties of the products are primarily affected by the way the monomer units are linked, i.e., with *cis*-1,4-, *trans*-1,4-, *iso*-3,4- (or 1,2), *syndio*-3,4- (or 1,2), or *atactic*-3,4- (or 1,2) regularities, and great efforts have been made to effect control over which form predominates. Rare-earth catalysts have been used

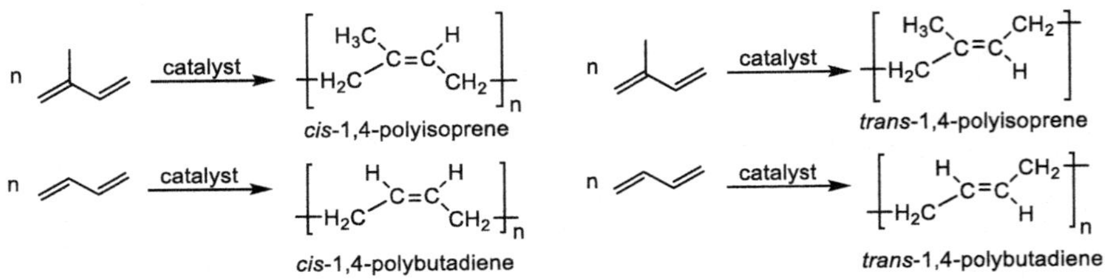

Scheme 13 Polymerization of isoprene and butadiene to yield the *cis*-1,4- and *trans*-1,4-products, respectively

with terrific results in the polymerization of 1,3-dienes, and research into rare-earth catalysts for these reactions remains very strong. A comprehensive review is beyond this scope of this chapter, so recent highlights in four key areas are covered here: *cis*-1,4-polymerization and *trans*-1,4-polymerization (discussed together), the 3,4-polymerization of isoprene, and copolymerization reactions of 1,3-dienes with unsaturated substrates.

3.2 Cis-*1,4-Polymerization and* Trans-*1,4-Polymerization*

The polymerization reactions of isoprene and butadiene are two of the most important industrial processes in chemistry, and great efforts have been made to produce viable rare-earth catalysts for these reactions. *Cis*-1,4- and *trans*-1,4-products are highly desirable, and both products find applications in the production of tires and other necessary products [70]. The polymerization reactions of isoprene and butadiene to yield the *cis*-1,4- and *trans*-1,4-products, respectively, are shown in Scheme 13.

A wide variety of rare-earth catalysts for these reactions have been prepared in the past several years, and a selection of them is shown in Scheme 14.

Interest in traditional metallocene catalysts remains strong, and solid results show that they still have an important part to play in rare-earth catalysis. Cui and coworkers used the substituted half-sandwich bis(allyl) yttrium complex **47** in tandem with $[Ph_3C][B(C_6F_5)_4]$ for a variety of polymerization reactions, including the highly *cis*-1,4-selective (99%) polymerization of butadiene (Table 6) [49]. Although unrelated to the scope of this review, **47** also promoted the highly syndiotactic polymerization of styrene (*rrrr* > 99%). Anwander and coworkers used a series of tetramethylcyclopentadienyl bis(tetramethylaluminate) Ln complexes (**48**–**52**) and various boron cocatalysts for the stereospecific polymerization of isoprene [50]. Moderate to excellent (55–95%) *trans*-1,4-selectivity was observed in most cases, with the nature of the boron cocatalyst affecting the outcome. However, when compared with previous results using the bulkier C_5Me_5 ligand [51], the reduction in the steric bulk from C_5Me_5 to C_5Me_4H was

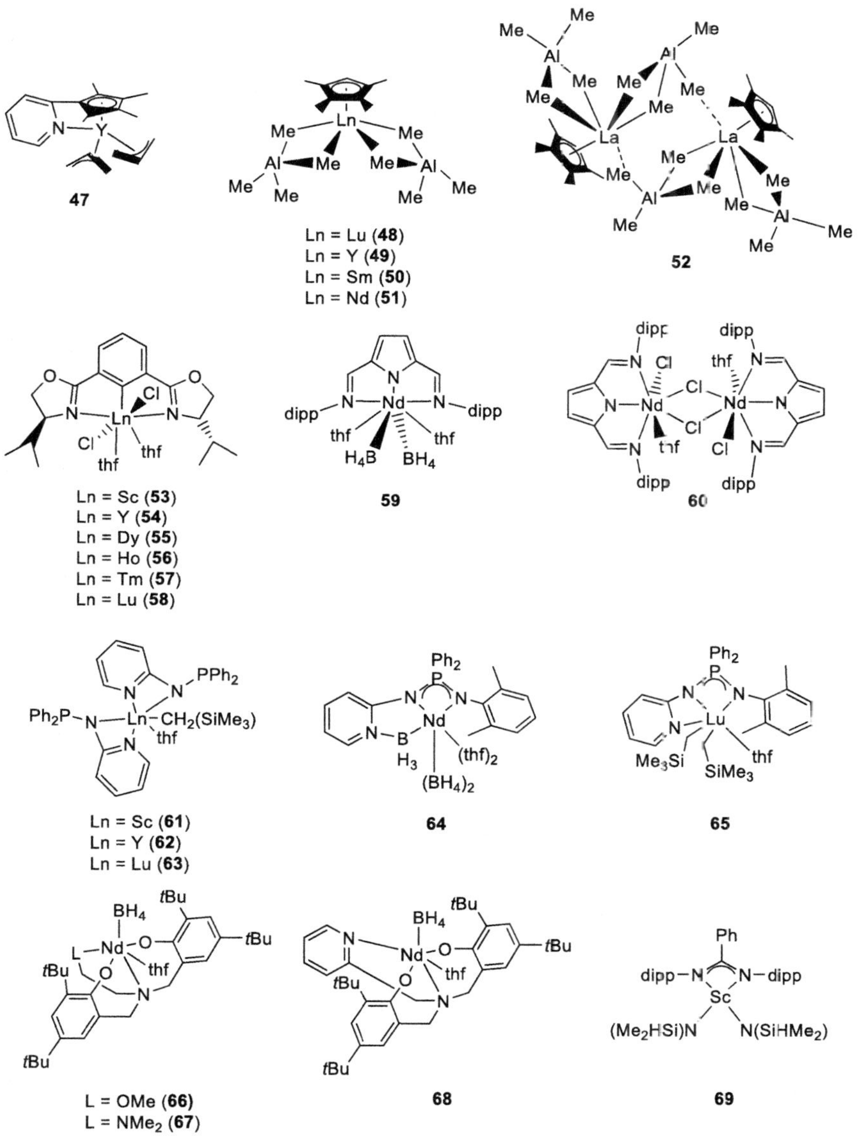

Scheme 14 A selection of rare-earth catalysts used for the polymerization of isoprene or butadiene to yield *cis*-1,4- or *trans*-1,4-products, respectively [49, 50, 57–60, 67]

accompanied by a loss in *trans*-1,4-selectivity and catalytic activity. Continuing the theme of cyclopentadienyl rare-earth bis(tetramethylaluminate) complexes, Anwander also prepared a series of dimethylanilium-functionalized cyclopentadienyl complexes that underwent intramolecular C–H activation to yield

Table 6 Highly *cis*-1,4-selective polymerization of butadiene by complex **47**/$[Ph_3C][B(C_6F_5)_4]$ [49]

Butadiene/ Lu	Time (min)	Conversion (%)	Activity polymer (kg mol_{Lu}^{-1} h^{-1})	*cis*-1,4 (%)	M_n ($\times 10^4$)	M_w/M_n
500	1	>99	1,620	97.1	3.1	1.24
1,000	1	>99	3,240	97.0	6.4	1.26
2,000	1	>99	6,480	97.0	13.2	1.25
500[a]	10	70	114	99.0	3.6	2.13

Conditions: Lu (10 μmol), $[Lu]:[Ph_3C][B(C_6F_5)_4] = 1:1$, $T = 20°C$, toluene/butadiene = 5:1 (v/v)
[a]Chlorobenzene (5 mL)

complexes of the form $[\{C_5Me_4C_6H_4NMe(\mu\text{-}CH_2)AlMe_3\}Ln(AlMe_4)]$ [52]. The lanthanum-based pre-catalysts and boron (or borate) cocatalysts were highly *trans*-1,4-selective (up to 95.6%) in the polymerization of isoprene. A living-polymerization mechanism was inferred from narrow molecular weight distributions ($M_n/M_w < 1.1$) and complete consumption of the isoprene monomers. It is also worth mentioning that Anwander and coworkers grafted monophosphacyclopentadienyl bis(tetramethylaluminate) complexes onto mesoporous silica SBA-15. These materials showed modest activity in the polymerization of isoprene, but spectacular *cis*-1,4-selectivity was observed (99%) [53]. Anwander continues to be prolific in the use of half-sandwich complexes, recently using half-sandwich tetramethylaluminate complexes with peripheral boryl ligands in the polymerization of isoprene [54]. He also used substituted tetramethylcyclopentadienyl bis(allyl) complexes in the polymerization of isoprene [55]. Using varying amounts of $[La(Cp^*)(BH_4)_2(thf)_2]$, Mg(*n*Bu)(Et), and a chain transfer agent, Zinck and coworkers were able to prepare *trans*-1,4-polyisoprene by chain transfer polymerization with very high selectivity values (up to 99%) [56]. However, the *trans*-1,4-selectivity decreased with increasing amounts of Mg(*n*Bu)(Et) to favor 3,4-selectivity.

Multidentate ligands are being increasingly utilized in all areas of catalysis. Complexes **53–60** are related in that they all have tridentate ancillary ligands. Xu and coworkers prepared a series of chiral bis(oxazolinyl)phenyl rare-earth dichloride complexes (**53–58**), which were then used in the polymerization of isoprene [57]. When used in conjunction with $[PhNHMe_2][B(C_6F_5)_4]$ and Al(*i*Bu)$_3$, complexes **54–57** afforded remarkable *cis*-1,4-selectivity values of up to 99.5% (Table 7). Complexes **53** and **58** performed very poorly, and this is a consequence of the small size of scandium and lutetium, respectively. Importantly, the selectivity of complex **54** dropped only marginally at 80°C while still maintaining high *cis*-1,4-selectivity (96.4%). Roesky and coworkers prepared a series of structurally related pyrrolyl Schiff base complexes (**59–60**) that were found to be highly active and reasonably *cis*-1,4-selective (up to 85%) catalysts for the polymerization of butadiene [58]. When used with $AlEt_3/B(C_6F_5)_3$ cocatalysts, the dimeric complex **60** showed exceptionally high activity, with 3.604 kg of polymer produced per mmol catalyst per hour, and this is significantly higher than other systems tested under comparable conditions.

Table 7 Highly *cis*-1,4-selective polymerization of isoprene by complexes **53–58**/[$PhNHMe_2$][$B(C_6F_5)_4$]/Al(*i*Bu)$_3$ [57]

Catalyst	Temp. (°C)	Time (min)	Yield (%)	M_n ($\times 10^4$)	M_w/M_n	*cis*-1,4 (%)	*trans*-1,4 (%)	3,4 (%)
53	30	60	Trace	–	–	–	–	–
54	−8	60	84	57.4	2.26	99.5	0	0.5
54	30	30	100	11.6	2.12	98.6	0.8	0.6
54	50	5	100	10.7	1.65	97.4	1.4	1.2
54	80	5	86	11.2	2.07	96.4	1.6	2.0
55	30	15	100	11.0	2.34	98.9	0.6	0.5
56	30	30	100	13.9	3.21	99.2	0.4	0.4
57	30	60	78	12.0	3.69	98.3	0.9	0.8
58	30	60	7	–	–	–	–	–

Conditions: C_6H_5Cl (3 mL), [Ln]:[$PhNHMe_2$][$B(C_6F_5)_4$]:Al(*i*Bu)$_3$ = 1:1:10, [isoprene]/[Ln] = 500

Table 8 Polymerization of isoprene by **61–65**/borate/Al(*i*Bu)$_3$[a] (or Mg(*n*Bu)$_2$[b]): differences in selectivity [59]

Catalyst	Temp. (°C)	Time (h)	Yield (%)	M_n ($\times 10^4$)	M_w/M_n	*cis*-1,4 (%)	*trans*-1,4 (%)	3,4 (%)
61 + **B**[a]	20	5 min	98	23.0	3.07	95.0	1.5	3.5
62 + **B**[a]	20	10 min	99	16.8	2.52	95.6	1.3	3.1
63 + **B**[a]	20	4	100	8.75	2.08	97.2	1.0	1.8
63 + **B**[a]	0	8	82	11.5	1.87	97.7	0	2.3
64 + **A**[a]	15	4	100	3.08	1.42	12.0		88.0
65 + **A**[b]	15	5	73	4.21	1.10	1.3	97.2	1.5
65 + **B**[b]	50	6	82	4.43	1.23	3.2	95.2	1.6

Conditions: toluene (5 mL), A = [$PhNHMe_2$][$B(C_6F_5)_4$], B = [Ph_3C][$B(C_6F_5)_4$], [Ln]:[borate]:[Al] = 1:1:10 (or [Ln]:[borate]:[Mg] = 1:1:1), [isoprene]/[Ln] = 1,000

[a] Al(*i*Bu)$_3$

[b] Mg(*n*Bu)$_2$

Cui and coworkers prepared a series of aminopyridinato rare-earth complexes (**61–63**) and some related NPN complexes (**64** and **65**) [59]. Remarkably, these complexes were found to promote either *cis*-1,4-, *trans*-1,4- or 3,4-selective polymerization of isoprene depending on which complex and cocatalysts were used (Table 8). Complexes **61–63** were highly *cis*-1,4 selective when used in conjunction with [Ph_3C][$B(C_6F_5)_4$] and Al(*i*Bu)$_3$, and selectivities of 95.0–97.7% were obtained. On the contrary, the neodymium NPN bis(borohydride) complex **64** afforded 3,4 selectivities of around 87–88% when used in conjunction with Al(*i*Bu)$_3$ and either [$PhNHMe_2$][$B(C_6F_5)_4$] or [Ph_3C][$B(C_6F_5)_4$]. The related lutetium NPN bis(neosilyl) complex **65** actually promoted the *trans*-1,4-specific polymerization of isoprene when used in concert with Mg(*n*Bu)$_2$ and either [$PhNHMe_2$][$B(C_6F_5)_4$], [Ph_3C][$B(C_6F_5)_4$], or $B(C_6F_5)_3$, and selectivities of 94.2–97.2% were achieved. The *trans*-selectivity of **65** was tentatively ascribed to the formation of a bimetallic intermediate.

Bonnet and others used the bis(phenolate)amine neodymium borohydride complexes **66–68** to promote the formation of *trans*-1,4-isoprene with very good selectivity values (ca. 93–95%) when combined with Mg(nBu)(Et) [60]. They were much more selective than analogous samarium complexes, as expected. The same systems also catalyzed the formation of atactic polystyrene. Unusually, the samarium catalysts were much more active than the neodymium ones.

Amidinate and β-diketiminate ligands continue to attract considerable attention in polymerization reactions [61–66], and a scandium benzamidinate system (**69**) provided some interesting results in the polymerization of polyisoprene [67]. When treated with $[Ph_3C][B(C_6F_5)_4]$, complex **69** showed good 3,4-selectivity (just under 90%). However, when $AlMe_3$ was added to the system, the regioselectivity changed and *cis*-1,4-selectivity (up to 94%) was observed after the reaction went to completion in only 2 min.

Although ligand design plays an important role in catalysis, it is always worth exploring well-established systems. Homoleptic [Ln(BTSA)$_3$] complexes have been known for many years, and they are easily prepared from commercially available precursors. With this in mind, Visseaux and Bonnet tested [Ln(BTSA)$_3$] (Ln=Y, Nd) in the polymerization of isoprene [68]. When used with $B(C_6F_5)_3$ and Al(iBu)$_3$, very high activities (up to 4,700 kg mol_{Ln}^{-1} h^{-1}) and *cis*-1,4-selectivities (up to 96%) were obtained.

Allyl complexes are popular polymerization catalysts, and Cui and Hou continued this trend with a series of $[Ln\{C_5Me_4(C_6H_4NMe_2\text{-}o)\}(C_3H_5)_2]$ complexes (Ln=Y (**70**); Ln=Nd (**71**); Ln=Gd (**72**); Ln=Dy (**73**)) [69]. These catalysts, in combination with Al(iBu)$_3$/$[PhMe_2NH][B(C_6F_5)_4]$, displayed varied activity in the *cis*-1,4-selective polymerization of isoprene. The activity of the catalysts followed the order Nd > Gd >> Dy >> Y. Although **71** was more active than **72**, the latter provided the highest selectivity (98.2% at 20°C; 99.2% at 0°C). Remarkably, in the presence of excess Al(iBu)$_3$ unprecedented living catalyzed chain growth was observed. The addition of a second equivalent of isoprene doubled the molecular weight without affecting the *cis*-1,4-selectivity or the molecular weight distribution, the same held true for increasing amounts of Al(iBu)$_3$. Moreover, with the sequential addition of further amounts of butadiene or isoprene, di- or triblock copolymers were produced.

3.3 3,4-Polymerization of Isoprene

Compared with the large number of *cis*-1,4-selective, and to a lesser degree *trans*-1,4-selective, rare-earth catalysts, the number of active and selective rare-earth catalysts for the 3,4- polymerization of isoprene is rather low due to unfavorable electronic factors [70]. However, 3,4-polyisoprene has a number of important uses, including incorporation into high-performance tires [70]. The 3,4-polymerization of isoprene is shown in Scheme 15.

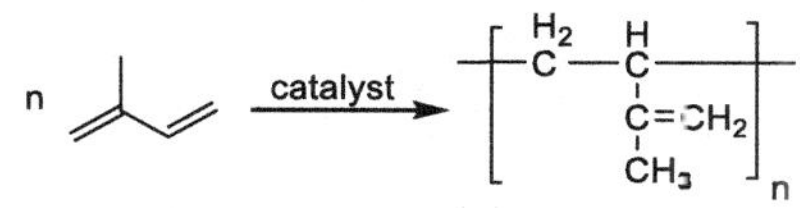

Scheme 15 Polymerization of isoprene to yield 3,4-polyisoprene

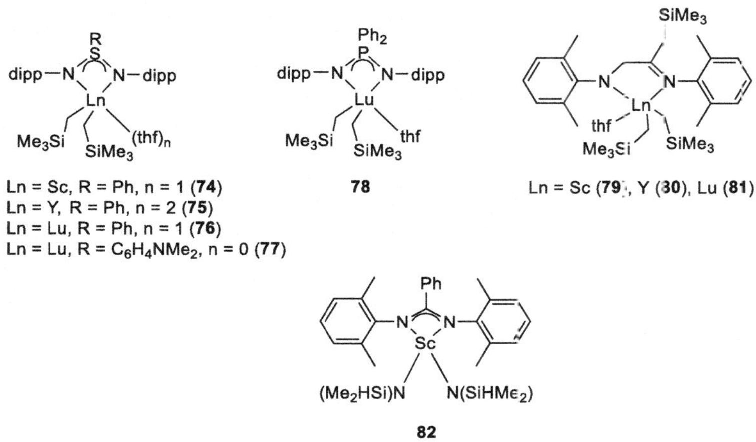

Scheme 16 Rare-earth catalysts for the 3,4-selective polymerization of isoprene [63, 71, 72]

Over the past several years, new rare-earth catalyst systems have emerged that promote the selective formation of 3,4-polyisoprene. A selection of the rare-earth pre-catalysts is shown in Scheme 16.

Cui and coworkers prepared a series of NSN- and NPN-ligated rare-earth complexes (**74–78**) that proved to be potent catalysts for the 3,4-selective polymerization of isoprene (Table 9) [71]. When used in conjunction with $[PhNHMe_2][B(C_6F_5)_4]$ and $Al(i Bu)_3$ cocatalysts at 10°C, all complexes gave very high 3,4-selectivities (91–98%), with most catalysts giving 100% conversion in 10 min. At −30°C, selectivities were even more impressive (98.2 to > 99%) but reaction times were much longer. Very good isospecific selectivities (*mmmm* > 99%) were achieved with **74–77**. The highest known T_m reported at that time (170°C) was also achieved. Surprisingly, using complex **78** switched the stereoselectivity and moderately syndiotactic polyisoprene was produced (*rr*/*rrrr* up to 66%/36%), and at this time, this was the most syndiotactic polyisoprene formed in a system using rare-earth catalysts.

The imino–amido bis(neosilyl) rare-earth complexes **79–81** also proved to be potent catalysts for the 3,4-selective polymerization of isoprene [72]. When used with an equimolar amount of $[Ph_3C][B(C_6F_5)_4]$, excellent selectivities (90–100%) were achieved, along with moderate to excellent yields (60–100%) and low polydispersity values (1.18–1.27).

Table 9 The 3,4-selective polymerization of isoprene by **74–78**/$[PhNHMe_2][B(C_6F_5)_4]/Al(i\mathrm{Bu})_3$ [71]

Catalyst	Time (min)	Temp. (°C)	Yield (%)	M_n ($\times10^4$)	M_w/M_n	3,4 (%)	*mm*/*mmmm* (%)	*mr* (%)	*rr*/*rrrr* (%)
74	10	10	100	13.7	2.1	95.0	93/58	7	2
75	10	10	100	13.5	2.3	97.1	90/60	8	2
76	10	10	100	14.7	2.3	98.0	92/68	5	3
77	60	10	100	11.6	2.1	93.1	74/19	24	2
78	10	10	82	7.9	1.7	97.2	15	29	56/27
74	360	−30	100	32.1	2.0	99	97/97	3	–
75	360	−30	100	36.7	2.1	>99	>99/>99	–	–
76	360	−30	100	35.0	2.3	>99	>99/>99	–	–
77	360	−30	100	37.1	2.0	>99	80/43	20	–
78[a]	1,440	−30	70	3.8	1.7	99.0	1	33	66/36

Conditions: C_6H_5Cl (5 mL), cat. (10 μmol), isoprene 1 mL, [cat.]:$[PhNHMe_2][B(C_6F_5)_4]$ = 1:1
[a]Toluene (5 mL)

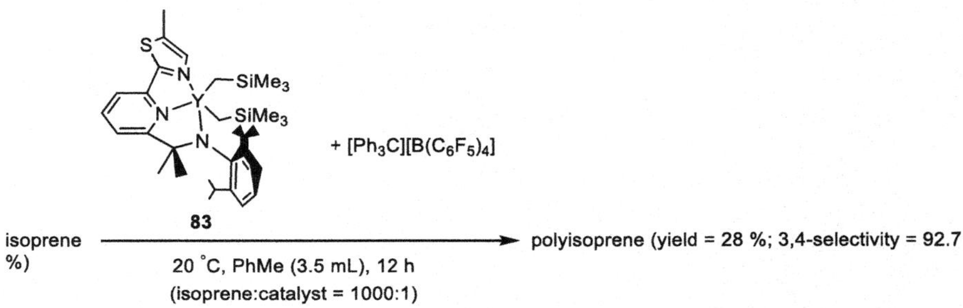

Scheme 17 3,4-Selective polymerization of isoprene by the thiazole-substituted amidopyridinate yttrium complex **83** [73]

The scandium complex **82** was prepared by Luo and coworkers [63], and it is structurally very similar to complex **69**. The binary system comprising complex **82** and $[Ph_3C][B(C_6F_5)_4]$ showed good 3,4-selectivity (up to 83%). On the other hand, when $AlMe_3$ was used as well, the selectivity was reversed and the system showed good *cis*-1,4-selectivity (83% after 5 min with a yield of 100%).

As seen in "Achiral Catalysts" of Section 2.1.2, thiazole-substituted amidopyridinate complexes can be useful catalysts for hydroamination reactions. Trifonov and coworkers prepared a related yttrium complex (**83**) that was used in the 3,4-selective polymerization of isoprene (Scheme 17) [73]. Complex **83** afforded high 3,4-selectivity (up to 92.7%) but poor conversion (28%). The conversion was not altered much by increasing the temperature to 50°C. Adding ten molar equivalents of $Al(i\mathrm{Bu})_3$ increased the yield to 81% but the 3,4-selectivity dropped to 61.7%. On the other hand, analogous complexes of Lu, Yb, and Er were moderately *trans*-1,4-selective (up to 76.4%).

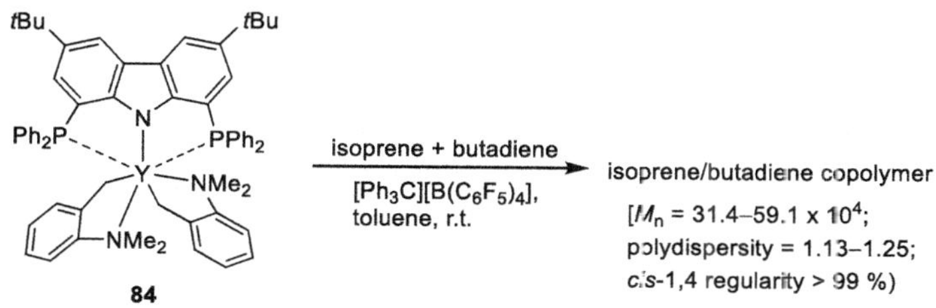

Scheme 18 Random copolymerization of isoprene and butadiene by **84**/$[Ph_3C][B(C_6F_5)_4]$ [74]

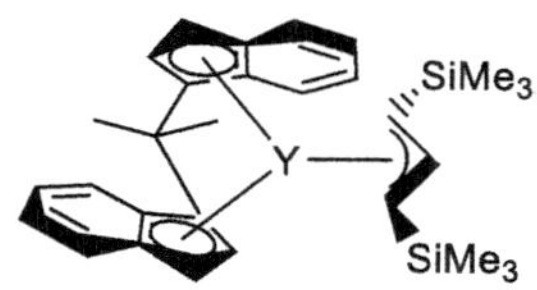

85

Fig. 11 An *ansa*-bis(indenyl) yttrium allyl complex (**85**) used in the simultaneous or sequential copolymerization reactions of isoprene and styrene [75]

3.4 Copolymerization Reactions of 1,3-Dienes with Alkenes

The copolymerization of dienes with alkenes provides an attractive route to functionalized polymers, which can be used for high-performance applications. However, low activity is often a distinctive feature of such reactions, but nevertheless rapid developments in rare-earth catalysis are beginning to address these problems.

Complex **84** could be used for the random copolymerization of butadiene and isoprene (Scheme 18) when used in conjunction with $[Ph_3C][B(C_6F_5)_4]$ [74]. Both the butenyl and isopentenyl fragments were found to have high *cis*-1,4-regularity (>99%), suggesting that this catalytic system could be used for the production of high-performance rubber. The same system was also efficient in the copolymerization of isoprene and ε-caprolactone.

The *ansa*-bis(indenyl) yttrium allyl complex **85** (Fig. 11) was used in both simultaneous and sequential copolymerization reactions of isoprene and styrene [75]. The simultaneous copolymerization reactions afforded copolymers with a large range of compositions (styrene molar fraction = 0.30–0.98) and stereoregular *trans*-1,4-polyisoprene units and isotactic polystyrene units. The catalyst system preferentially polymerized styrene, and this led to a tendency to form multiblock materials as opposed to materials with alternating units. Sequential copolymerization reactions yielded *trans*-1,4-polyisoprene-*block*-isotactic polystyrene featuring two crystalline blocks.

Visseaux and coworkers investigated the known neodymium borohydride complex, $[Nd(Cp^*)(BH_4)_2(thf)_x]$ (**86**), for the random copolymerization of butadiene and styrene [76]. Mg(*n*Bu)(Et) was used to activate the system and results are listed

Table 10 Results for the random copolymerization of butadiene and styrene by **86**/Mg(*n*Bu)(Et) [76]

[Mg]/[**86**]	Butadiene/styrene in the feed (mmol)	Butadiene/styrene in the copolymer (mol%)	Yield (%)	M_n ($\times 10^4$)	M_w/M_n	*trans*-1,4 (%)	*cis*-1,4 (%)	1,2 (%)
1	20/0	100/0	70	10.0	1.28	95.3	3.1	1.6
1	20/20	96.2/3.8	22	5.80	1.27	96.4	2.2	1.4
1	20/80	83.1/16.9	10	9.7	1.56	97.2	2.2	0.6
1	80/20	98.6/1.4	29	24.0	1.44	95.8	2.8	1.4
3	20/20	94.7/5.3	4	0.44	1.53	93.2	1.7	5.1
10	20/20	86.7/13.3	16	0.68	2.49	87.3	2.0	10.7

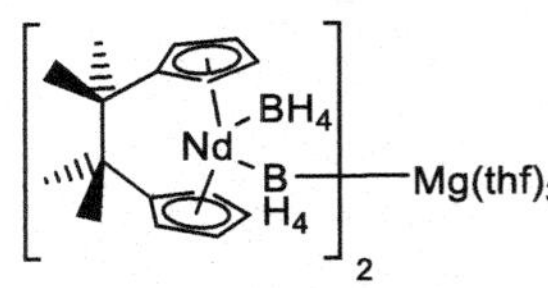

88

Fig. 12 Structure of complex **88** [77]

in Table 10. Increasing amounts of Mg(*n*Bu)(Et) negatively affected the yield but slightly increased the rate of styrene insertion. Increasing the amount of styrene also had a deleterious effect on the yield. However, in all cases, high *trans*-1,4-selectivity (87–97%) of the polybutadiene backbone was observed.

In a particularly nice piece of chemistry, Zinck and coworkers demonstrated the chain-shuttling copolymerization of isoprene and styrene using two rare-earth complexes and Mg(*n*Bu)(Et) [77]. The rare-earth complexes used were the half-sandwich complex, [La(Cp*)(BH_4)$_2$(thf)$_2$] (**87**), and the neodymium *ansa*-complex shown in Fig. 12 (**88**). Complex **87** afforded isoprene-rich copolymers with low T_g values. On the other hand, complex **88** gave copolymers with higher styrene content and higher T_g values. Chain-shuttling experiments were then conducted with mixtures of the catalysts. Depending on the relative amounts of the reagents, fully amorphous multiblock copolymers of poly(isoprene-co-styrene) containing soft blocks rich in isoprene and hard blocks rich in isoprene (very high *trans*-1,4-selectivity) were produced. These copolymers, which represent a new class of thermoplastic elastomers, have two separate T_g values that can be tuned by varying the amounts of reagents. However, increasing amounts of Mg(*n*Bu)(Et) led to the formation of more traditional statistical copolymers. Selected results are shown in Table 11.

Li and coworkers prepared a tropidinyl bis(alkyl) scandium catalyst (**89**) that was used in the random copolymerization of isoprene and norbornene (Scheme 19) [78]. Complex **89** was used with [Ph_3C][B(C_6F_5)$_4$]/Al(*i*Bu)$_3$, and by varying the amount of norbornene relative to a fixed amount of isoprene (5 mmol), random copolymers with 57–91 mol% of isoprene were produced with *cis*-1,4-

Table 11 Selected results in the copolymerization of isoprene and styrene by **87** and/or **88**/Mg(*n*Bu)(Et) [77]

87:**88** (%)	Time (h)	Mg (equiv.)	Isoprene/styrene (%) feed	Yield (%)	M_n ($\times 10^4$)	Isoprene/styrene (%) product	*trans*-1,4 (%)	*cis*-1,4 (%)	3,4 (%)	T_g (°C)
100:0	2	1	50:50	48	4.52	88:12	98	–	2	−50
0:100	2	1	50:50	5	0.94	50:50	100	–	–	n.d.
50:50	2	1	50:50	8	2.32	70:30	100	–	–	n.d.
50:50	26	10	50:50	72	1.00	62:38	92	–	8	−10
50:50	2	1	30:70	11	2.57	60:40	96	4	–	−10, 34
50:50	18	1	70:30	84	8.27	76:24	99	1	–	−35, 36
30:70	24	1	50:50	74	8.04	62:38	100	–	–	−10, 40
70:30	24	1	50:50	67	8.02	69:31	100	–	–	−27

Conditions: 50°C, toluene (1.5 mL), monomers/catalyst(s) ratio of 2000:1
n.d. not determined

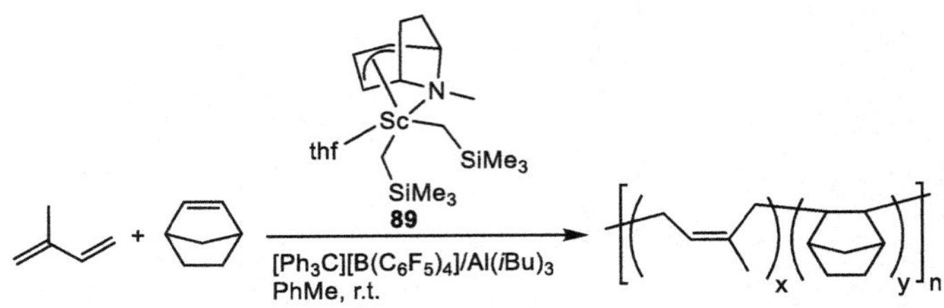

Scheme 19 Random copolymerization of isoprene and norbornene catalyzed by the **89**/$[Ph_3C][B(C_6F_5)_4]/Al(i\text{Bu})_3$ system [78]

configurations of up to 88%. Complex **89** was also used in the *cis*-1,4-selective polymerization of isoprene. Overall, it showed similar or better activity than traditional cationic (cyclopentadienyl)scandium catalysts.

3.5 Outlook

Although metallocene catalysts remain an important class of compounds, non-metallocene catalysts look like they will be increasingly favored in the search for catalysts that combine high activity with high selectivity. Important advances have been made in the development of *trans*-1,4-selective and 3,4-selective catalysts, but they are still scarcer than *cis*-1,4-selective catalysts. Bolstered by recent successes in achieving high *cis*-1,4-content in copolymers, it is likely that more intensive research efforts will be directed to this industrially important area.

Acknowledgment This work was supported by the DFG-funded transregional collaborative research center SFB/TRR 88 "3MET".

References

1. Watson PL (1983) Methane exchange reactions of lanthanide and early-transition-metal methyl complexes. J Am Chem Soc 105(21):6491–6493
2. Labinger JA, Bercaw JE (2002) Understanding and exploiting C-H bond activation. Nature 417(6888):507–514
3. Barros N, Eisenstein O, Maron L (2006) DFT studies of the methyl exchange reaction between $Cp_2M\text{-}CH_3$ or $Cp^*{}_2M\text{-}CH_3$ ($Cp = C_5H_5$, $Cp^* = C_5Me_5$, M = Y, Sc, Ln) and CH_4. Does M ionic radius control the reaction? Dalton Trans 25:3052–3057
4. Sadow AD, Tilley TD (2005) Synthesis and characterization of scandium silyl complexes of the type $Cp^*{}_2ScSiHRR'$. σ-Bond metathesis reactions and catalytic dehydrogenative silation of hydrocarbons. J Am Chem Soc 127(2):643–656
5. Ziegler T, Folga E, Berces A (1993) A density functional study on the activation of hydrogen-hydrogen and hydrogen-carbon bonds by $Cp_2Sc\text{-}H$ and $Cp_2Sc\text{-}CH_3$. J Am Chem Soc 115 (2):636–646

6. Evans WJ, Grate JW, Doedens RJ (1985) Organolanthanide and organoyttrium hydride chemistry. 7. Reaction of the samarium-hydrogen bond in the organosamarium hydride $[(C_5Me_5)_2SmH]_2$ with carbon monoxide: formation, isomerization, and x-ray crystallographic characterization of the samarium complexes cis- and trans-$\{(C_5Me_5)_2[(C_6H_5)_3PO]Sm\}_2(\mu\text{-OCH CHO})$. J Am Chem Soc 107(6):1671–1679
7. Evans WJ, Miller KA, Ziller JW (2006) Synthesis of $(O_2CEPh)1$- ligands (E = S, Se) by CO_2 insertion into lanthanide chalcogen bonds and their utility in forming crystallographically characterizable organoaluminum complexes $[Me_2Al(\mu\text{-}O_2CEPh)]_2$. Inorg Chem 45 (1):424–429
8. Klementyeva SV, Gamer MT, Schmidt A-C, Meyer K, Konchenko SN, Roesky PW (2014) Activation of SO_2 with $[(\eta^5\text{-}C_5Me_5)_2Ln(THF)_2]$ (Ln=Eu, Yb) leading to dithionite and sulfinate complexes. Chem Eur J 20(42):13497–13500
9. Klementyeva SV, Arleth N, Meyer K, Konchenko SN, Roesky PW (2015) Dithionite and sulfinate complexes from the reaction of SO_2 with decamethylsamarocene. New J Chem 39:7589–7594
10. Evans WJ, Ulibarri TA, Ziller JW (1990) Reactivity of $(C_5Me_5)_2Sm$ and related species with alkenes: synthesis and structural characterization of a series of organosamarium allyl complexes. J Am Chem Soc 112(6):2314–2324
11. Kaita S, Koga N, Hou Z, Doi Y, Wakatsuki Y (2003) Theoretical study of coordination and insertion of butadiene to bis(pentamethylcyclopentadienyl)samarium hydride. Organometallics 22(15):3077–3082
12. Roesky PW (ed) (2010) Molecular catalysis of rare-earth elements, vol 137, Structure and bonding. Springer, Berlin/Heidelberg
13. Hong S, Marks TJ (2004) Organolanthanide-catalyzed hydroamination. Acc Chem Res 37 (9):673–686
14. Müller TE, Hultzsch KC, Yus M, Foubelo F, Tada M (2008) Hydroamination: direct addition of amines to alkenes and alkynes. Chem Rev 108(9):3795–3892
15. Reznichenko A, Hultzsch K (2013) Early transition metal (group 3–5, lanthanides and actinides) and main group metal (group 1, 2, and 13) catalyzed hydroamination. In: Ananikov VP, Tanaka M (eds) Hydrofunctionalization, vol 43, Topics in organometallic chemistry. Springer, Berlin/Heidelberg, pp 51–114
16. Koshti V, Gaikwad S, Chikkali SH (2014) Contemporary avenues in catalytic PH bond addition reaction: a case study of hydrophosphination. Coord Chem Rev 265:52–73
17. Trifonov AA (2007) Non-metallocene rare-earth organometallic derivatives: synthesis, structure and application in the catalysis of transformations of unsaturated substrates. Russ Chem Rev 76(11):1122
18. Rodriguez-Ruiz V, Carlino R, Bezzenine-Lafollee S, Gil R, Prim D, Schulz E, Hannedouche J (2015) Recent developments in alkene hydro-functionalisation promoted by homogeneous catalysts based on earth abundant elements: formation of C-N, C-O and C-P bond. Dalton Trans 44(27):12029–12059
19. Reznichenko AL, Nguyen HN, Hultzsch KC (2010) Asymmetric intermolecular hydroamination of unactivated alkenes with simple amines. Angew Chem Int Ed 49 (47):8984–8987
20. Reznichenko AL, Hultzsch KC (2013) C1-symmetric rare-earth-metal aminodiolate complexes for intra- and intermolecular asymmetric hydroamination of alkenes. Organometallics 32(5):1394–1408
21. Basalov IV, Roşca SC, Lyubov DM, Selikhov AN, Fukin GK, Sarazin Y, Carpentier J-F, Trifonov AA (2014) Divalent heteroleptic ytterbium complexes – effective catalysts for intermolecular styrene hydrophosphination and hydroamination. Inorg Chem 53 (3):1654–1661
22. Hannedouche J, Collin J, Trifonov A, Schulz E (2011) Intramolecular enantioselective hydroamination catalyzed by rare earth binaphthylamides. J Organomet Chem 696 (1):255–262

23. Lovick HM, An DK, Livinghouse TS (2011) Structure-activity relationships in group 3 metal catalysts for asymmetric intramolecular alkene hydroamination. An investigation of ligands based on the axially chiral 1,1′-binaphthyl-2,2′-diamine motif. Dalton Trans 40 (30):7697–7700
24. Queffelec C, Boeda F, Pouilhès A, Meddour A, Kouklovsky C, Hannedouche J, Collin J, Schulz E (2011) Enantioselective intramolecular hydroamination of secondary amines catalyzed by easily accessible ate and neutral rare-earth complexes. ChemCatChem 3(1):122–126
25. Aillaud I, Olier C, Chapurina Y, Collin J, Schulz E, Guillot R, Hannedouche J, Trifonov A (2011) Alkyl complexes of lanthanum and samarium supported by a chiral binaphthylamido ligand. Syntheses, structures, and catalytic activity in hydroamination of amino-1,3-dienes. Organometallics 30(12):3378–3385
26. Chapurina Y, Guillot R, Lyubov D, Trifonov A, Hannedouche J, Schulz E (2013) LiCl-effect on asymmetric intramolecular hydroamination catalyzed by binaphthylamido yttrium complexes. Dalton Trans 42(2):507–520
27. Benndorf P, Jenter J, Zielke L, Roesky PW (2011) Chiral lutetium benzamidinate complexes. Chem Commun 47(9):2574–2576
28. Benndorf P, Kratsch J, Hartenstein L, Preuss CM, Roesky PW (2012) Chiral benzamidinate ligands in rare-earth-metal coordination chemistry. Chem Eur J 18(45):14454–14463
29. Zhang Y, Yao W, Li H, Mu Y (2012) Rare-earth-metal complexes supported by new chiral tetra-azane chelating ligands: synthesis, characterization, and catalytic properties for intramolecular asymmetric hydroamination. Organometallics 31(13):4670–4679
30. Bennett SD, Core BA, Blake MP, Pope SJA, Mountford P, Ward BD (2014) Chiral lanthanide complexes: coordination chemistry, spectroscopy, and catalysis. Dalton Trans 43 (15):5871–5885
31. Liu H, He J, Liu Z, Lin Z, Du G, Zhang S, Li X (2013) Quasi-living trans-1,4-polymerization of isoprene by cationic rare earth metal alkyl species bearing a chiral (S, S)-bis (oxazolinylphenyl)amido ligand. Macromolecules 46(9):3257–3265
32. Manna K, Kruse ML, Sadow AD (2011) Concerted C–N/C–H bond formation in highly enantioselective yttrium(III)-catalyzed hydroamination. ACS Catal 1(11):1637–1642
33. Wang Q, Zhang F, Song H, Zi G (2011) Synthesis, structure, and catalytic activity of organolanthanides with chiral biphenyl-based tridentate amidate ligands. J Organomet Chem 696(10):2186–2192
34. Vitanova DV, Hampel F, Hultzsch KC (2011) Linked bis(β-diketiminato) yttrium and lanthanum complexes as catalysts in asymmetric hydroamination/cyclization of aminoalkenes (AHA). J Organomet Chem 696(1):321–330
35. Hangaly NK, Petrov AR, Rufanov KA, Harms K, Elfferding M, Sundermeyer J (2011) Cyclopentadienylphosphazene (CpPN) complexes of rare-earth metals: synthesis, structural characterization, and hydroamination catalysis. Organometallics 30(17):4544–4554
36. Xu X, Chen Y, Feng J, Zou G, Sun J (2010) Half-sandwich indenyl rare earth metal dialkyl complexes: syntheses, structures, and catalytic activities of (1,3-$(SiMe_3)2C_9H_5$)Ln $(CH_2SiMe_3)_2$(THF). Organometallics 29(3):549–553
37. Chai Z, Hua D, Li K, Zhou S, Chu J, Yang G (2014) Rare-earth metal complexes with tridentate linked amido-indenyl ligand: synthesis, characterization, and catalytic properties for intramolecular hydroamination. J Organomet Chem 768:136–139
38. Trambitas AG, Panda TK, Jenter J, Roesky PW, Daniliuc C, Hrib CG, Jones PG, Tamm M (2010) Rare-earth metal alkyl, amido, and cyclopentadienyl complexes supported by imidazolin-2-iminato ligands: synthesis, structural characterization, and catalytic application. Inorg Chem 49(5):2435–2446
39. Trambitas Alexandra G, Panda Tarun K, Tamm M (2010) Rare earth metal complexes supported by ancillary imidazolin-2-iminato ligands. Z Anorg Allg Chem 636(12):2156–2171
40. Roux EL, Liang Y, Storz MP, Anwander R (2010) Intramolecular hydroamination/cyclization of aminoalkenes catalyzed by $Ln[N(SiMe_3)_2]_3$ grafted onto periodic mesoporous silicas. J Am Chem Soc 132(46):16368–16371

41. Lyubov DM, Luconi L, Rossin A, Tuci G, Cherkasov AV, Fukin GK, Giambastiani G, Trifonov AA (2014) Metal-to-ligand alkyl migration inducing carbon–sulfur bond cleavage in dialkyl yttrium complexes supported by thiazole-containing amidopyridinate ligands: synthesis, characterization, and catalytic activity in the intramolecular hydroamination reaction. Chem Eur J 20(12):3487–3499
42. Reznichenko AL, Hultzsch KC (2010) Catalytic σ-bond metathesis. Struct Bond 137:1–48
43. Behrle AC, Schmidt JAR (2013) Insertion reactions and catalytic hydrophosphination of heterocumulenes using α-metalated N, N-dimethylbenzylamine rare-earth-metal complexes. Organometallics 32(5):1141–1149
44. Liu B, Roisnel T, Carpentier J-F, Sarazin Y (2013) Heteroleptic alkyl and amide iminoanilide alkaline earth and divalent rare earth complexes for the catalysis of hydrophosphination and (cyclo)hydroamination reactions. Chem Eur J 19(40):13445–13462
45. Trifonov A, Kissel A, Mahrova T, Lyubov D, Cherkasov A, Fukin GK, Maron L, Del Rosal I (2015) Metallacyclic yttrium alkyl and hydrido complexes: synthesis, structures and catalytic activity in intermolecular olefin hydrophosphination and hydroamination. Dalton Trans 44(27):12137–12148
46. LeBlanc FA, Piers WE, Parvez M (2014) Selective hydrosilation of CO_2 to a bis(silylacetal) using an anilido bipyridyl-ligated organoscandium catalyst. Angew Chem Int Ed 53 (3):789–792
47. Abinet E, Martin D, Standfuss S, Kulinna H, Spaniol TP, Okuda J (2011) Rare-earth metal allyl and hydrido complexes supported by an (NNNN)-type macrocyclic ligand: synthesis, structure, and reactivity toward biomass-derived furanics. Chem Eur J 17(52):15014–15026
48. Lyubov DM, Shavyrin AS, Kurskii YA, Trifonov AA (2010) Hydrosilylation of non-1-ene with phenylsilane in the presence of yttrium and lutetium bisguanidinate hydride complexes. Russ Chem Bull 59(9):1765–1770
49. Jian Z, Tang S, Cui D (2010) A lutetium allyl complex that bears a pyridyl-functionalized cyclopentadienyl ligand: dual catalysis on highly syndiospecific and cis-1,4-selective (co) polymerizations of styrene and butadiene. Chem Eur J 16(47):14007–14015
50. Zimmermann M, Volbeda J, Törnroos KW, Anwander R (2010) Tetramethylcyclopentadienyl-supported rare-earth metal bis(tetramethyl)aluminate complexes: synthesis, structural chemistry, cation formation, and isoprene polymerization. C R Chim 13(6–7):651–660
51. Zimmermann M, Törnroos KW, Anwander R (2008) Cationic rare-earth-metal half-sandwich complexes for the living trans-1,4-isoprene polymerization. Angew Chem Int Ed 47 (4):775–778
52. Jende LN, Maichle-Mössmer C, Anwander R (2013) Rare-earth-metal alkylaluminates supported by N-donor-functionalized cyclopentadienyl ligands: C–H bond activation and performance in isoprene polymerization. Chem Eur J 19(48):16321–16333
53. Roux EL, Liang Y, Törnroos KW, Nief F, Anwander R (2012) Heterogenization of lanthanum and neodymium monophosphacyclopentadienyl bis(tetramethylaluminate) complexes onto periodic mesoporous silica SBA-15. Organometallics 31(18):6526–6537
54. Dettenrieder N, Hollfelder CO, Jende LN, Maichle-Mössmer C, Anwander R (2014) Half-sandwich rare-earth-metal alkylaluminate complexes bearing peripheral boryl ligands. Organometallics 33(7):1528–1531
55. Jende LN, Hollfelder CO, Maichle-Mössmer C, Anwander R (2015) Rare-earth-metal allyl complexes supported by the [2-(N, N-dimethylamino)ethyl]tetramethylcyclopentadienyl ligand: structural characterization, reactivity, and isoprene polymerization. Organometallics 34(1):32–41
56. Valente A, Zinck P, Vitorino MJ, Mortreux A, Visseaux M (2010) Rare earths/main group metal alkyls catalytic systems for the 1,4-trans stereoselective coordinative chain transfer polymerization of isoprene. J Polym Sci Part A Polym Chem 48(21):4640–4647

57. Pan Y, Xu T, Yang G-W, Jin K, Lu X-B (2013) Bis(oxazolinyl)phenyl-ligated rare-earth-metal complexes: highly regioselective catalysts for cis-1,4-polymerization of isoprene. Inorg Chem 52(6):2802–2808
58. Jenter J, Meyer N, Roesky PW, Thiele SKH, Eickerling G, Scherer W (2010) Borane and borohydride complexes of the rare-earth elements: synthesis, structures, and butadiene polymerization catalysis. Chem Eur J 16(18):5472–5480
59. Yang Y, Lv K, Wang L, Wang Y, Cui D (2010) Isoprene polymerization with aminopyridinato ligand supported rare-earth metal complexes. Switching of the regio- and stereoselectivity. Chem Commun 46(33):6150–6152
60. Bonnet F, Dyer HE, El Kinani Y, Dietz C, Roussel P, Bria M, Visseaux M, Zinck P, Mountford P (2015) Bis(phenolate)amine-supported lanthanide borohydride complexes for styrene and trans-1,4-isoprene (co-)polymerisations. Dalton Trans 44(27):12312–12325
61. Luo Y, Fan S, Yang J, Fang J, Xu P (2011) Rare earth metal bis(amide) complexes bearing amidinate ancillary ligands: synthesis, characterization, and performance as catalyst precursors for cis-1,4 selective polymerization of isoprene. Dalton Trans 40(12):3053–3059
62. Hamidi S, Jende LN, Martin Dietrich H, Maichle-Mössmer C, Törnroos KW, Deacon GB, Junk PC, Anwander R (2013) C–H bond activation and isoprene polymerization by rare-earth-metal tetramethylaluminate complexes bearing formamidinato N-ancillary ligands. Organometallics 32(5):1209–1223
63. Luo Y, Lei Y, Fan S, Wang Y, Chen J (2013) Synthesis of mono-amidinate-ligated rare-earth-metal bis(silylamide) complexes and their reactivity with $[Ph_3C][B(C_6F_5)_4]$, $AlMe_3$ and isoprene. Dalton Trans 42(11):4040–4051
64. Li L, Wu C, Liu D, Li S, Cui D (2013) Binuclear rare-earth-metal alkyl complexes ligated by phenylene-bridged β-diketiminate ligands: synthesis, characterization, and catalysis toward isoprene polymerization. Organometallics 32(11):3203–3209
65. Sun S, Ouyang H, Luo Y, Zhang Y, Shen Q, Yao Y (2013) Synthesis of β-diketiminate-ligated bimetallic and monometallic lanthanide amide complexes and their reactivity with isoprene and $AlMe_3$. Dalton Trans 42(46):16355–16364
66. Li D, Li S, Cui D, Zhang X (2010) β-Diketiminato rare-earth metal complexes. Structures, catalysis, and active species for highly cis-1,4-selective polymerization of isoprene. Organometallics 29(9):2186–2193
67. Chen F, Fan S, Wang Y, Chen J, Luo Y (2012) Unusual Si–H bond activation and formation of cationic scandium amide complexes from a mono(amidinate)-ligated scandium bis(silylamide) complex and their performance in isoprene polymerization. Organometallics 31(9):3730–3735
68. Martins N, Bonnet F, Visseaux M (2014) Highly efficient cis-1,4 polymerisation of isoprene using simple homoleptic amido rare earth-based catalysts. Polymer 55(20):5013–5016
69. Jian Z, Cui D, Hou Z, Li X (2010) Living catalyzed-chain-growth polymerization and block copolymerization of isoprene by rare-earth metal allyl precursors bearing a constrained-geometry-conformation ligand. Chem Commun 46(17):3022–3024
70. Zhang Z, Cui D, Wang B, Liu B, Yang Y (2010) Polymerization of 1,3-conjugated dienes with rare-earth metal precursors. Struct Bond 137:49–108
71. Liu B, Li L, Sun G, Liu J, Wang M, Li S, Cui D (2014) 3,4-Polymerization of isoprene by using NSN- and NPN-ligated rare earth metal precursors: switching of stereo selectivity and mechanism. Macromolecules 47(15):4971–4978
72. Du G, Wei Y, Ai L, Chen Y, Xu Q, Liu X, Zhang S, Hou Z, Li X (2011) Living 3,4-polymerization of isoprene by cationic rare earth metal alkyl complexes bearing iminoamido ligands. Organometallics 30(1):160–170
73. Luconi L, Lyubov DM, Rossin A, Glukhova TA, Cherkasov AV, Tuci G, Fukin GK, Trifonov AA, Giambastiani G (2014) Organolanthanide complexes supported by thiazole-containing amidopyridinate ligands: synthesis, characterization, and catalytic activity in isoprene polymerization. Organometallics 33(24):7125–7134

74. Wang L, Cui D, Hou Z, Li W, Li Y (2011) Highly cis-1,4-selective living polymerization of 1,3-conjugated dienes and copolymerization with ε-caprolactone by bis(phosphino) carbazolide rare-earth-metal complexes. Organometallics 30(4):760–767
75. Annunziata L, Duc M, Carpentier J-F (2011) Chain growth polymerization of isoprene and stereoselective isoprene–styrene copolymerization promoted by an ansa-bis(indenyl)allyl–yttrium complex. Macromolecules 44(18):7158–7166
76. Ventura A, Chenal T, Bria M, Bonnet F, Zinck P, Ngono-Ravache Y, Balanzat E, Visseaux M (2013) Trans-stereospecific polymerization of butadiene and random copolymerization with styrene using borohydrido neodymium/magnesium dialkyl catalysts. Eur Polym J 49 (12):4130–4140
77. Valente A, Stoclet G, Bonnet F, Mortreux A, Visseaux M, Zinck P (2014) Isoprene–styrene chain shuttling copolymerization mediated by a lanthanide half-sandwich complex and a lanthanidocene: straightforward access to a new type of thermoplastic elastomers. Angew Chem Int Ed 53(18):4638–4641
78. Liu S, Du G, He J, Long Y, Zhang S, Li X (2014) Cationic tropidinyl scandium catalyst: a perfectly acceptable substitute for cationic half-sandwich scandium catalysts in cis-1,4-polymerization of isoprene and copolymerization with norbornene. Macromolecules 47 (11):3567–3573

Struct Bond (2016) 172: 119–138
DOI: 10.1007/430_2015_207

Published online: 14 February 2016

The Renaissance of Fullerene Superconductivity

Yasuhiro Takabayashi and Kosmas Prassides

Abstract Unconventional high-T_c superconductivity, defined both in terms of the magnitude of the superconducting transition temperature, T_c, and the key role played by electronic correlations, not only is the realm of atom-based low-dimensional layered systems such as the cuprates or the iron pnictides but is also accessible in molecular systems such as the cubic alkali fullerides with stoichiometry A_3C_{60} (A=alkali metal). In fulleride superconductors, isotropic high-T_c superconductivity occurs in competition with electronic ground states resulting from a fine balance between electron correlations and electron–phonon coupling in an electronic phase diagram strikingly similar to those of unconventional superconductors such as the cuprates and the heavy fermions. Superconductivity at the highest T_c (38 K) known for any molecular material emerges from the antiferromagnetic insulating state solely by changing an electronic parameter – the overlap between the outer wave functions of the constituent molecules – and T_c scales universally in a structure-independent dome-like relationship with proximity to the Mott metal–insulator transition (quantified by V, the volume/C_{60}, or equivalently by (U/W), the ratio of the on-site Coulomb energy, U, to the electronic bandwidth, W), a hallmark of electron correlations characteristic of high-T_c superconductors other than fullerides. The C_{60} molecular electronic structure plays a key role in the Mott–Jahn–Teller (MJT) insulator formed at large V, with the on-molecule dynamic Jahn–Teller (JT) effect distorting the $C_{60}{}^{3-}$ units and quenching the t_{1u} orbital degeneracy responsible for metallicity. As V decreases, the MJT insulator transforms first into an unconventional correlated JT metal (where localised electrons coexist with metallicity and the on-molecule distortion persists) and then into a Fermi liquid with a less prominent molecular electronic signature. This normal state crossover is mirrored in the evolution of the

Y. Takabayashi and K. Prassides (✉)
World Premier International – Advanced Institute for Materials Research (WPI-AIMR), Tohoku University, 2-1-1 Katahira, Aoba-ku, Sendai 980-8577, Japan
e-mail: k.prassides@wpi-aimr.tohoku.ac.jp

superconducting state, with the highest T_c found at the boundary between unconventional correlated and conventional weak-coupling BCS superconductivity, where the interplay between extended and molecular aspects of the electronic structure is optimised to create the superconductivity dome.

Keywords Antiferromagnetism · Electron correlation · Fullerenes · Jahn–Teller effect · Metal–insulator transition · Mott insulator · Superconductivity

Contents

Abbreviations

AFM	Antiferromagnetic
bcc	Body-centred cubic
bco	Body-centred orthorhombic
BCS	Bardeen–Cooper–Schrieffer
EPR	Electron paramagnetic resonance
fcc	Face-centred cubic
fco	Face-centred orthorhombic
HOMO	Highest occupied molecular orbital
IR	Infrared
JT	Jahn–Teller
LRO	Long-range order
LUMO	Lowest unoccupied molecular orbital
MIT	Metal–insulator transition
MJT	Mott–Jahn–Teller
NMR	Nuclear magnetic resonance
ZFC	Zero field cooled
μSR	Muon spin relaxation

1 Introduction

The discovery of C_{60} (buckminsterfullerene, a quasi-spherical molecule of icosahedral (I_h) symmetry with dimensions of ~1 nm and the parent of the family of the hollow closed-cage all-carbon molecules named fullerenes) in 1985 marked the beginning of the modern era of carbon chemistry, physics and materials science research [1]. This was subsequently followed by the discovery of other novel forms of carbon such as carbon nanotubes [2] and graphene [3], equally exciting on their own accord. The development of fullerene research was addressed in volume 109 of the *Structure and Bonding* series published in 2004. This volume included contributions from both theorists and experimentalists and highlighted many of the interdisciplinary issues – ranging from the appearance of key physical properties such as superconductivity and ferromagnetism to promise for future applications in batteries, transistors, solar cells and sensors among others – then at the forefront of fullerene research. In particular, the structural and electronic properties of fulleride salts were specifically reviewed [4], with particular emphasis placed on the members of this family of materials which become superconducting with remarkably high transition temperatures, T_c (up to 33 K for the $RbCs_2C_{60}$ composition [5]).

Metallic behaviour and superconductivity were first discovered in potassium-doped C_{60} films with $C_{60}{}^{3-}$ charge in 1991 [6]. This was followed by the isolation of bulk alkali fulleride salts with stoichiometries K_3C_{60} and Rb_3C_{60} and T_c of 19 and 29 K, respectively [7–9]. The superconducting compositions adopted expanded fcc structures, in which the three cations occupy the available octahedral and tetrahedral interstitial sites in the fcc structure of solid C_{60} (Fig. 1a) [10]. Charge transfer between the alkali metals and the C_{60} molecules is essentially complete, and the conduction band of C_{60}, which arises from its lowest unoccupied molecular orbital (LUMO) of t_{1u} symmetry, is half filled (Fig. 1b). Changing the size of the alkali dopants affects sensitively the superconducting properties of the materials [11]. In the *fcc* structure of A_3C_{60}, there are octahedral (one per C_{60} unit with a radius of 2.06 Å) and tetrahedral (two per C_{60} unit with a radius of 1.12 Å) interstices, which are occupied by the alkali cations (Fig. 1a). The C_{60}–C_{60} distance is directly controlled by the alkali metal ionic size. As the dopant size increases, the interfullerene separation and fcc lattice constant increase, and T_c also increases monotonically (Fig. 2a).

These experimental observations gave credence to models for the origin of fulleride superconductivity based on conventional BCS electron–phonon coupling mechanisms in which pair binding is dominated by intramolecular on-ball vibrational modes [12]. In brief, as the interfullerene spacing increases, the overlap between the fulleride units decreases, leading to a reduced bandwidth, W. As the band filling across the A_3C_{60} series is precisely fixed at ½, this results in an increased density of states at the Fermi level, $N(\varepsilon_F)$ (Fig. 1b), which in turn leads to an increased T_c through the BCS-type relationship: $T_c \propto (\hbar\omega_{ph})\exp[-1/N(\varepsilon_F)v]$. Here $\hbar\omega_{ph}$ is the average intramolecular phonon energy, and v is the electron–phonon coupling strength. Therefore the observed T_c could be understood within a

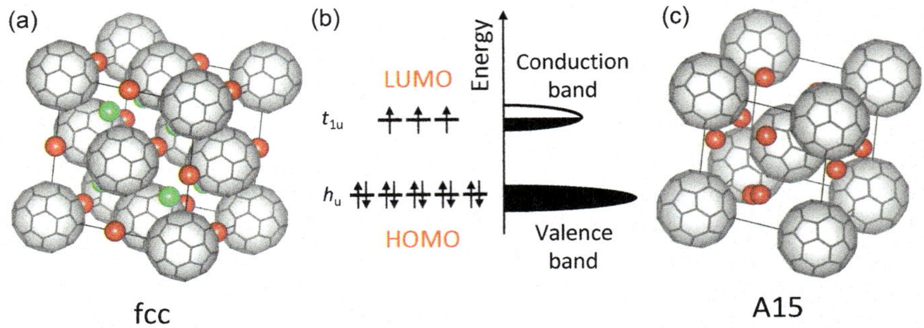

Fig. 1 Crystal and electronic structures for A_3C_{60} (A=alkali metal). (**a**) Fcc unit cell (space group $Fm\overline{3}m$). Two orientations related by 90° rotation about 100 exist in a disordered manner – only one of these is shown at each site for clarity. The A^+ ions are depicted as red and green spheres to signify symmetry inequivalent positions in the unit cell – octahedral and tetrahedral sites, respectively. (**b**) Molecular orbital energy scheme for the $C_{60}{}^{3-}$ anion and the corresponding valence and conduction bands for solid A_3C_{60}. (**c**) A15 unit cell (space group $Pm\overline{3}n$) based on bcc anion packing. One unique orientation of the $C_{60}{}^{3-}$ anions is present. The lower density of the body-centred packing offers more spacious sites for the A^+ cations with four fulleride neighbours

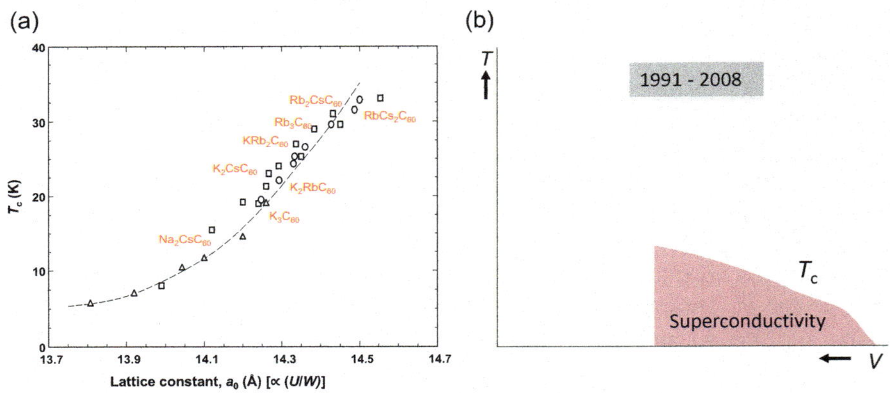

Fig. 2 (**a**) Variation of the superconducting transition temperature, T_c (K), with fcc lattice parameter, a_o (Å), for various compositions of A_3C_{60} fullerides at both ambient and elevated pressures. The dashed line is the T_c–a relationship expected from BCS theory using $N(\varepsilon_F)$ values obtained by LDA calculations. (**b**) Schematic electronic phase diagram of A_3C_{60} fullerides, T_c vs. V (volume occupied per fulleride anion), until the discovery of Cs_3C_{60} in 2008

BCS-like phenomenology, in terms of (1) a high-average phonon frequency, ω_{ph} (of comparable magnitude to the Fermi energy), resulting from the light carbon mass and the large force constants associated with the intramolecular modes; (2) a moderately large electron–phonon coupling constant, v, with contributions from both radial and tangential C_{60} vibrational modes; and (3) a high $N(\varepsilon_F)$ arising from the weak intermolecular interactions and narrow t_{1u} bandwidth. Therefore a consensus on the superconductivity mechanism as conventional phonon mediated was rapidly established and was almost universally accepted. Purely electronic models for pair binding [13], which were proposed very early after the discovery of fulleride superconductivity, were essentially considered inappropriate.

2 A_3C_{60} Fullerides and Electron Correlations

Although a BCS-type approach nicely provided a convincing rationalisation of the observed T_c of A_3C_{60} alkali fullerides and placed them near the top-performing conventional phonon-mediated superconductors, there were some features of their electronic structures which did not fit into the phenomenology developed. These were related to the relative magnitude of the t_{1u}-derived conduction bandwidth, W, and the on-site Coulomb repulsion, U. Estimates of U for the C_{60} molecule are on the order of 3 eV. This value is significantly reduced in the solid to ~1 eV [14]. At the same time, the t_{1u} orbital degeneracy is retained in the solid fcc fullerides because of the cubic symmetry, and typical values of W are on the order of 0.5 eV. Therefore the ratio (U/W) is much larger than 1 for all A_3C_{60}, implying that, in these systems, the gain in energy by delocalization of the outer electrons does not exceed the cost of putting two electrons on the same lattice site. As a result, instead of the experimentally observed metallicity and superconductivity, fcc A_3C_{60} should have been Mott–Hubbard insulators as the t_{1u} conduction band should split into two, a lower Hubbard sub-band which is fully occupied and an upper Hubbard sub-band which is empty.

The observed metallic behaviour of A_3C_{60} was accounted for theoretically by considering the triple orbital degeneracy of the LUMO states that survives in the solid state. It was shown that as a result of the additional possible hopping channels available to the carriers because of the orbital degeneracy compared to standard nondegenerate situations, the boundary of the metal to Mott–Hubbard insulator transition as monitored by the (U/W) ratio shifts to values higher than 1 – the critical value, $(U/W)_c$, scales roughly with the square root of the degeneracy [15]. These considerations naturally account for the stabilisation of the metallic state of A_3C_{60} in terms of the retention of the t_{1u} orbital degeneracy despite the extreme narrowing of the conduction band. They also introduce a new ingredient, namely, that of the enhanced electron correlations beyond the standard BCS-like phenomenology, and raise the issue of what the role of the correlations is in the superconductivity mechanism and in the determination of the electronic properties in the normal and superconducting states [16]. Moreover, in the special case of fcc-structured A_3C_{60}, there is the extra ingredient of geometric frustration – intrinsic to fcc packing – for the development of antiferromagnetic order. It can be shown that this sustains the highly correlated metallic state to even smaller electronic bandwidths [15], raising $(U/W)_c$ compared to bipartite topologies to the even higher value of ~2.3 (Fig. 3, [17]). Many early accounts of the electronic and superconducting properties of intercalated alkali fullerides exist as a function of

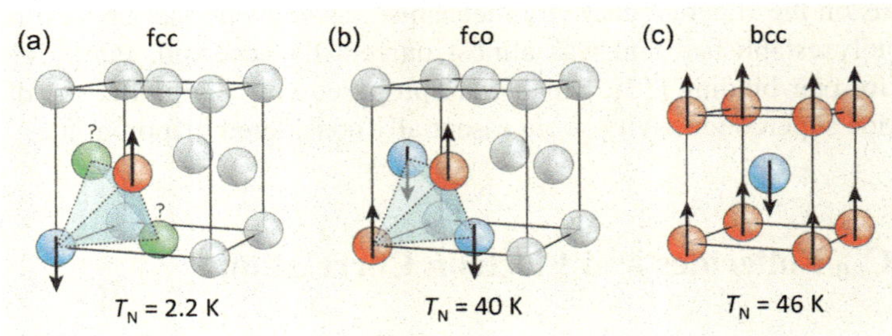

Fig. 3 Packing arrangements of C_{60}^{3-} anions in antiferromagnetic fullerides [17]. (**a**) Fcc lattice variant of A_3C_{60}. If the spin orientation of two anions (*blue and red spheres*) is specified, the remaining two (*green*) are left frustrated. (**b**) Fco lattice of $(NH_3)A_3C_{60}$. Spin frustration is suppressed due to the structural distortion and all spin orientations can be uniquely determined. (**c**) A15 lattice variant of A_3C_{60}. If the spin orientation of the central anion is defined, those of the remaining eight anions are automatically determined

interfullerene separation, orientational order/disorder, valence state, band filling, orbital degeneracy, low-symmetry distortions and metal–C_{60} interactions [18–26].

3 Lattice Expansion and Symmetry Lowering

An important issue arising from the correlation between T_c and interfullerene separation or (U/W), depicted in Fig. 2, is the availability of new materials with large lattice parameters. In principle, this may allow one to address the question of how high a T_c can be experimentally achieved in the fullerides before a transition to a Mott insulating state due to the accompanying band narrowing eventually leads to electron localization. In comparison with the electronic phase diagrams of other families of superconducting materials such as the high-T_c cuprates [27], the Fe pnictides/chalcogenides [28], the heavy fermions [29] and the organic superconductors [30], Fig. 2b indicates that there is a vast area of the alkali metal–C_{60} phase field, which has remained chemically and physically unexplored.

The way this issue was originally addressed experimentally was to attempt to access expanded fulleride structures by introducing large structural spacers comprising alkali ions solvated with neutral molecules, such as ammonia and methylamine, into the interstitial sites of the fullerene structure. As W depends sensitively on the interfullerene separation and U is a molecular quantity and does not vary across the A_3C_{60} series, (U/W) should increase markedly as the lattice constant of the expanded fulleride structures increases. This strategy successfully led to the isolation of non-superconducting compositions such as $(NH_3)K_3C_{60}$ [31, 32] and $(CH_3NH_2)K_3C_{60}$ [33] with interfullerene separations significantly exceeding those of the $RbCs_2C_{60}$ superconductor (Fig. 4). However, this also resulted in the reduction of the crystal symmetry from cubic to orthorhombic. This lifts the degeneracy of the t_{1u} orbitals, and the experimentally observed suppression of

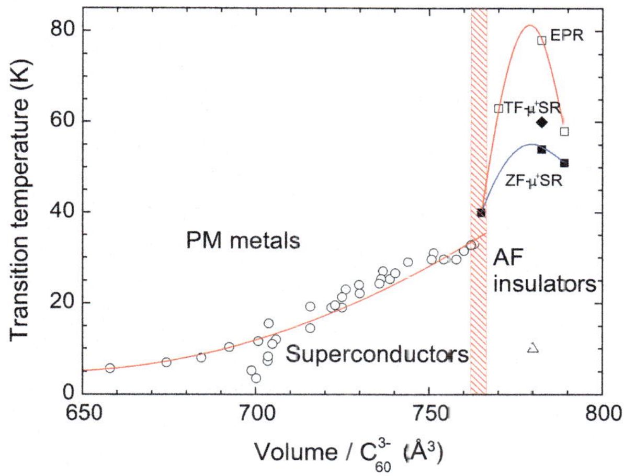

Fig. 4 Electronic phase diagram of fullerides [36] including the T_c values for fcc-structured A_3C_{60} superconductors (*open circles*) and the T_N values for the fco-structured antiferromagnets, $(NH_3)A_3C_{60}$ (*open/solid squares, diamond*) and $(CH_3NH_2)K_3C_{60}$ (*triangle*). The *shaded area* denotes the metal (PM)–insulator (AF) boundary

superconductivity can be traced to the decrease of the critical value of (U/W) for the transition to the Mott insulating state. Nonetheless the electronic ground states of these expanded fullerides were authenticated as those of $S = ½$ LRO AFM insulators bordering the metallic and superconducting phases (Fig. 4) [34–36]. This provided for the first time the hint of an intriguing commonality with the phenomenology in organic and high-T_c superconductors, namely, the proximity of the superconductivity to a magnetic insulating state.

4 Lattice Expansion with Retention of Cubic Symmetry

Despite the insight provided by the expanded ammoniated/aminated alkali fullerides through the suppression of the metallic state and the emergence of AFM insulators at large interfullerene separations [37], there was no definitive experimental evidence for a non-BCS origin for superconductivity in fullerides, where correlation would play a role. The origin of the metal–insulator transition could not be necessarily traced to bandwidth control but could be associated with the symmetry lowering, which lifted the orbital degeneracy. Therefore what has been still missing since 1992 at this stage was an ideal material in which the MIT could be traversed in a purely electronic manner without the complications of structural transitions (Fig. 4). This necessitated that the cubic site symmetry required for orbital degeneracy to survive in all competing electronic states is maintained across the transition. Such a material was finally discovered in 2008 in the form of the A15-structured polymorph of the most expanded, yet binary, cubic fulleride, Cs_3C_{60} (Fig. 1c) [38].

4.1 A15-Structured Cs_3C_{60}

Although Cs_3C_{60} had been a key target composition since the discovery of superconductivity in alkali fullerides, it remained elusive despite numerous attempts to synthesise it by many groups worldwide. The large mismatch between the ionic radius of the large Cs^+ ion and the small size of the tetrahedral interstice in the fulleride fcc structure renders Cs_3C_{60} thermodynamically unstable towards disproportionation into the stable CsC_{60} and Cs_4C_{60} phases. Indeed there were assertions in the literature that fcc Cs_3C_{60} is a nonexistent phase and cannot be formed whatever the method of preparation. Trace superconductivity (shielding fraction $<<0.1\%$) at 40 K under pressure was reported in multiphase samples with nominal composition Cs_3C_{60} in 1995. These were prepared by a low-temperature ammonia solution route, followed by heat treatment of the ammoniated precursors [39]. However, these results remained unconfirmed, and the identity, composition and structure of the putative high-T_c superconducting phase remained unidentified.

Access to nearly phase-pure samples of the Cs_3C_{60} fulleride was achieved by employing a variant of the previously used solution-based routes to alkali fulleride salts. The major difference lied in the use of methylamine as the solvent used to access the nano-sized methylaminated precursor, $(CH_3NH_2)_xCs_3C_{60}$ [38]. Controlled desolvation of the precursor by mild heating afforded the isolation of a polymorph of Cs_3C_{60} with the bcc-based A15 structure (Fig. 1c). The hyperexpanded A15 structure is less dense and incorporates larger distorted tetrahedral interstices than the fcc structural analogue ($r_{A15}(T_d) = 1.46$ Å vs. $r_{fcc}(T_d) = 1.12$ Å) rendering their occupation by the Cs^+ ion less unfavourable. Investigation of its electronic properties reveals that unlike the other A_3C_{60} fullerides, it is not superconducting at ambient pressure. Application of moderate hydrostatic pressure leads to the emergence of bulk superconductivity without any accompanying crystal structure change (Fig. 5). As the applied pressure increases, T_c increases until it reaches a broad maximum at ~7 kbar and then continues to decrease upon further pressurisation. The maximum T_c at 38 K is the highest observed in a bulk molecular material [38]. These results provided the first example of a superconducting $C_{60}{}^{3-}$ fulleride with non-fcc sphere packing. Very importantly, they marked the beginning of the transformation of the entire field of fullerene superconductivity as they revealed for the first time the existence of a key commonality with other unconventional superconductors – the presence of a conduction-bandwidth-controlled superconductivity dome in their electronic phase diagram (Fig. 6, Fig. 7a). DFT calculations in the absence of electron correlations find a smooth evolution of $N(\varepsilon_F)$ and W throughout the region of the phase diagram where the V dependence of T_c changes sign [40]. This suggests that superconducting pairing in fullerides, at least near the MIT, is more complex than BCS-like electron–phonon coupling.

These were followed by the observation that the parent state of the A15 Cs_3C_{60} superconductor is an antiferromagnetic ($S = ½$) insulator with a Néel temperature, T_N of 46 K [41]. The AFM insulator-to-superconductor (SC) transition is purely

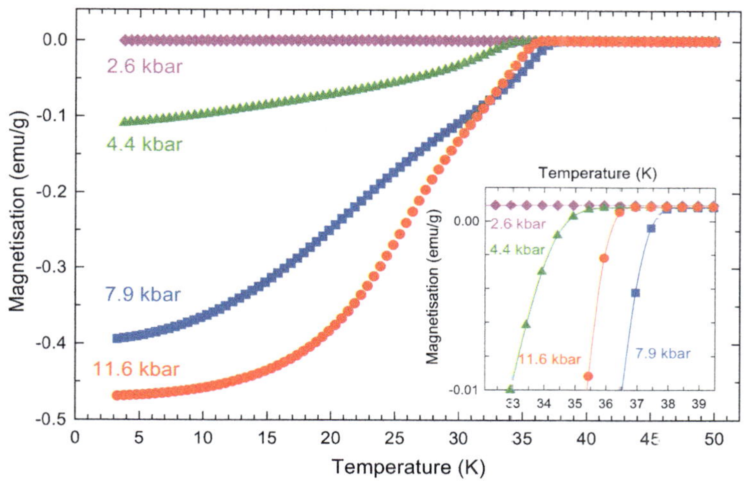

Fig. 5 Superconductivity under pressure in A15-structured Cs_3C_{60} [38]. Temperature dependence of the magnetization, M (ZFC, 20 Oe), at selected pressures. The *inset* shows an expanded view of the data near T_c

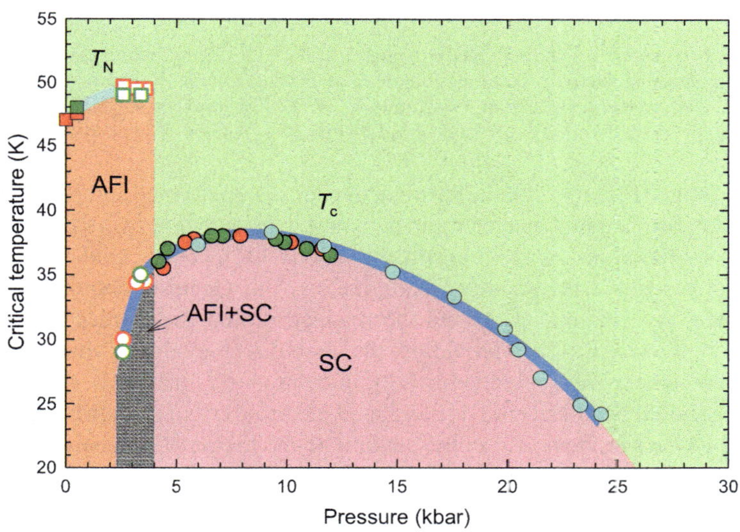

Fig. 6 Electronic phase diagram of A15 Cs_3C_{60} [41] showing the evolution of the Néel temperature, T_N (*squares*) and T_c (*circles*), and thus the isosymmetric transition from the ambient-pressure AFM-insulating state to the high-pressure superconducting state with change in pressure

electronically driven without any crystal structure change by pressure and in the absence of any positional, chemical or orientational disorder. It is of first order as evidenced by the coexistence of the two competing electronic states over a finite

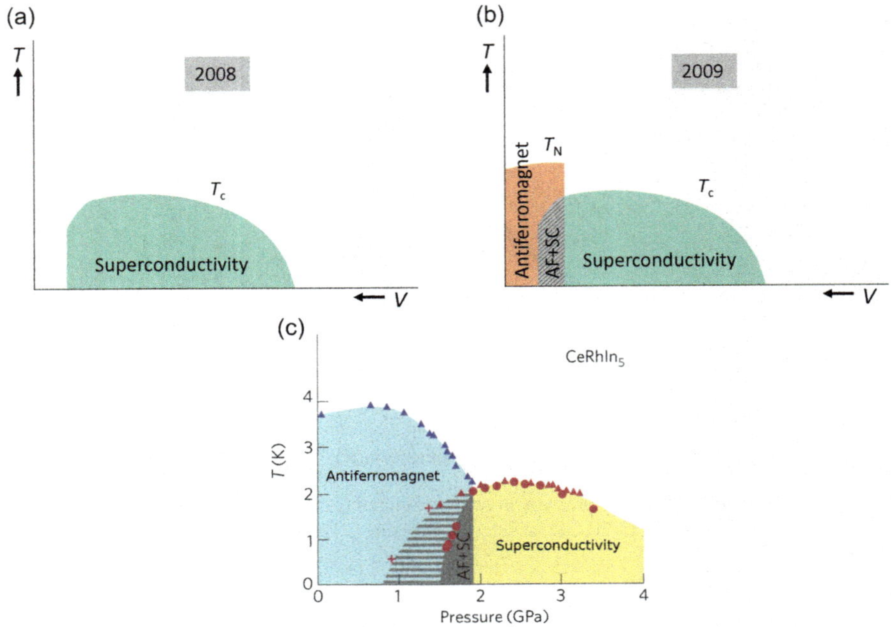

Fig. 7 (**a**, **b**) Schematic electronic phase diagrams of A15 Cs_3C_{60} following (**a**) the first observation of a superconductivity dome in fullerides in 2008 [38] and (**b**) the first evidence that superconductivity emerges out of an isosymmetric LRO AFM insulator in 2009 [41]. (**c**) Phase diagram of the heavy fermion superconductor, $CeRhIn_5$ as a function of applied pressure [29]

pressure range (Fig. 6), and antiferromagnetism is transformed into superconductivity solely by changing an electronic parameter: the extent of overlap between the outer wave functions of the constituent anions [41]. Thus, unlike in the less expanded systems with smaller alkali metals, the electron correlation effects represented by U dominate despite the retained orbital degeneracy in the cubic symmetry of A15 Cs_3C_{60}, providing the first direct evidence that correlation directly competes with superconductivity in cubic A_3C_{60} materials. Therefore the fullerides should be considered as correlated electron systems controlled by Mott–Hubbard physics in the region corresponding to the highest T_c. Theoretical models of the fullerides beyond conventional BCS theory which treated electron correlation effects and electron–phonon coupling on an equal footing predicted correlation enhancement of superconductivity near the MIT and provided significant steps forward in understanding the parent-insulating and correlated high-T_c superconducting states [42–46].

4.2 Fcc-Structured Cs_3C_{60}

Access to nearly phase-pure samples of the long-sought second polymorph of Cs_3C_{60} – fcc Cs_3C_{60} – isostructural with all the other superconducting A_3C_{60} fcc fullerides was then achieved [47] by developing an alternative multistep low-temperature synthesis to that which afforded A15 Cs_3C_{60}. This new synthetic protocol first included reaction between stoichiometric amounts of Cs and C_{60} in liquid ammonia, followed by removal of ammonia under vacuum and controlled thermal treatment. Formation of other competing fulleride was then further minimised by exposure of the intermediate to methylamine vapour, followed by additional thermal treatment at high temperature. The most expanded fcc-structured Cs_3C_{60} fulleride mimics the electronic response of the A15 Cs_3C_{60} polymorph. It is not a superconductor [47, 48] but an AFM ($S = ½$) insulator at ambient pressure and becomes superconducting without any structural change under pressure (Fig. 8a)

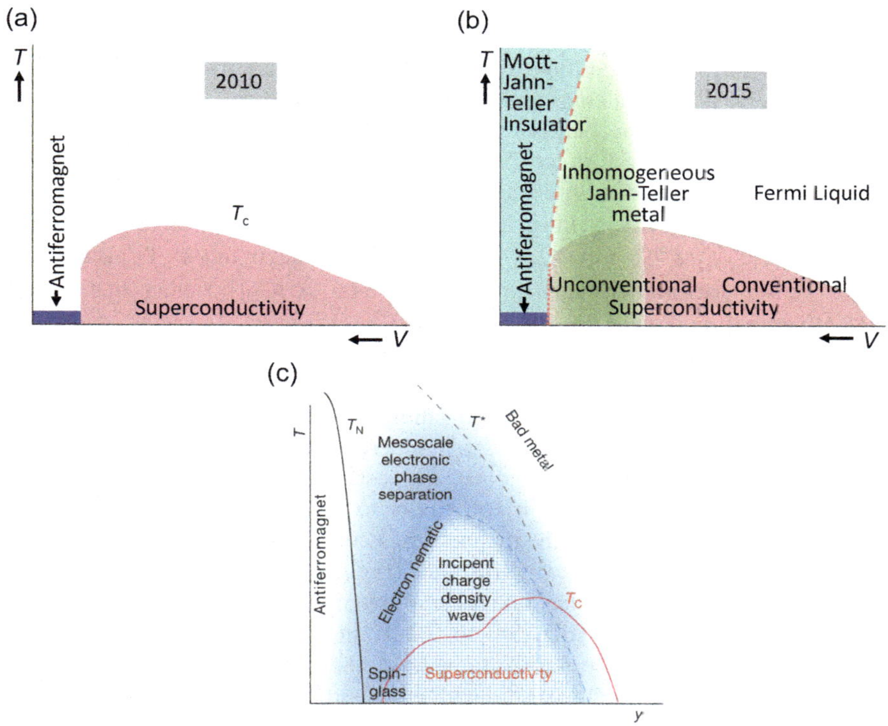

Fig. 8 (**a**, **b**) Schematic electronic phase diagrams of fcc A_3C_{60} showing (**a**) the existence of a LRO AFM insulator bordering the superconductivity dome as discovered in 2010 [47] and (**b**) the presence of a central JT metallic state bordering both the MJT insulating and Fermi liquid states and capping the unconventional part of the superconductivity dome found experimentally in 2015 [54]. (**c**) Schematic phase diagram of the high-T_c superconducting cuprates as a function of hole doping [53]

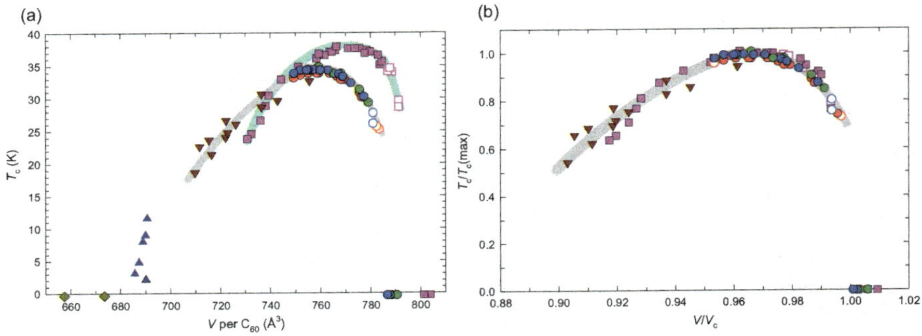

Fig. 9 (**a**) Superconducting transition temperature, T_c, as a function of volume occupied per fulleride anion, V, at low temperature for the fcc (*grey line*) and A15 (bcc packing, *blue line*) A_3C_{60} fullerides [47]. (**b**) Normalised superconducting transition temperature, $T_c/T_c(\max)$, as a function of V divided by the critical value V_c required to produce localization in the A_3C_{60} fulleride structures with fcc- and bcc-sphere packings

[47]. It therefore allowed us to investigate for the first time the role of electron correlations and the competition of the AFM and superconducting states in two high-symmetry lattice packings at the same doping level.

The different lattice packings of $C_{60}{}^{3-}$ change the maximum attainable T_c from 38 K in A15 Cs_3C_{60} to 35 K in fcc Cs_3C_{60} (the highest found in the fcc A_3C_{60} family of materials). Superconductivity emerges out of the insulating state in the fcc phase at a critical value of the packing density, $V_c = 786.0$ Å^3/$C_{60}{}^{3-}$, smaller than that for A15 Cs_3C_{60} (at 796.6 Å^3/$C_{60}{}^{3-}$) (Fig. 9a). The maximum in T_c also occurs for a denser packing arrangement ($V_{max} = 759.6$ Å^3/$C_{60}{}^{3-}$) than that in the A15-structured analogues (at 770.9 Å^3/$C_{60}{}^{3-}$). However, T_c scales identically with packing density for both polymorphs when it is normalised to the critical packing density, V_c, of each polymorph at the MIT. The maximum T_c in both cases occurs when $(V/V_c) \approx 0.967$ and normalisation to the maximum T_c in each family maps all of the known A15 and fcc $C_{60}{}^{3-}$ superconductors onto a single-packing-independent universal curve (Fig. 9b), suggesting that the same pair binding mechanism, which is controlled by proximity to the MIT, is operational in both cases [47]. On the other hand, the magnetism is controlled by the lattice packing – the fcc polymorph is geometrically frustrated (Fig. 3a) – and thus the onset of AFM order is suppressed by more than one order of magnitude from 46 K in the bipartite bcc-packed A15 structure (Fig. 3c) to 2.2 K in the fcc structure. The anion orientational disorder in the fcc case quenches the frustration and allows access to an inhomogeneous highly disordered magnetic state at low temperature as revealed by μSR [47] and also confirmed by specific heat measurements [49].

The two superconducting polymorphs of Cs_3C_{60} have shown that the same electronically active unit ($C_{60}{}^{3-}$) can be packed in distinct arrays, in contrast to the atom-based correlated systems such as the cuprates where the packing of the electronically active ions is essentially identical (two-dimensional square) in all known superconducting families. This has allowed the identification of structure-independent factors controlling the electronic behaviour which are associated with the electronic structure of the molecule itself.

4.3 Jahn–Teller Distortion of C_{60}^{3-} in Cs_3C_{60}

Approaching the electronic properties of Cs_3C_{60} from the molecular side, the C_{60}^{3-} ions are potentially unstable towards JT deformations since the three electrons in the t_{1u} molecular orbitals (Fig. 1b), whose threefold degeneracy is preserved in the solid state because of the high lattice (cubic) symmetry, can couple to the eight H_g vibrations of the molecular units. As the JT-active vibrations are the ones which mediate the electron–phonon coupling, they are intrinsically relevant to the study of the competing insulating and metallic/superconducting states [43]. The JT effect (static or dynamic) grows in importance when charges are localised on the C_{60}^{3-} units, thereby making expanded fullerides, such as the two polymorphs of Cs_3C_{60}, excellent systems to pursue the investigation of the interplay between the molecular JT effect and Mott electron localization. Indeed indirect evidence of the relevance of the JT effect in Mott insulating Cs_3C_{60} is provided by the observation of a low-spin $S = ½$ ground state [41, 47] despite the retention of cubic symmetry which should have favoured a high-spin $S = 3/2$ state for the C_{60}^{3-} anion (Fig. 10a).

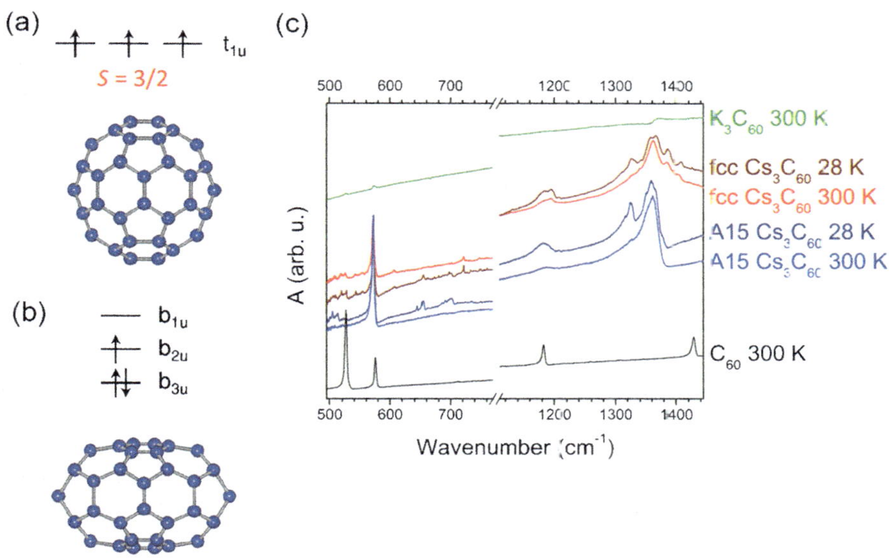

Fig. 10 Molecular orbitals of C_{60}^{3-} for (**a**) undistorted I_h symmetry with an unsplit triply degenerate t_{1u} LUMO and (**b**) JT-distorted D_{2h} symmetry with threefold splitting (b_{1u}, b_{2u} and b_{3u}) of the LUMO. 'Inverted Hund's rule coupling' leads to a low-spin ($S = ½$) state and the splitting of vibrational bands. (**c**) IR spectra of insulating A15 and fcc Cs_3C_{60} at 28 and 300 K. The spectra of insulating C_{60} and metallic K_3C_{60} at 300 K are also shown for comparison [50]

This purely molecular aspect of the C_{60}^{3-} superconductivity story was first explored experimentally by employing infrared spectroscopy, which can detect subtle changes in the shape of the C_{60}^{3-} ion due to the JT distortion [50]. The temperature dependence of the IR spectra in the insulating phases of both fcc and A15 Cs_3C_{60} polymorphs (Fig. 10c) can be explained by the gradual transformation from two temperature-dependent solid-state conformers to a single one, typical for JT systems. The JT effect is dynamic, but the distortion is seen as static on the timescale (10^{-11} s) of the IR measurements. This implies that the interconversion rate between JT conformers is less than 10^{11} s^{-1}. Complementary low-temperature magic-angle-spinning ^{13}C and ^{133}Cs NMR experiments [51] on the fcc Cs_3C_{60} polymorph have set a lower limit of 10^5 s^{-1} for this interconversion rate. These experiments are consistent with the freezing out of the C_{60}^{3-} JT dynamics at low temperature and the development of a t_{1u} electronic orbital glass state in fcc Cs_3C_{60} guided by the merohedral disorder of the fcc structure.

A simple interpretation of the experimental findings to date is that the Coulomb exchange energy, J_H, which favours the highest total spin and orbital angular momentum (Hund's rule) is smaller than the JT interaction, J_{JT}, caused by coupling of the electrons to symmetry-lowering molecular vibrations [43]. This leads to a physical regime of an 'inverted Hund's rule coupling' and accounts for the occurrence of a low-spin state according to the JT effect (Fig. 10b). Therefore the JT effect appears strong enough to overcome Hund's rule to enforce a low-spin state, but at the same time it is not strong enough to break the global cubic symmetry, notwithstanding the merohedral disorder of the fcc phase. Both Cs_3C_{60} polymorphs can be then classified as magnetic MJT insulators with the on-molecule dynamic distortion creating the $S = ½$ ground state which produces the magnetism from which the superconductivity emerges.

A prevailing feature of the theoretical literature on correlated fulleride superconductivity [16, 42–46] has been the necessity to implicate the (dynamic) JT effect in order to account for the robust experimental finding of a low-spin electronic ground state in the Mott insulating precursor state of the superconductors. This rests on the assumption that electron–electron interactions per se always favour the high-spin electronic configuration. It is therefore intriguing that recent theoretical calculations [52] of the t–J model of short-range repulsions on a C_{60} molecule show that Hund's first rule is violated without the necessity of electron–phonon interactions (dynamic Jahn–Teller effect). Moreover, for an average of three electrons per molecule (C_{60}^{3-}), an effective attraction (pair binding) occurs making it favourable to place four electrons on one molecule (C_{60}^{4-}) and two on a second (C_{60}^{2-}) rather than putting three on each. A dominantly electronic mechanism of superconductivity in the fullerides is therefore possible and not precluded by the currently available experimental findings.

4.4 Traversing the Superconductivity Dome in fcc $Rb_xCs_{3-x}C_{60}$

Understanding unconventional superconductivity and identifying the interactions which are important in the superconducting pair binding mechanism go beyond the study of the parent-insulating and superconducting states and crucially rest on their interrelationship with the normal metallic state above T_c. Such an approach has recently dictated the physics of the high-T_c cuprate superconductors with special focus on the 'bad metal' phase which dominates the phase diagram at temperatures well above T_c. Various forms of electronic order, including charge- and spin-density wave orders and nematic charge order are found to coexist in this part of the phase diagram and are inextricably intertwined with high-T_c superconductivity (Fig. 8c) [53]. Although the electronic phase diagram adopted by molecular superconductors such as both polymorphs of the cubic alkali fulleride, Cs_3C_{60} [38, 41, 47, 48], which are continuously tunable by pressure control of the bandwidth W via outer wave function overlap of the constituent molecules resembles those of unconventional superconductors, there has been little attempt to probe the normal metallic state above T_c in the fullerides especially close to the MJT insulator boundary.

The Mott metal–insulator transition has been traversed up to now by applying moderate physical pressure to the AFM parent Cs_3C_{60} insulators. However, substitution of the smaller Rb^+ for the Cs^+ cation in Cs_3C_{60} to afford the family of materials with composition, $Rb_xCs_{3-x}C_{60}$ ($0 \leq x \leq 3$), mimics the effect of physical pressure and shifts the transition to ambient pressure [54]. This permits the use of a variety of techniques to probe the electronic and crystal structure changes associated with the transition in a way that is not possible for Cs_3C_{60} where high pressures are needed. The electronic states of the trivalent molecular fullerides can be tracked with temperature at ambient pressure as the C_{60} packing density, V, varies across the full range of $Rb_xCs_{3-x}C_{60}$ compositions, allowing quasi-continuous access to the entire bandwidth-controlled phase diagram (Figs. 8b and 11) [54]. This extends from a strongly correlated AFM MJT insulator (Cs_3C_{60}) to a conventional metal (Rb_3C_{60}).

Contraction of the $x = 0$ MJT insulator first destroys the Mott insulator and yields an unconventional metal in which correlations sufficiently slow carrier hopping to allow the molecular JT distortions to survive; the local heterogeneities then gradually disappear with recovery of conventional metallic behaviour. The states at the two extremes of the phase diagram (Fig. 11) are thus straddled by an intermediate metallic phase where short-range quasi-localised electron behaviour associated with the intramolecular JT effect coexists with metallicity. This new state of matter, which was termed a JT metal [54], is characterised by both molecular (dynamically JT-distorted $C_{60}{}^{3-}$ ions) and free-carrier (electronic continuum) features as revealed by IR spectroscopy (Fig. 12). The JT metal exhibits a strongly enhanced spin susceptibility relative to that of a conventional Fermi liquid, characteristic of the importance of strong electron correlations. Finally there is a

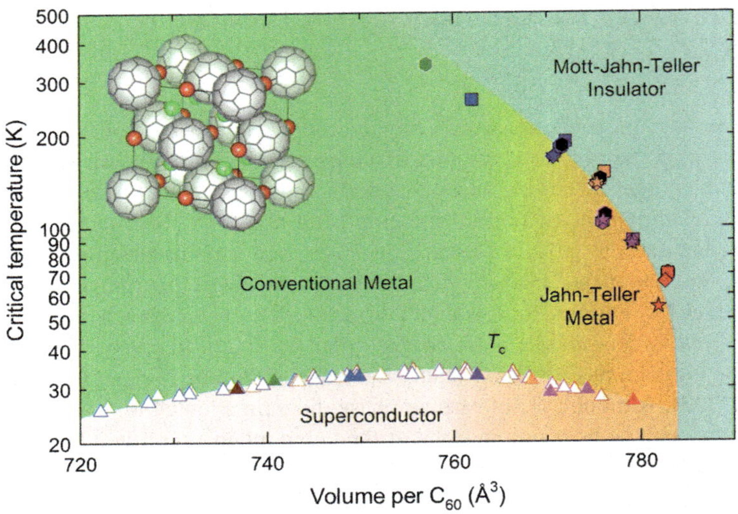

Fig. 11 Global electronic phase diagram of fcc-structured (inset) $Rb_xCs_{3-x}C_{60}$ showing the evolution of T_c (ambient P, *solid triangles*; high P, *unfilled triangles*) and the MJT insulator-to-JT metal crossover temperature, T' (synchrotron X-ray diffraction, *squares*; $\chi(T)$, *stars*; ^{13}C, ^{87}Rb and ^{133}Cs NMR spectroscopy, hexagon with *white*, coloured and *black* edges, respectively; IR spectroscopy, *diamonds*), as a function of V [54]. Within the metallic (superconducting) regime, gradient shading from *orange* to *green* schematically illustrates the JT metal to conventional metal (unconventional to weak-coupling BCS conventional superconductor) crossover

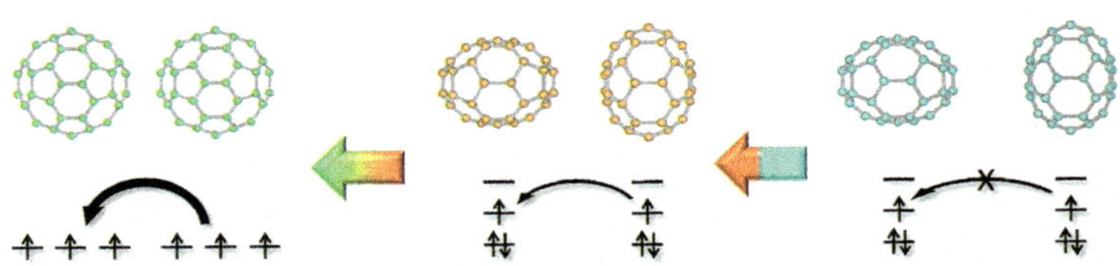

Fig. 12 Change in electronic state of the fulleride solids with change in volume per C_{60} [54]. Schematic depictions of the JT molecular distortion of the fullerene units in the MJT insulator (*blue molecules*) and the JT metal (*yellow molecules*), their respective molecular electronic structure (lifting of the orbital degeneracy due to the distortion) and the resulting intermolecular hopping of the electrons (prohibited in the insulator, weak hopping in the JT metal). This situation contrasts with the behaviour of the conventional metal where hopping is allowed, the orbital degeneracy is retained and the molecules are undistorted (*green molecules*)

crossover from the JT metal to a conventional Fermi liquid upon moving from the Mott boundary towards the underexpanded regime with the distortion arising from JT effect disappearing and the electron mean-free path extending to more than a few intermolecular distances.

The boundary with the JT metal directly affects superconductivity: the (multi-band single-gap) s-wave fulleride superconductors forming from the

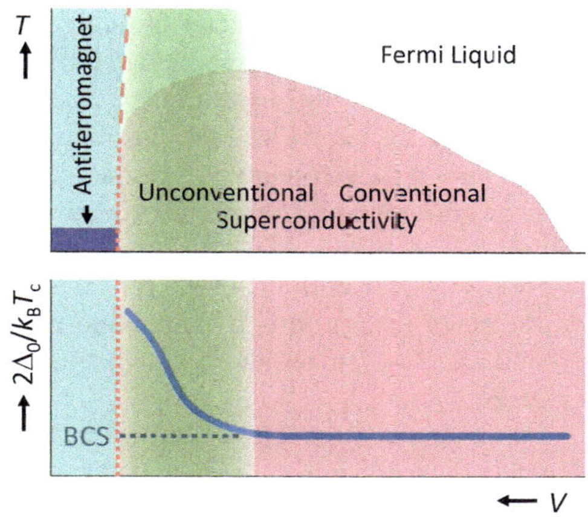

Fig. 13 Contrasting dependence of the normalised superconducting gap, $2\Delta_0/k_BT_c$, and of the superconducting transition temperature, T_c, on fulleride packing density, V

underexpanded conventional metals where effects of the JT metal are minimal display ratios of superconducting gaps to T_c, $2\Delta_0/k_BT_c$, characteristic of weak-coupling BCS superconductors (i.e. ~3.52); on the other hand, although *s*-wave symmetry is retained, the coupling strength is anomalously large, with this ratio approaching and exceeding 5, for the optimally and overexpanded superconductors emerging from JT metals upon cooling (Fig. 13) [54]. Moreover, the superconducting gap does not correlate with T_c in the overexpanded regime, where molecular features play a dominant role in producing the unconventional superconductivity. In contrast to the dome-shaped dependence of T_c on packing density, V, the gap increases monotonically with interfullerene separation. Notably, the maximum T_c occurs at the crossover between the two types of gap behaviour. Such a picture is also encountered as the gap magnitude varies with V upon the application of external pressure for both Cs_3C_{60} polymorphs [55, 56] but contrasts sharply with the situation established in the (single-band single-gap *d*-wave) cuprates and (multi-band multi-gap *s*-wave) iron pnictides. In the both the latter families, the gap magnitudes generally first increase with doping in the underdoped regime, varying in a similar manner to T_c, but then they soon reach a maximum and remain at the same size for larger doping levels [57, 58].

5 Conclusions

Understanding how electron–electron interactions are controlled near the metal–insulator crossover region is of generic significance for all strongly correlated systems, and, in this regard, the study of hyperexpanded Cs_3C_{60} in which fulleride frontier orbital degeneracy and geometrical frustration may be pressure tuned to access insulating, metallic and superconducting states is of paramount importance. Contrary to long-held beliefs, the fullerides emerged as simple members of the unconventional superconductivity family, i.e. the pairing interaction is something other than simply the conventional BCS electron–phonon interaction. This was demonstrated by showing that in Cs_3C_{60}, superconductivity emerges upon applied pressure out of an AFM Mott insulating state and displays an unconventional behaviour – a superconductivity dome – explicable by the prominent role of strong electron correlations. The parent-insulating state involves JT distortion of the anion, driven by coupling of the localised t_{1u}^3 electrons to intramolecular phonons, implying a controlling role for the molecular electronic structure. The on-molecule distortion is dynamic and may be responsible in creating the $S = ½$ ground state which produces the magnetism and from which superconductivity emerges.

However, although the similarities with the electronic phase diagrams of high-T_c superconductors such as the cuprates are striking, there existed no information on the metallic state out of which both the insulator and superconductor emerge. The relation of this parent state to both the pairing mechanism and the proximate antiferromagnetic insulator is at the heart of any attempt to understand high-temperature unconventional superconductivity and has been feverishly pursued for decades in the cuprates. The high symmetry and structural simplicity of the fullerides are here advantageous in pursuing a direct connection from (molecular) building unit to extended structure properties. The critical synthetic advance achieved was the identification of a new family of fullerides ($Rb_xCs_{3-x}C_{60}$) that allowed one to traverse the Mott metal–insulator transition at ambient pressure, thus enabling the deployment of a panoply of techniques to probe the electronic and crystal structure changes associated with this transition. This led to the discovery of an anomalous correlated metallic phase – the Jahn–Teller metal – which dominates the phase diagram at temperatures well above T_c and is the normal state at the maximum in T_c, where the superconducting pairing crosses over from conventional weak coupling to unconventional strong coupling. The JT metal is a dynamic, microscopically heterogeneous coexistence of itinerant metallic electrons with localised electrons, which produce JT on-molecule distortions. This electronic state is the parent of an unconventional strongly coupled superconductor and fades away into a conventional Fermi liquid metal and weak-coupling BCS superconductor when the molecular signatures disappear with lattice contraction. The optimal T_c in the fullerides – highest for any molecular material – associated with a strongly coupled (or extremely stable) Cooper pair is found at the boundary between unconventional and conventional behaviours, where the balance between molecular (JT distortion) and extended lattice (itinerant electrons) features of the electronic structure is optimised.

Acknowledgements This work was sponsored by the 'World Premier International (WPI) Research Center Initiative for Atoms, Molecules and Materials', Ministry of Education, Culture, Sports, Science, and Technology of Japan.

References

1. Kroto HW, Heath JR, O'Brien SC, Curl RF, Smalley RE (1985) Nature 318:162
2. Iijima S (1991) Nature 354:56
3. Novoselov KS, Geim AK, Morozov SV, Jiang D, Zhang Y, Dubonos SV, Grigorieva IV, Firsov AA (2004) Science 306:666
4. Margadonna S, Iwasa Y, Takenobu T, Prassides K (2004) Struct Bond 109:127
5. Tanigaki K, Ebbesen TW, Saito S, Mizuki J, Tsai JS, Kubo Y, Kuroshima S (1991) Nature 352:222
6. Haddon RC, Hebard AF, Rosseinsky MJ, Murphy DW, Duclos SJ, Lyons KB, Miller B, Rosamilia JM, Fleming RM, Kortan AR, Glarum SH, Makhija AV, Muller AJ, Eick RH, Zahurak SM, Tycko R, Dabbagh G, Thiel FA (1991) Nature 350:320
7. Hebard AF, Rosseinsky MJ, Haddon RC, Murphy DW, Glarum SH, Palstra TTM, Ramirez AP, Kortan AR (1991) Nature 350:600
8. Holczer K, Klein O, Huang SM, Kaner RB, Fu KJ, Whetten RL, Diederich F (1991) Science 252:1154
9. Rosseinsky MJ, Ramirez AP, Glarum SH, Murphy DW, Haddon RC, Hebard AF, Palstra TTM, Kortan AR, Zahurak SM, Makhija AV (1991) Phys Rev Lett 66:2830
10. Stephens PW, Mihaly L, Lee PL, Whetten RL, Huang SM, Kaner R, Deiderich F, Holczer K (1991) Nature 351:632
11. Fleming RM, Ramirez AP, Rosseinsky MJ, Murphy DW, Haddon RC, Zahurak SM, Makhija AV (1991) Nature 352:787
12. Schluter M, Lannoo M, Needels M, Baraff GA, Tomanek D (1992) Phys Rev Lett 68:526
13. Chakravarty S, Gelfand MP, Kivelson S (1991) Science 254:970
14. Lof RW, VanVeenendaal MA, Koopmans B, Jonkman HT, Sawatzky GA (1992) Phys Rev Lett 68:3924
15. Koch E, Gunnarsson O, Martin RM (1999) Phys Rev Lett 83:620
16. Capone M, Fabrizio M, Castellani C, Tosatti E (2002) Science 296:2364
17. Iwasa Y (2010) Nature 466:191
18. Gunnarsson O (1997) Rev Mod Phys 69:575
19. Prassides K (1997) Curr Opin Solid State Mater Sci 2:433
20. Rosseinsky MJ (1998) Chem Mater 10:2665
21. Kosaka M, Tanigaki K, Prassides K, Margadonna S, Lappas A, Brown CM, Fitch AN (1999) Phys Rev B 59:R6628
22. Forro L, Mihaly L (2001) Rep Prog Phys 64:649
23. Margadonna S, Prassides K (2002) J Solid State Chem 168:639
24. Iwasa Y, Takenobu T (2003) J Phys Condens Matter 15:R495
25. Gunnarsson O (2004) Alkali-doped fullerides: narrow-band solids with unusual properties. World Scientific, Singapore
26. Gunnarsson O, Han JE, Koch E, Crespi VH (2005) Struct Bond 114:71
27. Keimer B, Kivelson SA, Norman MR, Uchida S, Zaanen J (2015) Nature 518:179
28. Hashimoto K, Cho K, Shibauchi T, Kasahara S, Mizukami Y, Katsumata R, Tsuruhara Y, Terashima T, Ikeda H, Tanatar MA, Kitano H, Salovich N, Giannetta RW, Walmsley P, Carrington A, Prozorov R, Matsuda Y (2012) Science 336:1554
29. Knebel G, Aoki D, Braithwaite D, Salce B, Flouquet J (2006) Phys Rev B 74:020501
30. Ardavan A, Brown S, Kagoshima S, Kanoda K, Kuroki K, Mori H, Ogata M, Uji S, Wosnitza J (2012) J Phys Soc Jpn 81:011004
31. Rosseinsky MJ, Murphy DW, Fleming RM, Zhou O (1993) Nature 364:425
32. Takenobu T, Muro T, Iwasa Y, Mitani T (2000) Phys Rev Lett 85:381
33. Ganin AY, Takabayashi Y, Bridges CA, Khimyak YZ, Margadonna S, Prassides K, Rosseinsky MJ (2006) J Am Chem Soc 128:14784
34. Prassides K, Margadonna S, Arcon D, Lappas A, Shimoda H, Iwasa Y (1999) J Am Chem Soc 121:11227

35. Takabayashi Y, Ganin AY, Rosseinsky MJ, Prassides K (2007) Chem Commun 2007:870
36. Arvanitidis J, Papagelis K, Takabayashi Y, Takanobu T, Iwasa Y, Rosseinsky MJ, Prassides K (2007) J Phys Condens Matter 19:386235
37. Uemura YJ (2009) Nat Mater 8:253
38. Ganin AY, Takabayashi Y, Khimyak YZ, Margadonna S, Tamai A, Rosseinsky MJ, Prassides K (2008) Nat Mater 7:367
39. Palstra TTM, Zhou O, Iwasa Y, Sulewski PE, Fleming RM, Zegarski BR (1995) Solid State Commun 93:327
40. Darling GR, Ganin AY, Rosseinsky MJ, Takabayashi Y, Prassides K (2008) Phys Rev Lett 101:136404
41. Takabayashi Y, Ganin AY, Jeglič P, Arčon D, Takano T, Iwasa Y, Ohishi Y, Takata M, Takeshita N, Prassides K, Rosseinsky MJ (2009) Science 323:1585
42. Han JE, Gunnarsson O, Crespi VH (2003) Phys Rev Lett 90:167006
43. Capone M, Fabrizio M, Castellani C, Tosatti E (2009) Rev Mod Phys 81:943
44. Akashi R, Arita R (2013) Phys Rev B 88:054510
45. Murakami Y, Werner P, Tsuji N, Aoki H (2013) Phys Rev B 88:125126
46. Nomura Y, Sakai S, Capone M, Arita R (2015) Sci Adv 1, e1500568
47. Ganin AY, Takabayashi Y, Jeglič P, Arčon D, Potočnik A, Baker PJ, Ohishi Y, McDonald MT, Tzirakis MD, McLennan A, Darling GR, Takata M, Rosseinsky MJ, Prassides K (2010) Nature 466:221
48. Ihara Y, Alloul H, Wzietek P, Pontiroli D, Mazzani M, Riccó M (2010) Phys Rev Lett 104:256402
49. Kasahara Y, Takeuchi Y, Itou T, Zadik RH, Takabayashi Y, Ganin AY, Arcon D, Rosseinsky MJ, Prassides K, Iwasa Y (2014) Phys Rev B 90:014413
50. Klupp G, Matus P, Kamarás K, Ganin AY, McLennan A, Rosseinsky MJ, Takabayashi Y, McDonald MT, Prassides K (2012) Nat Commun 3:912
51. Potocnik A, Ganin AY, Takabayashi Y, McDonald MT, Heinmaa I, Jeglic P, Stern R, Rosseinsky MJ, Prassides K, Arcon D (2014) Chem Sci 5:3008
52. Jiang H-C, Kivelson S (2015) arXiv:1510.04704
53. Fradkin E, Kivelson SA (2012) Nat Phys 8:864
54. Zadik RH, Takabayashi Y, Klupp G, Colman RH, Ganin AY, Potočnik A, Jeglič P, Arčon D, Matus P, Kamarás K, Kasahara Y, Iwasa Y, Fitch AN, Ohishi Y, Garbarino G, Kato K, Rosseinsky MJ, Prassides K (2015) Sci Adv 1, e150059
55. Potočnik A, Krajnc A, Jeglič P, Takabayashi Y, Ganin AY, Prassides K, Rosseinsky MJ, Arčon D (2014) Sci Rep 4:4265
56. Wzietek P, Mito T, Alloul H, Pontiroli D, Aramini M, Riccò M (2014) Phys Rev Lett 112:066401
57. Tanaka K, Lee WS, Lu DH, Fujimori A, Fujii T, Terasaki I, Scalapino DJ, Devereaux TP, Hussain Z, Shen Z-X (2006) Science 314:1910
58. Xu Y-M, Richard P, Nakayama K, Kawahara T, Sekiba Y, Qian T, Neupane M, Souma S, Sato T, Takahashi T, Luo H-Q, Wen H-H, Chen G-F, Wang N-L, Wang Z, Fang Z, Dai X, Ding H (2011) Nat Commun 2:392

Struct Bond (2016) 172: 139–158
DOI: 10.1007/430_2015_212

Published online: 10 March 2016

Synthetic Bioinorganic Chemistry: Scorpionates Turn 50

Daniel Rabinovich

Abstract The coordination chemistry of scorpionates, anionic tripodal boron-centered ligands that readily bind to virtually all main group, transition, and f-block metals, has had a profound effect in the development of synthetic bioinorganic chemistry during the past 50 years. Highlighted in this article are some of the most meaningful results published in the area, primarily in the preparation of structural and functional model compounds for the active sites in metalloproteins and metalloenzymes, a theme that encompasses topics as diverse as nitrogen fixation and heavy metal poisoning. It has also been 50 years since the genesis of *Structure and Bonding*, so let this account also be a celebration of the many contributions that the staunch serial has had on the entire field of inorganic chemistry during the past half century.

Keywords Bioinorganic chemistry • Carbonic anhydrase • Copper • Gold • Iron • Manganese • Mercury • Molybdenum • Nickel • Scorpionates • Silver • Synthesis • Thiolate complexes • Tris(carbene)borates • Tris(mercaptoimidazolyl)borates • Tris(phosphino)borates • Tris(pyrazolyl)borate • Tris(thioether)borates • Tris (triazolyl)borates • Tungsten • Vanadium • Zinc

Contents

The original version of this chapter was revised: Figure 10 of this chapter was displayed incorrect. The erratum to this chapter is available at DOI: 10.1007/430_2016_213.

D. Rabinovich (✉)
Department of Chemistry, The University of North Carolina at Charlotte, 9201 University City Boulevard, Charlotte, NC 28223, USA
e-mail: drabinov@uncc.edu

Abbreviations

Ad	Adamantyl
ALAD	5-Aminolevulinate dehydratase
Bp	Bis(pyrazolyl)borate
Bu^t	*Tert*-butyl
CA	Carbonic anhydrase
Cp_2Co	Bis(cyclopentadienyl)cobalt (cobaltocene)
Cys	Cysteine
DMSO	Dimethyl sulfoxide
Et	Ethyl
His	Histidine
HSAB	Hard and soft acids and bases
Im	Imidazol-2-ylidene
IR	Infrared
LADH	Liver alcohol dehydrogenase
Me	Methyl
Mes	Mesityl (2,4,6-trimethylphenyl)
MRI	Magnetic resonance imaging
NHC	N-Heterocyclic carbene
NHSe	N-Heterocyclic selone
NHT	N-Heterocyclic thione
Ph	Phenyl
Pr^i	Isopropyl
THF	Tetrahydrofuran
Tm^R	Tris(mercaptoimidazolyl)borate
To^R	Tris(oxoimidazolyl)borate
$Tp^{RR'}$	Tris(pyrazolyl)borate
Tse^R	Tris(selenoimidazolyl)borate
Tt^R	Tris(thioether)borate
$Ttz^{RR'}$	Tris(triazolyl)borate

1 Introduction

1.1 Early Days of Bioinorganic Chemistry

In 1966, when the publication of the *Structure and Bonding* started, the field of bioinorganic chemistry was in its early stages of development. Only 2 years earlier Dorothy Crowfoot Hodgkin (1910–1994) had received the Nobel Prize in Chemistry "for her determinations by X-ray techniques of the structures of important biochemical substances," including that of vitamin B_{12} [1], and it was also only a few months after Barnett Rosenberg (1926–2009) had published his seminal paper on the anticancer properties of cisplatin [2]. Relatively little was known at the time about the mechanism of biological nitrogen fixation, and the first dinitrogen complex, $[Ru(NH_3)_5(N_2)]^{2+}$, had just been reported by Allen and Senoff [3]. Medicinal inorganic chemistry was also in an embryonic state: the use of gadolinium complexes as magnetic resonance imagining (MRI) agents [4–6] and the production of technetium-99m radiopharmaceuticals [7, 8] were still more than a decade away. Interestingly, the term "bioinorganic" wasn't even in the public domain back then, although things were bound to change dramatically within a few years!

A recent search of *Chemical Abstracts* revealed that the first documented use of the term "bioinorganic" was at an international symposium held at Virginia Tech (in Blacksburg, Virginia, USA) on 22–25 June 1970 (Fig. 1). The event showcased the state of the art in the field and led to the publication of a volume entitled *Bioinorganic Chemistry* in the *Advances in Chemistry Series* published by the American Chemical Society [9]. The book's 19 chapters included insightful reviews on nitrogen fixation [10, 11] and cobalamin chemistry [10], the uptake of oxygen by cobalt(II) complexes [12], the structure and function of metalloenzymes [13], and structural models for iron and copper proteins [14]. In a similar vein, the first peer-reviewed publication to mention the word "bioinorganic" appears to be a 1972 paper by Rayner-Canham and Lever describing simple models of iron sites in important biological systems such as hemes, rubredoxin, and ferredoxins [15]. Bioinorganic chemistry as a discipline was maturing at an accelerated pace, and the first few monographs on the subject appeared during the 1970s [16], including a couple with the word bioinorganic in their titles [17, 18].

1.2 Synthetic Bioinorganic Chemistry

The design and synthesis of small molecules is at the core of many areas of interest in bioinorganic chemistry, from the preparation of chelating agents for metal ion sensors and the diagnostic or treatment of diseases to the generation of model compounds for the active sites in metal-containing proteins and enzymes. Indeed, the synthesis and investigation of coordination complexes that mimic the structure and function of metallobiomolecules is currently one of the most active areas of

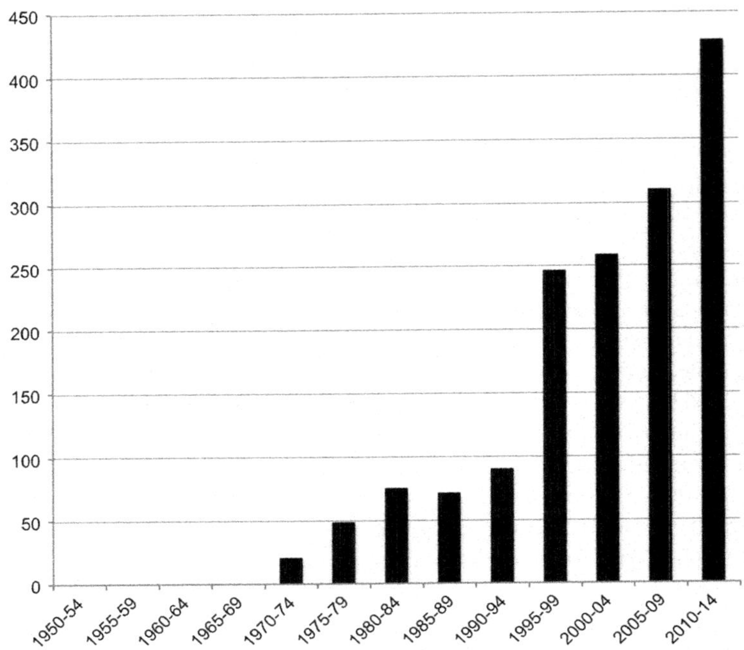

Fig. 1 Results of a search for the term "bioinorganic" in Chemical Abstracts (1950–2014) using the SciFinder Scholar search engine

research in bioinorganic chemistry. The synthetic analogue approach, outlined by Holm some 40 years ago [19], is still today a valuable and effective strategy to investigate the active sites of biological systems that rely on the use of metal ions as prosthetic groups. In essence, the method involves the preparation of low molecular weight metal complexes whose structures, spectroscopic properties, and reactivity closely match those of the actual biological sites of interest. In this regard, the distinction between *corroborative* and *speculative* model compounds synthesized to verify or predict the nature of the metal centers in metallobiomolecules, originally expounded by Hill [20], was subsequently refined by Ibers, Holm, and others [21–23].

1.3 Scorpionate Ligands

Trofimenko's tris(pyrazolyl)borates (Fig. 2), which are also celebrating in 2016 their 50th anniversary [24], are among the most versatile and pervasive ancillary ligands used in inorganic, bioinorganic, and organometallic chemistry. More than 200 different $Tp^{RR'}$ ligands have been synthesized to date, and coordination complexes thereof have been prepared for almost every metal and metalloid in the

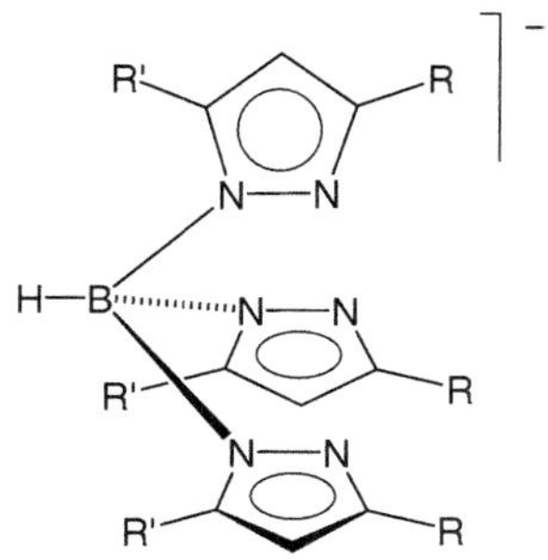

Fig. 2 The original tris(pyrazolyl)borate ($Tp^{RR'}$) ligand system (R, R′=alkyl, aryl)

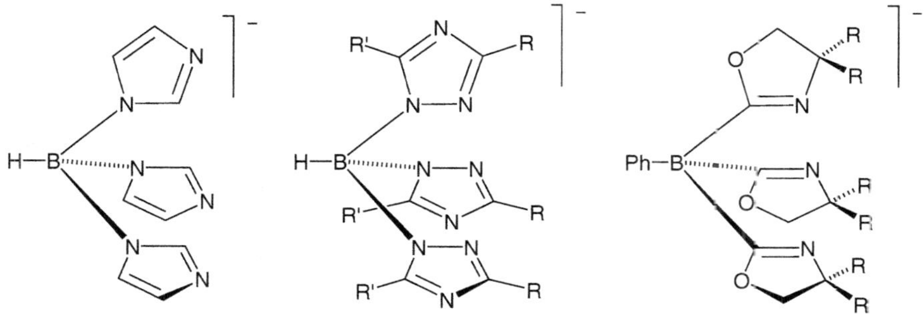

Fig. 3 Tris(imidazolyl)borate, tris(triazolyl)borate, and tris(oxazolinyl)borate ligands

periodic table [25–27]. Their widespread use likely stems from the ease with which they can be usually prepared and how their steric and electronic effects can be altered simply by changing the number and nature of the substituents on the pyrazolyl rings or the boron atom [28–30].

Structural variations on the $Tp^{RR'}$ theme have been readily implemented by incorporating nitrogen donor groups other than pyrazole in the assembly of the borate ligands. For example, the syntheses and reactivity of tris(azolyl)borate ligands derived from imidazole [31], oxazolines [32], and triazoles [33] have been reported (Fig. 3). In addition, related ligands based on carbene [34, 35], phosphine [36, 37], thioether [38], thione [39–41], and selone [39] donor moieties have been studied during the past 20 years or so and will be discussed in more detail below (Fig. 4).

This article will highlight certain facets of the bioinorganic chemistry of prototypical scorpionate ligands, i.e., those that are boron centered, anionic, and tridentate and exhibit threefold symmetry (C_3 or C_{3v}) when bound to a metal center. Thus, largely excluded from the discussion below are bidentate and tetradentate ligands (i.e., heteroscorpionates). Likewise, metal complexes of tris(pyrazolyl)aluminates [42] and those derived from neutral analogues of scorpionates centered on carbon

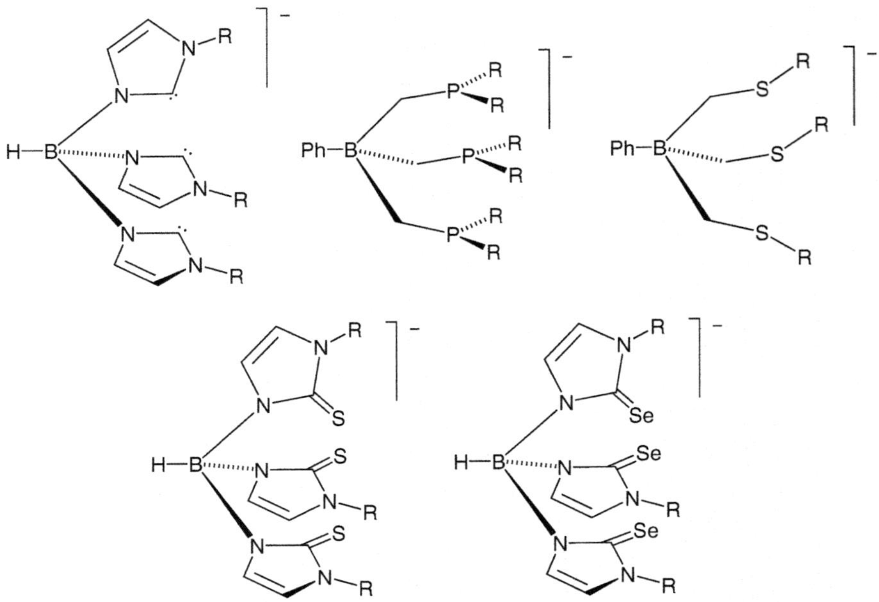

Fig. 4 Scorpionate ligands derived from N-heterocyclic carbene (NHC), tertiary phosphine, thioether, N-heterocyclic thione (NHT), and N-heterocyclic selone (NHSe) donor groups

[i.e., tris(pyrazolyl)methanes] [43–45] or silicon [i.e., tris(pyrazolyl)silanes] [46] will not be considered, even though such ligands (especially the former) have an extensive coordination chemistry, including applications in bioinorganic chemistry.

2 Hard Scorpionates

2.1 Tris(pyrazolyl)borates

Tris(pyrazolyl)borates and, to a lesser degree, the closely related bis(pyrazolyl) borate ($Bp^{RR'}$) ligands have played a pivotal role in the development of bioinorganic chemistry during the past few decades, particularly in the area of enzyme modeling. Every biologically relevant first-row transition metal has been the subject of studies in which a $Tp^{RR'}$ ligand provided an N_3 donor set resembling the imidazolyl groups of three histidine residues. For example, vanadyl bis(phenolate) complexes of general formula $(Tp^{Me2})VO(OAr)_2$ [47] have been synthesized as potential models for the active site in bromoperoxidase, a vanadium(V)-dependent enzyme that catalyzes the bromination of a variety of organic substrates [48].

Molybdenum, the only second-row transition metal recognized as an essential trace element in human nutrition [49], is present as a cofactor in most nitrogenases

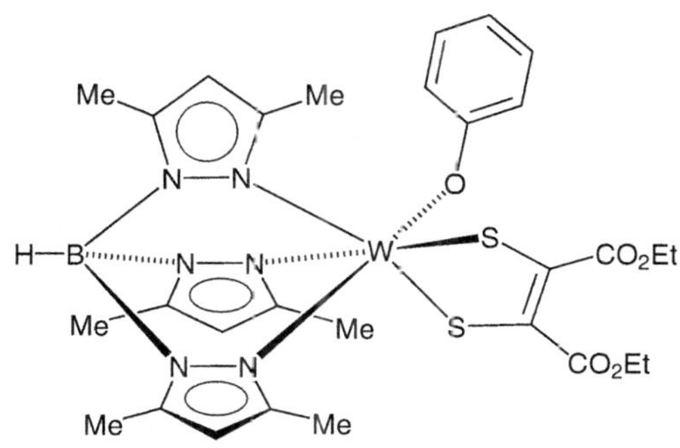

Fig. 5 Representation of the tungsten(IV) complex $(Tp^{Me2})W(OPh)\{S_2C_2(CO_2Et)_2\}$

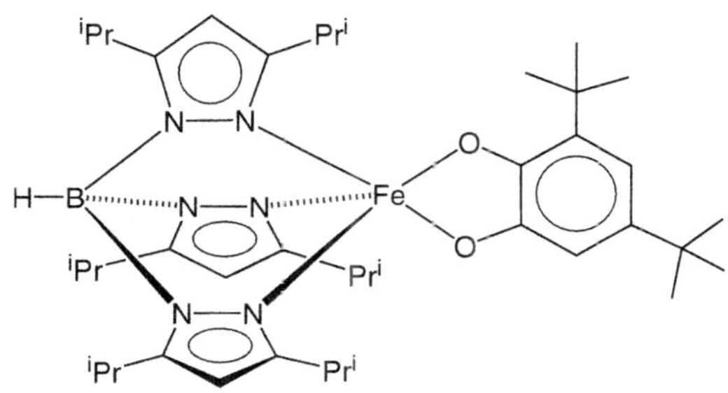

Fig. 6 Structural and functional model of cathecol dehydrogenase

and in more than 50 different redox-active enzymes that contain a molybdopterin unit and catalyze several hydroxylation and oxygen atom-transfer reactions [50, 51]. Multiple studies have been conducted over the years whereby $Tp^{RR'}$ ligands are applied to model the molybdenum centers in sulfite oxidase, xanthine oxidase/dehydrogenase, nitrate reductase, and dimethyl sulfoxide (DMSO) reductase enzymes [52–54]. In a similar vein, the active centers in tungsten-containing enzymes such as aldehyde oxidoreductase and formate dehydrogenase [55, 56] have been modeled using a variety of $Tp^{RR'}$ complexes [52, 57], as exemplified by an ene-1,2-dithiolate complex (Fig. 5) described by Young and coworkers [58].

The catalytic activity of manganese superoxide dismutase and other dinuclear redox-active manganese enzymes [59–61] has been investigated using complexes such as the benzoate derivative $(Tp^{iPr2})Mn(O_2CPh)$ [62]. Similarly, the formation of a stable peroxomanganese(III) species derived from $[(Tp^{iPr2})Mn(\mu\text{-}O)]_2$, which mimics certain aspects of the binding of molecular oxygen to the "distal" histidine residue in hemoglobin/myoglobin, was reported a few years ago [63].

One of the earliest uses of scorpionate ligands in biomimetic chemistry involved the synthesis of oxo-bridged diiron complexes of formula $[TpFe]_2(\mu\text{-}O)(\mu\text{-}O_2CR)_2$ [64, 65] as synthetic models for hemerythrin, a nonheme oxygen-binding protein present in several marine invertebrates [66]. Similarly, the cathecolato complex $(Tp^{iPr2})Fe(O_2C_6H_2Bu^t{}_2)$ (Fig. 6) was prepared as a structural and functional model of catechol dehydrogenase, and its reactivity towards O_2 was investigated [67].

Nonheme iron oxygenases [68] catalyze a variety of complex oxidation processes such as the oxidative cleavage of carbon–carbon bonds and several monohydroxylation and dihydroxylation reactions [69–73]. Mononuclear iron complexes of α-ketocarboxylates and α-hydroxycarboxylates supported by various Tp $^{RR'}$ ligands [as well as tetradentate tris(pyridyl)amines] have been synthesized as functional models for nonheme iron oxygenases that mediate the oxidative decarboxylation of such substrates in the presence of O_2 [74]. Significantly, mechanistic studies suggest that iron(III)-superoxo and iron(IV)-oxo species are implicated as key oxidants in these reactions.

A large number of copper scorpionate complexes have been prepared as model compounds for the binding sites in proteins and enzymes involved in dioxygen binding and activation (e.g., tyrosinase, hemocyanin, catechol oxidase, galactose oxidase) as well as in blue copper proteins and copper nitrite reductase [75–78]. For example, the use of sterically demanding Tp ligands bearing isopropyl or phenyl substituents in the 3- and 5-positions has allowed the isolation of peroxo-bridged dicopper complexes (Fig. 7) [79]. These complexes show remarkably similar spectroscopic properties to oxyhemocyanin and oxytyrosinase and suggest that dioxygen may bind in a side-on fashion in the enzymes.

The bulky "second-generation" $Tp^{RR'}$ ligands (R=Bu^t, R′=H; R=R′=Ph) [80] were also used by Tolman and coworkers to isolate the first examples of mononuclear copper nitrosyl complexes, ($Tp^{RR'}$)Cu(NO) [81]. These complexes, formally $\{CuNO\}^{11}$ in the Enemark–Feltham notation [82], are attractive model compounds for a proposed intermediate in the reduction of nitrite ions by copper nitrite reductase enzymes [83, 84].

The bioinorganic chemistry of zinc has benefitted tremendously from the application of scorpionate ligands to the preparation of synthetic analogues of zinc enzymes, including carbonic anhydrase (CA), carboxypeptidase, thermolysin, alcohol dehydrogenase, and matrix metalloproteinases [85–88]. In particular, carbonic anhydrase, one of the fastest enzymes known and the first one where the presence of zinc was confirmed [89], catalyzes the reversible hydration of carbon dioxide and therefore plays a key role in respiration and CO_2/HCO_3^- equilibration in biological systems (Scheme 1). The tetrahedral zinc center is coordinated by the imidazole

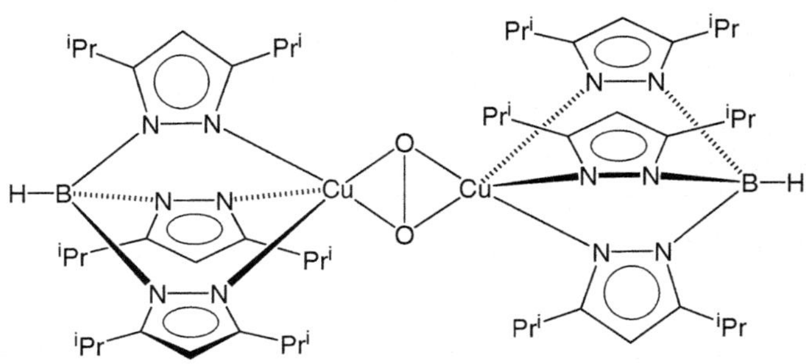

Fig. 7 Structurally characterized peroxo-bridged dicopper complex

Scheme 1 Catalytic cycle for the hydration of CO_2 to bicarbonate (modified from [87])

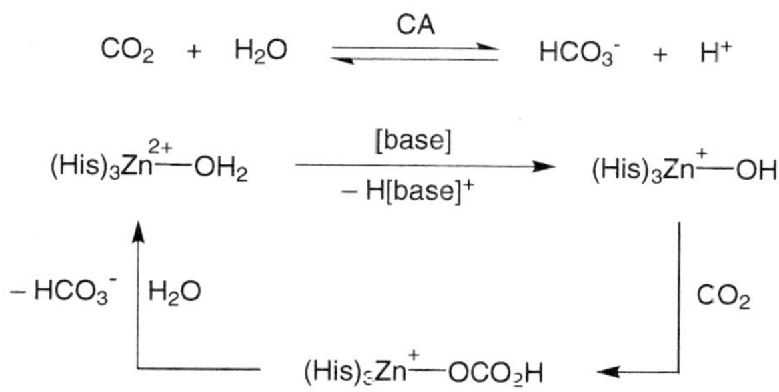

groups of three histidine (His) residues and a water molecule or (in the presence of an external base) a hydroxide ion. Significantly, the isolation of the first monomeric terminal zinc hydroxide complex, $(\mathrm{Tp}^{t\mathrm{Bu,Me}})\mathrm{ZnOH}$, was facilitated by the use of a highly substituted tris(pyrazolyl)borate ligand [90]. Furthermore, the zinc hydroxide complex reacts reversibly with CO_2 to generate a bicarbonate complex that has been characterized by IR spectroscopy [91] and is effectively a functional model for CA since it catalyzes the exchange of oxygen atoms between CO_2 and H_2O.

2.2 *Tris(triazolyl)borates*

The tris(triazolyl)borates ($\mathrm{Ttz}^{\mathrm{RR'}}$), derivatives of 1,2,4-triazoles rather than pyrazoles as the original $\mathrm{Tp}^{\mathrm{RR'}}$ ligands, were also pioneered by Trofimenko [92]. However, their chemistry remains underdeveloped relative to the latter, both in terms of ligand diversity and reactivity studies [33, 93]. The two salient properties of $\mathrm{Ttz}^{\mathrm{RR'}}$ ligands are a consequence of the presence of the additional nitrogen atom in the 4-position, namely the possibility of binding additional metal ions (leading to an increased tendency to form polymeric species) and an enhanced solubility in water.

The first applications of $\mathrm{Ttz}^{\mathrm{RR'}}$ ligands to bioinorganic chemistry involved the preparation of copper coordination polymers that possess some structural features reminiscent of those of hemocyanin (i.e., binding of copper to three His residues) [94]. A copper(I) complex that incorporates both the parent Ttz ligand (R=R′=H) and a water-soluble phosphine, $(\mathrm{Ttz})\mathrm{Cu}\{\mathrm{PPh_2}(p\text{-}\mathrm{C_6H_4CO_2H})\}$, exhibits moderate superoxide scavenging activity [95]. The bulky $\mathrm{Ttz}^{t\mathrm{Bu,Me}}$ ligand has been used to synthesize functional models for copper nitrite reductase, including mononuclear five-coordinate nitrite and nitrate complexes (Fig. 8) that have been structurally characterized [96].

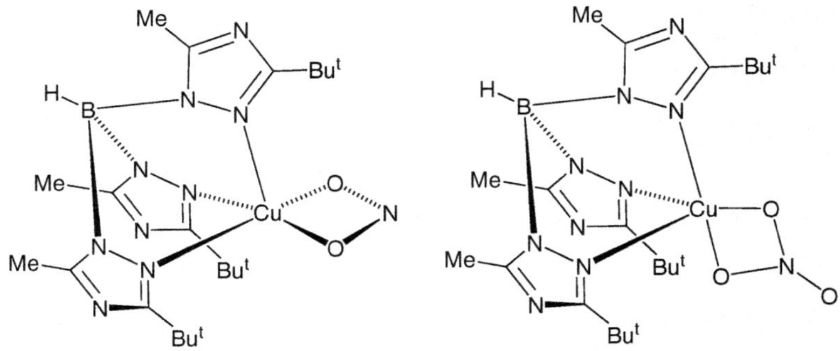

Fig. 8 Distorted square pyramidal (Ttz^{tBu,Me})Cu(NO$_2$) and distorted trigonal bipyramidal (Ttz ^{tBu,Me})Cu(NO$_3$) complexes

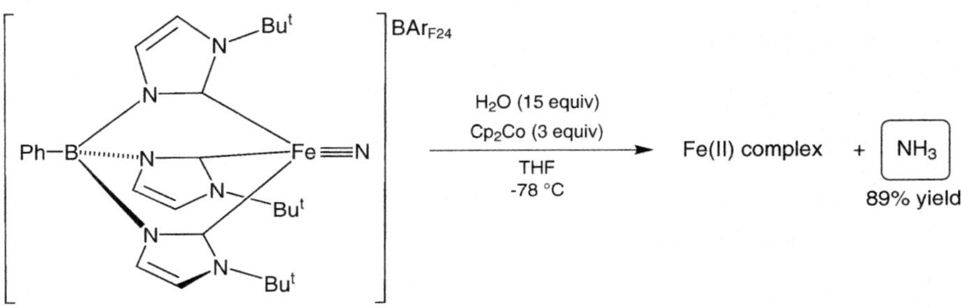

Scheme 2 Release of ammonia from an iron(V) nitrido complex

2.3 Tris(carbene)borates

The coordination chemistry of the tris(carbene)borates, first reported by Fehlhammer in 1996 [34], was very limited [97] until Smith et al. described in 2005 an improved synthetic route to analogous ligands bearing bulkier alkyl and aryl substituents [98]. These ligands, like the ubiquitous N-heterocyclic carbenes (NHCs) [99, 100], are primarily strong σ-donors and only weak π-acceptors [101]. It has also been estimated that, based on comparative studies using IR spectroscopy and cyclic voltammetry, they are even more electron donating than most tertiary alkyl phosphines [35] and, as such, should have the ability to stabilize transition metals in high oxidation states. Accordingly, Smith and coworkers have prepared a range of novel transition metal nitride complexes [102], some of which may represent key intermediates in the catalytic cycle of the nitrogenase enzyme. Significantly, the iron (IV) nitrido complex {PhB(MesIm)$_3$}FeN releases ammonia upon reaction with a hydrogen-atom donor [103]. Subsequently, the same researchers isolated and structurally characterized the remarkable cationic iron(V) nitrido complex [{PhB(^{t}BuIm)$_3$} FeN]$^+$, which also yields ammonia upon treatment with an excess of water under reducing conditions at low temperature (Scheme 2) [104].

3 Bioinorganic Chemistry of Soft Scorpionates

3.1 Tris(thioether)borates

The tetrakis(thioether)borate ligand $[B(CH_2SMe)_4]^-$ and its first two coordination compounds, the anionic molybdenum(0) derivative $[\{B(CH_2SMe)_4\}Mo(CO)_3]^-$ and the homoleptic iron(II) complex $Fe\{B(CH_2SMe)_4\}_2$, were described by Riordan and coworkers in 1994 [105]. Two years later, the first tris(thioether)borate ($PhTt^R$) ligand (R=Me), explicitly designed as a more polarizable, soft version of tris(pyrazolyl)borates, was reported together with its Fe(II), Co(II), and Ni (II) complexes [106]. In agreement with Pearson's hard and soft acids and bases (HSAB) principle [107], these ligands should have a good affinity for late transition metals in low oxidation states. Hence, many interesting nickel(I) and copper (I) complexes thereof have been isolated since then, including a tetranuclear compound $[Cu(PhTt^{Me})]_4$ [108] and pseudotetrahedral derivatives $(PhTt^R)Ni(L)$ (R=Bu^t, Ad; L=CO, PMe_3) [109, 110].

In the context of bioinorganic chemistry, the zinc phenylthiolate complex $(PhTt^{tBu})ZnSPh$ [111] was synthesized and characterized by X-ray diffraction as a structural model for the active site of the homocysteine-bound forms of methionine synthase enzymes [112, 113]. In a similar vein, a couple of elegant heterodinuclear nickel–copper complexes (Fig. 9) were prepared as synthetic analogues of the catalytic site of acetyl coenzyme A synthase [114].

The activation of dioxygen using monovalent nickel has also been investigated by Riordan et al., resulting in the generation and spectroscopic characterization of a metastable bis-μ-oxo dinickel(III) species (Fig. 10) in which the O_2 molecule has been reduced by four electrons with the simultaneous cleavage of the O=O bond [115].

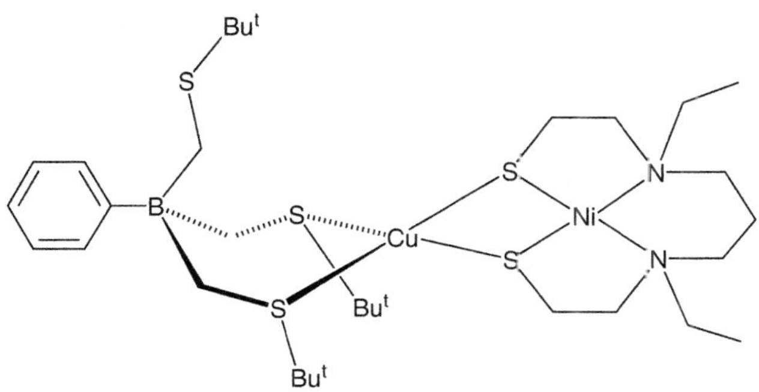

Fig. 9 Structural model of the catalytic site of acetyl coenzyme A synthase (ACS)

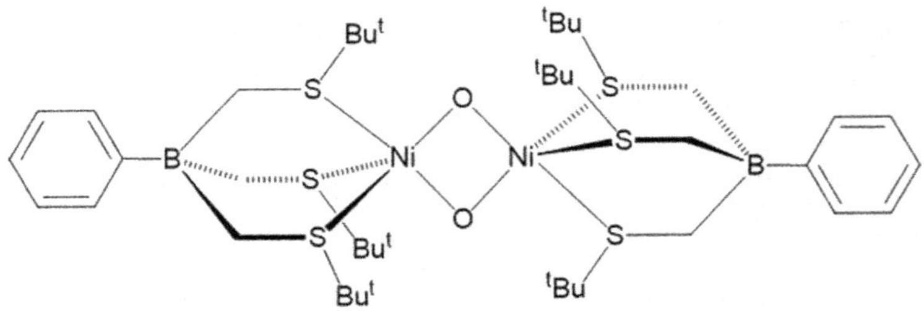

Fig. 10 Dinuclear Ni(III) complex with a $Ni_2(\mu\text{-}O)_2$ core

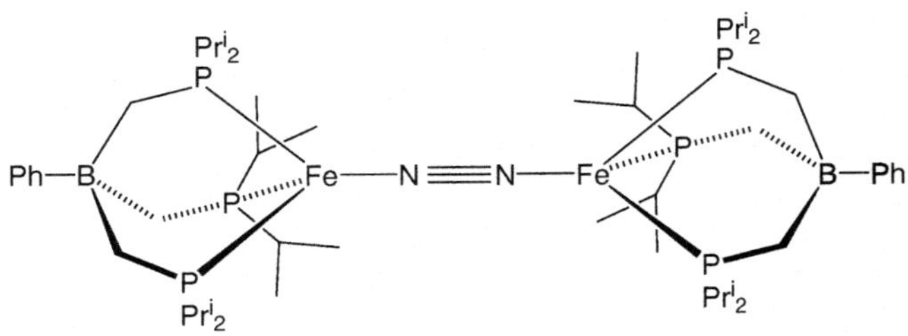

Fig. 11 Structurally characterized dinuclear iron(I) complex stabilized by a bulky tris(phosphino) borate ligand

3.2 *Tris(phosphino)borates*

The anionic tris(phosphino)borates $[PhB(CH_2PR_2)_3]^-$, originally reported in 1999 almost simultaneously by the research groups of Tilley [36] and Nocera [116], are excellent σ-donor ligands and weak π-acceptors that, like the tris(carbene)borates described above, are well poised to stabilize transition metals in unusually high oxidation states. Thus, a number of noteworthy complexes of iron that can be regarded as putative intermediates in the process of nitrogen fixation have been synthesized and studied by a variety of spectroscopic, structural, and computational methods [117]. For example, the use of the sterically demanding $[PhB(CH_2PPr^i{}_2)_3]^-$ ligand has allowed the generation of a mononuclear terminal iron(IV) nitride complex that readily converts to a dinitrogen-bridged diiron(I) species (Fig. 11) [118, 119].

3.3 *Tris(mercaptoimidazolyl)borates*

The tris(mercaptoimidazolyl)borates (Tm^R) constitute yet another class of soft anionic borate-type ligands in which three mercaptoimidazolyl groups provide

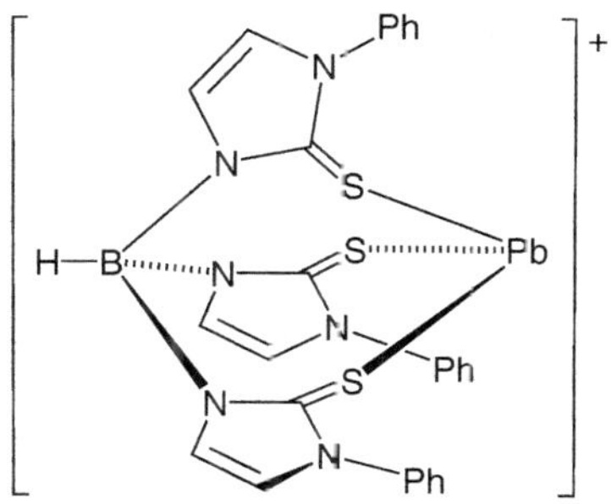

Fig. 12 Structural representation of the complex cation present in $[Pb(Tm^{Ph})]ClO_4$

the sulfur donor moieties. Introduced by Reglinski in 1996 [120], they contain thione (C=S) donor units which, as part of acyclic thioureas and a variety of N-heterocyclic thiones (NHTs), have a well-established coordination chemistry [121, 122]. An important difference between most scorpionates and Tm^R ligands is the size (and therefore shape) of the metallacycles that form upon coordination to a metal center. Whereas coordinated $Tp^{RR'}$ typically forms $B(–N–N–)^2M$ six-membered chelating rings in a boat conformation, Tm^R usually generates twisted $B(–N–C–S–)^2M$ eight-membered rings.

The coordination chemistry of Tm^R ligands has been reviewed [39–41] and includes the preparation of synthetic analogues for the active sites in various known sulfur-rich metalloenzymes [123]. For example, the first structurally characterized zinc hydroxide complex supported by a tridentate sulfur ligand, $(Tm^{Ph})ZnOH$, was prepared as a structural model for the zinc-containing enzyme 5-aminolevulinate dehydratase (ALAD) [124]. A couple of Tm^{Me} complexes of molybdenum stoichiometrically convert sulfite to sulfate and thereby serve as functional models for sulfite oxidase [125, 126]. The unusual trigonal pyramidal cation $[(Tm^{Ph})Pb]^+$ (Fig. 12) [127] was one of the key species that prompted a careful reevaluation of lead(II) coordination preferences in sulfur-rich environments [128], a study aimed at understanding the molecular basis of lead poisoning.

A fairly large number of Tm^R complexes of zinc, e.g., $(Tm^R)ZnX$ (X=Cl, Br, I), have been synthesized as straightforward starting materials for further reactivity studies [129]. Zinc thiolate complexes such as $(Tm^{Ph})Zn(SPh)$ [130] have been prepared to model the reactivity of tetrahedral zinc in a sulfur-only environment (i.e., $[ZnS^4]$), which can be found in liver alcohol dehydrogenase (LADH), farnesyltransferase, and methionine synthases [85–88, 131–133]. The Ada DNA repair protein, which restores damaged DNA by abstracting a methyl group from a methylated nucleic base residue or phosphate group and transfers it to one of its own cysteine thiolate groups, also has a $Zn(Cys)^4$ core [134, 135]. In this regard, the alkylation of well-defined zinc thiolate complexes has been studied by a number of researchers, including Lippard [136, 137], Darensbourg [138], Vahrenkamp [139–146], Parkin [130, 147–149], Riordan [111, 150], Carrano [151–154], Anders [155, 156], and Ibrahim [157, 158]. There appears to be mounting evidence that the methylation of zinc thiolates occurs by two distinct mechanisms: dissociative, in which a zinc-bound thiolate is released and then reacts with the electrophile, a pathway taking place in anionic complexes such as $[Zn(SPh)^4]^{2-}$ [137], and associative, in which the methylating agent directly attacks the thiolate ligand still bound to the metal [112, 159, 160]. In related work, the methylation of monothiolate complexes of cadmium and mercury, $(Tm^R)MSR$ (M=Cd, Hg), has also been explored [129].

Mercury alkyl complexes $(Tm^{tBu})HgR$ (R=Me, Et) have also been investigated by Melnick and Parkin, who observed the facile protolytic cleavage of the mercury–carbon bonds using thiophenol (PhSH) as the Brønsted acid [161]. These experiments suggest that such reactivity may be facilitated by the sulfur-rich environment surrounding the mercury ion in these complexes, an observation with potential implications in the detoxification of organomercury compounds.

4 Conclusions and Outlook

Fifty years on, the chemistry of scorpionate ligands and complexes continues to thrive, as are the range of applications they partake on. I do not know if anyone is planning to write a "Scorpionates III" book, a logical extension to "Scorpionates II" [26], but hundreds (if not more) of new relevant papers have appeared since the publication of the latter. It still amazes me to read Trofimenko's original 1966 communication [24], which had the rather short and unpretentious title "Boron-Pyrazole Chemistry," and see where the field stands today!

Given the popularity of Tm^R ligands since their inception 20 years ago, it was only a matter of time until their oxygen and selenium analogues would emerge. Predictably, Parkin and coworkers have already reported the syntheses and initial reactivity studies of the tris(oxoimidazolyl)borate (To^R) [162] and tris (selenoimidazolyl)borate (Tse^R) [163–165] ligand systems (Fig. 13) in 2006 and 2011, respectively. These ligands, particularly the oxygen donors, should be particularly useful for binding "hard" early transition metal and lanthanide ions.

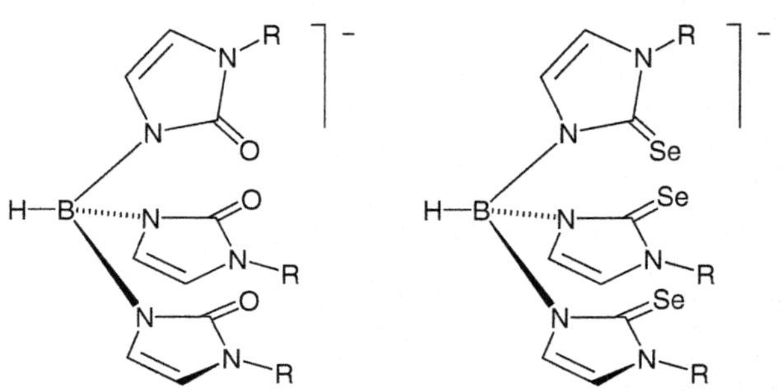

Fig. 13 Tris(oxoimidazolyl)borate (To^R) and tris(selenoimidazolyl)borate (Tse^R) ligands

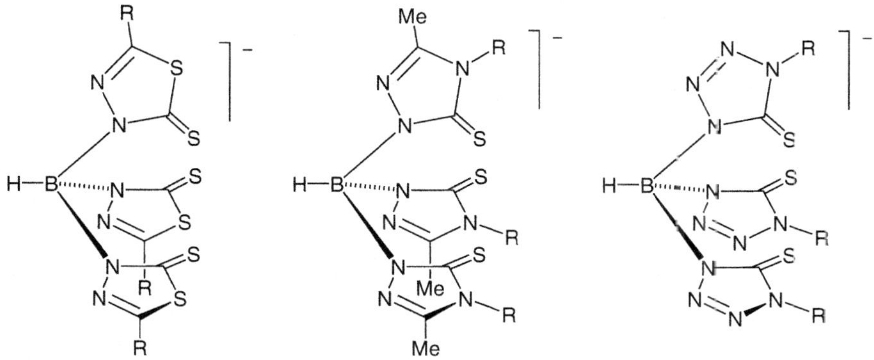

Fig. 14 The three basic types of Janus scorpionates reported to date

The synthesis and reactivity of ambidentate scorpionate ligands featuring both hard (N) and soft (S) donor groups, the so-called Janus scorpionates (Fig. 14), is another underdeveloped area of research that is bound to continue to grow in the near future [166–172].

I look forward to seeing what new developments arise in scorpionate chemistry in the near future, and I hope that this overview of the topic in *Structure and Bonding* serves as a source of inspiration for new generations of chemists willing to make their own contributions to the field.

Acknowledgments Insightful discussions with Professors Joshua Telser (Roosevelt University), Brian Hoffman (Northwestern University), and Bill Jensen (University of Cincinnati) regarding the first public use of the term "bioinorganic" are sincerely appreciated. The National Science Foundation (Award No. CHE-0911407) and The University of North Carolina at Charlotte are gratefully acknowledged for support of our research in synthetic bioinorganic chemistry since 1996.

References

1. Hodgkin DC (1965) Science 150:979
2. Rosenberg B, Van Camp L, Krigas T (1965) Nature 205:698
3. Allen AD, Senoff CV (1965) Chem Commun 621
4. De Haën C (2001) Top Magn Reson Imaging 12:221
5. Caravan P, Ellison JJ, McMurry TJ, Lauffer RB (1999) Chem Rev 99:2293
6. Maiocchi A (2003) Mini Rev Med Chem 3:845
7. Schwochau K (1994) Angew Chem Int Ed Engl 33:2258
8. Liu S, Chakraborty S (2011) Dalton Trans 40:6077
9. Dessy R, Dillard J, Taylor L (eds) (1971) Bioinorganic chemistry, a symposium co-sponsored by the Division of Inorganic Chemistry of the American Chemical Society and the Chemical Institute of Canada at Virginia Polytechnic Institute and State University, Blacksburg, Va (Advances in chemistry series, No 100), June 22–25, 1970

10. Schrauzer GN (1971) Dessy R, Dillard J, Taylor L (eds) Bioinorganic chemistry, a symposium co-sponsored by the Division of Inorganic Chemistry of the American Chemical Society and the Chemical Institute of Canada at Virginia Polytechnic Institute and State University, Blacksburg, Va (Advances in chemistry series, No 100), June 22–25, 1970, Chap. 1, p 1
11. Allen AD (1971) Dessy R, Dillard J, Taylor L (eds) Bioinorganic chemistry, a symposium co-sponsored by the Division of Inorganic Chemistry of the American Chemical Society and the Chemical Institute of Canada at Virginia Polytechnic Institute and State University, Blacksburg, Va (Advances in chemistry series, No 100), June 22–25, 1970, Chap. 4, p 79
12. Wilkins RG (1971) Dessy R, Dillard J, Taylor L (eds) Bioinorganic chemistry, a symposium co-sponsored by the Division of Inorganic Chemistry of the American Chemical Society and the Chemical Institute of Canada at Virginia Polytechnic Institute and State University, Blacksburg, Va (Advances in chemistry series, No 100), June 22–25, 1970, p 111
13. Ulmer DD, Vallee BL (1971) Dessy R, Dillard J, Taylor L (eds) Bioinorganic chemistry, a symposium co-sponsored by the Division of Inorganic Chemistry of the American Chemical Society and the Chemical Institute of Canada at Virginia Polytechnic Institute and State University, Blacksburg, Va (Advances in chemistry series, No 100), June 22–25, 1970, Chap. 10, p 187
14. Gray HB (1971) Dessy R, Dillard J, Taylor L (eds) Bioinorganic chemistry, a symposium co-sponsored by the Division of Inorganic Chemistry of the American Chemical Society and the Chemical Institute of Canada at Virginia Polytechnic Institute and State University, Blacksburg, Va (Advances in chemistry series, No 100), June 22–25, 1970, Chap. 17, p 365
15. Rayner Canham GW, Lever ABP (1972) J Chem Educ 49:656
16. Williams DR (1971) The metals of life. Van Nostrand Reinhold, London
17. Williams DR (1976) An introduction to bio-inorganic chemistry. Charles C. Thomas, Springfield
18. Ochiai EI (1977) Bioinorganic chemistry: an introduction. Allen and Bacon, Boston
19. Holm RH (1977) Acc Chem Res 10:427
20. Hill HAO (1976) Chem Brit 12:119
21. Ibers JA, Holm RH (1980) Science 209:223
22. Karlin KD (1993) Science 261:701
23. Lee SC, Holm RH (2003) Proc Natl Acad Sci U S A 100:3595
24. Trofimenko S (1966) J Am Chem Soc 88:1842
25. Trofimenko S (1999) Scorpionates: the coordination chemistry of polypyrazolylborate ligands. Imperial College, London
26. Pettinari C (2008) Scorpionates II: chelating borate ligands. Imperial College, London
27. Santini C, Pellei M, Gioia Lobbia G, Papini G (2010) Mini Rev Org Chem 7:84
28. Sadimenko AP (2001) Adv Heterocycl Chem 81:167
29. Akita M, Hikichi S (2002) Bull Chem Soc Jpn 75:1657
30. Trofimenko S (2005) J Chem Educ 82:1715
31. Janiak C, Temizdemir S, Röhr C (2000) Z Anorg Allg Chem 626:1265
32. Dunne JF, Su J, Ellern A, Sadow AD (2008) Organometallics 27:2399
33. Papish ET, Dixon NA, Kumar M (2014) Struct Bond 160:115
34. Kernbach U, Ramm M, Luger P, Fehlhammer WP (1996) Angew Chem Int Ed Engl 35:310
35. Smith JM (2008) Comments Inorg Chem 29:189
36. Peters JC, Feldman JD, Tilley TD (1999) J Am Chem Soc 121:9871
37. Barney AA, Heyduk AD, Nocera DG (1999) Chem Commun 2379
38. Riordan CG (2010) Coord Chem Rev 254:1815
39. Parkin G (2007) New J Chem 31:1996
40. Spicer MD, Reglinski J (2009) Eur J Inorg Chem 1553
41. Reglinski J, Spicer MD (2015) Coord Chem Rev 297–298:181
42. Snyder CJ, Heeg MJ, Winter CH (2011) Inorg Chem 50:9210
43. Reger DL (1999) Comments Inorg Chem 21:1
44. Pettinari C, Pettinari R (2005) Coord Chem Rev 249:525

45. Bigmore HR, Lawrence SC, Mountford P, Tredget CS (2005) Dalton Trans 635
46. Pullen EE, Rabinovich D, Incarvito CD, Concolino TE, Rheingold AL (2000) Inorg Chem 39:1561
47. Holmes C, Carrano CJ (1991) Inorg Chem 30:1231
48. Butler A, Baldwin AH (1997) Struct Bond 89:109
49. Rajagopalan KV (1988) Annu Rev Nutr 8:401
50. McMaster J, Enemark JH (1998) Curr Opin Chem Biol 2:201
51. Feng C, Tollin G, Enemark JH (2007) Biochim Biophys Acta 1774:527
52. Enemark JH, Cooney JJA, Wang JJ, Holm RH (2004) Chem Rev 104:1175
53. Romão MJ, Huber R (2004) Chem Rev 104:1175
54. Romão MJ, Huber R (1998) Struct Bond 90:69
55. Hagen WR, Arendsen AF (1998) Struct Bond 90:161
56. Bevers LE, Hagedoorn PL, Hagen WR (2009) Coord Chem Rev 253:269
57. Thomas S, Tiekink ERT, Young CG (1996) Organometallics 15:2428
58. Eagle AA, Harben SM, Tiekink ERT, Young CG (1994) J Am Chem Soc 116:9749
59. Wieghardt K (1989) Angew Chem Int Ed Engl 28:1153
60. Yocum CF, Pecoraro VL (1999) Curr Opin Chem Biol 3:182
61. Miriyala S, Spasojevic I, Tovmasyan A, Salvemini D, Vujaskovic Z, St Clair D, Batinic-Haberle I (2012) Biochim Biophys Acta 1822:794
62. Kitajima N, Osawa M, Tamura N, Moro-oka Y, Hirano T, Hirobe M, Nagano T (1993) Inorg Chem 32:1879
63. Singh UP, Sharma AK, Hikichi S, Komatsuzaki H, Moro-oka Y, Akita M (2006) Inorg Chim Acta 359:4407
64. Armstrong WH, Spool A, Papaefthymiou GC, Frankel RB, Lippard SJ (1984) J Am Chem Soc 106:3653
65. Feig AL, Lippard SJ (1994) Chem Rev 94:759
66. Stenkamp RE (1994) Chem Rev 94:715
67. Ogihara T, Hikichi S, Akita M, Moro-oka Y (1998) Inorg Chem 37:2614
68. Ryle MJ, Hausinger RP (2002) Curr Opin Chem Biol 6:193
69. Bugg TDH, Ramaswamy S (2008) Curr Opin Chem Biol 12:134
70. Abu-Omar MM, Loaiza A, Hontzeas N (2005) Chem Rev 105:2227
71. Kovaleva EG, Lipscomb JD (2008) Nat Chem Biol 4:187
72. Oloo WN, Que L Jr (2015) Acc Chem Res 48:2612
73. Sallmann M, Limberg C (2015) Acc Chem Res 48:2734
74. Paine TK, Que L Jr (2014) Struct Bond 160:39
75. Gennari M, Marchiò L (2009) Curr Bioact Compd 5:244
76. Kitajima N, Tolman WB (1995) Prog Inorg Chem 43:419
77. Kitajima N, Moro-oka Y (1994) Chem Rev 94:737
78. Blackman AG, Tolman WB (2000) Struct Bond 97:179
79. Kitajima N, Fujisawa K, Fujimoto C, Moro-oka Y, Hashimoto S, Kitagawa T, Toriumi K, Tatsumi K, Nakamura A (1992) J Am Chem Soc 114:1277
80. Trofimenko S, Calabrese JC, Thompson JS (1987) Inorg Chem 26:1507
81. Ruggiero CE, Carrier SM, Antholine WE, Whittaker JW, Cramer CJ, Tolman WB (1993) J Am Chem Soc 115:11285
82. Enemark JH, Feltham RD (1974) Coord Chem Rev 13:339
83. Suzuki S, Kataoka K, Yamaguchi K, Inoue T, Kai Y (1999) Coord Chem Rev 190–192:245
84. Merkle AC, Lehnert N (2012) Dalton Trans 41:3355
85. Kimura E, Koike T, Shionoya M (1997) Struct Bond 89:1
86. Vahrenkamp H (1999) Acc Chem Res 32:589
87. Parkin G (2000) Chem Commun 1971
88. Parkin G (2004) Chem Rev 104:699
89. Keilin D, Mann T (1940) Biochem J 34:1163
90. Alsfasser R, Trofimenko S, Looney A, Parkin G, Vahrenkamp H (1991) Inorg Chem 30:4098

91. Looney A, Han R, McNeill K, Parkin G (1993) J Am Chem Soc 115:4690
92. Trofimenko S (1967) J Am Chem Soc 89:3170
93. Janiak C, Scharmann TG, Green JC, Parkin RPG, Kolm MJ, Riedel E, Mickler W, Elguero J, Claramunt RM, Sanz D (1996) Chem Eur J 2:992
94. Janiak C, Scharmann TG, Günther W, Hinrichs W, Lentz D (1996) Chem Ber 129:991
95. Santini C, Pellei M, Gioia Lobbia G, Fedeli D, Falcioni G (2003) J Inorg Biochem 94:348
96. Kumar M, Dixon NA, Merkle AC, Zeller M, Lehnert N, Papish ET (2012) Inorg Chem 51:7004
97. Fränkel R, Kernbach U, Bakola-Christianopoulou M, Plaia U, Suter M, Ponikwar W, Nöth H, Moinet C, Fehlhammer WP (2001) J Organomet Chem 617–618:530
98. Nieto I, Cervantes-Lee F, Smith JM (2005) Chem Commun 3811
99. Hahn FE, Jahnke MC (2008) Angew Chem Int Ed 47:3122
100. Hopkinson MN, Richter C, Schedler M, Glorius F (2014) Nature 510:485
101. Díez-González S, Nolan SP (2007) Coord Chem Rev 251:874
102. Smith JM (2014) Prog Inorg Chem 58:417
103. Scepaniak JJ, Young JA, Bontchev RP, Smith JM (2009) Angew Chem Int Ed 48:3158
104. Scepaniak JJ, Vogel CS, Khusniyarov MM, Heinemann FW, Meyer K, Smith JM (2011) Science 331:1049
105. Ge P, Haggerty BS, Rheingold AL, Riordan CG (1994) J Am Chem Soc 116:8406
106. Ohrenberg C, Ge P, Schebler P, Riordan CG, Yap GPA, Rheingold AL (1996) Inorg Chem 35:749
107. Pearson RG (1963) J Am Chem Soc 85:3533
108. Ohrenberg C, Saleem MM, Riordan CG, Yap GPA, Verma AK, Rheingold AL (1996) Chem Commun 1081
109. Schebler P, Mandimutsira BS, Riordan CG, Liable-Sands LM, Incarvito CD, Rheingold AL (2001) J Am Chem Soc 123:331
110. Fujita K, Rheingold AL, Riordan CG (2003) Dalton Trans 2004
111. Chiou SJ, Innocent J, Riordan CG, Lam KC, Liable-Sands L, Rheingold AL (2000) Inorg Chem 39:4347
112. Penner-Hahn J (2007) Curr Opin Chem Biol 11:166
113. Matthews RG (2001) Acc Chem Res 34:681
114. Krishnan R, Voo JK, Riordan CG, Zahkarov L, Rheingold AL (2003) J Am Chem Soc 125:4422
115. Mandimutsira BS, Yamarik JL, Brunold TC, Gu W, Cramer SP, Riordan CG (2001) J Am Chem Soc 123:9194
116. Barney AA, Heyduk AF, Nocera DG (1999) Chem Commun 2379
117. Mehn MP, Peters JC (2006) J Inorg Biochem 100:634
118. Betley TA, Peters JC (2004) J Am Chem Soc 126:6252
119. Hendrich MP, Gunderson W, Behan RK, Green MT, Mehn MP, Betley TA, Lu CC, Peters JC (2006) Proc Natl Acad Sci U S A 103:17107
120. Garner M, Reglinski J, Cassidy I, Spicer MD, Kennedy AR (1996) Chem Commun 1975
121. Raper ES (1985) Coord Chem Rev 61:115
122. Akrivos PD (2001) Coord Chem Rev 213:181
123. Reglinski J, Spicer MD (2009) Curr Bioact Compd 5:264
124. Bridgewater BM, Parkin G (2001) Inorg Chem Commun 4:126
125. Tran BL, Carrano CJ (2007) Inorg Chem 46:5429
126. Wallace D, Gibson LT, Reglinski J, Spicer MD (2007) Inorg Chem 46:3804
127. Bridgewater BM, Parkin G (2000) J Am Chem Soc 122:7140
128. Magyar JS, Weng TC, Stern CM, Dye DF, Rous BW, Payne JC, Bridgewater BM, Mijovilovich A, Parkin G, Zaleski JM, Penner-Hahn JE, Godwin HA (2005) J Am Chem Soc 127:9495
129. Rabinovich D (2006) Struct Bond 120:143
130. Bridgewater BM, Fillebeen T, Friesner RA, Parkin G (2000) J Chem Soc Dalton Trans 4494

131. Auld DS (2001) BioMetals 14:271
132. Lipscomb WN, Sträter N (1996) Chem Rev 96:2375
133. Hightower KE, Fierke CA (1999) Curr Opin Chem Biol 3:176
134. Sedgwick B, Lindahl T (2002) Oncogene 21:8886
135. Lindahl T, Sedgwick B, Sekiguchi M, Nakabeppu Y (1988) Annu Rev Plant Physiol Plant Mol Biol 57:133
136. Wilker JJ, Lippard SJ (1995) J Am Chem Soc 117:8682
137. Wilker JJ, Lippard SJ (1997) Inorg Chem 36:969
138. Grapperhaus CA, Tuntulani T, Reibenspies JH, Darensbourg MY (1998) Inorg Chem 37:4052
139. Brand U, Rombach M, Vahrenkamp H (1998) Chem Commun 2717
140. Rombach M, Vahrenkamp H (2001) Inorg Chem 40:6144
141. Brand U, Rombach M, Seebacher J, Vahrenkamp H (2001) Inorg Chem 40:6151
142. Seebacher J, Ji M, Vahrenkamp H (2004) Eur J Inorg Chem 409
143. Ji M, Vahrenkamp H (2005) Eur J Inorg Chem 1398
144. Ji M, Benkmil B, Vahrenkamp H (2005) Inorg Chem 44:3518
145. Ibrahim MM, He G, Seebacher J, Benkmil B, Vahrenkamp H (2005) Eur J Inorg Chem 4070
146. Ibrahim MM, Seebacher J, Steinfeld G, Vahrenkamp H (2005) Inorg Chem 44:8531
147. Docrat A, Morlok MM, Bridgewater BM, Churchill DG, Parkin G (2004) Polyhedron 23:481
148. Morlok MM, Janak KE, Zhu G, Quarless DA, Parkin G (2006) J Inorg Biochem 100:1147
149. Melnick JG, Zhu G, Buccella D, Parkin G (2005) J Am Chem Soc 127:14039
150. Chiou SJ, Riordan CG, Rheingold AL (2003) Proc Natl Acad Sci U S A 100:3695
151. Warthen CR, Hammes BS, Carrano CJ, Crans DC (2001) J Biol Inorg Chem 6:82
152. Hammes BS, Carrano CJ (2001) Inorg Chem 40:919
153. Smith JN, Shirin Z, Carrano CJ (2003) J Am Chem Soc 125:868
154. Smith JN, Hoffman JT, Shirin Z, Carrano CJ (2005) Inorg Chem 44:2012
155. Notni J, Görls H, Anders E (2006) Eur J Inorg Chem 1444
156. Notni J, Günther W, Anders E (2007) Eur J Inorg Chem 985
157. Ibrahim MM (2010) J Sulfur Chem 31:395
158. Ibrahim MM, Mosa A (2014) Arab J Chem 7:672
159. Picot D, Ohanessian G, Frison G (2008) Inorg Chem 47:8167
160. Isaac M, Latour JM, Sénèque O (2012) Chem Sci 3:3409
161. Melnick JG, Parkin G (2007) Science 317:225
162. Al-Harbi A, Sattler W, Sattler A, Parkin G (2011) Chem Commun 47:3123
163. Minoura M, Landry VK, Melnick JG, Pang K, Marchiò L, Parkin G (2006) Chem Commun 3990
164. Landry VK, Pang K, Quan SM, Parkin G (2007) Dalton Trans 820
165. Landry VK, Buccella D, Pang K, Parkin G (2007) Dalton Trans 866
166. Bailey PJ, Lanfranchi M, Marchiò L, Parsons S (2001) Inorg Chem 40:5030
167. Cammi R, Gennari M, Giannetto M, Lanfranchi M, Marchiò L, Mori G, Paiola C, Pellinghelli MA (2005) Inorg Chem 44:4333
168. Silva RM, Gwengo C, Lindeman SV, Smith MD, Gardinier JR (2006) Inorg Chem 45:10998
169. Silva RM, Gwengo C, Lindeman SV, Smith MD, Long GJ, Grandjean F, Gardinier JR (2008) Inorg Chem 47:7239
170. Imran M, Neumann B, Stammler HG, Monkowius U, Ertl M, Mitzel NW (2013) Dalton Trans 42:15785
171. Imran M, Mix A, Neumann B, Stammler HG, Monkowius U, Bleckenwegner P, Mitzel NW (2014) Dalton Trans 43:10956
172. Imran M, Mix A, Neumann B, Stammler HG, Monkowius U, Bleckenwegner P, Mitzel NW (2014) Dalton Trans 43:14737

Struct Bond (2016) 172: 159–200
DOI: 10.1007/430_2015_194

Published online: 10 November 2015

Recent Advances in Phthalocyanine-Based Functional Molecular Materials

Yongzhong Bian and Jianzhuang Jiang

Abstract In this chapter, we wish to review the recent progress in the application of phthalocyanines as functional molecular materials including (1) semiconducting materials for organic photovoltaic cells and organic field effect transistors, (2) functional organic dyes as photosensitizers for photodynamic therapy and dye-sensitized solar cells, and (3) single-molecule magnets. The structure–function relationship has been highlighted on the bases of selected examples since 2010, which hopefully will be informative for the future developments of phthalocyanine chemistry and related materials science.

Keywords Dye-sensitized solar cell (DSSC) · Organic field effect transistor (OFET) · Organic photovoltaic cell (OPVC) · Photodynamic therapy (PDT) · Phthalocyanine · Single-molecule magnet (SMM)

Contents

Y. Bian and J. Jiang (✉)
Beijing Key Laboratory for Science and Application of Functional Molecular and Crystalline Materials, Department of Chemistry, University of Science and Technology Beijing, Beijing 100083, China
e-mail: jianzhuang@ustb.edu.cn

1 Introduction

Phthalocyanines (Pc's) are conjugated macrocyclic tetrapyrrole compounds of 18 π-electron aromaticity [1–3], which are analogous to naturally occurring porphyrins (Por's) [4]. The rigid coplanar structural features endow Pc's with intriguing physicochemical properties such as narrow and tunable molecular band gaps and related photo- and electro-activities [5], versatile substitutions around the Pc ring and strong coordination ability of the inner cavity [3], and highly thermal and chemical stability as well. In addition to the traditional applications as dyestuffs, Pc's have attracted great research interests in the recent several decades for their potentials in functional molecular materials including organic semiconductors for organic photovoltaic cells (OPVCs) [6], organic field effect transistors (OFETs) [7] and organic light-emitting diodes (OLEDs) [8], photosensitizers for photodynamic therapy (PDT) [9] and dye-sensitized solar cells (DSSCs) [10], single-molecule magnets (SMMs) [11], nonlinear optical (NLO) materials [12], and catalysts for chemical and photo-redox reactions [13–15] among others [16–19].

Along with the development of function-oriented strategies for molecular design and synthesis, in this new century various Pc derivatives with novel structural and electrical features have been obtained, for example, π-expanded Pc molecules in the horizontal [20] and vertical [21] directions of Pc planes, electron-donor–acceptor (D–A) systems of Pc's hybridized with other electro-active subunits [22, 23], and even covalent organic frameworks (COFs) of Pc's [24]. On the other hand, the advances in controllable assembly techniques and related nanotechnologies enable Pc building blocks to be incorporated into supramolecular architectures [25], nanoparticles [26], thin films [27], and other ordered condensed matters and then can be fabricated to prototype devices and evaluated for the functional parameters. This research direction has achieved, recently, exciting progress not only for the practical and potential applications such as organic photodetectors [28] and supramolecular spin valves [29] but also for their operation mechanisms of the fundamental photo-, electro-, and magneto-physics.

Many excellent reviews and monographs about functional phthalocyanine materials have been published in recent years with emphases on molecular photovoltaics [22, 30–38], thin-film electronics [25] and sensors [39–41], PDT sensitizers [42–50], molecular magnets [51], NLO materials [52, 53], catalysts [13, 54, 55], photochemical and photophysical properties [56], and synthetic and coordination chemistry of Pc's [57–64]. We also edited and published in 2010 the volume of "Functional Phthalocyanine Molecular Materials" for the series of Structure and Bonding [65], with nine chapters covering the chemistry of near-infrared absorbing

[66], photochromic [67], and ball-type Pc's [68], the supramolecular arrays [69] and nanostructures [70, 71] of Pc's, and their applications in OFETs [72], DSSCs [73], and molecular magnets [74]. As a follow-up of the above volume, the present chapter is intended to briefly review the advances in functional Pc's since 2010, with selected topics of semiconducting materials for OPVCs and OFETs, functional photosensitizers for PDT and DSSCs, and molecular magnetic materials of SMMs. On the bases of the most recent examples, we wish to highlight the new perspectives as well as the structure–function relationships for each of the selected topics.

2 Organic Semiconducting Materials

As a class of typical small-molecule semiconducting materials, Pc's are superior for the high charge-carrier mobility and tunable highest occupied molecular orbital (HOMO) and lowest unoccupied molecular orbital (LUMO) energy levels and thus tunable molecular band gaps, which enable them to be stable and high-performance *p*-channel, *n*-channel, and even ambipolar charge-transporting materials. In the case of OFETs, the charge carriers are injected into the semiconductors from metal electrodes; however, for OPVCs the charge carriers are generated by photon-induced charge separation (CS) between electron-donor and electron-acceptor materials, where Pc's can be good electron-donor materials due to the highly destabilized HOMOs by the large π-conjugation and the largely minimized reorganization energy by the rigid planar structure. In addition, the strong absorption of Pc's, covering from the ultraviolet to the visible and near-IR regions, leads them to be suitable for efficient solar light-harvesting components in OPVCs.

2.1 *Organic Photovoltaic Cells*

Organic photovoltaic cells, based on organic semiconductors of small molecules or polymers, provide great opportunities for low-cost and high-efficiency solar energy conversion, due to several advantages of organic materials such as intense absorption of visible and near-infrared (NIR) light, synthetic variability, low-cost fabrication process, and flexibility of devices, among others. The generation of photocurrent in OPVCs is basically a cascade of four steps: exciton generation upon photon absorption, exciton diffusion to the heterojunction, exciton dissociation into free charge carriers, and charge transport and collection at the external electrodes. The performance of an OPVC is usually assessed by analyzing the current–voltage (I–V) curves, the key parameters including the open-circuit voltage (V_{OC}, determined by the energy gap between the LUMO level of the acceptor and the HOMO level of the donor), short-circuit current density (J_{sc}, proportional to photon absorption), fill factor (FF), and power conversion efficiency (PCE). On the bases of the understanding of mechanistic picture for solar energy harvesting and photoinduced charge-carrier generation, a remarkable enhancement of the PCE has been achieved recently by the combined efforts of (1) the development of new light-harvesting and electron-donating materials, (2) the fine control of the morphology of active layer, and (3) the rationally optimization of the device architectures.

2.1.1 Vacuum-Deposited Pc's for OPVCs

It has been widely accepted that the organic bilayer planar heterojunction (PHJ) device reported by Tang in 1986 [75] is one of the most significant breakthroughs for OPVCs, in which a layer of copper phthalocyanine complex (CuPc) (**1**) (Fig. 1) was used as the electron donor to form a donor–acceptor heterojunction with a layer of perylene diimide derivative as the electron acceptor, and PCE of 1% was obtained. After that, metal phthalocyanine complexes received more and more research attention as small molecular donors in OPVCs, not only for the prominent charge-transporting and photon absorption properties but also for the synthetic accessibility and commercial viability of Pc's [76]. Great progress has been achieved for the field of Pc-based OPVCs in the recent 5 years. For example, Xu et al. [77] reported bulk heterojunction (BHJ) cells of P3HT:CuMePc:$PC_{61}BM$ (1:1:2, P3HT = poly(3-hexythiophene), CuMePc = tetramethyl-substituted phthalocyaninato Cu(II) complex (**2**), PCBM = [6,6]-phenyl C61-butyric acid methyl ester), which showed improved photovoltaic performance with a PCE of 5.3%, as a result of the enhancement of the charge-carrier mobility and photon harvesting by incorporating CuMePc nanocrystals in P3HT:$PC_{61}BM$ BHJ solar cells. In another trial, Qu et al. [78] fabricated PHJ cells using the same molecular donor of CuMePc (**2**) (Fig. 1) and a different acceptor of C_{60} by vacuum-deposition technique. With MoO_3 as the electron-blocking buffer layer and after a thermal annealing treatment, an enhanced PCE of 3.53% was achieved.

Fluorinated phthalocyaninato zinc complexes can also serve as an efficient electron donor for BHJ cells. As demonstrated by Meiss et al. [79], the single BHJ cell with F_4ZnPc (**3**) (Fig. 1) as an electron donor and C_{60} as an electron acceptor (F_4ZnPc:C_{60}) showed highly improved V_{OC} in comparison to the device based on the unsubstituted phthalocyaninato zinc complex and C_{60} (ZnPc:C_{60}); accordingly, the PCE values were increased from 3.3% (ZnPc:C_{60}) to 4.6% (F4ZnPc:C_{60}). Very recently, Brendel et al. [80] systematically investigated the

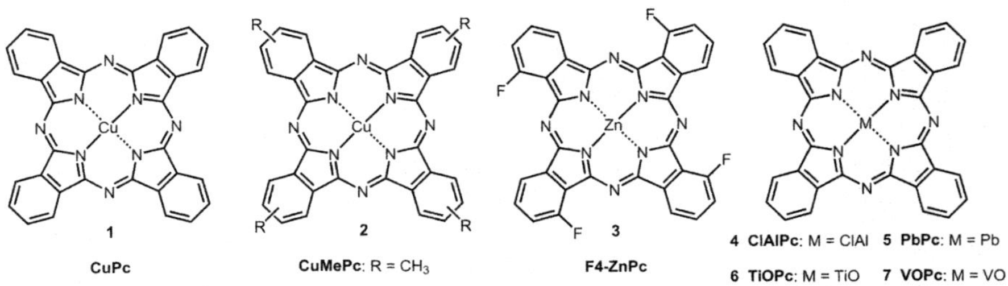

Fig. 1 Vacuum-deposited Pc's for OPVCs

effect of fluorination of the ZnPc molecules on the OPV performance, they found that upon the gradual fluorination of F_nZnPc ($n = 0, 4, 8$), the effective band gap at the donor–acceptor interface and in turn the V_{OC} for bilayer BHJ cells were increased.

Chloroaluminum phthalocyanine (ClAlPc) (**4**) (Fig. 1) [81, 82] is another promising donor material for OPVCs, due to its enhanced photon absorption in the NIR region and stabilized HOMO in comparison to CuPc (**1**). Particularly, the existence of axial chlorine is feasible for the optimization of donor/acceptor and donor/electrode interfaces by vacuum-deposited techniques in heterojunction-based cells; as a consequence, high device performance with a PCE of 4.8% can be obtained for a BHJ cell of ClAlPc:C_{60} by Verreet et al. [83]. Some other nonplanar Pc's, such as lead phthalocyanine (PbPc) (**5**) (Fig. 2) [84], titanyl phthalocyanine

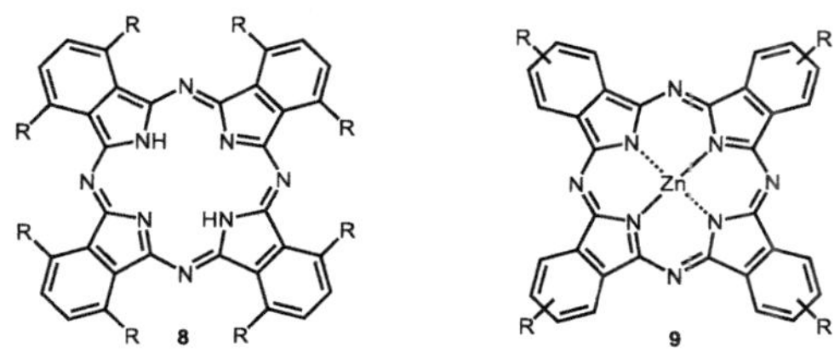

8 C_6PcH_2: R = C_6H_{13}

9 **ZnPc*t*Bu₄**: R = *t*Bu

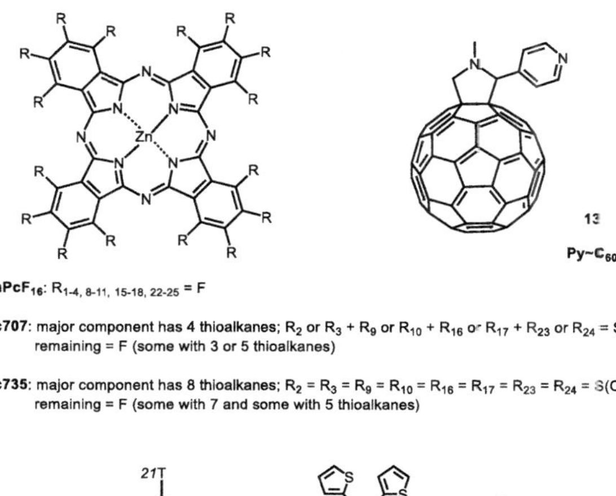

13 **Py-C_{60}**

10 **ZnPcF$_{16}$**: $R_{1-4,\,8-11,\,15-18,\,22-25}$ = F

11 **Pc707**: major component has 4 thioalkanes; R_2 or R_3 + R_9 or R_{10} + R_{16} or R_{17} + R_{23} or R_{24} = $S(CH_2)_{11}CH_3$ remaining = F (some with 3 or 5 thioalkanes)

12 **Pc735**: major component has 8 thioalkanes; $R_2 = R_3 = R_9 = R_{10} = R_{16} = R_{17} = R_{23} = R_{24} = S(CH_2)_{11}CH_3$ remaining = F (some with 7 and some with 5 thioalkanes)

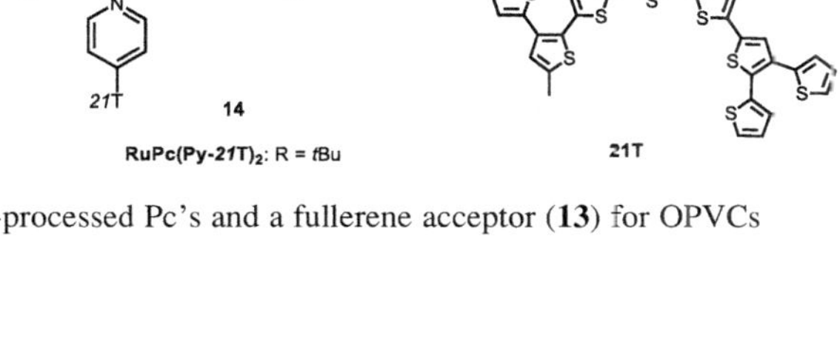

14 **RuPc(Py-*21T*)$_2$**: R = *t*Bu

21T

Fig. 2 Solution-processed Pc's and a fullerene acceptor (**13**) for OPVCs

(TiOPc) (**6**) (Fig. 1) [85], and vanadyl phthalocyanine (VOPc) (**7**) (Fig. 1) [86], have also been explored as the donor materials to extend the spectral coverage and to control the film morphology and packing structures. It was found that, for these polymorphic Pc's, the optical and electrical characteristics could be regulated by templating or annealing-induced phase transition.

2.1.2 Solution-Processed Pc's for OPVCs

Because most of the above Pc's are insoluble in common organic solvents, the related devices can only be fabricated by vacuum evaporating techniques. In order to produce Pc-based OPVCs by low-cost solution-processed methods, some significant efforts have been carried out.

Hori et al. [87] reported a soluble metal-free phthalocyanine with long alkyl substitutes, 1,4,8,11,15,18,22,25-octahexylphthalocyanine (C6PcH2) (**8**) (Fig. 2), which was fabricated to HBJ cells with PCBM donor by spin coating of the mixed solution. A PCE of 3.2% was obtained for the optimized devices. Varotto et al. [88] demonstrated BHJ cells with self-organized blends of Pc derivatives, which include ZnPc*t*Bu_4 (**9**), and the mixed products (Pc707 (**11**) and Pc735 (**12**)) of the peripheral fluorine substitution reaction between hexadecafluorophthalocyaninatozinc (II) ($ZnPcF_{16}$) (**10**) and 1-dodecanethiol. The compounds Pc707 (**11**) and Pc735 (**12**) have successively decreased molecular band gaps while the HOMO energy levels are destabilized more than the LUMOs, resulting in a suitable internal energy alignment of frontier molecular orbitals among the Pc derivatives. When the Pc blends were fabricated to BHJ cells with a complementary fullerene acceptor (**13**) (Fig. 2), significant increase in J_{sc} and PCE relative to any given Pc components was achieved for the solution-processed devices. In a follow-up study, Jurow et al. [89] found that, for this class of soluble Pc donors, the peripheral alkane substituents can also be used to control the morphology and molecular packing of the active layer in BHJ cells. Fischer et al. [90] reported a series of highly soluble tetra-*tert*-butyl phthalocyaninato ruthenium(II) complexes (RuPc-Py-*n*T) (**14**) (Fig. 2) with dendritic oligothiophene (DOT) axial ligands. The DOT ligands can enhance the absorption in the range of 380–550 nm, where RuPc core does not possess a strong absorption. These RuPc-Py-*n*T complexes were applied to solution-processed BHJ solar cells with fullerene PCBM as the acceptor; PCE values of up to 1.6% were achieved.

The water-soluble tetra-sulfonic phthalocyaninato copper(II) complex (TSCuPc) (**15**) (Fig. 3) was fabricated to PHJ cells with C_{60} as the acceptor layer by Schumann et al. [91]. A PCE value up to 0.32% was obtained for this water-soluble molecular semiconductor. Later, Ryan et al. [92] studied the photovoltaic performance of a series of water-soluble ZnPc derivatives, which have two, three, or four sulfonate substituents at the periphery of Pc rings, noted as ZnPc-S2 (**16**), ZnPc-S3 (**17**), and ZnPc-S4 (**18**), respectively (Fig. 3). It was revealed that the frontier orbital energies could be shifted by changing the number of substituents, thus altering the

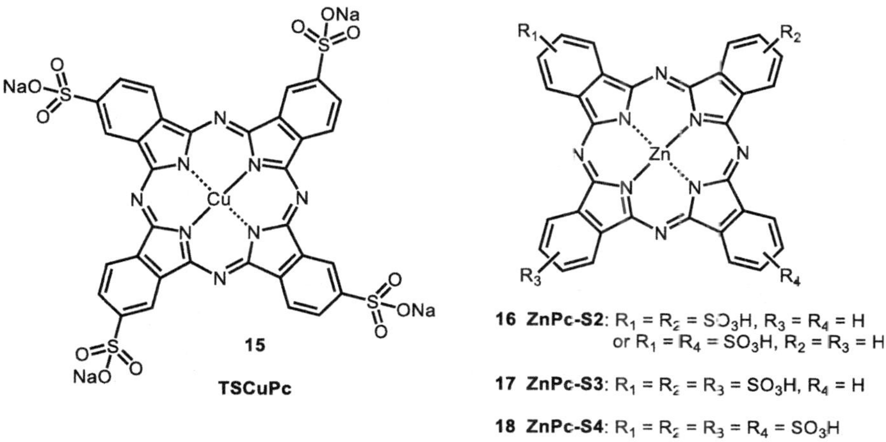

Fig. 3 Water-soluble Pc's for OPVCs

performance of the PHJ cells. The disulfonated species ZnPc-S2 (**16**) showed the most improved PCE (0.46%) under standard conditions.

Sandwich-type phthalocyaninato rare-earth metal complexes (SPRMCs) [57, 93] are narrow band-gap molecular semiconductors with broad absorption and good solubility over corresponding monomeric Pc derivatives. However, there are very few examples for the application of SPRMCs in small-molecule OPVCs [94–98]. In a recent trial, Li et al. [98] fabricated solution-processed organic–inorganic hybrid BHJ solar cells using a series of rare-earth double-decker complexes as the electron donors. By choosing PDI, with complementary absorptions to the double-decker donors, as the primary electron acceptor, blended with nanoporous TiO_2 as the secondary electron acceptor to suppress the fatal back charge transfer, an optimal PCE of up to 0.82% was achieved, which is comparable to some vacuum-deposited cells of monomeric Pc's.

2.2 *Organic Field Effect Transistors*

Organic field effect transistors (OFETs) were first reported in 1980s [99], since then they have attracted considerable attention due to the great potentials for the fields of integrated circuits, flexible displays, sensors, and other low-cost large-area electronic devices. In principle, when suitable gate voltage (V_G) and drain-source voltage (V_{DS}) are applied to an OFET device, the injected charges (hole for *p*-channel and electron for *n*-channel) from the source electrode will gather at the semiconductor/gate insulator interface and move to the drain electrode; as a result, a tunable drain-source current (I_{DS}) is generated in the semiconductor layer. The performance of an OFET device can be evaluated by three key parameters: (1) charge-transfer mobility (μ/cm^2 v^{-1} s^{-1}), which quantifies the average charge-carrier drift velocity per unit electric field; (2) on/off current ratio (I_{on}/I_{off}), which is the ratio of the drain currents between the on and off states; and (3) threshold voltage (V_T/v), which expresses the minimum gate voltage applied to turn the transistor on. Although the charge transport mechanism in OFETs remains undefined yet, it is believed that, in addition to the semiconductor layer, other factors such as the device configuration, the properties of the electrodes, the gate insulator, and the interfaces are all crucial in determining the performance.

Over the years, the research field of OFETs [100–103] has achieved great progress as a result of the advances in material design and device engineering. In particular, due to the prominent charge transport properties, chemical stability, and synthetic versatility of Pc's, OFET devices with Pc's as the active materials [72, 104, 105] have also experienced much research efforts and great development, as exemplified by the highly improved charge-transfer mobility for Pc-based OFETs from the beginning of $1.5–7.0 \times 10^{-4}$ cm^2 v^{-1} s^{-1} (NiPc films in 1989) [106] to the most high values of 1.0–10.0 cm^2 v^{-1} s^{-1} (TiOPc films in 2007) [107] and emerging applications of sensors [108–110], photodetectors [111] and ring oscillators [112], etc. In the following part of this section, we will briefly survey the development of Pc-based OFETs in the past 5 years, in the order of *p*-type, *n*-type, and ambipolar OFETs.

2.2.1 *p*-Channel OFETs

CuPc (**1**) is one of the most intensively studied *p*-type semiconductors for OFETs. The recent interests for this classical organic semiconductor (OSC) material are focused on the modulating of the molecular packing, crystalline structure, and interfacial properties [113], which have shown crucial role to realize high-performance OFETs. Particularly, Jung et al. prepared α-phase CuPc nanowires which then were transformed to β-phase rectangular nanotubes through hydrothermal and annealing process. Comparative OFET studies showed that the charge-transfer mobility and on/off current ratio were improved to 8 and 10 times, respectively, upon the transformation from α-phase nanowire to β-phase nanotube [114]. Alternatively, in order to develop low-voltage OFETs, Su et al. introduced a solution-processed bilayer Al_2O_y/TiO_x system as the high-κ gate dielectric for CuPc-based OFETs, which exhibited high hole mobility of 0.06 cm^2 v^{-1} s^{-1} at an operation voltage of −1.5 V [115]. Recently, Deng et al. developed a simple and efficient etching-assisted transfer printing (ETP) method to fabricate CuPc nanowire array-based flexible OFETs, by which a high hole mobility (2.0 cm^2 v^{-1} s^{-1}), a large on/off ratio (>10^4), and a low threshold voltage (−2 v) were achieved, in addition to the excellent bend stability [116]. The high crystal quality of the nanowires is believed to be responsible for the high mobility, while the use of high-κ dielectric layer (Si_3N_4, $\varepsilon = 7.5$) optimizes the current on/off ratio and the operating voltages. On the other hand, it is also important to select preferred material for the source and drain electrodes to optimize the work function match with the channel material. For example, Liu et al. demonstrated that graphene source and drain electrodes could perform higher hole injection efficiency to CuPc nanoribbons than that of gold electrodes [117]. Moreover, Luo et al. reported that doping gold nanoparticles on the top surface of the CuPc layer could improve the performance of CuPc-based OFETs [118].

Miniaturized electronic devices are highly desired in modern technology. In this regard, Cao et al. developed molecular scale OFETs based on a specially designed thiolated phthalocyaninato Cu(II) complex (2thioPc, **19**) [119] (Fig. 4). The molecular gap electrodes were fabricated by a "top-down" molecular lithography approach, which were then connected by a "bottom-up" approach of self-assembling of the 2thioPc molecules through Au–S bonding. The obtained

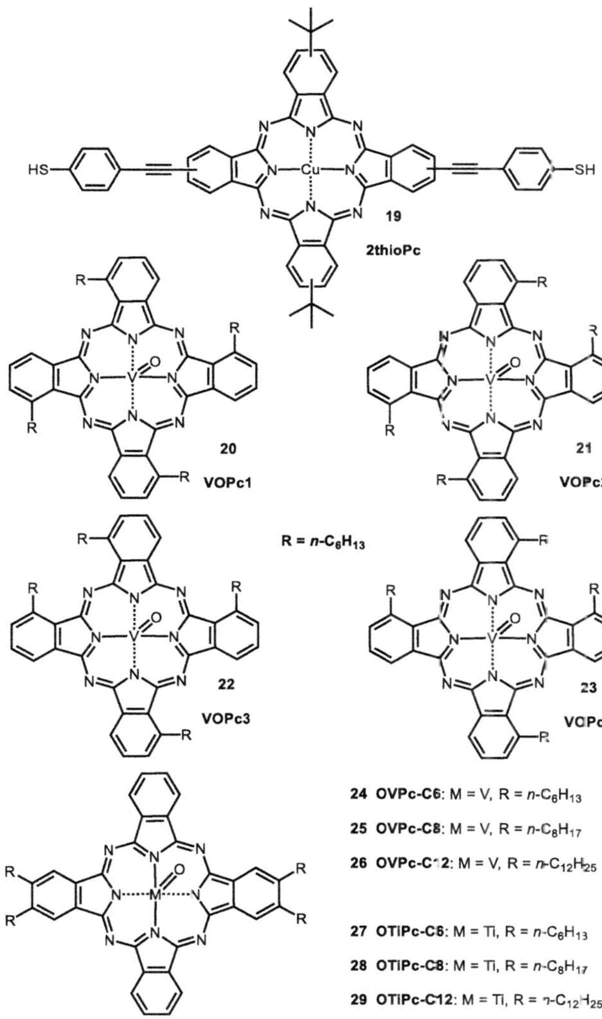

Fig. 4 Pc's for *p*-channel OFETs

molecular scale OFETs showed high hole mobility of 0.16 $cm^2 v^{-1} s^{-1}$. Later, Jiang et al. developed a facile approach of "molecular crystal lithography" to prepare nano-gap electrodes for thermally evaporated CuPc (**1**) layers with tunable channel lengths from 20 μm down to 50 nm [120].

Solution processible organic semiconductors are of great importance for the printable and low-cost organic electronic devices [121]. As for OFETs, Langmuir–Blodgett (LB) monolayers of CuPc (**1**) were applied by Cao et al. as *p*-channel materials with single-layer graphene as planar contacts [122]; the devices showed carrier mobility of up to 0.04 $cm^2 v^{-1} s^{-1}$, on/off ratio of $>10^6$, and ultrasensitive photoresponsive behavior as well.

Non-planar metal phthalocyanine complexes (MPc's) are also promising materials for OFETs because of the easily controllable intermolecular π–π staking and the highly improved solubility relative to planar MPc's. For instance, upon epitaxy growth of vanadyl phthalocyanine (VOPc) (**7**) on proper inducing layers [123, 124], high hole mobility up to 2.6 $cm^2 v^{-1} s^{-1}$ could been obtained for VOPc-based OFETs. Interestingly, Dong et al. successfully synthesized and separated four structural isomers of non-peripheral tetrahexyl-substituted vanadyl phthalocyanine complexes VOPc1–VOPc4 (**20–23**) (Fig. 4) [125]. They found that the substitution pattern significantly influences the intermolecular π–π stacking and electronic coupling. As a result, the OFET mobilities in the range of 1.2×10^{-4} to 0.13 $cm^2 v^{-1} s^{-1}$ were obtained for the four isomers, in which the most abundant isomer (VOPc3, **22**) showed the best performance in the series. Later, these authors designed new series of titanyl and vanadyl phthalocyanines with ABAB symmetry (OVPc-Cn, $n=6$ (**24**), 8 (**25**), 12 (**26**); and OTiPc-Cn, $n=6$ (**27**), 8 (**28**), 12 (**29**)) (Fig. 4) [126]. The specially designed molecular substitution pattern could help to keep tight cofacial π–π stacking of Pc cores by placing all of the alkyl substituents between the Pc core layers (Fig. 5). The slipped cofacial π–π stacking motif of the Pc cores is obviously favorable for intermolecular π-orbital overlap and thus is beneficial for efficient charge transporting. As expected, the OFET mobilities of

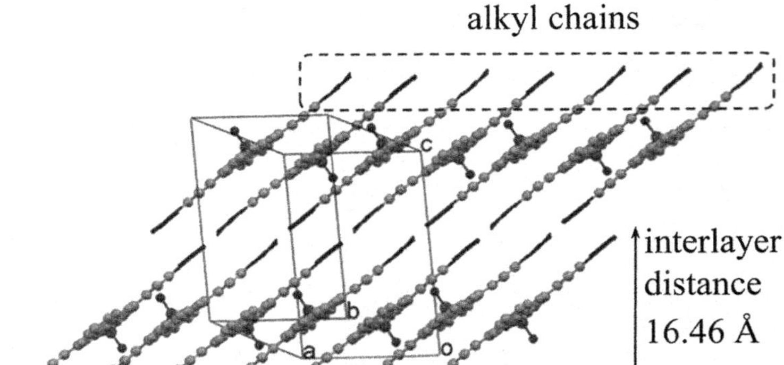

Fig. 5 Crystal packing structure of OTiPc-C6 (**27**) along the *c*-axis. Reprinted with permission from [126]. Copyright 2013 Wiley-VCH Verlag GmbH & Co. KGaA, Weinheim

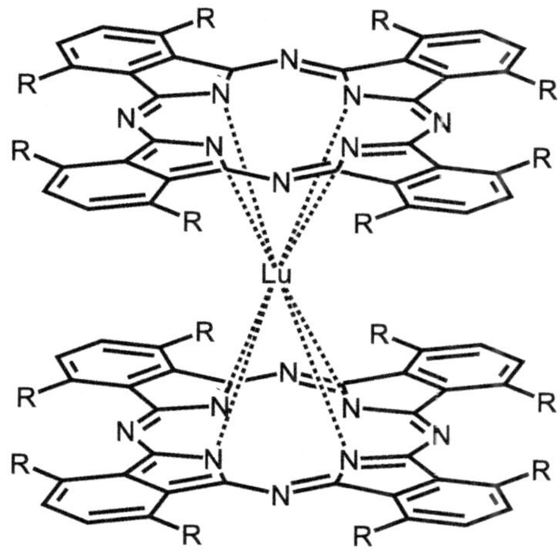

R = n-C_8H_{17}

30

Fig. 6 Schematic molecular structure of Lu(Pc[α-C_8H_{17}-*n*]$_8$)$_2$ (**30**)

27–29 have been highly improved to the range of 0.18–0.96 $cm^2\ v^{-1}\ s^{-1}$, in which OTiPc-C8 (**28**) showed the best device performance. These examples demonstrate the significant effect of rational molecular design on the properties of Pc-based semiconductors and OFET devices.

As stated above in the Sect. 2.1.2, sandwich-type phthalocyaninato rare-earth metal complexes (SPRMCs) [57, 93] are mostly solution-processable semiconductors with narrow molecular band gaps, which allows for the fabrication of a family of SPRMCs-based thin-film OFETs by using Langmuir–Blodgett (LB) technique coupled with amphiphilic molecular design [127–132]. Recently, Chaure et al. synthesized a liquid crystalline bisphthalocyaninate lutetium(III) sandwich complex Lu(Pc[α-C_8H_{17}-*n*]$_8$)$_2$ (**30**) (Fig. 6) and fabricated to OFETs through spin-coating technique [133]. They found that annealing the devices above the temperature of crystal to mesophase transition could significantly improve the performance of OFETs, as such the hole mobility increased to five times, the on/off ratio increased by two orders of magnitude, while the threshold voltage decreased to half of the values upon annealing the as-deposited films. The improvement in performance was attributed to the increase in grain sizes and the decrease in mean surface roughness of the spin-coated films upon annealing.

2.2.2 *n*-Channel OFETs

In order to construct integrated organic circuits, air-stable *n*-channel OFET devices are highly desired to complement *p*-channel devices. In this respect, hexadecafluorinated metallophthalocyanines (F_{16}MPc, M = Zn (**10**), Cu (**31**), Co (**32**), and

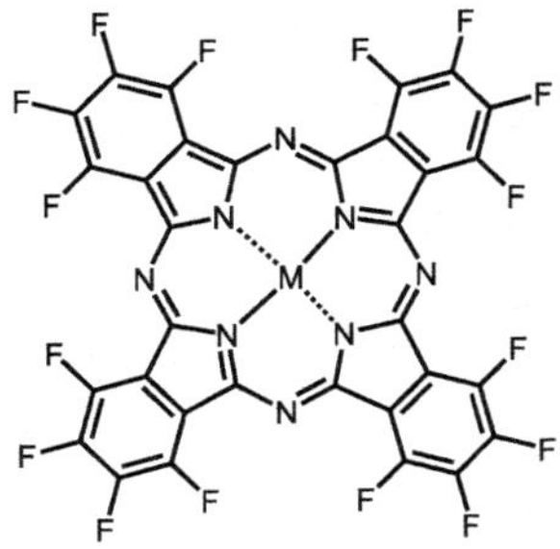

31 F_{16}**CuPc**: M = Cu

32 F_{16}**CoPc**: M = Co

33 F_{16}**FePc**: M = Fe

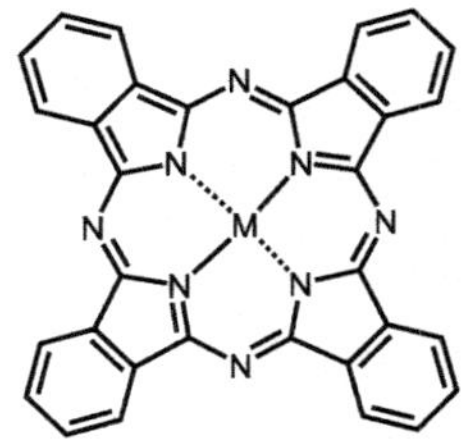

34 SnOPc: M = SnO

35 $SnCl_2$**Pc**: M = $SnCl_2$

Fig. 7 Pc's for *n*-channel OFETs

Fe (**33**)) (Fig. 7) are one of the first classes of high-performance air-stable *n*-channel semiconductors [134, 135], in which the vacuum-deposited films of copper complex **31** showed the best performance with electron mobility up to 0.03 $cm^2\ v^{-1}\ s^{-1}$. However, in a recent work [136], Jiang et al. reported that the single crystal OFETs of F_{16}ZnPc (**10**) showed higher electron mobility (1.1 $cm^2\ v^{-1}\ s^{-1}$) than those of F_{16}CoPc (**32**, 0.8 $cm^2\ v^{-1}\ s^{-1}$) and F_{16}CuPc (**31**, 0.6 $cm^2\ v^{-1}\ s^{-1}$), which was attributed to higher electron transfer integrals and weaker electron–phonon coupling strength of F_{16}ZnPc (**10**) than those of the other two. In addition to the above hexadecafluorinated Pc's, tin(IV) phthalocyanate complexes with axial oxygen (SnOPc, **34**) (Fig. 7) [137] or dichloride ($SnCl_2$Pc, **35**) (Fig. 7) [138] groups are another kind of high-performance *n*-type semiconductors. Recently, Obaidulla et al. [139] reported that vacuum-deposited $SnCl_2$Pc (**35**) OFETs exhibited good *n*-channel behavior with electron mobility of 0.01 $cm^2\ v^{-1}\ s^{-1}$, low threshold voltages of 4 V, on/off current ratio of 10^4, and high operation stability as well.

The first solution-processible Pc-based *n*-type semiconductor was reported in 2011 [140]. By stepwisely replacing electron-donating 15-crown-5 units with electron-withdrawing octyloxycarbonyl groups, the transporting feature of the series of phthalocyaninato copper complexes (**36–41**, Fig. 8) were successfully tuned from *p*-type to *n*-type, along with the decreasing of the LUMO levels in the same order.

2.2.3 Pc's-Based Ambipolar OFETs

Ambipolar OFET devices can accumulate and conduct both holes and electrons, thus providing both *n*- and *p*-channel behaviors, which are highly interested for the application of low-cost complementary circuits and light emission devices [141, 142]. Usually three approaches have been adapted to achieve ambipolar transporting properties in OFETs: (1) *n*- and *p*-type semiconductors in a bilayer architecture (or *p–n* junction) [143–145], (2) blends of *n*- and *p*-type materials in a

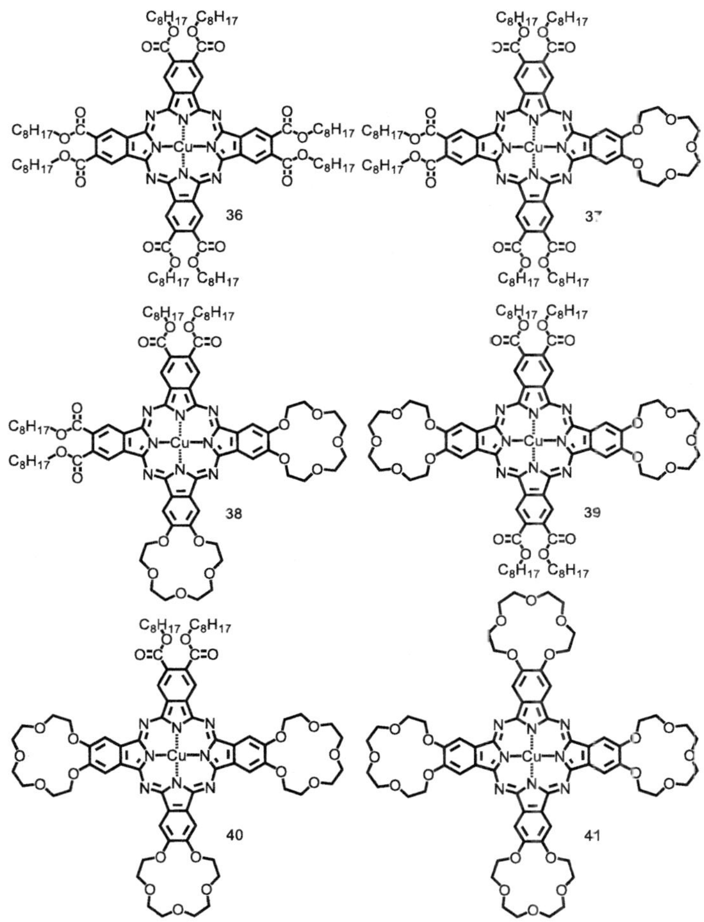

Fig. 8 Solution-processible Pc's for *n*-channel OFETs

single layer [145, 146], and (3) single component in a single layer. Here the last approach will be primarily discussed to highlight the superiority of Pc derivatives for ambipolar transport, where the tunable and narrow band-gap nature of Pc's is crucial for hole and electron injection.

Opitz et al. systematically investigated the ambipolar transporting properties of CuPc (**1**) [147–149]. They found that modifying the substrate (gate insulator) with suitable passivation layers could realize bipolar charge transport in typical *p*-type semiconductor of CuPc (**1**). When tetratetracontane (TTC) was used as the passivation layer, balanced charge-carrier mobilities up to 0.03 $cm^2\ v^{-1}\ s^{-1}$ can be achieved in polycrystalline thin films of CuPc (**1**) [148]. Another typical example is PbPc (**5**), due to its narrow molecular band gap and unique π–π stacking mode; high-performance ambipolar charge-transporting properties were discovered by

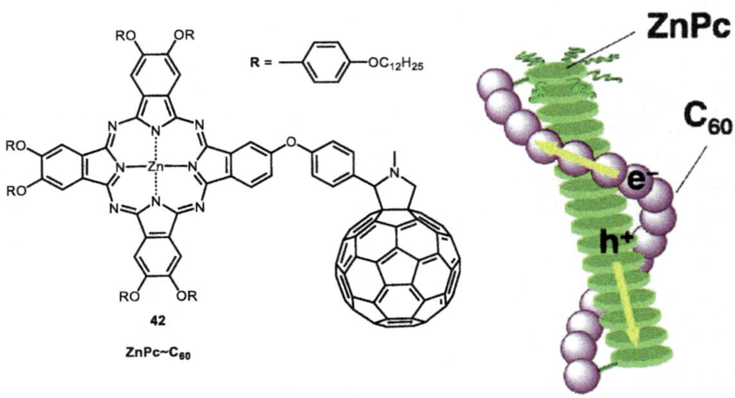

Fig. 9 Phthalocyanine–fullerene dyad (**42**) for ambipolar charge transporting. Reprinted with modification and permission from [153]. Copyright 2011 American Chemical Society

experimental measurements [150] and theoretical modeling [151] as well. Nevertheless, Peng et al. recently reported that, upon light illumination with proper wavelength and power density, PbPc-based *p*-channel OFET devices could be converted to ambipolar ones with balanced carrier mobilities [152].

Hayashi et al. reported an attractive strategy for achieving single-component ambipolar charge transport based on a mesogenic phthalocyanine–fullerene dyad (**42**, Fig. 9) [153]. The segregated columnar liquid crystals (LCs) result in bicontinuous donor–acceptor structures, which exhibit efficient ambipolar charge-transporting properties as measured by time-of-flight (TOF) and time-resolved microwave conductivity (TRMC) methods.

It is clear that the π-conjugation extension can lead to narrow band-gap organic semiconductors, which is favor of matching with the work function of given electrodes for both of hole and electron injection. In this regard, Shi and Li designed π-extended binuclear phthalocyaninato copper (II) complex, $Cu_2[Pc(COOC_8H_{17})_6]_2$ (**43**, Fig. 10), with electron-withdrawing octyl formate groups at the periphery positions [154]. The new compound is soluble in conventional organic solvents, and more significantly, the HOMO and LUMO energy levels have been successfully adjusted to the range for an air-stable ambipolar organic semiconductor. By a solution-based quasi-Langmuir–Shäfer (QLS) method, thin-film OFET devices were fabricated and air-stable ambipolar properties were proved with electron mobilities up to 0.17 $cm^2\ v^{-1}\ s^{-1}$ and hole mobilities to $2.3 \times 10^{-4}\ cm^2\ v^{-1}\ s^{-1}$.

Another approach to extend the π-conjugation system of Pc's is the formation of sandwich-type phthalocyaninato rare-earth metal complexes (SPRMCs) [57, 93], in which the unique intramolecular π–π interactions can lead to notable decreasing of the LUMO energy levels and increasing of the HOMO energy levels in comparison with the corresponding monomeric Pc derivatives. As a result, prominent ambipolar transporting properties were identified for bis(phthalocyaninato) rare-earth

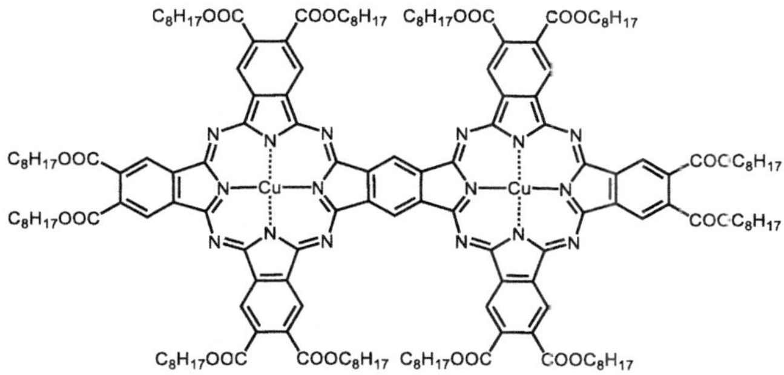

43

$Cu_2[Pc(COOC_8H_{17})_6]_2$

Fig. 10 Structure of the binuclear phthalocyaninato copper (II) complex, $Cu_2[Pc(COOC_8H_{17})_6]_2$ (**43**)

complexes ($RE(Pc)_2$) by experimental [155] and theoretical studies [156]. Furthermore, Chen et al. found that the electronic property and number of peripheral substituents on the Pc rings could be effective factors for tuning the semiconducting nature among *p*-, *n*-, and even ambipolar types for $RE(Pc)_2$ [157, 158].

Triple-decker phthalocyaninato rare-earth complexes ($RE_2(Pc)_3$) have further extended conjugation system and in turn further decreased molecular band gaps than the corresponding double deckers ($RE(Pc)_2$). The outstanding *p*-type OFET properties of triple deckers had already been demonstrated till 2009 [127, 129–131]. Recently, Jiang, Chen, and coworkers developed a series of air-stable and solution-processable ambipolar OFETs on the bases of Pc-involved triple deckers [159–163]. By incorporating suitable electron-withdrawing groups (phenoxyl, *p*-fluorophenoxyl, or naphthoxyl) at the periphery of Pc ligands in the triple deckers, the HOMO and LUMO energy levels were precisely adjusted to the window for air-stable ambipolar transporting. Long-range ordered thin films of the triple deckers were prepared by a solution-based QLS method. High-performance air-stable ambipolar OFET devices were obtained for triple decker $Eu_2[Pc(OPhF)_8]_3$ (**44**, Fig. 11) with hole and electron mobilities of 0.24 and 0.042 $cm^2\ v^{-1}\ s^{-1}$, respectively [159]. In a subsequent study, homoleptic triple decker $Eu_2[Pc(OPh)_8]_3$ (**45**)-based devices showed lower mobilities of 0.041 and 0.0026 $cm^2\ v^{-1}\ s^{-1}$ for electron and hole, respectively, in comparison with those of heteroleptic triple decker $(Pc)Eu[Pc(OPh)_8]Eu[Pc(OPh)_8]$ (**46**)-based ones, for which improved mobilities of 0.68 and 0.014 $cm^2\ v^{-1}\ s^{-1}$ for electron and hole, respectively, were obtained [160]. The improvement of charge mobilities was attributed to the change in molecular packing from *J*-aggregation for $Eu_2[Pc(OPh)_8]_3$ (**45**) to *H*-aggregation for $(Pc)Eu[Pc(OPh)_8]Eu[Pc(OPh)_8]$ (**46**) in the QLS films. Notably, a new version of heteroleptic triple-decker complex, $(Pc)Eu[Pc(ONPh)_8]Eu[Pc(ONPh)_8]$ (**47**) [161] showed high potential for air-stable ambipolar OFETs. Upon solvent vapor annealing (SVA) over the QLS films of $(Pc)Eu[Pc(ONPh)_8]Eu[Pc(ONPh)_8]$ (**47**), high and balanced mobilities of 1.71 and

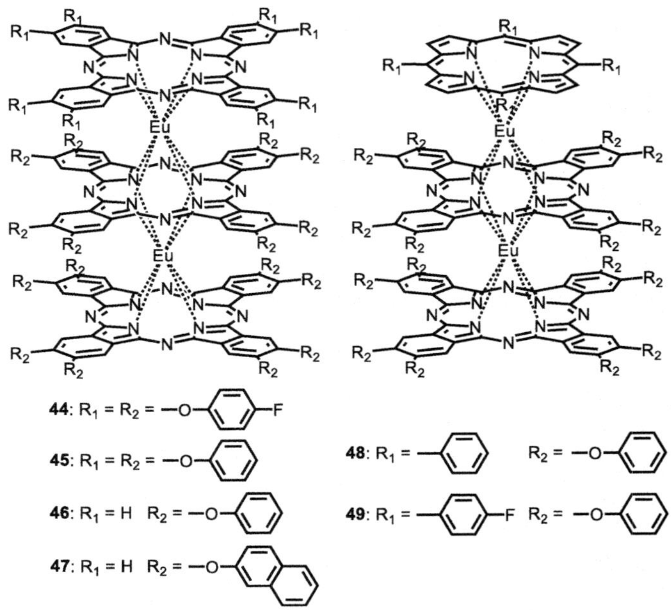

Fig. 11 Triple-decker rare-earth complexes for ambipolar OFETs

1.25 $cm^2\ v^{-1}\ s^{-1}$ for holes and electrons, respectively, could be obtained with high on/off ratios of 10^6 [161]. In addition, mixed (phthalocyaninato)(porphyrinato) triple-decker complexes (TPP)Eu[Pc(OPh)$_8$]Eu[Pc(OPh)$_8$] (**48**) [162] and (TFPP) Eu[Pc(OPh)$_8$]Eu[Pc(OPh)$_8$] (**49**, Fig. 11) [163] were developed and fabricated to OFETs by the solution-based QLS technique. Ambipolar transporting properties were also revealed for the single-component devices [162, 163] and the two-component bilayer heterojunction devices [163] as well, in which the heterojunction transistors of (TFPP)Eu[Pc(OPh)$_8$]Eu[Pc(OPh)$_8$]/CuPc showed competent performance with mobilities of 0.16 and 0.30 $cm^2\ v^{-1}\ s^{-1}$ for holes and electrons, respectively.

3 Organic Functional Dyes

3.1 Dye-Sensitized Solar Cells

Dye-sensitized solar cells (DSSCs) [164] are promising solar energy transformation devices due to the high efficiency, cost-effectiveness, and ease of fabrication [165]. The photophysical processes of DSSCs are carried by the main components of the light-harvesting dye, the wide-gap mesoporous semiconductor for electron transporting, and the redox couple in electrolyte for hole transporting. The

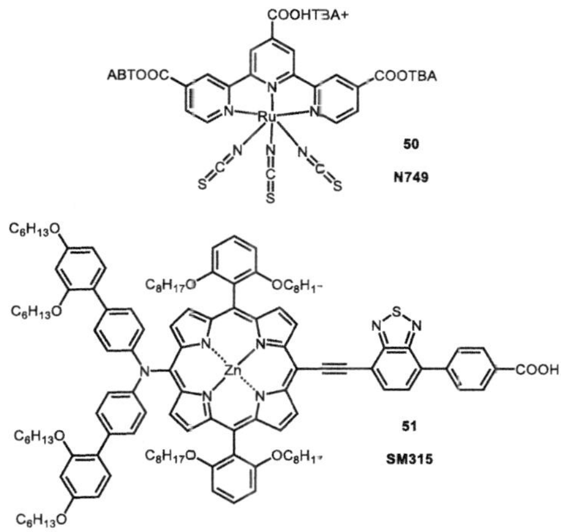

Fig. 12 Structures of the polypyridyl ruthenium (II) complex N749 (**50**) and the Por-based sensitizer SM315 (**51**)

light-harvesting dye, which absorbs solar photons and injects electrons to the semiconductor working electrode, plays a key role in achieving high PCE for the device. Despite the high-cost and environmental issues of polypyridyl ruthenium (II) complexes, PCEs of larger than 11% have been obtained, for example, by the use of black dye (N749, **50**) (Fig. 12), and complemented with D-π-A type organic co-sensitizers [166]. In order to explore low-cost and safe sensitizers, great efforts have been paid to ruthenium-free organic dyes; in particular, numerous sophisticatedly designed porphyrin (Por) dyes [167, 168] have been tested for DSSC applications and proved to be highly efficient sensitizers, for example, by the Por-based sensitizer SM315 (**51**) (Fig. 12); a record PCE of 13.0% was achieved in the absence of any co-sensitizer [168].

As one class of the most important synthetic analogues of Por's, Pc's are also attractive sensitizers because of the intense absorbance down to near-IR region and the excellent thermal and chemical stabilities [32, 34]. However, it has been revealed that the PCEs of Pc-based DSSCs are still lower than those of Por-based ones, mainly due to the weak absorption from 400 to 600 nm, the π–π aggregation between planar Pc rings, and the lack of electron D–A systems [35, 36].

In 1999 Nazeeruddin et al. reported the first examples of Pc-based DSSCs [169], in which an optimal dye of tetracarboxyphthalocyanine zinc(II) (**52**) (Fig. 13) was identified with PCE of 1.00%. After that, especially in recent years, significant progress has been achieved in Pc-based DSSCs through rational molecular design. At first, proper anchoring groups are important for strong binding of dyes to the TiO_2 surface, and short and conjugated spacers are suitable for efficient electron injection. In this regard, carboxyethynyl or carboxylic acid groups [170, 171] could provide "push–pull" directionality to the dye molecules, resulting in efficient electron injection from the excited dye to the conduction band (CB) of TiO_2. Secondly, dye aggregation should be minimized to avoid the rapid deactivation of the excited state. Bulk peripheral substituents and/or axial ligands [172–174] are useful to suppress the π–π staking of Pc dyes. Thirdly, the HOMO and LUMO

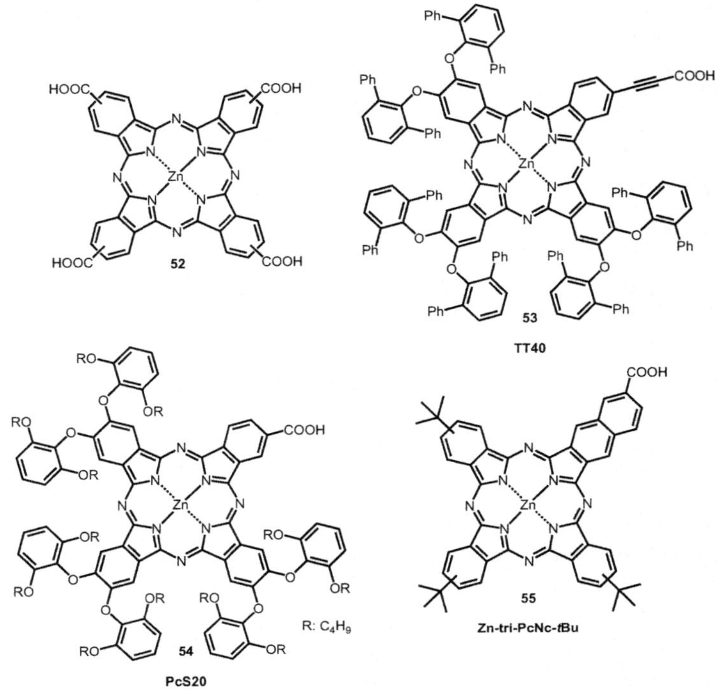

Fig. 13 Pc's for DSSCs

energy levels [175, 176] should be carefully adjusted for the dye regeneration from the electrolyte and the efficient electron injection into TiO_2. Finally, the molar absorption coefficient in a wide range of sunlight defines the light-harvesting property, which is also crucial for high-performance DSSCs.

The above concept of molecular design has fruited many elegant Pc dyes, for example, Torres, Nazeeruddin, and coworkers developed the TT*x* ($x = 1$–43) series [177–181], in which TT40 (**53**) (Fig. 13) showed the best performance with a PCE up to 6.01% under standard solar conditions [182]. Meanwhile, Kimura, Mori, and coworkers invented the PcS*X* ($X = 1$–21) series of Pc dyes [183–186], in which PcS20 (**54**) (Fig. 13) was found to be the most successful Pc dye to date with a record PCE of 6.40% [186]. Introduction of bulk diphenylphenoxy (TT40 (**53**)) and dibutoxylphenoxy (PcS20 (**54**)) substituents in the periphery of Pc rings was proved effective to suppress aggregation phenomena and to reduce the recombination at the TiO_2/electrolyte interface. Furthermore, Peng, Li, and coworkers designed the asymmetrical tribenzonaphtho-condensed Pc dyes [187–190]. The asymmetrical fusion of a benzo ring to one of the four benzopyrrole units of the Pc core induced not only the red shift of absorptions but also the directionality of the electronic orbitals in the excited state of the Pc dyes. As a result, high-performance DSSCs was obtained for Zn-tri-PcNc-*t*Bu (**55**) (Fig. 13) [187], with PCE of 3.65% under standard solar conditions, and the efficiency could be improved to 6.61% by a co-sensitization technique [191]. In addition, subphthalocyanines [192], hyperbranched phthalocyanine oligomers [193], and phthalocyanine–porphyrin dyads [194] were also tested for DSSCs and showed promising potentials.

3.2 Sensitizers for Photodynamic Therapy

Photodynamic therapy (PDT) is a highly controllable treatment for cancer in comparison with traditional surgery, radiation therapy, and chemotherapy. Generally, PDT process involves three key elements: a photosensitizer, light, and molecular oxygen. Upon photo irradiation, the photosensitizer is populated to the lowest excited singlet state (S_1), which then undergoes spin inversion and transforms to the lowest excited triplet state (T_1) via intersystem crossing (ISC). The T_1 state can trigger two kinds of processes defined as type I and type II reactions. Type I reactions feature directly charge transfer between the T_1 state of photosensitizer and local substrates, whereas type II reactions feature energy transfer from the T_1 state of photosensitizer to the ground state of molecular oxygen ($^3O_2$), producing highly reactive singlet oxygen ($^1O_2$) as the cytotoxic agent. The efficiency of PDT is strongly dependent on the properties of photosensitizers, particularly for type II reactions; the lifetime and quantum yield (τ_T and Φ_T) of the T_1 state are decisive for PDT efficiency. Other factors, such as an intense absorption in 600–900 nm region, water-soluble and amphiphilic characters, selective affinity toward tumor cells, and low dark toxicity, are also crucial for clinical application of PDT.

Conjugated cyclic tetrapyrrole compounds are one of the most successful class of PDT photosensitizers [9, 195, 196], in which Photofrin® (**56**) (Fig. 14), a mixture of porphyrin dimers and higher oligomers, is the first clinically approved photosensitizer. However, the first-generation photosensitizers, as represented by Photofrin®, suffer from the long-term skin photosensitivity and the limited light penetration depth. In order to overcome these problems, the second- and third-generation photosensitizers, with characteristics of long-wavelength absorptions

Fig. 14 Photofrin® (**56**)

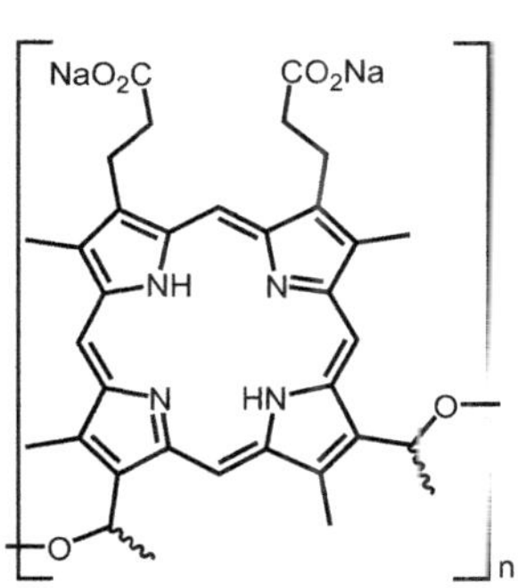

56

Photofrin: n = 1-9

and tumor-targeting functions, respectively, have emerged as the focus of PDT research in recent years.

Though Pc's usually have strong absorption in the NIR region due to the large π-conjugated system, their tendency of aggregation and low water solubility is unfavorable for PDT sensitizers. In this regard, various water-soluble substituents have been connected to the α- and/or β-positions of Pc rings, such as solketal [197], amine and its protonation salt [198, 199], and carboxylate acid groups [200, 201]. Supramolecular assembling with cyclodextrin also could improve the water solubility of hydrophobic Pc's [202]. In the above symmetrical Pc photosensitizers, the heavy-atom effect of platinum(II) [197] and hydriodate [198, 199] was suggested to enhance the intersystem crossing (ISC) and thus singlet oxygen generation, whereas the α-substitution [199] and π-extension [201] could shift the NIR absorption maxima to about 750 nm. In particular, the quaternized α-substituted Zn(Pc) derivative (2Zn-Et, **57**) (Fig. 15) [199] showed high singlet oxygen quantum yield (Φ_Δ) up to 0.91 and superb anticancer activity with the half-maximum inhibitory concentration (IC_{50}) value as low as 0.22 μM against the SK-MEL-28 cells.

Unsymmetrical modification of Pc ring is an effective way to produce Pc-based amphiphilic photosensitizers [203, 204]; more significantly, unsymmetrical Pc's can provide feasible platform for tumor-targeting and selectively activatable [205] strategies. In recent examples, Ke et al. connected linear and cyclic peptides, respectively, to the 1,4-disubstituted zinc(II) phthalocyanine by click chemistry. They found that the linear nuclear localization sequence (NLS) peptide (**58**, Fig. 16) could enhance the cellular uptake and tumor-retention properties [206], while the cyclic arginine–glycine–aspartic acid (RGD) sequence (**59**) could induce high selectivity toward integrin $\alpha_v\beta_3$ overexpressed cancer cells [207]. Alternatively, Zhang et al. adopted erlotinib, a small molecular anticancer drug, as the targeting moiety to modify Zn(Pc) core [208, 209]. The obtained Zn(Pc)-erlotinib conjugates (**60** and **61**) showed high specific affinity toward epidermal growth factor receptor (EGFR) overexpressed cancer cells. An oxaliplatin derivative was also linked to a

Fig. 15 The quaternized α-substituted Zn (Pc) derivative (2Zn-Et, **57**)

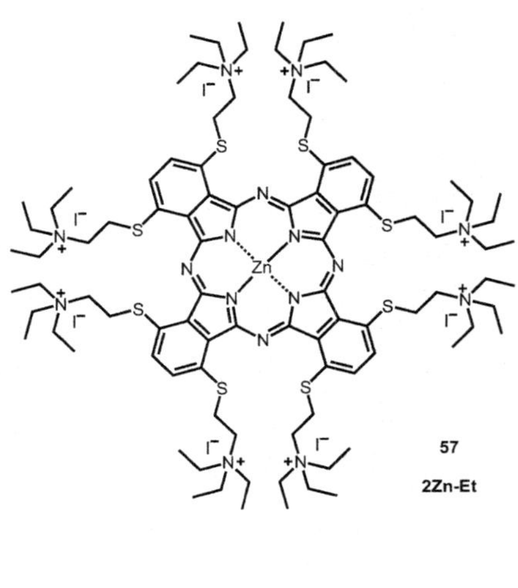

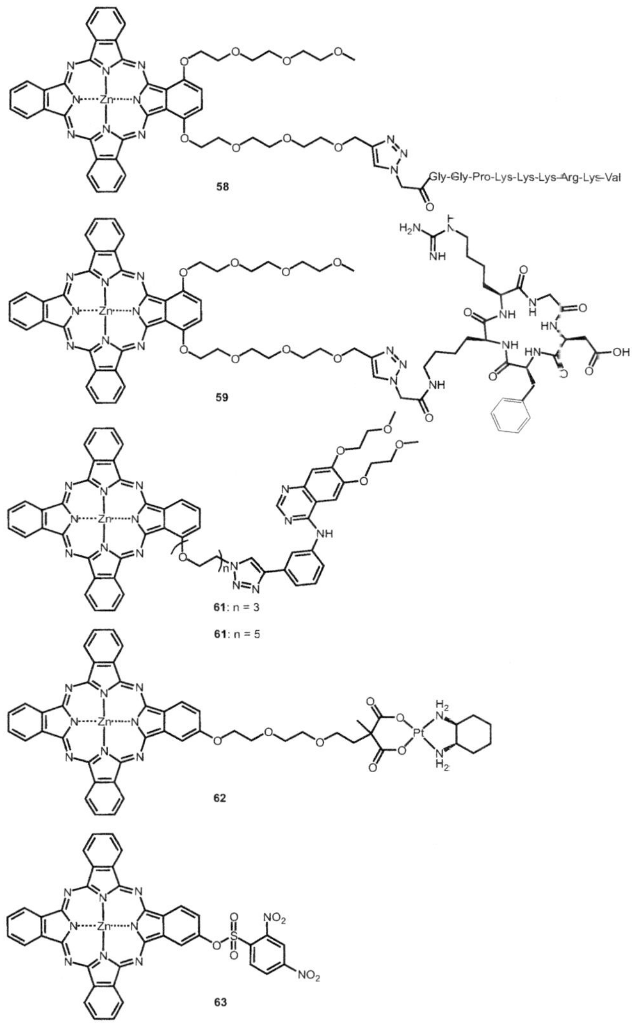

Fig. 16 Unsymmetrical modification of Pc's for PDT

Zn(Pc) core by Lau et al. [210] to construct a dual chemo- and photodynamic therapeutic agent (**62**); synergistic effects were observed with IC_{50} value of 0.11 μM toward the HT29 cells. Very recently, He et al. developed a "smart" photosensitizer by connecting a 2,4-dinitrobenzenesulfonate group (as a removable quencher) to the periphery of Zn(Pc) (**63**, Fig. 16) [205], which could be activated by glutathione (GSH) through cutting off the sulfonate ester linker.

Axial modification of silicon(IV) phthalocyanine (Si(Pc)) is another useful approach to Pc-based PDT sensitizers. The axial ligands can not only depress the tendency of aggregation and affect the photophysical properties [211–213] but also

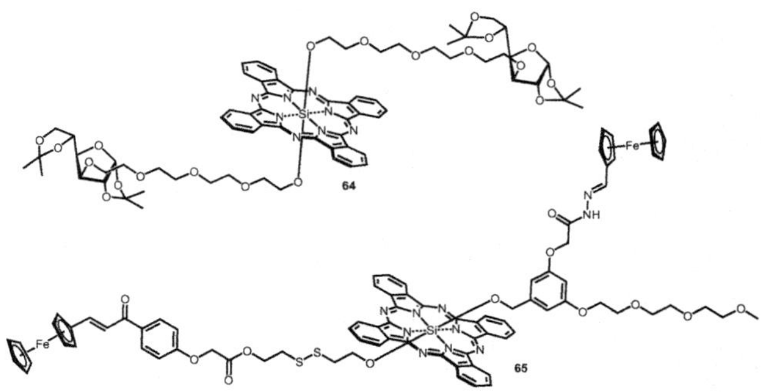

Fig. 17 Axial modified Si(Pc) for PDT

induce targeted delivery and stimuli-responsive vectors. For example, the gluco-conjugated Si(Pc) (**64**, Fig. 17) showed improved uptake and potent phototoxicity toward HT29 and HepG2 cells [214], and the polyamine-conjugated Si(Pc)s exhibited pH-responsive photodynamic activity [215–217]. Interestingly, Lau et al. reported an unsymmetrical bisferrocenyl-conjugated Si(Pc) (**65**, Fig. 17) [218], in which the disulfide and hydrazone connections can be cleaved by acid and dithiothreitol (DTT), respectively. The removal of the ferrocenyl quenchers switches on the singlet oxygen production of Si(Pc) core, leading to a dual-responsive photosensitizer.

In addition to the above molecular design strategies for Pc-based photosensitizers, current research interests have extended to the interdisciplines of chemistry, biology, and medical science to develop Pc-based pharmaceutical formulations. To increase the efficacy and reduce the side effects, the antibody-directed phototherapy (ADP) [219, 220] was first proposed on the basis of anti-M-1-hematoporphyrin conjugates. Recently, Kobayashi and coworkers first reported the Pc-based NIR photoimmunotherapy (PIT) formulations [221], in which monoclonal antibodies (mAb's) were employed to target epidermal growth factor receptors. Both in vitro and in vivo tests revealed prominent tumor specificity and potent PDT efficacy [222–225]. On the other hand, Huang and coworkers utilized an amino-terminal fragment (ATF) of urokinase as the targeting moiety for a Zn(Pc) dye [226]. The ATF-ZnPc photosensitizer showed high specificity toward cancer cells, where urokinase-type plasminogen activator receptors (uPARs) are overexpressed. In this direction, the recombinants of human serum albumin (HAS) and the ATF of urokinase were tested as tumor-targeting drug carriers for Zn(Pc) dyes, and positive results were obtained with highly specific phototoxicity on uPAR-expressing tumor cells [227, 228]. Another notable recent example of biomolecule-based targeting agents for Pc-based photosensitizers was reported by Liu and coworkers [229]. The formulation was constructed in such a way that the integrin $\alpha v\beta 6$-specific peptide HK was first linked to the biotin; thereafter the biotinylated HK peptide (B-HK) was connected to dye-labeled streptavidin (Dye-SA) through streptavidin–biotin binding chemistry. The resulted Dye-SA-B-HK (Fig. 18) showed specific integrin $\alpha v\beta 6$-targeting ability both in vitro and in vivo. In combination with the unique

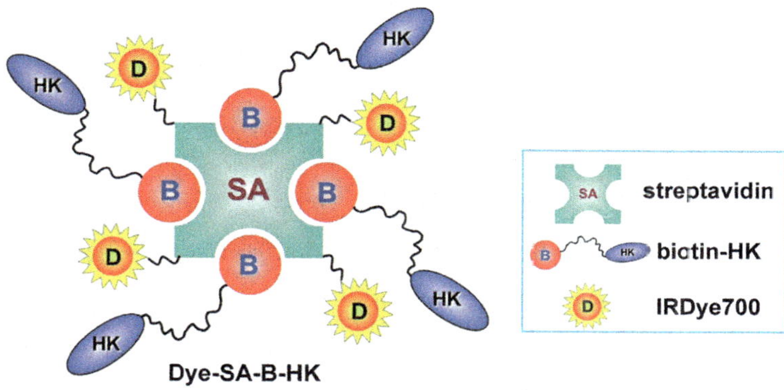

Fig. 18 Schematic drawing of Dye-SA-B-HK. Reproduced with modification and permission from [229]. Copyright 2015 Elsevier Ltd

photophysical properties of the SiPc-based NIR dye, optical imaging-guided surgery and phototherapy of pancreatic cancer were found to be successful in mouse models [229].

The progress of nanotechnology issued the concept of nanomedicine [230, 231], which was proved to be applicable for the Pc-based PDT field with virtues of targeted drug delivery, facile cellular uptake, and enhanced permeability and retention (EPR) properties. The current Pc-based PDT nanomedicine may broadly be classified to three categories: (1) Pc molecules self-assembled to nanometer-size objects, such as nanoparticles [232] and hollow nanospheres [233]; (2) Pc molecules encapsulated in the interior of nanocontainers [234], including the recent examples of polymer micelles [235, 236], dendrimers [237], and viruslike particles [238]; and (3) Pc molecules conjugated to the surface or shell of nanovehicles [239], which include the recent examples of gold nanoparticles [240–245], upconversion fluorescent nanoparticles [246–248], and magnetic nanoparticles [249]. These elegant examples demonstrate the great potential of nano-platform, by which improved PDT efficacy and multifunctional formulations can be realized for Pc-based photosensitizers.

At the end of this section, it should be noted that Pc-based dyes have also been involved in the applications of optical imaging [250], antimicrobial PDT [251–253], and photothermal therapy (PTT) of diseases [254], which have shown promising results and therefore will receive increasing research interests.

4 Single-Molecule Magnets

Single-molecule magnets (SMMs) are molecules that exhibit spontaneous magnetization in the absence of a collective magnetic order and show slow relaxation of magnetization below a characteristic blocking temperature (T_B) due to an energy barrier to magnetization reversal [255, 256]. Due to the unique quantum mechanics of SMMs, they become important prototype models for the understanding of modern magnetism and hold potentials in quantum computing, spintronics, and information storage applications.

4.1 Double-Decker Lanthanide SMMs

Since the first report of a bis(phthalocyaninato) lanthanide SMM by Ishikawa et al. in 2003 [11], the sandwich-type phthalocyaninato lanthanide complexes have emerged as an important research focus for SMMs. In the sandwich-type molecular structures, the neighboring Pc ligands form a C_4 symmetrical (*pseudo-*D_{4d}) ligand field (LF) (Fig. 19), which induces the zero-field splitting of the J ground state of lanthanide ion (Ln) into new sublevels. This LF-induced uniaxial magnetic anisotropy, together with the intrinsic large-spin ground state (S) and the strong spin-orbit interactions, leads to significant relaxation energy barriers for this class of SMMs [74, 257].

Recently, Fukuda et al. carefully investigated the magnetic relaxation dynamics of the anionic bis(phthalocyaninato) terbium SMM ($[Tb(Pc)_2]^-.TBA^+$, $TBA^+ = (C_4H_9)_4N^+$) in the nonthermally activated temperature range [258]. For the 1:49 diluted sample with the diamagnetic isostructural yttrium complex ([Y

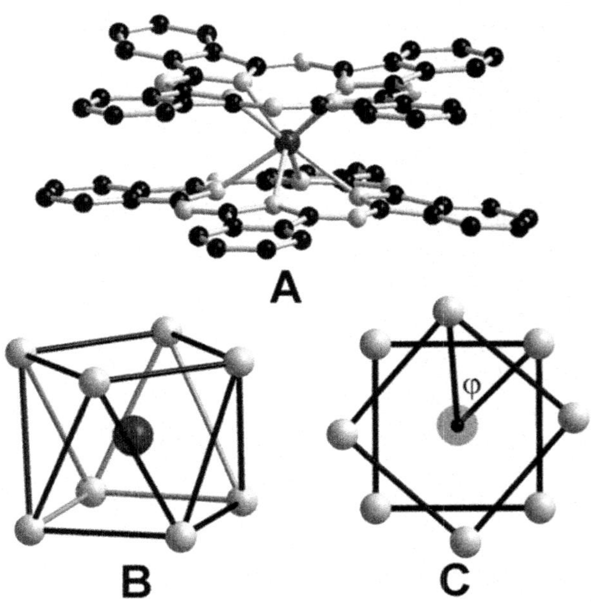

Fig. 19 (**a**) Molecular structure of bis (phthalocyaninato) lanthanide complexes, (**b**) *pseudo-*D_{4d} ligand field, and (**c**) the twist angle. Reproduced with permission from [257]. Copyright 2015 Elsevier B.V

$(Pc)_2]^{-}.TBA^{+}$), they found that the relaxation time depends on the externally applied direct current (dc) magnetic field and temperature, which suggests that the direct process dominates the magnetic relaxation pathway in the lower temperature range. Because the *pseudo*-D_{4d} LF of bis(phthalocyaninato) terbium complexes can be tuned by the redox states of molecules [259, 260] and the substituents on Pc ligands [261–267], these are useful approach for adjusting the sublevel structure of the lanthanide ions and in turn for modulating the SMM performance. Notably, Cao et al. reported an effective strategy to enhance the anisotropy and energy barrier of double-decker SMMs via atom replacement [268]. They incorporated a core-modified porphyrin, in which one of the four pyrrole nitrogen atoms was replaced by an oxygen or sulfur atom, into a mixed (phthalocyaninato) (porphyrinato) dysprosium complex to adjust the LF. This rational modification induced significant enhancement of the SMM performance, as such the relaxation energy barriers increased from 28 cm^{-1} for **66** to 94 and 135 cm^{-1} for **67** and **68**, respectively (Fig. 20). Impressively, the energy barrier of 135 cm^{-1} for **68** represents the highest one for all of the bis(tetrapyrrole) dysprosium SMMs till now. Furthermore, in a series of mixed (phthalocyaninato)(porphyrinato) dysprosium complexes [269], Wang et al. found that the twist angle between the two ligands showed clear effect on the quantum tunneling (QT) of the SMMs.

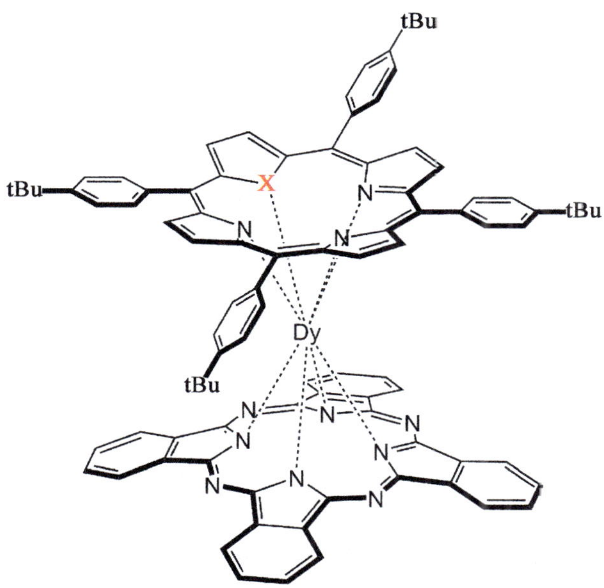

Fig. 20 Schematic molecular structure of the mixed (phthalocyaninato)(porphyrinato) dysprosium complexes: **66** (X = N), **67** (X = O), and **68** (X = O)

4.2 Triple-Decker Lanthanide SMMs

Heteroleptic tris(phthalocyaninato) lanthanide SMMs were reported as early as 2005 [270]. However, the SMM behavior of the homoleptic tris(phthalocyaninato) di-terbium triple-decker analogues was not studied until 2011 [271]. One of the most obvious structural features for these binuclear systems is the short intramolecular Tb–Tb distance (ca. 3.5 Å) [271], which is expected to lead to significant f–f interactions. Magnetic studies revealed dual relaxation processes for the dinuclear Tb–Tb complexes in the low-temperature region in the presence of a dc magnetic field, but a single relaxation process for the mononuclear Tb complexes. These results suggest that the magnetic relaxation depends on the intramolecular dipole–dipole interaction in the triple-decker Tb SMMs. In addition, a comparative investigation of the SMM properties of the mono- and dinuclear Dy phthalocyaninate complexes [272] showed that the relaxation time for the bis-Dy SMMs was enhanced relative to that of the mono-Dy species.

Furthermore, mixed (phthalocyaninato)(porphyrinato) triple-decker complexes of terbium [273] and dysprosium [274] were also investigated to clarify the effects of different coordination environments and intramolecular f–f interactions on the SMM properties. For the bis-Tb compound **69**, one Tb ion surrounded by two phthalocyanine ligands shares a square-antiprism (SAP) coordination field (with a twist angle of 43.6°), while the other Tb ion sandwiched between phthalocyanine and porphyrin ligands has a square-prism (SP) coordination geometry (with a twist angle of 1.2°) (Fig. 21). Similar to the heteroleptic tris(phthalocyaninato) system [270], the intramolecular interionic interaction in the bis-Tb compound **69** is ferromagnetic based on its static magnetic characteristics, and the quantum tunneling of magnetization (QTM) is highly suppressed by the intramolecular dipole–dipole (f–f) interactions. However, in the case of a bis-Dy triple decker [274], both Dy ions site in distorted SAP octa-coordination geometries with twist angles of ca. 25° and 10°, respectively. It was found that the coordination geometry of the spin carriers, instead of intramolecular ionic f–f interactions, plays a dominant role in determining the SMM properties [274].

4.3 Tetrakis- and Pentakis(Phthalocyaninato) Lanthanide SMMs

Since 2010, continuous progress has been achieved in the synthesis of sandwich-type multikis(phthalocyaninato) rare-earth complexes with more than three tetrapyrrole decks [275]. Thus, the availability of lanthanide multiple-decker complexes allowed the investigation of the effects of f–f interactions systematically.

The first tetrakis(phthalocyaninato) lanthanide SMM was reported in 2011 by Wang et al. [276]. It was found that the average twist angle of the $[(Pc^{12})Dy(Pc^{12})]^{-}$ subunits is 20° and the average separation between the intramolecular Dy(III) ions

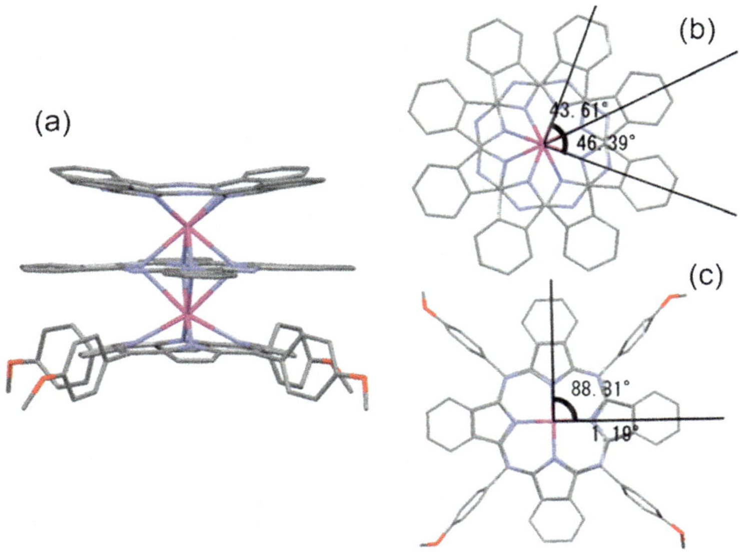

Fig. 21 Crystal structures of the mixed (phthalocyaninato)(porphyrinato) terbium triple-decker complex (**69**) from three perspectives: (**a**) *side*, (**b**) *top* and (**c**) *bottom*; protons are omitted for clarity. Reproduced with permission from [273]. Copyright 2012 the Royal Society of Chemistry

is 6.7 Å in the tetrakis(phthalocyaninato) dysprosium–cadmium quadruple decker **70** (Fig. 22). The frequency-dependent dynamic magnetization of this compound indicated the slow magnetization relaxation. The energy barriers are 15.6 cm^{-1} ($\tau_0 = 3.6 \times 10^{-7}$ s) and 14.4 cm^{-1} ($\tau_0 = 8.1 \times 10^{-7}$ s) in zero and 2,000 Oe dc magnetic fields, respectively, which are smaller than that for $[(Pc)Dy(Pc)]^{-}.TBA^{+}$ (28.0 cm^{-1}) due to the non-ideal D_{4d} local symmetry of the dysprosium ion and the presence of dipole–dipole interactions in the quadruple-decker compound.

In order to clarify the presence of intramolecular lanthanide–lanthanide (f–f) interactions and their effects on the SMM properties, a series of homoleptic and heteroleptic tetrakis(phthalocyaninato) terbium–cadmium quadruple deckers were studied [277]. The magnetic measurements confirmed the presence of inter-Tb ionic ferromagnetic interactions in the bis-Tb compounds where the intramolecular Tb ions are separated by 6.6 Å. The intramolecular f–f interactions lead to the effective suppression of QTM in the bis-terbium quadruple decker compared with the mono-terbium counterparts. This also results in an enhanced energy barrier of $U_{eff} = 228.6$ cm^{-1} ($\tau_0 = 5.5 \times 10^{-9}$) for the bis-terbium **SMM** relative to that of the mono-terbium counterparts.

Ishikawa et al. prepared another series of quadruple-decker complexes [(Pc)Ln(Pc)Cd(Pc)Ln(Pc)] (Ln-Cd-Ln = Tb-Cd-Tb, Dy-Cd-Dy, Er-Cd-Er) [278, 279]. As expected, comparative studies of their magnetic properties demonstrated their SMM nature and confirmed the presence of long-range f–f interactions in the quadruple-decker compounds. The effect of f–f interactions on suppressing the

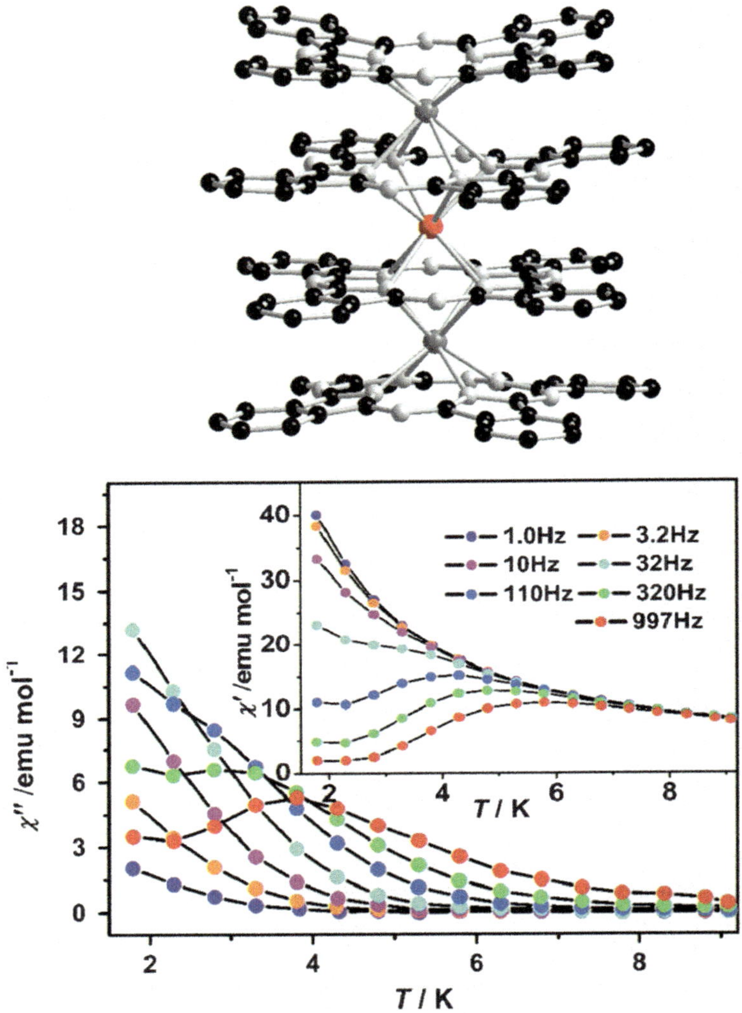

Fig. 22 Crystal structure of **70** (*top*) and plots of χ_M' and χ_M'' vs. T (*bottom*) under a zero dc magnetic field. Reproduced with permission from [276]. Copyright 2011 the Royal Society of Chemistry

QTM was also shown in the low-temperature region under a small dc magnetic field. However, under a higher magnetic field such as 1,000 Oe, the QTM was enhanced in the bis-lanthanide SMMs compared with their mono-lanthanide counterparts due to the enhanced nuclear spin-driven QTMs. Recently, Shang et al. investigated the SMM properties of tetrakis(phthalocyaninato) dysprosium (III)-cadmium(II) quadruple-decker compounds with a different extent of phthalocyanine peripheral substitution [280]. The results indicated that introducing different phthalocyanine ligands into quadruple-decker compounds is a simple and effective way to tune the magnetic properties.

In order to demonstrate the distance-dependent nature of the intramolecular dipole–dipole (f–f) interactions, Yamashita and coworkers studied the magnet properties of pentakis(phthalocyaninato) lanthanide–cadmium quintuple-decker compounds and compared with those of other multiple-decker complexes [281, 282]. The results showed that, along with the increase in the inter-terbium ionic separation from 3.5 Å for the triple decker, to 6.6 Å for the quadruple decker, to 9.9 Å for the quintuple decker, the magnetic relaxation mechanism was significantly changed, leading to the conclusion that the SMM properties of these dinuclear systems are sensitive to the dipole–dipole (f–f) interactions and thus to the inter-ion distances.

4.4 Fused Phthalocyanine Lanthanide Sandwich SMMs

Two bis(phthalocyaninato) dysprosium(III) units can be fused by a conjugated dimeric Pc ligand forming a novel sandwich-type tetrapyrrole lanthanide architecture **71** (Fig. 23) [283]. At the neutral state, two π-radicals exist in this novel sandwich molecule as elucidated by the spectroscopic results and theoretical calculations, where each π-radical electron delocalizes over an individual double-decker subunit. The static magnetic properties suggest that antiferromagnetic interactions present between the intramolecular radicals, the radical and Dy ion, as well as the intramolecular Dy ions in **71**. The dynamic magnetic behaviors of the neutral and reduced forms of **71** (denoted as **71a** and **71b**) also indicate the SMM nature of both forms of the di-dysprosium sandwich architecture. Nevertheless, the intramolecular biradical-involved antiferromagnetic interaction is helpful for suppressing the QTM in a zero dc magnetic field.

A tetra-terbium sandwich SMM **72** was prepared using the same Pc ligands [284], in which the two triple-decker subunits are connected by a common benzene ring (Fig. 24). Ferromagnetic interactions dominate among the four Tb^{3+} ions in this multiple-decker system, including strong intramolecular dipole–dipole interactions within the triple-decker subunit and weak ones between the two triple-decker subunits. As expected, this tetra-terbium sandwich skeleton also exhibits slow relaxation of magnetization with an energy barrier of 149 cm^{-1} ($\tau_0 = 2.7 \times 10^{-8}$ s). In addition, the slow relaxation behavior of **72** in a high temperature range is dominated by a spin–lattice interaction-induced Orbach process, which is independent of the dc magnetic field. By contrast, in a low-temperature range, the magnetic relaxation process is split into dual processes due to a direct or phonon-induced tunneling process.

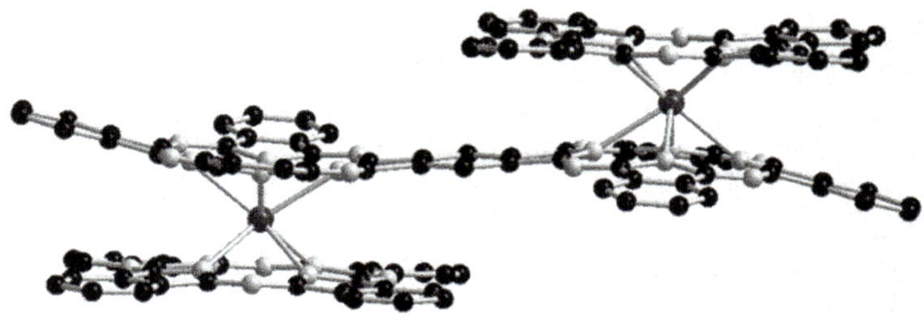

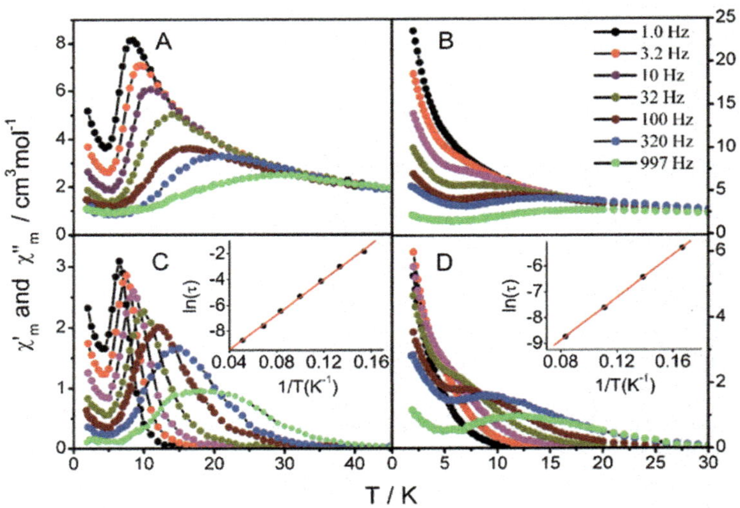

Fig. 23 Temperature dependency of the in-phase and out-of-phase ac magnetic susceptibility of **71a** (**a** and **c**) and **71b** (**b** and **d**) under a zero dc field. Reproduced with permission from [283]. Copyright 2013 Wiley-VCH Verlag GmbH&Co. KGaA, Weinheim

4.5 SMM Properties on Surfaces

Due to the relatively high blocking temperature and long magnetic relaxation time at low temperature, the sandwich-type phthalocyaninato lanthanide complexes are considered to be one of the most promising SMMs [257, 285]. Their unique molecule quantum effect, together with the tunable molecular band gap and supramolecular properties, endow promising potential applications in molecular spintronic devices [286–288]. Significant advances have been achieved in this area. For example, Yamashita and coworkers could observe and switch the Kondo signal of different double-decker molecules on an Au(111) surface by scanning tunneling microscopy (STM) and spectroscopy (STS) [289–292]. Vincent et al. performed the electronic transport studies for determining the four hyperfine nuclear spin states of Tb^{3+} in a $[(Pc)Tb(Pc)]^0$ SMM [293], while Thiele

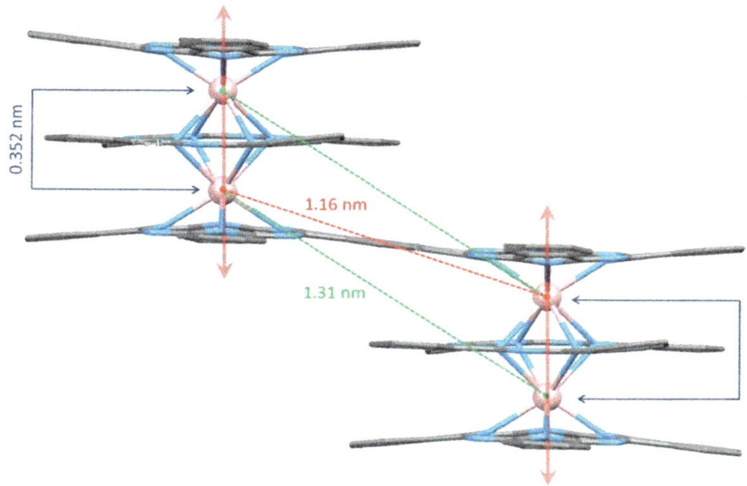

Fig. 24 Simulated structure of the tetra-terbium sandwich SMM **72**. Substituents are omitted for clarity. Reproduced with permission from [284]. Copyright 2013 American Chemical Society

et al. demonstrated coherent manipulation of the four-level nuclear spin qubit transistor of $[(Pc)Tb(Pc)]^0$ by using of the hyperfine Stark effect [294], which was suggested to be a general method for the electrical control of nuclear-spin-based devices.

In order to deeply understand the properties of SMMs on surfaces and the interaction between a substrate and the localized spin in SMMs, bis (phthalocyaninato) lanthanide SMMs, in particular $[(Pc)Tb(Pc)]^0$, were applied to various substrates including Cu(100) surface [295], Al foil [296], Si substance [297], carbon materials [29, 298–305], and others [306–308]. The experimental results demonstrate that the spin relaxation and magnetic anisotropy of double-decker SMMs are readily influenced by the nature of the substrate and the molecular arrangement. However, for the practical application of spintronics, more research is required to control over the interface effect and the spin transport properties as well. The interested readers in this field are addressed to the recent review papers [286–288].

5 Summary and Outlook

In this chapter, we try to outline the advances for the application of phthalocyanines as functional molecular materials in the recent 5 years. The scope is further converged to the topics of semiconducting materials for organic photovoltaic cells and organic field effect transistors, functional photosensitizers for photodynamic therapy and dye-sensitized solar cells, and single-molecule magnets. For each topic, selected examples have been emphasized to address the significant achievement. Although great progress has been made in the field of functional Pc's, theoretical and technical challenges still exist for the practical applications. The performance and efficiency need to be improved, which should rely on the deep understanding of the structure–function relationship of Pc's. The stability and the bio- and environmental compatibility might be new challenges for researchers involved in the phthalocyanine chemistry and related materials science.

Acknowledgement Financial support from the Natural Science Foundation of China, National Key Basic Research Program of China (Grant Nos. 2012CB224801 and 2013CB933400), Beijing Natural Science Foundation, and Fundamental Research Funds for the Central Universities is gratefully acknowledged.

References

1. Leznoff CC, Lever ABP (eds) (1989) Phthalocyanines: properties and applications. VCH, Cambridge
2. McKeown NB (1998) Phthalocyanine materials: structure, synthesis and function. Cambridge University Press, Cambridge
3. Kadish KM, Smith KM, Guilard R (eds) (2003) Phthalocyanines: synthesis, vol 15, The porphyrin handbook. Academic, San Diego
4. Kadish KM, Smith KM, Guilard R (eds) (2003) Phthalocyanines: structural characterization, vol 20, The porphyrin handbook. Academic, San Diego
5. Kadish KM, Smith KM, Guiland R (eds) (2003) Phthalocyanines: spectroscopic and electrochemical characterization, vol 16, The porphyrin handbook. Academic, San Diego
6. Peumans P, Uchida S, Forrest SR (2003) Nature 425(6954):158–162
7. Horowitz G (1998) Adv Mater 10(5):365–377
8. Huang Q, Evmenenko GA, Dutta P, Lee P, Armstrong NR, Marks TJ (2005) J Am Chem Soc 127(29):10227–10242
9. Bonnett R (1995) Chem Soc Rev 24(1):19–33
10. He J, Benkoe G, Korodi F, Polivka T, Lomoth R, Kermark B, Sun L, Hagfeldt A, Sundstroem V (2002) J Am Chem Soc 124(17):4922–4932
11. Ishikawa N, Sugita M, Ishikawa T, Koshihara S, Kaizu Y (2003) J Am Chem Soc 125 (29):8694–8695
12. de la Torre G, Vazquez P, Agullo-Lopez F, Torres T (2004) Chem Rev 104(9):3723–3750
13. Sorokin AB (2013) Chem Rev 113(10):8152–8191
14. Mele G, Del Sole R, Vasapollo G, Garcia-Lopez E, Palmisano L, Schiavello M (2003) J Catal 217(2):334–342
15. Reetz MT, Jiao N (2006) Angew Chem Int Ed 45(15):2416–2419
16. Zhang QM, Li H, Poh M, Xia F, Cheng ZY, Xu H, Huang C (2002) Nature 419 (6904):284–287
17. Somani PR, Radhakrishnan S (2003) Mater Chem Phys 77(1):117–133
18. Gryko D, Li J, Diers JR, Roth KM, Bocian DF, Kuhr WG, Lindsey JS (2001) J Mater Chem 11(4):1162–1180
19. van Nostrum CF, Picken SJ, Schouten A, Nolte RJM (1995) J Am Chem Soc 117 (40):9957–9965
20. Matsushita O, Derkacheva VM, Muranaka A, Shimizu S, Uchiyama M, Luk'Yanets EA, Kobayashi N (2012) J Am Chem Soc 134(7):3411–3418

21. Wang H, Qi D, Xie Z, Cao W, Wang K, Shang H, Jiang J (2013) Chem Commun 49 (9):889–891
22. Bottari G, de la Torre G, Guldi DM, Torres T (2010) Chem Rev 110(11):6768–6816
23. Bottari G, de la Torre G, Torres T (2015) Acc Chem Res 48(4):900–910
24. Spitler EL, Dichtel WR (2010) Nat Chem 2(8):672–677
25. Imahori H, Umeyama T, Kurotobi K, Takano Y (2012) Chem Commun 48(34):4032–4045
26. de la Torre G, Claessens CG, Torres T (2007) Chem Commun 43(20):2000–2015
27. Abel M, Clair S, Ourdjini O, Mossoyan M, Porte L (2011) J Am Chem Soc 133(5):1203–1205
28. Baeg K, Binda M, Natali D, Caironi M, Noh Y (2013) Adv Mater 25(31):4267–4295
29. Urdampilleta M, Klyatskaya S, Cleuziou JP, Ruben M, Wernsdorfer W (2011) Nat Mater 10 (7):502–506
30. Martinez-Diaz MV, de la Torre G, Torres T (2010) Chem Commun 46(38):7090–7108
31. Walter MG, Rudine AB, Wamser CC (2010) J Porphyrins Phthalocyanines 14(9):759–792
32. Wang X, Tamiaki H (2010) Energy Environ Sci 3(1):94–106
33. Bottari G, Trukhina O, Ince M, Torres T (2012) Coord Chem Rev 256(21–22):2453–2477
34. Ragoussi M, Ince M, Torres T (2013) Eur J Org Chem 2013(29):6475–6489
35. Singh VK, Kanaparthi RK, Giribabu L (2014) RSC Adv 4(14):6970–6984
36. Martin-Gomis L, Fernandez-Lazaro F, Sastre-Santos A (2014) J Mater Chem A 2 (38):15672–15682
37. Imahori H, Kurotobi K, Walter MG, Rudine AB, Wamser CC (2012) Porphyrin and phthalocyanine based solar cells. In: Kadish KM, Smith KM, Guilard R (eds) Handbook of porphyrin science, vol 18. World Scientific, Singapore, pp 57–121
38. Kesters J, Verstappen P, Kelchtermans M, Lutsen L, Vanderzande D, Maes W (2015) Adv Energy Mater. doi:10.1002/aenm.201500218
39. Trogler WC (2012) Struct Bond 142:91–118
40. Bill NL, Trukhina O, Sessler JL, Torres T (2015) Chem Commun 51(37):7781–7794
41. Rodriguez-Mendez ML, Apetrei C, Medina-Plaza C, Munoz R, de Saja JA (2014) Sensor arrays based on phthalocyanines: new developments on nanostructured and biomimetic electrochemical sensors. In: Lvova L, Kirsanov D, Di Natale C, Legin A (eds) Multisensor systems for chemical analysis. Pan Stanford, Singapore, pp 139–179. doi:10.1201/b15491-5
42. Josefsen LB, Boyle RW (2012) Theranostics 2(9):916–966
43. Ishii K (2012) Coord Chem Rev 256(15–16):1556–1568
44. Sekkat N, van den Bergh H, Nyokong T, Lange N (2012) Molecules 17:98–144
45. Ali H, van Lier JE (2010) Porphyrins and phthalocyanines as photosensitizers and radiosensitizers. In: Kadish KM, Smith KM, Guilard R (eds) Handbook of porphyrin science, vol 4. World Scientific, Singapore, pp 1–119
46. Nyokong T (2011) Pure Appl Chem 83(9):1763–1779
47. Pietrangeli D, Rosa A, Ristori S, Salvati A, Altieri S, Ricciardi G (2013) Coord Chem Rev 257(15–16):2213–2231
48. Jiang Z, Shao J, Yang T, Wang J, Jia L (2014) J Pharm Biomed 87:98–104
49. Sergeeva NN, Senge MO (2012) Photochemical transformations involving porphyrins and phthalocyanines. In: Griesbeck A, Oelgemoller M, Ghetti F (eds) CRC handbook of organic photochemistry and photobiology, vol 1, 3rd edn. CRC, Boca Raton, pp 831–879. doi:10.1201/b12252-35
50. Ding L, Luan L, Shi J, Liu W (2013) Wuji Huaxue Xuebao 29(8):1591–1598, In Chinese
51. Komeda T, Katoh K, Yamashita M (2014) Prog Surf Sci 89(2):127–160
52. Chen J, Wang S, Yang G (2015) Wuli Huaxue Xuebao 31(4):595–611, In Chinese
53. Venugopal Rao S (2012) Proc SPIE 8434(Nonlinear Optics and Applications VI):84341B
54. Zagal JH, Griveau S, Silva JF, Nyokong T, Bedioui F (2010) Coord Chem Rev 254 (23–24):2755–2791
55. Sorokin AB, Kudrik EV (2010) Catal Today 159(1):37–46
56. Kadish KM, Smith KM, Guilard R (eds) (2010) Electronic absorption spectra-phthalocyanines, vol 9, Handbook of porphyrin science. World Scientific, Singapore

57. Bian Y, Zhang Y, Ou Z, Jiang J (2011) Chemistry of sandwich tetrapyrrole rare earth complexes. In: Kadish KM, Smith KM, Guilard R (eds) Handbook of porphyrin science, vol 14. World Scientific, Singapore, pp 249–460
58. Ragoussi M, Torres T (2014) Chem Asian J 9(10):2676–2707
59. Mack J, Kobayashi N (2011) Chem Rev 111(2):281–321
60. Dumoulin F, Ahsen V (2011) J Porphyrins Phthalocyanines 15(7–8):481–504
61. Cook MJ, Chambrier I (2011) J Porphyrins Phthalocyanines 15(3):149–173
62. Gorbunova YG, Martynov AG, Tsivadze AY (2012) Crown-substituted phthalocyanines. From synthesis towards materials. In: Kadish KM, Smith KM, Guilard R (eds) Handbook of porphyrin science, vol 24. World Scientific, Singapore, pp 271–388
63. Figueira F, Pereira PMR, Silva S, Cavaleiro JAS, Tome JPC (2014) Curr Org Synth 11 (1):110–126
64. Kobayashi N (2012) Synthesis and characterization of chiral phthalocyanines. In: Kadish KM, Smith KM, Guilard R (eds) Handbook of porphyrin science, vol 23. World Scientific, Singapore, pp 373–440
65. Jiang J (2010) Functional phthalocyanine molecular materials. In: Mingos M (ed) Structure bonding, vol 135. Springer, Berlin. doi:10.1007/978-3-642-04752-7
66. Nyokong T (2010) Struct Bond 135:45–88
67. Luo Q, Liu Y, Tian H (2010) Struct Bond 135:89–104
68. Bekaroglu O (2010) Struct Bond 135:105–136
69. Liu J, Lo P, Ng DKP (2010) Struct Bond 135:169–210
70. de la Torre G, Bottari G, Hahn U, Torres T (2010) Struct Bond 135:1–44
71. Yoshimoto S, Kobayashi N (2010) Struct Bond 135:137–168
72. Zhang Y, Cai X, Bian Y, Jiang J (2010) Struct Bond 135:275–322
73. Li X, Wang H, Wu H (2010) Struct Bond 135:229–274
74. Ishikawa N (2010) Struct Bond 135:211–228
75. Tang CW (1986) Appl Phys Lett 48(2):183–185
76. Osedach TP, Andrew TL, Bulović V (2013) Energy Environ Sci 6(3):711–718
77. Xu Z, Roy VAL, Low K, Che C (2011) Chem Commun 47(34):9654–9656
78. Qu D, Guo R, Yue S, Wu Y, Yan P, Cheng G (2014) J Phys D 47(41):415104
79. Meiss J, Merten A, Hein M, Schuenemann C, Schäfer S, Tietze M, Uhrich C, Pfeiffer M, Leo K, Riede M (2012) Adv Funct Mater 22(2):405–414
80. Brendel M, Krause S, Steindamm A, Topczak AK, Sundarraj S, Erk P, Höhla S, Fruehauf N, Koch N, Pflaum J (2015) Adv Funct Mater 25(10):1565–1573
81. Beaumont N, Hancox I, Sullivan P, Hatton RA, Jones TS (2011) Energy Environ Sci 4 (5):1708–1711
82. Zhong S, Zhong JQ, Wang XZ, Huang MY, Qi DC, Chen ZK, Chen W (2012) J Phys Chem C 116(3):2521–2526
83. Verreet B, Müller R, Rand BP, Vasseur K, Heremans P (2011) Org Electron 12 (12):2131–2139
84. Vasseur K, Rand BP, Cheyns D, Froyen L, Heremans P (2011) Chem Mater 23(3):886–895
85. Vasseur K, Rand BP, Cheyns D, Temst K, Froyen L, Heremans P (2012) J Phys Chem Lett 3 (17):2395–2400
86. Suzuki A, Furukawa R, Akiyama T, Oku T (2015) AIP Conf Proc 1649(1, Irago Conference 2014):107–112
87. Hori T, Fukuoka N, Masuda T, Miyake Y, Yoshida H, Fujii A, Shimizu Y, Ozaki M (2011) Sol Energy Mater Sol Cells 95(11):3087–3092
88. Varotto A, Nam C, Radivojevic I, Tomé JPC, Cavaleiro JAS, Black CT, Drain CM (2010) J Am Chem Soc 132(8):2552–2554
89. Jurow MJ, Hageman BA, DiMasi E, Nam C, Pabon C, Black CT, Drain CM (2013) J Mater Chem A 1(5):1557–1565
90. Fischer MKR, López-Duarte I, Wienk MM, Martínez-Díaz MV, Janssen RAJ, Bäuerle P, Torres T (2009) J Am Chem Soc 131(24):8669–8676

91. Schumann S, Hatton RA, Jones TS (2011) J Phys Chem C 115(11):4916–4921
92. Ryan JW, Anaya-Plaza E, Escosura ADL, Torres T, Palomares E (2012) Chem Commun 48 (49):6094–6096
93. Jiang J, Ng DKP (2009) Acc Chem Res 42(1):79–88
94. Videlot C, Fichou D, Garnier F (1998) Mol Cryst Liq Cryst Sci Technol Sect A 322 (1):319–328
95. Liu L, Hu AT (2003) J Porphyrins Phthalocyanines 7(8):565–571
96. Liu MO, Hu AT (2004) J Organomet Chem 689(15):2450–2455
97. Wang Q, Li Y, Yan X, Rathi M, Ropp M, Galipeau D, Jiang J (2008) Appl Phys Lett 93 (7):73303
98. Li Y, Bian Y, Yan M, Thapaliya PS, Johns D, Yan X, Galipeau D, Jiang J (2011) J Mater Chem 21(30):11131–11141
99. Tsumura A, Koezuka H, Ando T (1986) Appl Phys Lett 49(18):1210
100. Gao X, Zhao Z (2015) Sci China Chem 58(6):947–968
101. Mei J, Diao Y, Appleton AL, Fang L, Bao Z (2013) J Am Chem Soc 135(18):6724–6746
102. Wang C, Dong H, Hu W, Liu Y, Zhu D (2012) Chem Rev 112(4):2208–2267
103. Murphy AR, Fréchet JMJ (2007) Chem Rev 107(4):1066–1096
104. Melville OA, Lessard BH, Bender TP (2015) ACS Appl Mater Interfaces 7(24):13105–13118
105. Li L, Tang Q, Li H, Hu W, Yang X, Shuai Z, Liu Y, Zhu D (2008) Pure Appl Chem 80 (11):2231–2240
106. Guillaud G, Madru R, Sadoun MA, Maitrot M (1989) J Appl Phys 66(9):4554
107. Li L, Tang Q, Li H, Yang X, Hu W, Song Y, Shuai Z, Xu W, Liu Y, Zhu D (2007) Adv Mater 19(18):2613–2617
108. Shaymurat T, Tang Q, Tong Y, Dong L, Liu Y (2013) Adv Mater 25(16):2269–2273
109. Huang W, Sinha J, Yeh M, Hardigree JFM, LeCover R, Besar K, Rule AM, Breysse PN, Katz HE (2013) Adv Funct Mater 23(33):4094–4104
110. Fernandes EGR, Vieira NCS, de Queiroz AAA, Guimarães FEG, Zucolotto V (2010) J Phys Chem C 114(14):6478–6483
111. Sun L, Zhang J, Zhao F, Luo X, Lv W, Li Y, Ren Q, Wen Z, Peng Y, Liu X (2015) Nanotechnology 26(18):185501
112. Zhang L, Wang H, Zhao Y, Guo Y, Hu W, Yu G, Liu Y (2013) Adv Mater 25(38):5455–5460
113. Maheshwari P, Mukherjee S, Bhattacharya D, Sen S, Tokas RB, Honda Y, Basu S, Padma N, Pujari PK (2015) ACS Appl Mater Interfaces 7(19):10169–10177
114. Jung JS, Lee JW, Kim K, Cho MY, Jo SG, Joo J (2010) Chem Mater 22(7):2219–2225
115. Su Y, Wang C, Xie W, Xie F, Chen J, Zhao N, Xu J (2011) ACS Appl Mater Interfaces 3 (12):4662–4667
116. Deng W, Zhang X, Wang J, Shang Q, Gong C, Zhang X, Zhang Q, Jie J (2014) Org Electron 15(7):1317–1323
117. Liu W, Jackson BL, Zhu J, Miao C, Chung C, Park YJ, Sun K, Woo J, Xie Y (2010) ACS Nano 4(7):3927–3932
118. Luo X, Li Y, Lv W, Zhao F, Sun L, Peng Y, Wen Z, Zhong J, Zhang J (2015) Nanotechnology 26(3):35201
119. Cao L, Chen S, Wei D, Liu Y, Fu L, Yu G, Liu H, Liu X, Wu D (2010) J Mater Chem 20 (12):2305–2309
120. Jiang L, Dong H, Meng Q, Tan J, Jiang W, Xu C, Wang Z, Hu W (2012) Adv Mater 24 (5):694–698
121. Ward JW, Lamport ZA, Jurchescu OD (2015) ChemPhysChem 16(6):1118–1132
122. Cao Y, Wei Z, Liu S, Gan L, Guo X, Xu W, Steigerwald ML, Liu Z, Zhu D (2010) Angew Chem Int Ed 49(36):6319–6323
123. Pan F, Tian H, Qian X, Huang L, Geng Y, Yan D (2011) Org Electron 12(8):1358–1363
124. Huang L, Zhu F, Liu C, Treske U, Grobosch M, Tian H, Zhang J, Geng Y, Knupfer M, Yan D (2012) Adv Funct Mater 22(21):4598–4607

125. Dong S, Tian H, Huang L, Zhang J, Yan D, Geng Y, Wang F (2011) Adv Mater 23 (25):2850–2854
126. Dong S, Bao C, Tian H, Yan D, Geng Y, Wang F (2013) Adv Mater 25(8):1165–1169
127. Chen Y, Su W, Bai M, Jiang J, Li X, Liu Y, Wang L, Wang S (2005) J Am Chem Soc 127 (45):15700–15701
128. Su W, Jiang J, Xiao K, Chen Y, Zhao Q, Yu G, Liu Y (2005) Langmuir 21(14):6527–6531
129. Chen Y, Li R, Wang R, Ma P, Dong S, Gao Y, Li X, Jiang J (2007) Langmuir 23 (25):12549–12554
130. Li R, Ma P, Dong S, Zhang X, Chen Y, Li X, Jiang J (2007) Inorg Chem 46(26):11397–11404
131. Gao Y, Ma P, Chen Y, Zhang Y, Bian Y, Li X, Jiang J, Ma C (2009) Inorg Chem 48(1):45–54
132. Ma P, Chen Y, Sheng N, Bian Y, Jiang J (2009) Eur J Inog Chem 2009(7):954–960
133. Chaure NB, Sosa-Sanchez JL, Cammidge AN, Cook MJ, Ray AK (2010) Org Electron 11 (3):434–438
134. Bao Z, Lovinger AJ, Brown J (1998) J Am Chem Soc 120(1):207–208
135. Keil C, Schlettwein D (2011) Org Electron 12(8):1376–1382
136. Jiang H, Ye J, Hu P, Wei F, Du K, Wang N, Ba T, Feng S, Kloc C (2014) Sci Rep 4:7573
137. Song D, Zhu F, Yu B, Huang L, Geng Y, Yan D (2008) Appl Phys Lett 75(14):143303
138. Song D, Wang H, Zhu F, Yang J, Tian H, Geng Y, Yan D (2008) Adv Mater 20 (11):2142–2144
139. Md Obaidulla SK, Goswami DK, Giri PK (2014) Appl Phys Lett 104(21):213302
140. Ma P, Kan J, Zhang Y, Hang C, Bian Y, Chen Y, Kobayshi N, Jiang J (2011) J Mater Chem 21(46):18552–18559
141. Zaumseil J, Sirringhaus H (2007) Chem Rev 107(4):1296–1323
142. Bisri SZ, Piliego C, Gao J, Loi MA (2014) Adv Mater 26(8):1176–1199
143. Zhang Y, Dong H, Tang Q, Ferdous S, Liu F, Mannsfeld SCB, Hu W, Briseno AL (2010) J Am Chem Soc 132(33):11580–11584
144. Fujimoto T, Matsushita MM, Awaga K (2013) J Phys Chem C 117(11):5552–5557
145. Nénon S, Kanehira D, Yoshimoto N, Fages F, Videlot-Ackermann C (2011) Synth Met 161 (17–18):1915–1920
146. Opitz A, Bronner M, Bruetting W, Himmerlich M, Schaefer JA, Krischok S (2007) Appl Phys Lett 90(21):212111–212112
147. Kraus M, Richler S, Opitz A, Brutting W, Haas S, Hasegawa T, Hinderhofer A, Schreiber F (2010) J Appl Phys 107(9):94503
148. Kraus M, Haug S, Bruetting W, Opitz A (2011) Org Electron 12(5):731–735
149. Opitz A, Horlet M, Kiwull M, Wagner J, Kraus M, Bruetting W (2012) Org Electron 13 (9):1614–1622
150. Yasuda T, Tsutsui T (2006) Jpn J Appl Phys 45(6L):L595
151. Zhong A, Bian Y, Zhang Y (2010) J Phys Chem C 114(7):3248–3255
152. Peng Y, Gao P, Lv W, Yao B, Fan G, Chen D, Xie J, Zhou M, Li Y, Wang Y (2013) IEEE Photon Technol Lett 25(22):2149–2152
153. Hayashi H, Nihashi W, Umeyama T, Matano Y, Seki S, Shimizu Y, Imahori H (2011) J Am Chem Soc 133(28):10736–10739
154. Shi Y, Li X (2014) Org Electron 15(1):286–293
155. Guillaud G, Al Sadoun M, Maitrot M, Simon J, Bouvet M (1990) Chem Phys Lett 167 (6):503–506
156. Zhang Y, Cai X, Qi D, Bian Y, Jiang J (2008) J Phys Chem C 112(37):14579–14588
157. Chen Y, Li D, Yuan N, Gao J, Gu R, Lu G, Bouvet M (2012) J Mater Chem 22 (41):22142–22149
158. Kong X, Jia Q, Wu F, Chen Y (2015) Dyes Pigments 115:67–72
159. Kan J, Chen Y, Qi D, Liu Y, Jiang J (2012) Adv Mater 24(13):1755–1758
160. Li D, Wang H, Kan J, Lu W, Chen Y, Jiang J (2013) Org Electron 14(10):2582–2589
161. Kong X, Zhang X, Gao D, Qi D, Chen Y, Jiang J (2015) Chem Sci 6(3):1967–1972
162. Zhang X, Chen Y (2014) Inorg Chem Commun 39:79–82

163. Gao D, Zhang X, Kong X, Chen Y, Jiang J (2015) ACS Appl Mater Interfaces 7 (4):2486–2493
164. O'Regan B, Gratzel M (1991) Nature 353(6346):737–740
165. Hardin BE, Snaith HJ, McGehee MD (2012) Nat Photonics 6(3):162–169
166. Han L, Islam A, Chen H, Malapaka C, Chiranjeevi B, Zhang S, Yang X, Yanagida M (2012) Energy Environ Sci 5(3):6057–6060
167. Li L, Diau EW (2013) Chem Soc Rev 42(1):291–304
168. Mathew S, Yella A, Gao P, Humphry-Baker R, Curchod BFE, Ashari-Astani N, Tavernelli I, Rothlisberger U, Nazeeruddin MK, Grätzel M (2014) Nat Chem 6(3):242–247
169. Nazeeruddin MK, Humphry-Baker R, Grätzel M, Wöhrle D, Schnurpfeil G, Schneider G, Hirth A, Ttombach N (1999) J Porphyrins Phthalocyanines 03(03):230–237
170. Zhang L, Cole JM (2015) ACS Appl Mater Interfaces 7:3427–3455
171. Sharma D, Steen G, Korterik JP, García-Iglesias M, Vázquez P, Torres T, Herek JL, Huijser A (2013) J Phys Chem C 117(48):25397–25404
172. Lim B, Margulis GY, Yum J, Unger EL, Hardin BE, Grätzel M, McGehee MD, Sellinger A (2013) Org Lett 15(4):784–787
173. Radivojevic I, Bazzan G, Burton-Pye BP, Ithisuphalap K, Saleh R, Durstock MF, Francesconi LC, Drain CM (2012) J Phys Chem C 116(30):15867–15877
174. Hayat A, Shivashimpi GM, Nishimura T, Fujikawa N, Ogomi Y, Yamaguchi Y, Pandey SS, Ma T, Hayase S (2015) Appl Phys Express 8(4):47001. doi:10.7567/APEX.8.047001
175. Barea EM, Ortiz J, Payá FJ, Fernández-Lázaro F, Fabregat-Santiago F, Sastre-Santos A, Bisquert J (2010) Energy Environ Sci 3(12):1985–1994
176. Li R, Zhang X, Zhu P, Ng DKP, Kobayashi N, Jiang J (2006) Inorg Chem 45(5):2327–2334
177. Cid J, Yum J, Jang S, Nazeeruddin MK, Martínez-Ferrero E, Palomares E, Ko J, Grätzel M, Torres T (2007) Angew Chem Int Ed 46(44):8358–8362
178. Cid J, García-Iglesias M, Yum J, Forneli A, Albero J, Martínez-Ferrero E, Vázquez P, Grätzel M, Nazeeruddin MK, Palomares E, Torres T (2009) Chem Eur J 15(20):5130–5137
179. Garcia-Iglesias M, Cid J, Yum J, Forneli A, Vazquez P, Nazeeruddin MK, Palomares E, Gratzel M, Torres T (2011) Energy Environ Sci 4(1):189–194
180. García-Iglesias M, Yum J, Humphry-Baker R, Zakeeruddin SM, Péchy P, Vázquez P, Palomares E, Grätzel M, Nazeeruddin MK, Torres T (2011) Chem Sci 2(6):1145–1150
181. Ragoussi M, Cid J, Yum J, de la Torre G, Di Censo D, Grätzel M, Nazeeruddin MK, Torres T (2012) Angew Chem Int Ed 51(18):4375–4378
182. Ragoussi M, Yum J, Chandiran AK, Ince M, de la Torre G, Grätzel M, Nazeeruddin MK, Torres T (2014) ChemPhysChem 15(6):1033–1036
183. Mori S, Nagata M, Nakahata Y, Yasuta K, Goto R, Kimura M, Taya M (2010) J Am Chem Soc 132(12):4054–4055
184. Kimura M, Nomoto H, Masaki N, Mori S (2012) Angew Chem Int Ed 51(18):4371–4374
185. Kimura M, Nomoto H, Suzuki H, Ikeuchi T, Matsuzaki H, Murakami TN, Furube A, Masaki N, Griffith MJ, Mori S (2013) Chem Eur J 19(23):7496–7502
186. Ikeuchi T, Nomoto H, Masaki N, Griffith MJ, Mori S, Kimura M (2014) Chem Commun 50 (16):1941–1943
187. Yu L, Zhou X, Yin Y, Liu Y, Li R, Peng T (2012) ChemPlusChem 77(11):1022–1027
188. Yu L, Shi W, Lin L, Liu Y, Li R, Peng T, Li X (2014) Dalton Trans 43(22):8421–8430
189. Lin L, Peng B, Shi W, Guo Y, Li R (2015) Dalton Trans 44(12):5867–5874
190. Yu L, Shi W, Lin L, Guo Y, Li R, Peng T (2015) Dyes Pigments 114:231–238
191. Yu L, Fan K, Duan T, Chen X, Li R, Peng T (2014) ACS Sustain Chem Eng 2(4):718–725
192. Ince M, Medina A, Yum J, Yella A, Claessens CG, Martínez-Díaz MV, Grätzel M, Nazeeruddin MK, Torres T (2014) Chem Eur J 20(7):2016–2021
193. Li Y, Lu P, Yan X, Jin L, Peng Z (2013) RSC Adv 3(2):545–558
194. Yamamoto S, Mori S, Wagner P, Mozer AJ, Kimura M (2015) Isr J Chem. doi:10.1002/ijch.201500023
195. Ethirajan M, Chen Y, Joshi P, Pandey RK (2011) Chem Soc Rev 40(1):340–362

196. Yuan A, Wu J, Tang X, Zhao L, Xu F, Hu Y (2013) J Pharm Sci 102(1):6–28
197. Zorlu Y, Dumoulin F, Durmuş M, Ahsen V (2010) Tetrahedron 66(17):3248–3258
198. Chen X, Li Y, Wang A, Zhou L, Lu S, Zhou J, Lin Y, Wei S (2015) Dyes Pigments 114:93–104
199. Machacek M, Cidlina A, Novakova V, Svec J, Rudolf E, Miletin M, Kučera R, Simunek T, Zimcik P (2015) J Med Chem 58(4):1736–1749
200. Liu W, Jensen TJ, Fronczek FR, Hammer RP, Smith KM, Vicente MGH (2005) J Med Chem 48(4):1033–1041
201. Luan L, Ding L, Zhang W, Shi J, Yu X, Liu W (2013) Bioorg Med Chem Lett 23 (13):3775–3779
202. Lu S, Ma YJ, Xuan HY, Wang A, Zhao B, Li XD, Zhou JH, Lin Y, Zhou L, Wei SH (2014) RSC Adv 4(104):59759–59763
203. Bai M, Lo P, Ye J, Wu C, Fong W, Ng DKP (2011) Org Biomol Chem 9(20):7028–7032
204. Duan W, Lo P, Duan L, Fong W, Ng DKP (2010) Bioorg Med Chem 18(7):2672–2677
205. He H, Lo P, Ng DKP (2014) Chem Eur J 20(21):6241–6245
206. Ke M, Yeung S, Fong W, Ng DKP, Lo P (2012) Chem Eur J 18(14):4225–4233, S4221-S4225
207. Ke M, Ng DKP, Lo P (2014) Chem Asian J 9(2):554–561
208. Zhang FL, Huang Q, Zheng K, Li J, Liu JY, Xue JP (2013) Chem Commun 49 (83):9570–9572
209. Zhang F, Huang Q, Liu J, Huang M, Xue J (2015) ChemMedChem 10(2):312–320
210. Lau JTF, Lo P, Fong W, Ng DKP (2012) J Med Chem 55(11):5446–5454
211. Luan L, Ding L, Shi J, Fang W, Ni Y, Liu W (2014) Chem Asian J 9(12):3491–3497
212. Lau JTF, Lo P, Fong W, Ng DKP (2011) Chem Eur J 17(27):7569–7577
213. Lau JTF, Lo P, Tsang Y, Fong W, Ng DKP (2011) Chem Commun 47(34):9657–9659
214. Chan CMH, Lo P, Yeung S, Ng DKP, Fong W (2010) Cancer Biol Ther 10(2):126–134
215. Jiang X, Lo P, Tsang Y, Yeung S, Fong W, Ng DKP (2010) Chem Eur J 16(16):4777–4783
216. Jiang X, Yeung S, Lo P, Fong W, Ng DKP (2011) J Med Chem 54(1):320–330
217. Jiang X, Lo P, Yeung S, Fong W, Ng DKP (2010) Chem Commun 46(18):3188–3190
218. Lau JTF, Lo P, Jiang X, Wang Q, Ng DKP (2014) J Med Chem 57(10):4088–4097
219. Mew D, Wat CK, Towers GH, Levy JG (1983) J Immunol 130(3):1473–1477
220. Hayley P, Stamati I, Yahioglu G, Butt M, Deonarain M (2013) Antibodies 2(2):270–305
221. Mitsunaga M, Ogawa M, Kosaka N, Rosenblum LT, Choyke PL, Kobayashi H (2011) Nat Med 17(12):1685–1691
222. Sano K, Mitsunaga M, Nakajima T, Choyke PL, Kobayashi H (2013) J Nucl Med 54 (5):770–775
223. Nakajima T, Sano K, Choyke PL, Kobayashi H (2013) Theranostics 3(6):357–365
224. Sato K, Nagaya T, Mitsunaga M, Choyke PL, Kobayashi H (2015) Cancer Lett 365 (1):112–121
225. Maawy AA, Hiroshima Y, Zhang Y, Heim R, Makings L, Garcia-Guzman M, Luiken GA, Kobayashi H, Hoffman RM, Bouvet M (2015) PLoS One 10(3):e121989
226. Chen Z, Xu P, Chen J, Chen H, Hu P, Chen X, Lin L, Huang Y, Zheng K, Zhou S, Li R, Chen S, Liu J, Xue J, Huang M (2014) Acta Biomater 10(10):4257–4268
227. Li R, Zheng K, Hu P, Chen Z, Zhou S, Chen J, Yuan C, Chen S, Zheng W, Ma E, Zhang F, Xue J, Chen X, Huang M (2014) Theranostics 4(6):642–659
228. Zhou X, Zheng K, Li R, Chen Z, Yuan C, Hu P, Chen J, Xue J, Huang M (2015) Acta Biomater 23:116–126
229. Gao D, Gao L, Zhang C, Liu H, Jia B, Zhu Z, Wang F, Liu Z (2015) Biomater 53:229–238
230. Moghimi SM, Hunter AC, Murray JC (2005) FASEB J 19(3):311–330
231. Kim BYS, Rutka JT, Chan WCW (2010) New Engl J Med 363(25):2434–2443
232. Zhao B, Duan W, Lo P, Duan L, Wu C, Ng DKP (2013) Chem Asian J 8(1):55–59
233. Hota R, Baek K, Yun G, Kim Y, Jung H, Park KM, Yoon E, Joo T, Kang J, Park CG, Bae SM, Ahn WS, Kim K (2013) Chem Sci 4(1):339–344

234. Jang W, Yim D, Hwang I (2014) J Mater Chem B 2(16):2202–2211
235. Master AM, Rodriguez ME, Kenney ME, Oleinick NL, Gupta AS (2010) J Pharm Sci 99 (5):2386–2398
236. Liang R, Ma L, Zhang L, Li C, Liu W, Wei M, Yan D, Evans DG, Duan X (2014) Chem Commun 50(95):14983–14986
237. Taratula O, Schumann C, Duong T, Taylor KL, Taratula O (2015) Nanoscale 7(9):3888–3902
238. Brasch M, de la Escosura A, Ma Y, Uetrecht C, Heck AJR, Torres T, Cornelissen JJLM (2011) J Am Chem Soc 133(18):6878–6881
239. Nyokong T, Antunes E (2013) Coord Chem Rev 257(15–16):2401–2418
240. Camerin M, Magaraggia M, Soncin M, Jori G, Moreno M, Chambrier I, Cook MJ, Russell DA (2010) Eur J Cancer 46(10):1910–1918
241. Jang B, Park J, Tung C, Kim I, Choi Y (2011) ACS Nano 5(2):1086–1094
242. Cheng Y, Samia AC, Li J, Kenney ME, Resnick A, Burda C (2010) Langmuir 26 (4):2248–2255
243. Obaid G, Chambrier I, Cook MJ, Russell DA (2012) Angew Chem Int Ed 51(25):6158–6162
244. Stuchinskaya T, Moreno M, Cook MJ, Edwards DR, Russell DA (2011) Photochem Photobiol Sci 10(5):822–831
245. Obaid G, Chambrier I, Cook MJ, Russell DA (2015) Photochem Photobiol Sci 14(4):737–747
246. Guo H, Qian H, Idris NM, Zhang Y (2010) Nanomed Nanotechnol Biol Med 6(3):486–495
247. Cui S, Chen H, Zhu H, Tian J, Chi X, Qian Z, Achilefu S, Gu Y (2012) J Mater Chem 22 (11):4861
248. Cui S, Yin D, Chen Y, Di Y, Chen H, Ma Y, Achilefu S, Gu Y (2013) ACS Nano 7 (1):676–688
249. Wang F, Chen X, Zhao Z, Tang S, Huang X, Lin C, Cai C, Zheng N (2011) J Mater Chem 21 (30):11244–11252
250. Abramczyk H, Brozek-Pluska B, Surmacki J, Musial J, Kordek R (2014) Analyst 139 (21):5547–5559
251. Giuliani F, Martinelli M, Cocchi A, Arbia D, Fantetti L, Roncucci G (2010) Antimicrob Agents Chemother 54(2):637–642
252. Ke M, Eastel JM, Ngai KLK, Cheung Y, Chan PKS, Hui M, Ng DKP, Lo P (2014) Eur J Med Chem 84:278–283
253. Ke M, Eastel JM, Ngai KLK, Cheung Y, Chan PKS, Hui M, Ng DKP, Lo P (2014) Chem Asian J 9(7):1868–1875
254. Mathew S, Murakami T, Nakatsuji H, Okamoto H, Morone N, Heuser JE, Hashida M, Imahori H (2013) ACS Nano 7(10):8908–8916
255. Sessoli R, Gatteschi D, Caneschi A, Novak MA (1993) Nature 365(6442):141–143
256. Gatteschi D, Sessoli R (2003) Angew Chem Int Ed 42(3):268–297
257. Wang H, Wang B, Bian Y, Gao S, Jiang J (2016) Coord Chem Rev 306:195–216. doi:10.1016/j.ccr.2015.07.004
258. Fukuda T, Shigeyoshi N, Yamamura T, Ishikawa N (2014) Inorg Chem 53(17):9080–9086
259. Gonidec M, Davies ES, McMaster J, Amabilino DB, Veciana J (2010) J Am Chem Soc 132 (6):1756–1757
260. Gonidec M, Krivokapic I, Vidal-Gancedo J, Davies ES, McMaster J, Gorun SM, Veciana J (2013) Inorg Chem 52(8):4464–4471
261. Gonidec M, Amabilino DB, Veciana J (2012) Dalton Trans 41(44):13632–13639
262. Waters M, Moro F, Krivokapic I, McMaster J, Slageren JV (2012) Dalton Trans 41 (4):1128–1130
263. Glebe U, Weidner T, Baio JE, Schach D, Bruhn C, Buchholz A, Plass W, Walleck S, Glaser T, Siemeling U (2012) ChemPlusChem 77(10):889–897
264. Katoh K, Isshiki H, Komeda T, Yamashita M (2011) Coord Chem Rev 255 (17–18):2124–2148
265. Katoh K, Umetsu K, Breedlove Brian K, Yamashita M (2012) Sci China Chem 55 (6):918–925

266. Ganivet CR, Ballesteros B, de la Torre G, Clemente-Juan JM, Coronado E, Torres T (2013) Chem Eur J 19(4):1457–1465
267. Gonidec M, Luis F, Vílchez À, Esquena J, Amabilino DB, Veciana J (2010) Angew Chem Int Ed 49(9):1623–1626
268. Cao W, Gao C, Zhang Y, Qi D, Liu T, Wang K, Duan C, Gao S, Jiang J (2015) Chem Sci 6(10):5947–5954
269. Wang H, Wang K, Tao J, Jiang J (2012) Chem Commun 48(24):2973–2975
270. Ishikawa N, Otsuka S, Kaizu Y (2005) Angew Chem Int Ed 44(5):731–733
271. Katoh K, Kajiwara T, Nakano M, Nakazawa Y, Wernsdorfer W, Ishikawa N, Breedlove BK, Yamashita M (2011) Chem Eur J 17(1):117–122
272. Katoh K, Asano R, Miura A, Horii Y, Morita T, Breedlove BK, Yamashita M (2014) Dalton Trans 43(21):7716–7725
273. Sakaue S, Fuyuhiro A, Fukuda T, Ishikawa N (2012) Chem Commun 48(43):5337–5339
274. Kan J, Wang H, Sun W, Cao W, Tao J, Jiang J (2013) Inorg Chem 52(15):8505–8510
275. Cao W, Wang H, Shang H, Jiang J (2013) Zhongguo Xitu Xuebao 31(5):513–521 (in Chinese)
276. Wang H, Qian K, Wang K, Bian Y, Jiang J, Gao S (2011) Chem Commun 47(34):9624
277. Wang H, Liu T, Wang K, Duan C, Jiang J (2012) Chem Eur J 18(25):7691–7694
278. Fukuda T, Kuroda W, Ishikawa N (2011) Chem Commun 47(42):11686–11688
279. Fukuda T, Matsumura K, Ishikawa N (2013) J Phys Chem A 117(40):10447–10454
280. Shang H, Zeng S, Wang H, Dou J, Jiang J (2015) Sci Rep 5:8838
281. Katoh K, Horii Y, Yasuda N, Wernsdorfer W, Toriumi K, Breedlove BK, Yamashita M (2012) Dalton Trans 41(44):13582–13600
282. Horii Y, Katoh K, Yasuda N, Breedlove BK, Yamashita M (2015) Inorg Chem 54 (7):3297–3305
283. Wang K, Qi D, Wang H, Cao W, Li W, Liu T, Duan C, Jiang J (2013) Chem Eur J 19 (34):11162–11166
284. Morita T, Katoh K, Breedlove BK, Yamashita M (2013) Inorg Chem 52(23):13555–13561
285. Blagg RJ, Ungur L, Tuna F, Speak J, Comar P, Collison D, Wernsdorfer W, McInnes EJL, Chibotaru LF, Winpenny REP (2013) Nat Chem 5(8):673–678
286. Auwaerter W, Ecija D, Klappenberger F, Barth JV (2015) Nat Chem 7(2):105–120
287. Katoh K, Isshiki H, Komeda T, Yamashita M (2012) Chem Asian J 7(6):1154–1169
288. Gottfried JM (2015) Surf Sci Rep 70(3):259–379
289. Komeda T, Isshiki H, Liu J, Zhang Y, Lorente N, Katoh K, Breedlove BK, Yamashita M (2011) Nat Commun 2:217
290. Robles R, Lorente N, Isshiki H, Liu J, Katoh K, Breedlove BK, Yamashita M, Komeda T (2012) Nano Lett 12(7):3609–3612
291. Komeda T, Isshiki H, Liu J, Katoh K, Shirakata M, Breedlove BK, Yamashita M (2013) ACS Nano 7(2):1092–1099
292. Komeda T, Isshiki H, Liu J, Katoh K, Yamashita M (2014) ACS Nano 8(5):4866–4875
293. Vincent R, Klyatskaya S, Ruben M, Wernsdorfer W, Balestro F (2012) Nature 488 (7411):357–360
294. Thiele S, Balestro F, Ballou R, Klyatskaya S, Ruben M, Wernsdorfer W (2014) Science 344 (6188):1135–1138
295. Stepanow S, Honolka J, Gambardella P, Vitali L, Abdurakhmanova N, Tseng T, Rauschenbach S, Tait SL, Sessi V, Klyatskaya S, Ruben M, Kern K (2010) J Am Chem Soc 132(34):11900–11901
296. Margheriti L, Chiappe D, Mannini M, Car P, Sainctavit P, Arrio M, de Mongeot FB, Cezar JC, Piras FM, Magnani A, Otero E, Caneschi A, Sessoli R (2010) Adv Mater 22 (48):5488–5493
297. Mannini M, Bertani F, Tudisco C, Malavolti L, Poggini L, Misztal K, Menozzi D, Motta A, Otero E, Ohresser P, Sainctavit P, Condorelli GG, Dalcanale E, Sessoli R (2014) Nat Commun 5:4582

298. Malavolti L, Mannini M, Car P, Campo G, Pineider F, Sessoli R (2013) J Mater Chem C 1 (16):2935–2942
299. Candini A, Klyatskaya S, Ruben M, Wernsdorfer W, Affronte M (2011) Nano Lett 11 (7):2634–2639
300. Lopes M, Candini A, Urdampilleta M, Reserbat-Plantey A, Bellini V, Klyatskaya S, Marty L, Ruben M, Affronte M, Wernsdorfer W, Bendiab N (2010) ACS Nano 4(12):7531–7537
301. Wang H, Qian K, Qi D, Cao W, Wang K, Gao S, Jiang J (2014) Chem Sci 5(8):3214–3320
302. Gonidec M, Biagi R, Corradini V, Moro F, De Renzi V, Del Pennino U, Summa D, Muccioli L, Zannoni C, Amabilino DB, Veciana J (2011) J Am Chem Soc 133 (17):6603–6612
303. Klar D, Candini A, Joly L, Klyatskaya S, Krumme B, Ohresser P, Kappler J, Ruben M, Wende H (2014) Dalton Trans 43(28):10686–10689
304. Kyatskaya S, Mascarós JRG, Bogani L, Hennrich F, Kappes M, Wernsdorfer W, Ruben M (2009) J Am Chem Soc 131(42):15143–15151
305. Ganzhorn M, Klyatskaya S, Ruben M, Wernsdorfer W (2013) ACS Nano 7(7):6225–6236
306. Krull C, Balashov T, Kavich JJ, Mugarza A, Miedema PS, Thakur PK, Sessi V, Klyatskaya S, Ruben M, Stepanow S, Gambardella P, Lodi Rizzini A (2011) Phys Rev Lett 107(17):177205
307. Lodi Rizzini A, Krull C, Balashov T, Mugarza A, Nistor C, Yakhou F, Sessi V, Klyatskaya S, Ruben M, Stepanow S, Gambardella P (2012) Nano Lett 12(11):5703–5707
308. Malavolti L, Poggini L, Margheriti L, Chiappe D, Graziosi P, Cortigiani B, Lanzilotto V, de Mongeot FB, Ohresser P, Otero E, Choueikani F, Sainctavit P, Bergenti I, Dediu VA, Mannini M, Sessoli R (2013) Chem Commun 49(98):11506–11508

Struct Bond (2016) 172: 201–290
DOI: 10.1007/430_2015_204

Published online: 10 March 2016

Transition Metal-Based Photofunctional Materials: Recent Advances and Potential Applications

Hok-Lai Wong, Margaret Ching-Lam Yeung, and Vivian Wing-Wah Yam

Abstract This chapter highlights the importance of structure–property relationships in transition metal complexes for the construction of molecular- and supramolecular-based photofunctional materials and summarizes the recent advancements of this class of complexes with potential applications in the areas of energy, catalysis, materials, biology, and diagnostics.

Keywords Artificial photosynthesis • Biological sensors • Carbon dioxide reduction • Dye-sensitized solar cells • Hydrogen generation • Imaging • Light-harvesting • Molecular machines • Organic light-emitting diodes • Oxygen generation • Photochromism • Photosensitizer • Supramolecular chemistry • Transition metal complexes

Contents

H.-L. Wong, M.C.-L. Yeung, and V.W.-W. Yam (✉)
Institute of Molecular Functional Materials (Areas of Excellence Scheme, University Grants Committee, Hong Kong) and Department of Chemistry, The University of Hong Kong, Pokfulam Road, Hong Kong, People's Republic of China
e-mail: wwyam@hku.hk

Abbreviations

$[Ir(ppy)_3]$	Tris(2-phenylpyridine)iridium(III)
4-ppy	4-Phenylpyridine
bda	2,2′-Bipyridine-6,6′-dicarboxylate
BIBP	4,4′-Dialkylbipyridinium
bpm	2,2′-Bipyrimidine
bpy	2,2′-Bipyridine
bpz	2,2′-Bipyrazine
bzimpy	2,6-Bis(benzimidazol-2-yl)pyridine
CBPQT	Cyclobis(paraquat-*p*-phenylene)
CD	*β*-Cyclodextrin
CIE	Commission Internationale de L'Eclairage
CRI	Color rendering index
DABCO	1,4-Diazabicyclo[2.2.2]octane
DAE	1,2-Diarylethene
DIP	4,7-Diphenyl-1,10-phenanthroline
$dmgBF_2$	(Difluoroboryl)dimethylglyoxime
dppee	1,2-Bis(diphenylphosphino)ethene
DSSCs	Dye-sensitized solar cells
e.r.	Enantiomeric ratio
EQE	External quantum efficiency
FIrpic	Bis[(4,6-difluorophenyl)pyridinato-N,C^2](picolinato)iridium(III)
FRET	Förster resonance energy transfer
H_2bda	2,2′-Bipyridine-6,6′-dicarboxylic acid
HOMO	Highest occupied molecular orbital
HTI	Hemithioindigo
ILCT	Intraligand charge transfer
IQE	Internal quantum efficiency
LF	Ligand field
LLCT	Ligand-to-ligand charge transfer
LUMO	Lowest unoccupied molecular orbital
MC	Merocyanine
^3MLCT	Triplet metal-to-ligand charge transfer
MLCT	Metal-to-ligand charge transfer

^{3}MMLCT	Triplet metal–metal-to-ligand charge transfer
MOFs	Metal-organic frameworks
NHCs	*N*-Heterocyclic carbenes
NIR	Near-infrared
NLO	Nonlinear optical
NPB	4,4′-Bis[*N*-(1-naphthyl)-*N*-phenylamino]biphenyl
OEC	Oxygen-evolving complex
OLEDs	Organic light-emitting diodes
PCE	Power conversion efficiency
PCET	Proton-coupled electron transfer
PET	Photo-induced electron transfer
phen	1,10-Phenanthroline
PHOLEDs	Phosphorescent organic light-emitting diodes
PLEDs	Polymer light-emitting diodes
PMMA	Poly(methyl methacrylate)
PPE-SO_3^-	Poly(phenylene ethynylene sulfonate)
SMMs	Single-molecule magnets
TADF	Thermally activated delayed fluorescence
TEOA	Triethanolamine
TONs	Turnover numbers
TTF	Tetrathiafulvalene
WOLEDs	White organic light-emitting diodes
WPLEDs	White polymer light-emitting diodes

1 Introduction

Coordination of transition metal centers to organic frameworks provides enormous synthetic diversity, which originates from the variation in the nature of the metal centers, bonding modes of metal complexation as well as ligand designs, and constitutes a representative class of building blocks. Rational design of transition metal complexes with the incorporation of versatile functionalities provides not only the tuning of electronic absorption, luminescence, and excited-state redox properties but also their diverse photofunctional properties, which leads to their employment in recent advanced technologies. These also indicate the importance of the structure–property relationships of this class of complexes for the design of molecular-based photofunctional materials. The accessibility of triplet excited states in this class of heavy metal-containing complexes provided by efficient intersystem crossing via enhanced spin–orbit coupling allows their utilization as triplet emitters in PHOLEDs with IQEs of unity. Functionalization of the transition metal complexes with custom-made ligands imparts them with unique and rich photophysical properties that lead to the development of PHOLEDs of whole-visible-spectrum colors as well as photosensitizers and light-harvesting dyes that are panchromatic. High flexibility of coordination modes and diversity of oxidation states give rise to the metal-based artificial photosynthetic systems with rich excited-state chemistry, which could not be offered by the simple organic counterparts. Their improved absorptivity in the visible region also demonstrates the importance of metal complexation to the organic framework for the construction of efficient artificial photosynthetic systems. Photo-responsive systems and molecular machines constructed from transition metal complexes have also benefited from the ease of step-by-step synthesis, tunability, and the use of less destructive, lower-energy triggering light sources. Moreover, long-lived phosphorescence along with large Stokes shifts favor the employment of this class of complexes in the design of sensing probes applicable in biological environments because of the minimized interferences from short-lived background autofluorescence. Prominent sensitivities of their triplet excited-state properties toward microenvironmental changes give rise to the sensitive response of these biological sensors. Apart from that, supramolecular non-covalent metal–metal interactions are observed for some of the transition metal complexes, which govern their photophysical as well as structural properties. The utilization of supramolecular assembly behaviors for the design of photofunctional materials has expanded the research dimension to supramolecular-based materials beyond the molecular levels, attributed to the diversity originating from their hierarchical complexity which offers unlimited possibilities. In this chapter, recent advancements of the class of transition metal complexes since 2007 with diverse ranges of photofunctional properties being utilized in the applications of OLEDs, DSSCs, artificial photosynthesis, photo-switchable functional materials, molecular machines, as well as biological detection, imaging, and diagnostics will be discussed. In addition, the potential applications of the supramolecular-based materials with photofunctional properties have also been highlighted.

2 Highlights of Recent Developments of Highly Efficient Transition Metal-Based OLEDs

The phosphorescence nature arising from enhanced intersystem crossing as a result of large spin–orbit coupling attributed to the presence of heavy metal center endows the classes of transition metal complexes with attractive features to serve as promising candidates for the fabrication of PHOLEDs based on triplet harvesting effect [1]. Both singlet and triplet excitons formed by the recombination of electro-generated hole and electron can be harvested to achieve an IQE of unity. The ease of tuning emission energy via judicious structural modification on the transition metal complexes leads to the construction of desirable emitters of monochromatic and white OLEDs, which can be potentially utilized in flat-panel displays and solid-state lightings. Systematic studies reported by Yersin and coworkers have also indicated that increasing metal contribution in the lowest-lying triplet excited state of transition metal complexes would result in significant zero-field splitting of triplet state into substates [1]. Such high metal characters would give rise to not only the reduction of energy difference between the singlet and the lowest-lying triplet excited states which facilitates singlet-triplet intersystem crossing but also the relaxation of spin forbiddenness of the transition from the lowest-lying triplet excited state to singlet ground state which leads to an increase in the radiative decay rate and a shortening of emission lifetime [1]. In contrast to the millisecond-range lifetime of pure organic phosphors, the microsecond to submicrosecond regime of radiative lifetimes of transition metal complexes contributes critical significance in PHOLED fabrication because the problems of efficiency roll-off ascribed to triplet-triplet annihilation and saturation effect at increased current density can be effectively avoided [2].

$[Ir(ppy)_3]$ has been serving as the prototype for triplet emitters in monochromatic green PHOLEDs because of its high photoluminescence quantum yield, microsecond-range lifetime, as well as pronounced thermal stability and robustness [3]. Recent reports indicated its significant zero-field splitting of 170 cm^{-1}, illustrative of the substantial involvement of ^{3}MLCT character in the emissive excited states, and this accounts for its optimal properties being utilized as the triplet emitter in PHOLEDs [3]. Extensive studies on the construction of devices based on the class of iridium(III) complexes have blossomed and led to a versatile library of whole-visible-spectrum monochromatic triplet emitters, in which PHOLEDs for the three primary colors with EQEs exceeding the predicted upper limit of 20% have been achieved [2, 4]. Investigations on the utilization of platinum(II), osmium (II), and ruthenium(II) complexes for the fabrication of monochromatic PHOLEDs have also been well documented [2, 4–7].

WOLEDs can be constructed by mixing the emitters of three primary colors (blue, green, and red) or two complementary colors (blue and orange/yellow). In contrast to the versatile range of high-efficiency red and green PHOLEDs, design of triplet emitters for blue PHOLEDs remains challenging in terms of color purity, stability, and quantum efficiency. This has been attributed to the large energy gap between HOMO and LUMO required for the emitters as well as the presence of the relatively low-lying d-d LF states which would provide a non-radiative decay pathway [7, 8]. Construction of efficient triplet emitters with blue electroluminescence based on the utilization of iridium(III) complexes functionalized with high triplet energy-containing or electron-withdrawing ligands has been extensively explored [4, 5], while only few examples have been demonstrated with the use of other metal centers. Li and coworkers have utilized a difluoro-containing cyclometalated platinum(II) complex for the fabrication of blue PHOLEDs with a maximum EQE of 16% (Fig. 1a) [9]. A tetradentate platinum(II) complex reported by Li and coworkers has been fabricated as deep-blue PHOLEDs, exhibiting narrow electroluminescence spectral bandwidth and high spectral purity with CIE coordinates of (0.15, 0.08) as well as a maximum EQE of 24.8% (Fig. 1b) [10]. Recently, a demonstration on the use of a platinum(II) complex with spiro-acridine-containing tetradentate ligand for sky-blue PHOLED fabrication has been reported through the collaborative efforts of Chi, Chang, Chou, and Robertson, with a maximum EQE of 15.3% achieved (Fig. 1c) [11]. However, challenges associated with energy transfer between multiple emitters leading to voltage-dependent color changes and differential aging arisen from the variation in lifetimes of each emitter would be of critical concern for the WOLED fabrication upon the mixing of emitters of different colors [12]. In contrast to the d^6 counterparts of octahedral geometry, coordinate unsaturated d^8 platinum(II) complexes with sterically

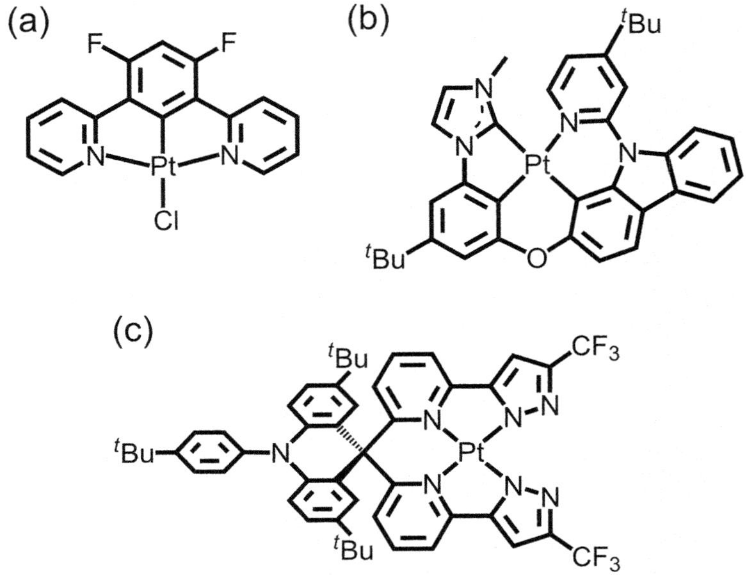

Fig. 1 Chemical structures of platinum(II) complexes with (**a**) difluoro-containing cyclometalating ligand [9], (**b**) phenylimidazole-containing tetradentate ligand [10], and (**c**) spiro-acridine-containing tetradentate ligand [11], which are utilized for blue PHOLED fabrication

undemanding ligands have been shown to exhibit strong propensity to assemble into oligomers in the solid state, along with red-shifted emission originated from the excited state associated with non-covalent metal–metal and/or π–π stacking interactions [13–16]. Such propensity has led to the exploitation of single-dopant WOLEDs based on the utilization of the simultaneous emission arisen from the monomers and aggregates or excimers of this class of complexes. Initiated by the first successful demonstration by Thompson and coworkers on the use of the excimer-forming platinum(II) complexes for the construction of phosphorescent single-dopant WOLED by judicious tuning of the dopant concentrations (Fig. 2a) [12, 17], increasing attention and continuous efforts have been made to improve the performance of platinum(II)-based single-dopant WOLEDs. By making use of the blue emission from the hole-transporting layer of NPB, Zhou, Wong, Ma and coworkers have prepared single-dopant WOLEDs based on the green-emitting platinum(II) complexes (Fig. 2b) [18]. Owing to the matched LUMO levels of the NPB and these green emitters, transfer of electrons from the emitting layer to the NPB layer followed by the combination with the holes injected from the anode would give rise to the blue electroluminescence. With the combination of the blue color from NPB and green and orange-red electroluminescence originating from the monomeric and excimeric forms of the triplet emitters, a well-balanced white illumination has been achieved. Introduction of two dimethylamino moieties to a blue-emissive cyclometalated platinum(II) complex has been shown by the group of Cocchi and Williams to control the excimer formation of this complex, with the emission of the aggregates covering the midrange of visible spectrum instead of

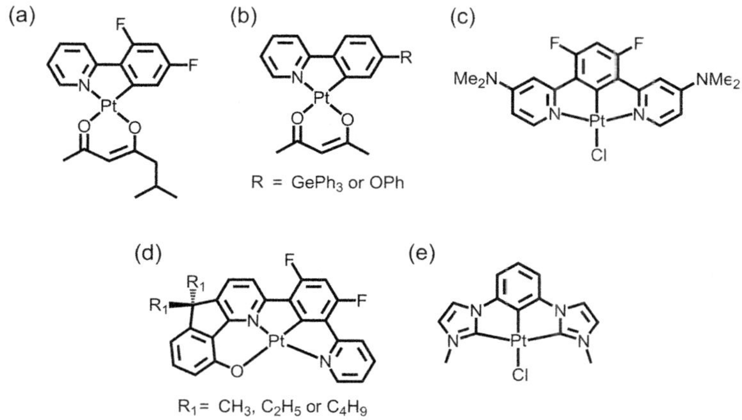

Fig. 2 Chemical structures of platinum(II) complexes reported by the groups of (**a**) Thompson [12, 17], (**b**) Zhou, Wong, and Ma [18], (**c**) Cocchi and Williams [19], (**d**) Che [20, 21], and (**e**) Li [22], which are utilized for single-dopant WOLED fabrication

extending to the NIR region (Fig. 2c) [19]. The good coverage of this complex in the visible spectral region upon the modulation of monomer/excimer formation has given rise to WOLEDs with CIE coordinates of (0.36, 0.37) and CRI of 87. Recently, a series of tetradentate platinum(II) complexes reported by Che and coworkers has been shown to exhibit strong dependence of their intermolecular interactions as well as the ratio of monomer/excimer emission from the corresponding device on the length of alkyl chains (Fig. 2d) [20, 21]. A vacuum-deposited single-dopant device based on the butyl chain-containing complex with well-balanced white electroluminescence has been constructed with maximum EQE of 16.5% achieved [20]. In addition, solution-processable WPLEDs based on the complex with ethyl chains were demonstrated with a maximum EQE of 12.7%, of which the efficiency is the highest among the reported single-dopant WPLEDs and comparable to the WPLEDs of highest performance based on multi-emitters [21]. Li and coworkers have utilized a blue-emitting platinum(II) bis(*N*-methyl-imidazolyl)benzene complex for the fabrication of monomer/excimer-based WOLEDs with a maximum EQE of 20.1% (Fig. 2e) [22]. In addition, the device has been shown to exhibit CIE coordinates of (0.33, 0.33) and CRI of 80 with its electroluminescence properties independent of the current density, which is desirable for high-quality white illumination.

Exploration on the use of other transition metal complexes as emitters in PHOLEDs has also been initiated. Incorporation of strong σ-donating alkynyl ligands to bis-cyclometalated gold(III) moiety has been demonstrated by Yam and coworkers to enhance the luminescence properties of the resultant complexes (Fig. 3a), owing to the destabilization of the non-emissive d-d LF states [23]. Such charge-neutral complexes have been found to be thermally stable and employed as phosphorescent emitters for the fabrication of PHOLEDs by vacuum deposition. Upon increasing voltages, devices with the phenylethynyl-containing gold(III)

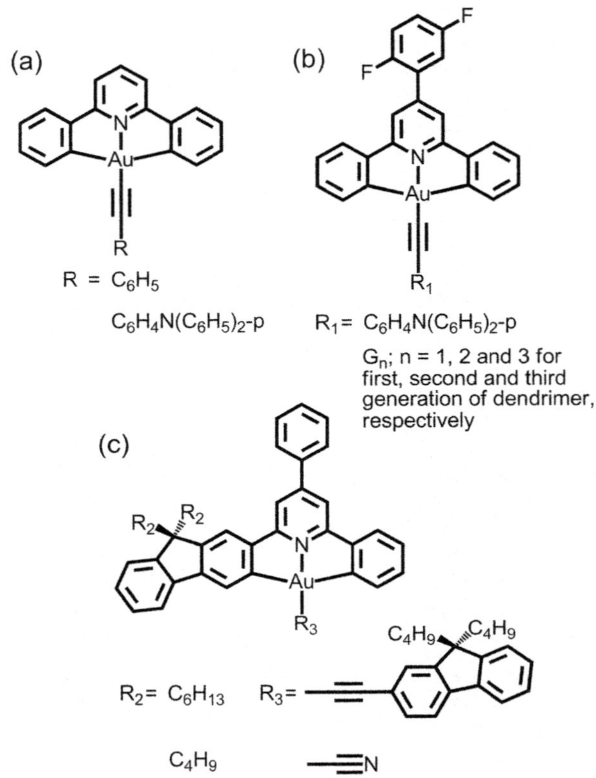

Fig. 3 Chemical structures of gold(III) complexes with (**a**) bis-cyclometalating ligand [23], (**b**) 2,5-difluorophenyl-containing bis-cyclometalating ligand [24, 25], and (**c**) fluorene moiety-containing bis-cyclometalating ligand [26], which are utilized for PHOLED or PLED fabrication

complex as emitter have been found with emission colors changed from orange to blue, which are originated from the oligomeric/excimeric emission arising from the π-stacking of the bis-cyclometalating ligand and the emission from adjacent hole-transporting layer, respectively. Extended conjugation on bis-cyclometalating ligand has been shown by the same group to further enhance the luminescence properties of this class of complexes. The photoluminescence quantum yield of the thin film of the 2,6-diphenyl-4-(2,5-difluorophenyl)pyridine gold(III) complex (Fig. 3b) doped in PMMA has been found to increase with respect to that of the control complex in Fig. 2a [24]. In addition, a high-performance device with such complex functioned as the emitter with optimized dopant concentration of 4% has been achieved, exhibiting green electroluminescence with a maximum EQE of 11.5% that is comparable to that of the [Ir(ppy)$_3$]-based devices. Advantaged by the enhanced solubility upon the incorporation of dendritic structure with carbazole as dendrons and *tert*-butyl functionalities as surface groups, dendritic alkynylgold (III) complexes have been fabricated by Yam and coworkers as relatively low-cost solution-processable PHOLEDs (Fig. 3b) [25]. High-efficiency devices based on the use of an alkynylgold(III) complex containing the first generation of dendrimer has been determined with an EQE of 7.8%. More importantly, reduced bathochromic shift has been observed for the device emission with the use of higher generations of dendrimers upon increasing dopant concentration. This is indicative of an effective control on the degree of intermolecular interactions and hence the fine-tuning of emission color of these solution-processable PHOLEDs. Che and coworkers have reported color-tunable PLEDs constructed from a yellow-emitting fluorene moiety-containing bis-cyclometalated alkynylgold(III) complex and the blue-emitting FIrpic with a maximum EQE of 13.2% achieved (Fig. 3c) [26]. Yellow electroluminescence has been observed at low driving voltage due to efficient energy transfer from FIrpic to the gold(III) emitter. Increasing voltage would result in the observation of white color owing to an increase in the blue component of the electroluminescence of the devices. Further increase in voltage would lead to the blue electroluminescence of FIrpic due to the saturation of the triplet excited state of the gold(III) emitter and hence the blocking of energy transfer pathway. Similar strategy has also been employed for the fabrication of color-tunable vacuum-deposited OLEDs based on the cyanide-containing analogue with a maximum EQE of 22.0% achieved (Fig. 3c) [26].

Non-porphyrin-containing palladium(II) complexes with luminescence properties are rarely reported, attributed to the non-emissive thermally accessible d-d LF states leading to effective non-radiative decay [27]. Upon coordination of strong σ-donating tetradentate ligands to the palladium(II) center, destabilization of the d-d LF states as well as the increased rigidity of the structural scaffold led to the reduction of non-radiative decay resulting from ligand distortion, leading to the enhanced luminescence properties of the palladium(II) complex as shown in Fig. 4a [27]. The high thermal stability of this complex enables its employment as the emitter in PHOLEDs with green emission by vacuum deposition. A maximum EQE of 7.4% has been determined for the device, which is comparable to those reported for PHOLEDs based on platinum(II) complexes. Recent work by Li and coworkers has demonstrated the preparation of a tetradentate palladium(II) complex for vacuum-deposited OLED fabrication, with green electroluminescence and maximum EQE of 20.4% (Fig. 4b) [28].

Research dimensions in OLEDs have also been directed toward the exploration on the use of earth-abundant and low-cost metals, e.g., copper(I) and zinc (II) complexes, as the alternative candidates for OLED fabrication. This has been attributed to the lack of emission quenching pathways via the population of d-d LF excited states for this class of metal complexes of d^{10} electronic configuration, as

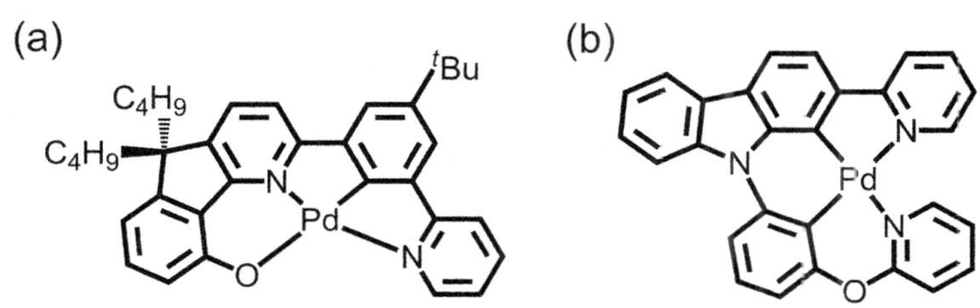

Fig. 4 Chemical structures of palladium(II) complexes reported by the groups of (**a**) Che [27] and (**b**) Li [28], which are utilized for PHOLED fabrication

well as their exhibition of TADF, arising from the occurrence of reverse intersystem crossing at ambient temperature owing to the small energy separation between the lowest-lying singlet and triplet excited states [29, 30]. Extensive investigations on the temperature-dependent luminescence properties of the class of copper(I) complexes have been performed by Yersin and coworkers, which have indicated that TADF emitters would allow the singlet harvesting of both singlet and triplet excitons by the fluorescent excited state in the electroluminescence process [29, 30]. This would lead to enhanced luminescence at ambient temperature and shortened radiative decay time that is advantageous for minimizing the efficiency roll-off upon increasing current densities. Thermally stable bis(phosphine)-diarylamido dinuclear copper(I) complex [31], bis(ditolylphosphino)benzene copper(I) complex [32], and bis(hydroxylphenylbenzoxazole)zinc(II) complex [33] reported by the groups of Peters, Osawa, and Adachi, respectively, have been found to exhibit TADF as confirmed by temperature-dependent time-resolved photoluminescence and lifetime measurements (Figs. 5 and 6a, b). They have been demonstrated to function as emitters in vapor-deposited OLEDs with green emission and maximum EQEs of 16.1%, 21.3%, and 19.6% attained, respectively. Recent work by Yersin and coworkers has reported a photoluminescence quantum yield of a dinuclear copper(I) complex in powder form of 92% at ambient temperature with short luminescence lifetime (Fig. 6c) [34]. Its emission has been assigned to originate from the lowest-lying triplet state due to direct phosphorescence in addition to that from the singlet excited state due to TADF. Such additional phosphorescence decay pathway has been attributed to the efficient spin–orbit coupling. This has been supported by the large zero-field splitting of the lowest-lying triplet excited state and a relatively fast radiative decay rate from the lowest-lying triplet excited state to singlet ground state with respect to that of the copper (I) complexes with normally millisecond-ranged phosphorescence lifetimes due to the relatively weak spin–orbit coupling [34]. The unusual feature of this complex

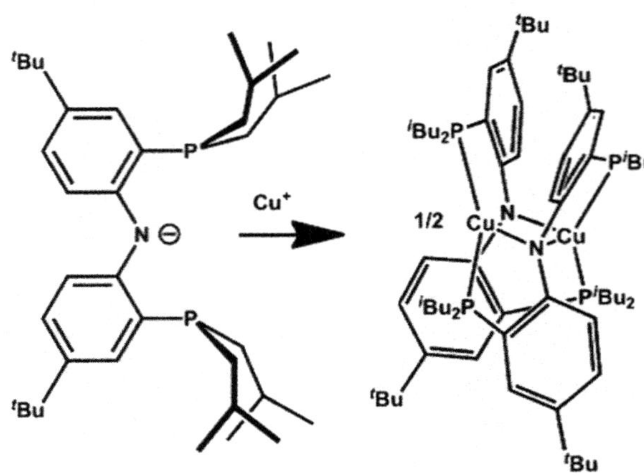

Fig. 5 Formation of bis-(phosphine)diarylamido dinuclear copper(I) complex. Reprinted with permission from Peters and coworkers [31]. Copyright 2010 American Chemical Society

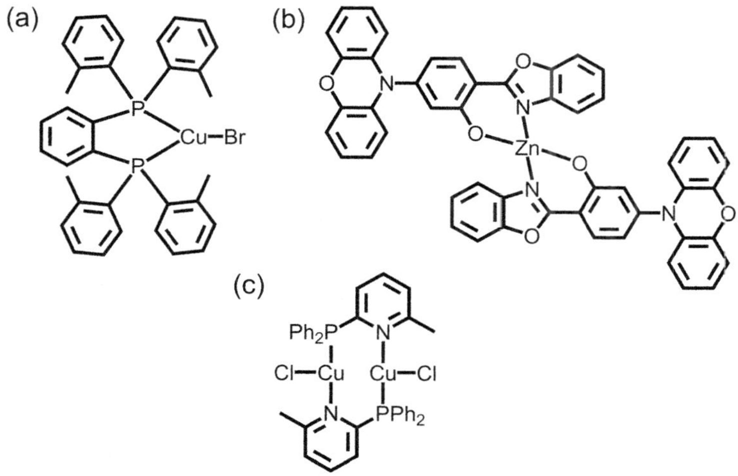

Fig. 6 Chemical structures of copper(I) and zinc(II) complexes reported by the groups of (**a**) Osawa [32], (**b**) Adachi [33], and (**c**) Yersin [34], which have been found to exhibit TADF

with additional phosphorescence decay pathway has opened up a new avenue for the design of efficient copper(I)-based OLED emitters with an overall radiative lifetime that has been remarkably shortened.

3 Recent Advances of Transition Metal Complexes as Sensitizers for Light-Harvesting and Artificial Photosynthetic Systems

Material development for solar energy harvesting has received enormous attention recently to meet the increasing global energy demand. Researchers have been trying to understand and mimic how photosynthesis is done by Mother Nature, which has proven to be a successful process for millions of years. Materials featuring long-lived charge-separated state, wide light absorption range from visible to NIR region, high absorptivity, low cost, ease of tuning of photophysical properties and spatial arrangements, as well as high photo- and electrochemical stability are highly desirable. Inspired by chlorophyll and other plant pigments, transition metal complexes offer unique advantages of structural diversity and long-lived excited states over their organic counterparts. These facilitate efficient and fine-tuning of energy levels and promote controlled energy transfer, electron transfer, and PCET after photo-excitation. In this section, recent advances in various selected transition metal complexes for DSSCs [35–95] and artificial photosynthetic systems [96–235] will be introduced.

3.1 Transition Metal Complexes for Dye-Sensitized Solar Cells

Among different materials for solar energy harvesting, ruthenium(II) polypyridine complexes for DSSCs, pioneered by Grätzel in 1991 [49], represent an important class of transition metal-containing sensitizers. The operation principle involves light absorption by the photosensitizer to generate the excited state. Electron injection occurs subsequently to the conduction band of the titanium dioxide, resulting in a charge-separated state, oxidized sensitizer, and photocurrent that flow through the external circuit. The presence of redox mediator facilitates the regeneration of the photosensitizer to continue the solar-harvesting cycle. The schematic diagram for the operation of DSSC is shown in Fig. 7a. Optimization of DSSC materials and device structures to achieve better electron injection, electron collection, dye regeneration, solar spectrum coverage, device stability, and minimizing unfavorable charge recombination process will be essential. Ruthenium(II)-based dyes show various intrinsic advantages in the area of DSSCs, including strong MLCT transition in the visible to NIR region, good energy match with the conduction band of titanium dioxide and I^-/I_3^- redox couple, and good stability. One particular class of ruthenium(II)-based sensitizers with general formula $[Ru(L)(L')(SCN)_2]$, where L and L′ are diimine ligands, shows impressive performance. Due to its synthetic versatility, a lot of work have been done to structurally modify the sensitizer and optimize the DSSC device performance at the molecular level. By fixing one ligand (L) with 2,2′-bipyridine-4,4′-dicarboxylic acid as anchoring group and varying the substituents on another diimine ligand (L′), structure–property relationship has been systematically investigated and compared with the ruthenium(II) dye, **N3**, reported by Grätzel and coworkers with PCE of 10% in 1993 [50]. In 2007, Zakeeruddin, Grätzel and coworkers functionalize the bipyridine ligand with methoxyphenylethene group and reported a highly efficient DSSC (up to 9.5%) exhibiting long-term stability (up to 1,000 h) [51]. Transient absorption spectroscopic experiment was done to understand the dye regeneration kinetics in the presence and absence of the redox mediator [51]. Other than

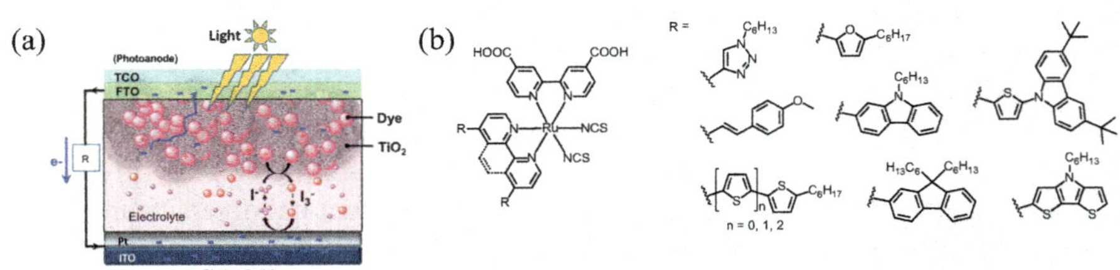

Fig. 7 (**a**) Schematic diagram of an operative TiO_2-DSSC using the I_3^-/I^- redox couple in the electrolyte (*ITO* indium tin oxide, *Pt* platinum, *FTO* fluorine-doped tin oxide, *TCO* transparent conductive oxide). Reprinted with permission from Nazeeruddin, Torres and coworkers [47]. Copyright 2014 American Chemical Society. (**b**) Selected examples of ruthenium(II) diimine-based dyes [49–65]

carbocycles [51–54], heterocyclic rings with solubilizing groups including oligothiophene [55–62], thienothiophene [59], ethylene-diethoxythiophene [55], furan [57], and others [63–65] have also been introduced to enhance the DSSC performance and the processability. Wu, Ho and coworkers reported the use of the electron-rich hole-transporting thienylcarbazole group to construct ruthenium(II)-based DSSC [57]. Under the same experimental condition, this sensitizer shows better PCE (9.72%) than the **N3** dye (8.51%), probably due to the more red-shifted and stronger absorption in the visible to NIR region. Lin and coworkers introduced alkyl heterocycles, such as fluorene, carbazole, and dithieno[3,2-*b*:2′,3′-*d*]pyrrole, in the ancillary bipyridine ligand [65]. Improved processability and bathochromic shift was observed due to the presence of alkyl chains and extended π-conjugation, respectively. This class of complexes showed PCEs from 5.94% to 6.91% (versus **N719** 7.13%) [65]. Selected examples of ruthenium(II) diimine-based DSSCs are shown in Fig. 7b.

To extend the lifetime of the DSSCs, another approach is to use tridentate terpyridine ligands instead of bidentate bipyridine ligands to achieve more stable complexation. Trithiocyanato-4,4′,4″-tricarboxy-2,2′:6′,2″-terpyridine ruthenium (II) complex, known as black dye, is a well-known example reported by Nazeeruddin, Grätzel and coworkers, showing overall conversion efficiency of 10.4% in 2001 [66]. Han and coworkers functionalize the terpyridine ligand with the 4-methylstyryl group (Fig. 8) [67]. This π-extended dye shows a more intense absorption in the UV, visible, and NIR region when compared to the black dye, resulting in a better light-harvesting ability, better photocurrent, and improved PCE of 11.1%. Chou, Chi and coworkers replaced two thiocyanate ligands with mono-anonic, bidentate pyridine–pyrazolate ligands (Fig. 8) and obtained a remarkable PCE of 10.05% (versus black dye, 9.07%) [68]. The replacement of the thiocyanate ligand is considered to be beneficial toward the device lifespan due to the relatively weaker bond of Ru–NCS, which shows notable decomposition during the device operation. The same research group later replaced all three thiocyanate ligands by another tridentate, dianionic pyridine–bis(pyrazolate) ligand to yield a heteroleptic ruthenium(II) sensitizer (Fig. 8) [69]. The device shows impressive performance of 10.7%, ruling out the absolute need of the thiocyanate ligands in ruthenium(II) sensitizers. In addition, the use of two tridentate ligands also eliminates the problematic isomerization from the synthetic point of view, resulting in easier purification and better product yields.

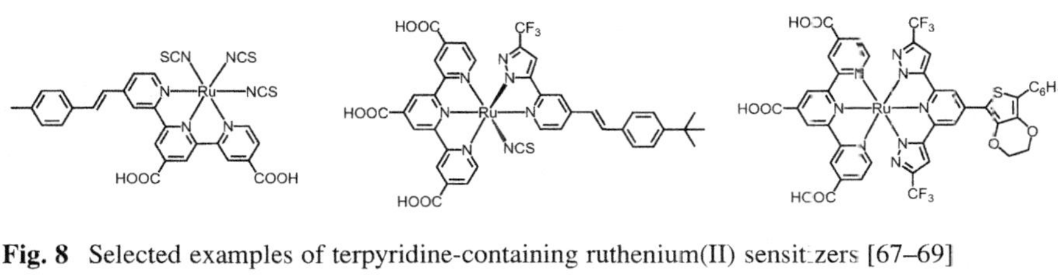

Fig. 8 Selected examples of terpyridine-containing ruthenium(II) sensitizers [67–69]

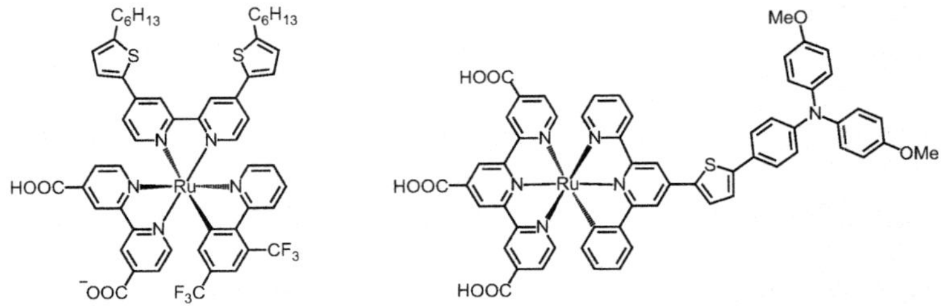

Fig. 9 Selected examples of cyclometalated ruthenium(II) sensitizers [73, 77]

To replace the labile Ru–NCS bond, versatile and diversified phenylpyridine-type cyclometalating ligands could also be employed [70–77]. In 2009, Nazeeruddin, Grätzel and coworkers replaced the NCS ligands by cyclometalating 2-(2,4-difluorophenyl)pyridine ligand and obtained a PCE of 10.1% [72]. Later, Berlinguette and coworkers reported a series of NCS-free tris-heteroleptic cyclometalated ruthenium(II) sensitizers (Fig. 9) that showed a broader and more intense absorption relative to the **N3** dye [73]. Structural engineering through installation of electron-withdrawing groups on the phenyl ring resulted in higher $Ru^{III/II}$ reduction potential to match with that of the redox mediator. Alkylthiophene was also introduced to suppress charge recombination. A PCE of 7.3% at AM1.5 sunlight was achieved (versus **N3**, 6.3%) [73]. A similar approach by using 3-(2′-pyridyl)-1,8-naphthalimide as the cyclometalating ligand was utilized by the same research group [74]. The structure–property relationship was investigated via systematic variation of the substituents on the heterocyclic rings. A PCE of 7.0% was achieved (versus black dye, 5.7%) [74]. Cyclometalated ruthenium(II) sensitizers with two tridentate ligands that are analogues of the ruthenium black dye were first documented by Koten and coworkers [76]. A PCE of 2.3% was achieved. Later, a comprehensive study was done by Berlinguette and coworkers (Fig. 9) [77]. With judicious choices of ancillary groups and cyclometalating positions, the complexes exhibited broad and intense absorption band to 800 nm, arising from MLCT and triphenylamine-based ILCT transitions. It was found that through structural modification, the HOMO and HOMO–1 could be positioned on the triphenylamine and ruthenium metal center, respectively, resulting in movement of photo-induced holes toward the triphenylamine unit and subsequent MLCT event to achieve high DSSC efficiency [77].

Apart from the ruthenium(II)-based sensitizers, metalloporphyrin-based sensitizers also show impressive performance in the field of DSSCs, owing to their strong absorption in the visible to NIR region, high stability, and synthetic versatility. Early examples involved the use of β-substituted metalloporphyrin (Cu and Zn, Fig. 10) reported by Nazeeruddin, Grätzel and coworkers, showing a PCE of up to 4.8% [78]. Sessler, Kim and coworkers introduced an amino group to the β-anchored metalloporphyrin, showing a comparable performance to that of the ruthenium-based **N3** dye [79]. Visible and NIR absorbing π-extended, fused, or oligomeric metalloporphyrin sensitizers have recently been reported and demonstrated to potentially function as efficient DSSC active materials [80–87]. In 2011,

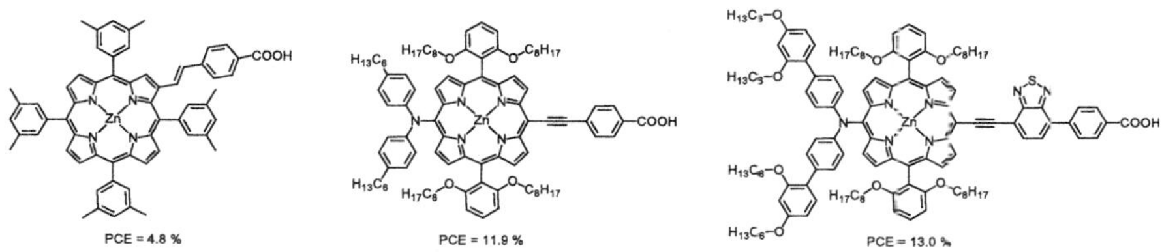

Fig. 10 High-performance DSSC light-harvesting materials reported by Grätzel and coworkers [78, 88, 93]

Yeh, Zakeeruddin, Grätzel and coworkers reported a breakthrough by replacing the iodide/triiodide redox couple with the $Co^{III/II}$ tris(bipyridine) redox electrolyte and by using a custom designed donor-π-bridge-acceptor zinc porphyrin (Fig. 10) which absorbs the entire visible to NIR region [88]. A $Co^{III/II}$ tris(diimine)-based redox mediator was first documented by Moser and coworkers using cobalt(II) bis[2,6-bis (1′-butylbenzimidazol-2′-yl)pyridine] in 2001 [89]. Under 94 Wcm^{-2} sunlight illumination and with the use of Z316 photosensitizer, the device showed respectable performance, with a fill factor of 0.68 and a PCE of 5.2%. The reduced performance (PCE: 2.2%) under full sunlight illumination was attributed to the deficient diffusive charge transport by the oxidized mediator [89]. In 2002, a series of cobalt(II) complexes using different substituted bipyridines, terpyridines, and phenanthrolines as ligands have been systematically investigated as redox mediators by Elliott, Bignozzi and coworkers [90]. Although the resulting performance is low, which is probably due to the mismatch between the non-bulky sensitizer **N3** and the cobalt complex [91], two important advantages over the I^-/I_3^- redox couple have been unveiled [90, 92]. Firstly, cobalt(II) complexes show less absorptivity in the visible region, resulting in less competitive light harvesting with the sensitizer and subsequently higher photocurrent. Secondly, the $Co^{III/II}$ redox potential could be deliberately positioned via structural modifications to match the sensitizer so as to minimize the overpotential loss to maximize the photovoltage. However, due to the ion-pairing effect by negatively charged dye and the positively charged cobalt complexes, the regeneration of the oxidized mediator was hindered, resulting in back electron transfer process, poor charge collection, and low device performance. One approach is to introduce bulky substituents on the sensitizer to function as a blocking layer [88, 91]. The tailor-made zinc porphyrin sensitizer reported by Yeh, Zakeeruddin, Grätzel and coworkers possesses hindered alkoxy chains to impair back electron transfer from the conduction band of the nanocrystalline titanium dioxide to the oxidized cobalt redox mediator [88]. The resulting photovoltage approaches 1 V. These, together with the high photocurrent generated, have produced a high PCE of 11.9%, reflecting the importance of molecular engineering toward improving device performance [88]. Grätzel and coworkers further improved the system by introducing a stronger acceptor, benzothiadiazole, to broaden and to red shift both the Soret and Q-bands, giving rise to larger photocurrent and higher PCE of 13% at full-sun illumination (Fig. 10) [93]. Investigations on the structure–property relationship of this class of porphyrin sensitizers have also been reported by variation of π-conjugated spacer [94, 95] and donor moiety [95], showing PCEs of up to 12.75% under simulated one-sun illumination [94].

3.2 *Catalytic and Photocatalytic Oxygen Generation*

To make good use of renewable solar energy, it must be properly stored and be released when needed. The success and understanding of natural photosynthesis have inspired us on the research and development in photo-induced water splitting, providing valuable hydrogen gas and oxygen gas as solar fuels for clean sources of energy [96–98]. In nature, oxidation of water proceeds in photosystem II, initiated by the sensitization of light-harvesting chlorophyll P680 and electron transfer to benzoquinone [99–106]. The resulting $P680^+$ oxidizes tyrosine by long-range electron transfer, subsequently activating the OEC oxidatively, where the actual water oxidation is taken place. It involves four sequential PCET processes (Kok cycle) [107], and electrons are collected in photosystem I, where reduction occurs to convert protons to a "natural form" of hydrogen by conversion of NADP to NADPH. Overall, nature harvests sunlight and stores it in the form of chemical energy. Construction of artificial photosynthetic systems would require better understanding of these biological redox systems and PCET processes, which have been systematically investigated by research groups such as those of Meyer [100], Mayer [108, 109], Saveant [110, 111], and Fontecave [112–114]. In addition, it is equally important to develop robust light-harvesting sensitizers that show more positive reduction potential than the redox couple of water oxidation [(1.23–0.059pH) V vs NHE] and oxygen-evolving catalysts that are capable of successive PCET [99–106]. Noticeably, ruthenium(II) polypyridine complexes with their ^{3}MLCT excited states exhibit good redox properties to undergo excited-state PET to give the oxidized sensitizer in the presence of sacrificial acceptor or TiO_2, generating charge-separated states. More importantly, the $Ru^{III/II}$ redox couples of ruthenium(II) polypyridines [102] are thermodynamically feasible to drive the artificial oxygen-evolving catalyst for further water oxidation. A schematic diagram of a photosynthetic model system was discussed by Lewis and Nocera and is depicted in Fig. 11 [99]. To construct such kind of solar-driven water-splitting

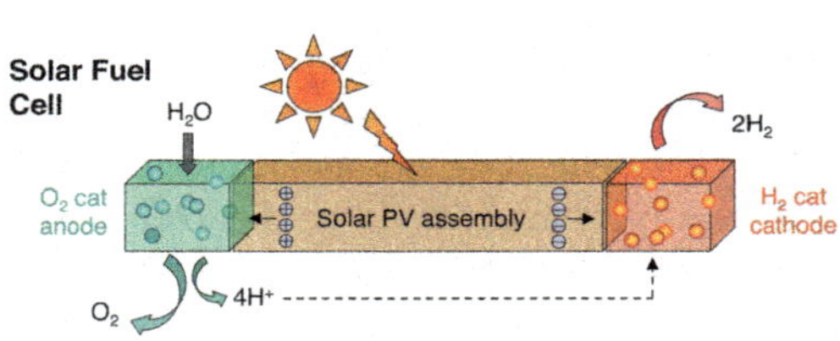

Fig. 11 Schematic diagram of molecular assembly of photo-induced water-splitting system. Reprinted with permission from Lewis and Nocera [99]. Copyright (2006) National Academy of Sciences, U.S.A

device, proper choices of solar absorbers, catalysts, and membranes that are efficient, robust, and scalable would be essential [99, 115–118]. Recently, developments of inorganic light-harvesting semiconductors and catalysts for artificial photosynthesis have received tremendous attention [119–145]. A number of inspiring works toward future production of clean sources of fuels are reported. Various light absorbers and catalysts that are potentially applicable in solar-driven water-splitting devices as photoanodes or photocathodes, such as $[WO_3/Ag^+]$ by Gray, Lewis and coworkers [119], $[BiVO_4]$ by Choi and coworkers [120, 121], transparent Pt-coated *p*-InP by Heller [122], Pt/Ru-coated *p*-WSe_2 by Gray, Lewis and coworkers [123], MoS_2 by Chorkendorff and coworkers [124], Ni_2P by Lewis, Schaak and coworkers [125], $Ba_{0.5}Sr_{0.5}Co_{0.8}Fe_{0.2}O_{3-\delta}$ perovskite by research groups of Halle [126] and Gasteiger [127], cobalt–phosphate assembly reported by Nocera and coworkers [128], $Rb_8K_2\{[Ru_4O_4(OH)_2(H_2O)_4](\gamma\text{-}SiW_{10}O_{36})_2\}\cdot 25H_2O$ reported by Botar, Hill and coworkers [129], $Cs_{10}[Ru_4(\mu\text{-}O)_4(\mu\text{-}OH)_2(H_2O)_4(\gamma\text{-}SiW_{10}O_{36})_2]$ reported by Sartorel, Geremia, Bonchio and coworkers [130], nickel–borate assembly reported by Nocera and coworkers [131], $[Co_4(H_2O)_2(PW_9O_{34})_2]^{10-}$ reported by Hill and coworkers [132], and others [133–135], have been investigated. Synthetic systems that mimic the manganese–calcium–oxygen containing OEC have also been documented by various groups [136–141], including $[Mn_3CaO_4]^{6+}$ core by Agapie and coworkers [138–140] and Mn_4Ca cluster by Zhang, Dong, Dau and coworkers [141]. By combining the ruthenium(II)-based sensitizers and different inorganic oxygen-evolving catalysts, different photocatalytic systems that parallel natural photosynthesis with chlorophyll-sensitized charge separation and oxidation to oxidize water molecules could be realized. Examples include the IrO_2-immobilized ruthenium diimine sensitizers reported by Mallouk and coworkers [142], $\{Ru(bpy)_2[bpy\text{-}(COO)_2]\}/\{Mn_4O_4[(MeOPh)_2PO_2]_6\}^+$ reported by Spiccia and coworkers [143], $[Ru(bpy)_3]^{2+}/[Mn^{III}{}_3Mn^{IV}O_3(CH_3COO)_3(A\text{-}\alpha\text{-}SiW_9O_{34})]^{6-}$ reported by Scandola, Kortz, Bonchio and coworkers [144], $[Ru(bpy)_3]^{2+}/[(V^{IV}{}_5V^{V})O_7(OCH_3)_{12}]^-$ reported by Santoni, Scandola, Campagna and coworkers [145], as well as $[Ru(bpy)_3]^{2+}/[CoMo_6O_{24}H_6]^{3-}$ by Sakai and coworkers [146].

With judicious choice of the ligands and appropriate molecular design and synthesis, different transition metal complexes could also be able to function as oxygen-evolving catalyst. The first well-characterized and synthetic molecular catalyst for water oxidation is the ruthenium blue dimer, *cis,cis*-$[(bpy)_2(H_2O)Ru^{III}ORu^{III}(OH_2)(bpy)_2]^{4+}$, in the presence of Ce(IV) as oxidant reported by Meyer and coworkers in 1982 [147]. The mechanism was proposed to involve four successive PCET to give $[(bpy)_2(O)Ru^VORu^V(O)(bpy)_2]^{4-}$ as a reactive intermediate to oxidize water molecules (Fig. 12) [148, 149]. It is believed that the rate-limiting step is the oxidation of $[(HO_2)Ru^{III}ORu^V(O)]^{3+}$ to $[(HO_2)Ru^{IV}ORu^V(O)]^{4+}$ (k_5 in Fig. 12), which rapidly reacts with H_2O to form $[(HO)Ru^{IV}ORu^{III}(OH_2)]^{4+}$

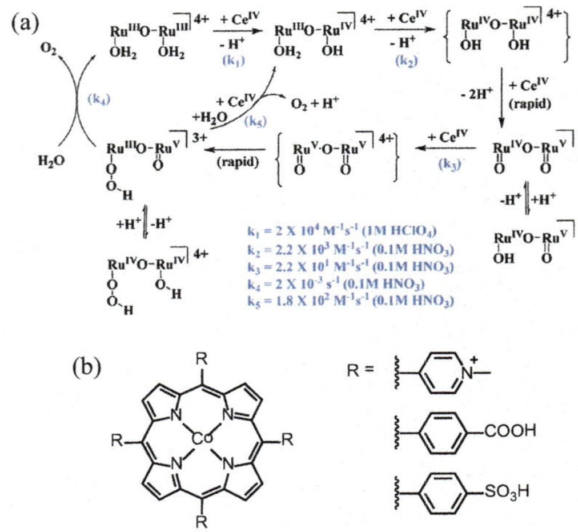

Fig. 12 (**a**) Mechanism of water oxidation by ruthenium blue dimer. Reprinted with permission from Meyer and coworkers [101]. Copyright 2009 American Chemical Society. (**b**) Chemical structures of cobalt porphyrin-based water oxidation catalysts [152]

and spontaneously releases oxygen gas as the final step in the catalytic cycle [149]. To improve the rate of water oxidation, the same research group systematically introduced different ruthenium-based redox mediators, $[Ru(bpy)_2(L)]^{2+}$, where L is bpm or bpz [149]. Due to the difference in the rate of oxidation of $[Ru(bpy)_3]^{2+}$ and $[(HO_2)Ru^{III}ORu^{V}(O)]^{3+}$ (k_5 in Fig. 12a) by Ce(IV) oxidant, $[Ru(bpy)_3]^{3+}$ produced by Ce(IV) serves as an intermediate oxidant for $[(HO_2)Ru^{III}ORu^{V}(O)]^{3+}$, resulting in the enhancement of oxidation rate of up to ~30 times [149]. However, as shown in Fig. 12, the oxidation of water does not necessarily involve two ruthenium metal centers, inspiring their later work to explore simpler mononuclear systems for the same purpose. In 2008, they reported the kinetic and mechanistic studies of catalytic water oxidation by $[Ru(tpy)(bpm)(OH_2)]^{2+}$ and $[Ru(tpy)(bpz)(OH_2)]^{2+}$, showing agreement with the DFT calculation [150] by Baik and coworkers predicting that one ruthenium is sufficient to carry out the process. Experimental evidence by Sakai and coworkers also revealed that $[Ru(tpy)(bpy)(OH_2)]^{2+}$ could be the active species and was fairly stable in strongly oxidizing condition, while $[Ru(tpy)(bpy)(Cl)]^{2+}$ showed no catalytic activity toward oxidation of water [151]. The same research group also examined the catalytic activities of water-soluble cobalt porphyrins (Fig. 12b) [152]. Conclusive evidence indicated that the molecular metal complex was the water oxidation catalyst, instead of being a pre-catalyst for the formation of cobalt oxides. In the presence of $[Ru(bpy)_3]^{2+}$ and $Na_2S_2O_8$, the combined system showed photochemical oxygen production with TON ranging from 88.7 to 121.8 [152].

$[Ru(L)(L')_2]$, where H_2L and L′ are H_2bda- and mono-imine-type ligand, respectively, is another class of complexes that show catalytic water oxidation behavior [153–157]. In 2010, Sun and coworkers reported a photoelectrochemical device with ruthenium(II) diimine sensitizer immobilized on TiO_2 as the photoanode, platinum as the cathode, and $[Ru(bda)(4\text{-picoline})_2]$ as the catalyst (Fig. 13a) [153]. Visible-light-driven water splitting was successfully demonstrated. The proposed catalytic mechanism is shown in Fig. 13b [154]. Similar to the enhanced electron transfer by immobilizing IrO_2 nanoparticles on ruthenium(II) diimine complexes [142], Li, Sun and coworkers covalently linked the pyridine-functionalized ruthenium(II) diimine sensitizer to the axial position of the [Ru (bda)] catalyst (Fig. 13c) [155]. The catalytic system was found to show enhanced water oxidation performance relative to the multicomponent control system with $[Ru(bda)(4\text{-picoline})_2]$ as catalyst and $[Ru(bpy)_3]^{2+}$ as photosensitizer in the presence of sodium persulfate as the sacrificial electron acceptor. This has been attributed to the faster electron transfer process under controlled immobilization on the electrode surface. In addition, the assembled system also shows better TON of 38 TN as compared to the multicomponent counterpart (8 TN) [155]. Meyer and coworkers investigated electropolymerization containing a mixture of vinyl-functionalized $[Ru(vinyl\text{-}bpy)_2\{bpy\text{-}PO(OH)_2\}]^{2+}$ sensitizer and [Ru(bda) $(4\text{-vinyl-py})_2]$ catalyst [156]. Polymerization was done on planar electrode, resulting in the construction of polymer networks of sensitizer–catalyst photocatalytic assembly on mesoporous, nanostructured TiO_2. The thin-film photocatalytic device showed catalytic water oxidation properties in phosphate buffer at pH 7, demonstrating the feasibility of fabricating photocatalytic surfaces on metal oxide electrodes through electropolymerization. Apart from covalently linking the sensitizer and the catalyst, Sun and coworkers utilized the non-covalent interactions between CD and hydrophobic 4-ppy to connect the sensitizer and the catalyst [157]. The resulting 1:1 adduct, $[Ru(bpy)_2(bpy\text{-}CD)]^{2+}/[Ru(bda)(4\text{-}ppy)_2]$, displayed remarkable improvement of almost one order of magnitude over the multicomponent counterparts with a TON of 267 and quantum efficiency of 84%. These assembled photocatalytic systems [155–157] have provided a promising direction toward future developments of artificial photosynthesis to mimic the controlled spatial arrangements and the stepwise electron transfers, similar to that of the Z-scheme.

With appropriate design, iridium(III) complexes could also function as water oxidation catalysts [158–164]. In 2008, Bernhard and coworkers first reported a series of mononuclear cyclometalated iridium(III) aquo complexes [158]. The well-developed synthetic pathway enabled the selective functionalization of different substituents to yield complexes with a wide range of redox potentials for a systematic study. Water oxidation with evolution of oxygen gas was observed in aqueous solution. The unusual kinetic progression revealed a possible two-step process during water oxidation, suggesting the possible involvement of multiple active iridium catalysts [158]. Eisenstein, Brudvig, Crabtree and coworkers later postulated that using a more electron-donating ligand could improve the catalytic

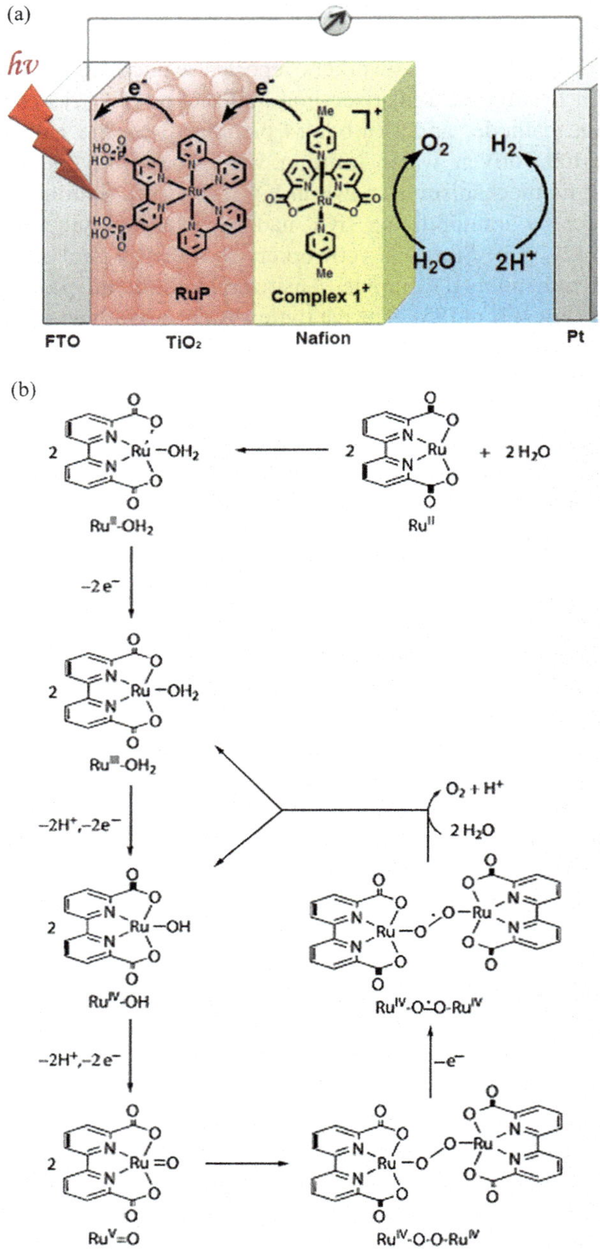

Fig. 13 (**a**) A photoelectrochemical device for visible-light-driven water-splitting reaction. Reproduced from Sun and coworkers [153] with permission from the Royal Society of Chemistry. (**b**) Proposed catalytic mechanism and axial ligands are omitted for clarity. Reprinted with permission from Sun and coworkers [154]. Copyright (2012) National Academy of Sciences, USA. (**c**) Chemical structure of the ruthenium assembly. Reproduced from Li, Sun and coworkers [155] with permission from Wiley-VCH

(c)

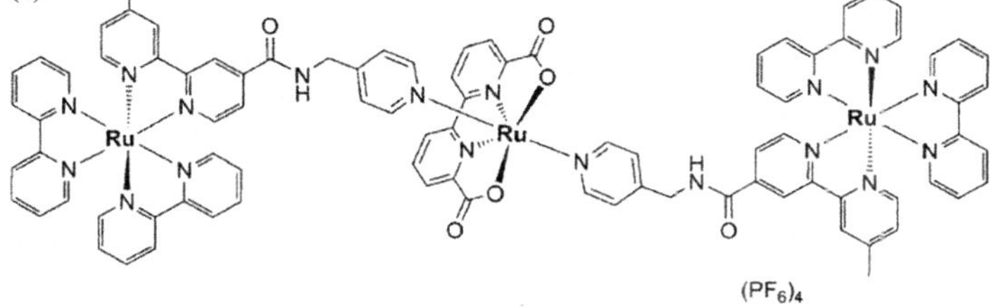

Fig. 13 (continued)

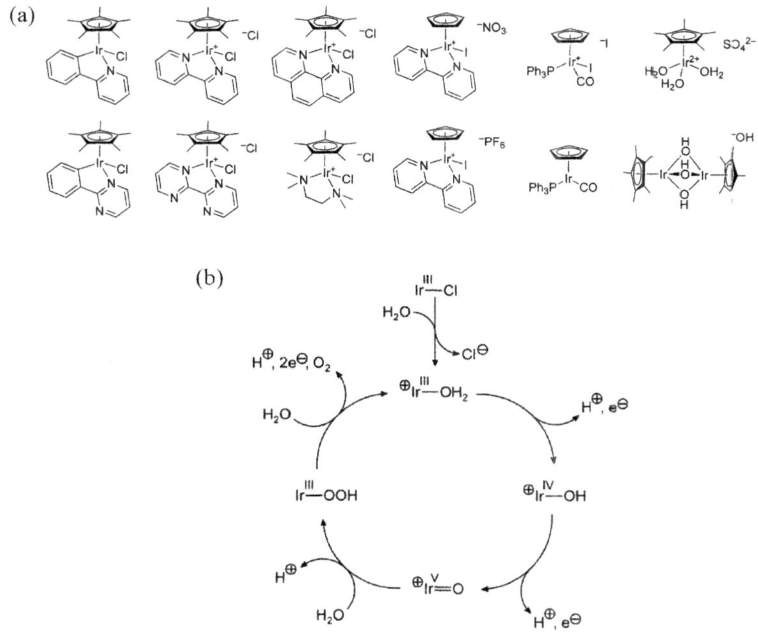

Fig. 14 (**a**) Chemical structures of the iridium(III) complexes studied for water oxidation and (**b**) the proposed reaction mechanism. Reprinted with permission from Eisenberg, Brudvig, Crabtree and coworkers [160]. Copyright 2010 American Chemical Society

performance, thereby replacing the diimine ligands with the pentamethylcyclopentadiene (Fig. 14a) [159–161, 163]. Initial work in 2009 observed a significant improvement on catalytic activity over the previous example as well as the ruthenium(II) counterparts [159]. A library of this kind of half-sandwich iridium complexes has been studied by the same research group [160]. It was found that [Cp*Ir $(H_2O)_3]SO_4$ and $[(Cp*Ir)_2(\mu\text{-}OH)_3]OH$ showed high turnover frequencies of up to 20 min^{-1} at pH 0.89. A reaction mechanism was proposed, as shown in Fig. 14b. The rate showed dependence on the cerium(IV) concentration in low Ce

(IV) loading, and the rate determining step was believed to be the oxidation of catalyst to a high-valent metal-oxo species. On the other hand, when high loading of the Ce(IV) oxidant was added, the rate determining step may be the formation of the dioxygen bond [160]. Immobilization of this class of iridium complex onto the MOF has been done by Lin and workers [162]. This offered advantages in the mechanistic study by isolating the catalysts to prohibit the degradation pathway involving multiple metal complexes. After carrying out the reaction, the water oxidation catalyst could also be unambiguously characterized since the MOF catalyst could be readily and easily separated from the reaction mixture. The result indicated that oxidative modification of Cp* was observed and the active catalyst was [(bpy-dicarboxylate)Ir(H_2O)(X)Cl], where X was likely to be formate or acetate, supported by various spectroscopic analyses [162]. Further optimization of this system by attaching a strongly electron-donating carbene ligand has been done by Bernhard, Albrecht and coworkers, showing a TON of up to 10,000 in 5 days, which is the highest reported to date [163]. In the presence of zinc porphyrin as photosensitizer, Schmuttenmaer, Crabtree, Brudvig and coworkers reported a visible-light-driven water-splitting photoanode by codeposition of the iridium complex with the sensitizer onto the TiO_2 [164]. High photocurrent was observed under visible-light irradiation. This indicated that the photo-generated zinc porphyrin underwent an electron transfer to the TiO_2 and the resulting cation radical was thermodynamically capable to activate the iridium water oxidation catalyst [164].

Oxygen gas not only could be utilized as clean fuel but also could function as an oxidizing agent in the presence of transition metal catalysts [165–168]. In nature, cytochrome P450, methane monooxygenase, and Rieske dioxygenases are enzymes that use dioxygen as oxidant [165–168]. Thus, complexes that could react with dioxygen have received attention for many years [165–172]. Tolman and coworkers initially observed the reversible formation and breaking of oxygen–oxygen bond in a dinuclear copper complex, supported by a tripodal ligand (1,4,7-triisopropyl-1,4,5-triazacyclononane) [169]. An equilibrium between $[Cu_2(\mu\text{-}\eta^2\text{:}\eta^2\text{-}O_2)]^{2+}$ and $[Cu_2(\mu\text{-}O)_2]^{2+}$ has been determined by kinetics spectroscopic and crystallographic studies [169]. In 2011, an important finding has been reported by Schwarz and coworkers [173]. Diatomic $[CuO]^+$ has been successfully synthesized in gas phase, and it was found that the C–H bond in methane could be activated though hydrogen abstraction to give CH_3 radical or oxygen transfer to give methanol [173]. Tolman and coworkers then reported a novel copper complex, bearing [$LCu^{III}OH$] ($L = N$, N'-bis(2,6-diisopropylphenyl)-2,6-pyridinedicarboxamide), to mimic a protonated form of $[CuO]^+$ (Fig. 15) [170, 171]. The resulting complex could react with dihydroanthracene to yield anthracene. Kinetic studies revealed a rapid H-atom abstraction process, with a second-order rate of 1.1 $M^{-1}s^{-1}$ at $-80°C$ [170]. Other organic molecules, such as fluorene, cyclohexene, tetrahydrofuran, and cyclohexane, could also react with the copper complex. In the case of THF, 2-hydroxytetrahydrofuran and γ-butyrolactone were found to be the products [171].

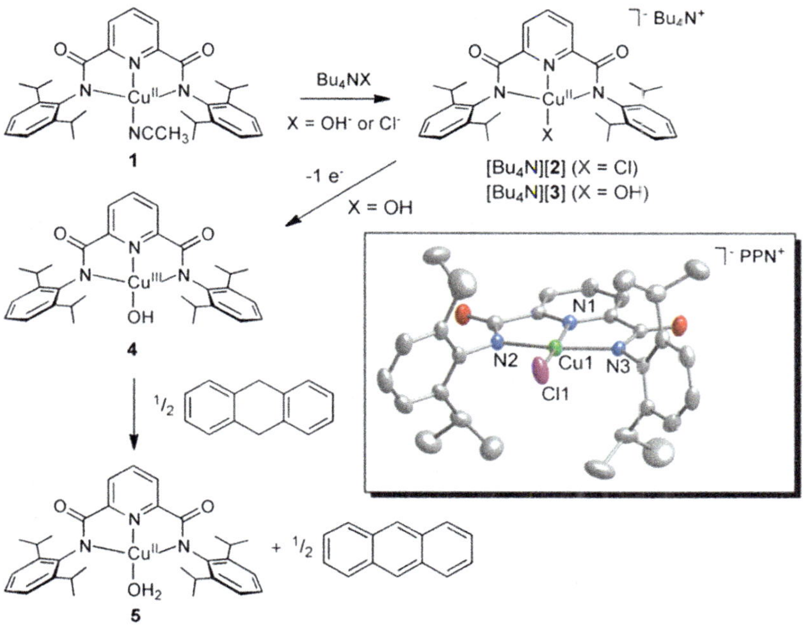

Fig. 15 Reaction scheme of the dehydrogenation of dihydroanthracene. Reprinted with permission from Tolman and coworkers [170]. Copyright 2011 American Chemical Society

3.3 *Catalytic and Photocatalytic Hydrogen Generation*

Apart from the oxygen-evolving catalyst, development of transition metal complexes as hydrogen-evolving catalysts is also important to the future generation of fuel and battery. An early example is the cobaloxime system $[Co(dmgBF_2)_2(H_2O)_2]$ (Fig. 16), which was documented to exhibit hydrogen reduction behavior in 1986 by Espenson and coworkers [174]. Since then, mechanistic studies and structural modifications of related cobaloxime complexes have been done by various research groups [175]. Lewis, Peters and coworkers reported the electrocatalytic hydrogen evolution by cobalt (difluoroboryl)diglyoximate complexes in the presence of strong acids in acetonitrile solution [176]. Electrocatalytic production of hydrogen gas with a Faradaic yield of ca. 90% was achieved under bulk electrolysis condition at −0.37 V vs SCE. Later on, they reported a $[Co(dmgBF_2)_2(MeCN)_2]$-modified glassy carbon electrode, which was active toward electrochemical hydrogen production in aqueous solution at $pH < 4.5$ [177]. The same research group also compared the cobaloxime system with the cobalt tetraamine complexes [178]. These two cobalt systems showed electrochemical hydrogen evolution between −0.55 and −0.20 V vs SCE in acetonitrile solutions. Mechanistic studies relating the hydrogen evolution potentials, the metal hydride pK_a value, and the redox properties of the cobalt complexes in the presence of different ligands have

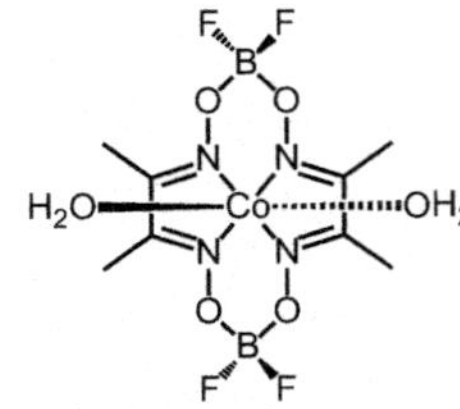

Fig. 16 Chemical structure of $[Co(dmgBF_2)_2(H_2O)_2]$ [174]

been studied [178]. Winkler, Gray and coworkers also revealed the proton transfer mechanism from brominated naphthol to Co^I-diglyoxime, forming cobalt(III) hydride by transient absorption spectroscopy [179]. Rate constants for $Co^{III}H$ formation ($3.5–4.7 \times 10^9\ M^{-1}s^{-1}$) and $Co^{III}H$ reduction to $Co^{II}H$ by excess Co^I-diglyoxime ($9.2 \times 10^6\ M^{-1}s^{-1}$) have been determined. Other related cobalt complexes, such as cobalt bis(iminopyridine) complex by Peters, Gray and coworkers [180] as well as dicobaloxime by Gray and coworkers [181], were also shown to possess electrocatalytic water reduction behaviors.

Besides working as sensitizers for water oxidation, ruthenium complexes could also sensitize catalytic hydrogen generation during artificial photosynthesis by conversion of protons to hydrogen gas [96–98, 182]. In 2008, Artero and coworkers covalently assembled the ruthenium(II) tris(diimine) sensitizer with the cobaloxime catalyst to achieve a photochemical hydrogen production system (Fig. 17a), showing the highest TON of 103 for such combination at that moment [183]. The high activity is believed to be due to three reasons. Firstly, the catalytic center is more resistant toward acid hydrolysis and hydrogenation. Secondly, the BF_2-bridged cobaloxime is easier to be reduced than the H-bridged analogue, resulting in more facile electron transfer process from the sensitizer. Lastly, the covalent linkage between the sensitizer and catalyst facilitates the electron transfer, showing 1.5–8.5 times efficiency enhancement over the multicomponent counterparts [183]. The photochemical system composed of $[Ru(bpy)]^{2+}/[Co(dbt)_2]^-$ (Fig. 17a) and ascorbic acid as sacrificial donor was investigated by Holland, Eisenberg and coworkers [184]. This system shows high catalytic activity of > 2,700 turnovers and an initial turnover rate of 880 mol H_2/mol catalyst/h after 12 h of illumination. The same research group further investigated the structure–property relationship on the catalytic activities by using different ligands attached to the cobalt center [185]. It is found that introduction of electron-withdrawing groups improves the TON, achieving ~ 9,000 TONs for $[Co(mnt)_2]_2^{2-}$ (Fig. 17a) with an initial turnover frequency of up to 3,450 h^{-1}. Besides using cobalt(II) as a hydrogen-evolving catalyst, platinum(II) complexes were also shown to exhibit catalytic properties in photochemical hydrogen production system [186]. Sakai and coworkers reported a covalently assembled system consisting of ruthenium(II) tris (diimine) sensitizer and platinum(II) diimine complex as the catalyst (Fig. 17a) [187]. Visible-light-driven hydrogen production was demonstrated in the presence of EDTA as sacrificial electron donor. In contrast, the multicomponent counterpart

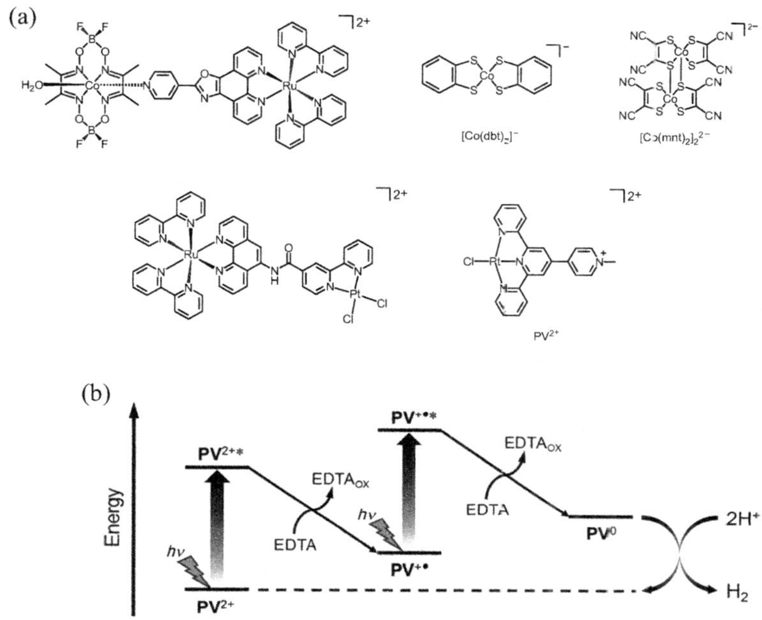

Fig. 17 (**a**) Chemical structures of various catalysts for proton reduction [183–185, 187, 188]. (**b**) Proposed Z-scheme photochemical hydrogen production. Reproduced from Sakai and coworkers [188] with permission from Wiley-VCH

did not show any activity unless in the presence of *N*,*N′*-dimethyl-4,4′-bipyridinium [187]. The same research group later developed another chloroplatinum(II) terpyridine complex functionalized with methylpyridinium (PV^{2+}, Fig. 17a) [188]. Mechanistic studies indicated that the assembly underwent successive reductive quenching after photo-excitation before leading to hydrogen generation (Fig. 17b). This stepwise photon absorption is reminiscent of the Z-scheme of natural photosynthesis [188].

Apart from using ruthenium complexes as photosensitizers, another metal–ligand chromophore, alkynylplatinum(II) terpyridine, which was first reported by Yam and coworkers in 2001 [236], could also be used. [Pt(4-tol-tpy)(C≡CPh)], documented by Wu, Tung and coworkers in 2002 [237], was reported as photosensitizer by Eisenberg and coworkers to sensitize colloidal platinum [189] and cobaloxime complex for hydrogen generation [190]. The photocatalytic system consisting of [Co(dmgH)$_2$(py)Cl] (dmgH = dimethylglyoximate; py = pyridine, Fig. 18), [Pt(4-tol-tpy)(C≡CPh)] (Fig. 18), and TEOA showed ~ 1,000 turnovers after 10 h of irradiation at pH 8.5 [190]. Later, variations on the photocatalytic system based on the choice of different cobalt catalysts by Eisenberg and coworkers [191] as well as different alkynylplatinum(II) terpyridine sensitizers by Castellano and coworkers [192] have been reported. On the other hand, a library of cyclometalated iridium(III) complexes have been investigated as sensitizers by Bernhard and coworkers [193, 194]. High quantum efficiencies up to 37 times of

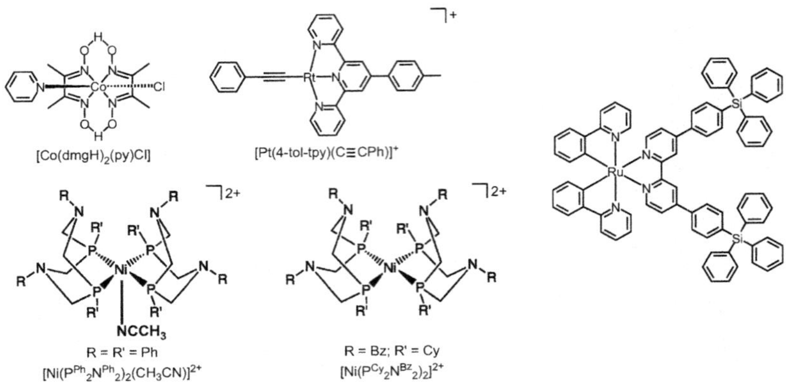

Fig. 18 Chemical structures of various catalysts and platinum-based sensitizer for proton reduction [189, 190, 195–199]

the ruthenium(II) tris(bipyridine) counterparts were observed. Further exploration of efficient iridium sensitizers has been done by Park and coworkers [195]. Introduction of bulky tetraphenylsilane onto the bipyridine ligands (Fig. 18) resulted in dramatic improvement in the durability and the catalytic property. TONs of 17,000 have been achieved, which was the highest value reported to date [195]. In addition to those transition metal complexes, nickel(II) phosphine complexes have also been studied as an electrocatalyst for hydrogen production [196–199]. As revealed by DuBois and coworkers, two closely related complexes with different substituents showed the opposite catalytic behavior with $[Ni(P^{Ph}{}_2N^{Ph}{}_2)_2(CH_3CN)](BF_4)_2$ being a hydrogen production catalyst and $[Ni(P^{Cy}{}_2N^{Bz}{}_2)_2](BF_4)_2$ being a hydrogen oxidation catalyst (Fig. 18) [196]. Initial work observed a turnover frequency of 350 s^{-1} for $[Ni(P^{Ph}{}_2N^{Ph}{}_2)_2(CH_3CN)](BF_4)_2$ catalyzed hydrogen production [197]. A structure–property relationship has been investigated by Roberts, DuBois and coworkers via different substituents on the phenyl ring and the presence of different acids [198]. The catalytic rate increased with a more electron-withdrawing *para*-substituent. However, the turnover frequencies did not show any correlation with the pKa of the acids, but they were influenced by the size of the acid and the presence of water [198]. Recently, $[Ni(P^{Ph}{}_2N^{Ph}{}_2)_2(CH_3CN)](BF_4)_2$ showed impressive turnover frequencies of 33,000 s^{-1} in dry acetonitrile and 106,000 s^{-1} in 1.2 M of water, using protonated dimethylformamide as the proton source [199].

3.4 Catalytic and Photocatalytic Reduction of Carbon Dioxide

During the past decades, we have been relying on the combustion of fossil fuels for the generation of electricity. Carbon dioxide has been massively produced as a major end product, which is a stable greenhouse gas. Recycling carbon dioxide to usable fuel or chemical intermediates has been a challenging topic for scientists for years as reduction of proton could be one major competitive reaction pathway [200–221]. In nature, one of the most active and selective reduction catalysts is carbon monoxide dehydrogenase from the anaerobic bacteria *Carboxydothermus hydrogenoformans* [209]. The active site was found to consist of a $NiFe_4S_4$ center, which allowed CO_2 to bind onto both iron and nickel centers with minimal reorganization energy. The enzyme catalyzed the reduction of carbon dioxide to carbon monoxide involving a two-electron process at a high turnover rate of 31,000 s^{-1} [209]. This has stimulated scientists to develop transition metal catalysts for reduction of carbon dioxide to carbon monoxide, which could be used with hydrogen gas to prepare liquid fuels during the Fischer–Tropsch process. In the past 30 years, a variety of transition metal complexes have been examined, including cobalt and nickel phthalocyanines by Meshitsuka and coworkers [210], cobalt and nickel tetraazomacrocyclic complexes by Eisenberg and coworkers [211], nickel cyclams by Sauvage and coworkers [212], iron porphyrins by Savéant and coworkers [213], $[Re(bpy)(CO)_3Cl]$ by Lehn and coworkers [214], $[Ru(bpy)_2(CO)_2]^{2+}$ by Tanaka and coworkers [215], rhodium and iridium diimine complexes by Meyer and coworkers [216], $[Rh(dppe)_2Cl]$ by Wagenknecht and workers [217], cobalt macrocycles by Fujita and coworkers [200, 218], palladium triphosphine complexes by DuBois and coworkers [219], nickel phosphine complexes by Kubiak and coworkers [220, 221], as well as cobalt and iron corroles by Fujita and coworkers [222]. Recently, electrocatalytic reduction of carbon dioxide at the mercury electrode by nickel(II) macrocycle-based catalysts have been reported by Schneider, Fujita and coworkers [223].

In 2008, Ishitani and coworkers extended the work to investigate the mechanism of the rhenium(I) ($[Re(bpy)(CO)_3L]$, L = SCN, Cl, CN)-catalyzed CO_2 reduction [224]. It was found that the monodentate ligand (L) had to be dissociated to leave a vacant site for binding of CO_2, which was the reason accounting for the inactivity of the $[Re(bpy)(CO)_3(CN)]$. A photochemical system based on *fac*-$[Re(bpy)(CO)_3(CH_3CN)]^+$/*fac*-$[Re\{4,4'\text{-}(MeO)_2bpy\}(CO)_3\{P(OEt)_3\}]^+$ showed photo-induced CO formation quantum yield of 0.59 [224]. Mechanistic studies of tricarbonylrhenium(I) diimine chloride complexes have also been reported by Kubiak and coworkers [225–227]. The two-electron reduced form, $[Re(bpy)(CO)_3]^-$, was prepared by a chemical method and was unambiguously characterized by X-ray crystallography [225]. Further experiments revealed that this anion exhibited higher selectivity to CO_2 than weak acid [226]. By X-ray absorption spectroscopy and computational quantum analysis, they attributed this to the involvement of $Re^0(bpy^-)$ as the two-electron reduced state, which was more favorable to react with CO_2 over H^+ because of smaller reorganization energy during the electron transfer process [227]. An electrocatalytic cycle was proposed in Fig. 19a [228]. In 2011, Chardon-Noblat, Deronzier and coworkers first reported the use of earth-abundant *fac*-tricarbonylmanganese(I) diimine complex as the electrocatalyst for carbon dioxide reduction [229]. The selectivity and the Faradaic efficiency for production of carbon monoxide were found to be similar to the rhenium counterparts, but the manganese system was shown to be more advantageous with a lower overpotential during the CO_2 reduction [229]. The tendency of dimerization of

(a)

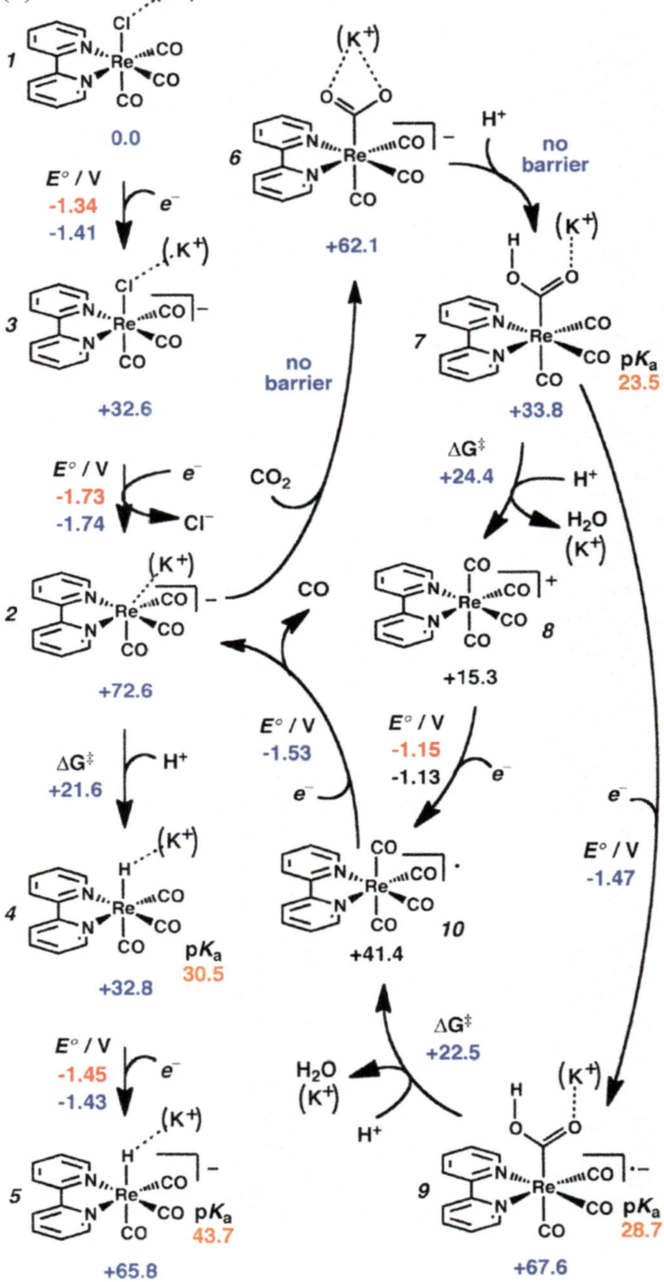

Fig. 19 (**a**) Electrocatalytic cycle for the reduction of carbon dioxide. Reprinted with permission from Carter and coworkers [228]. Copyright 2013 American Chemical Society. (**b**) Proposed catalytic CO_2 reduction mechanism of the manganese complex. Reprinted with permission from Kubiak and coworkers [230]. Copyright 2014 American Chemical Society

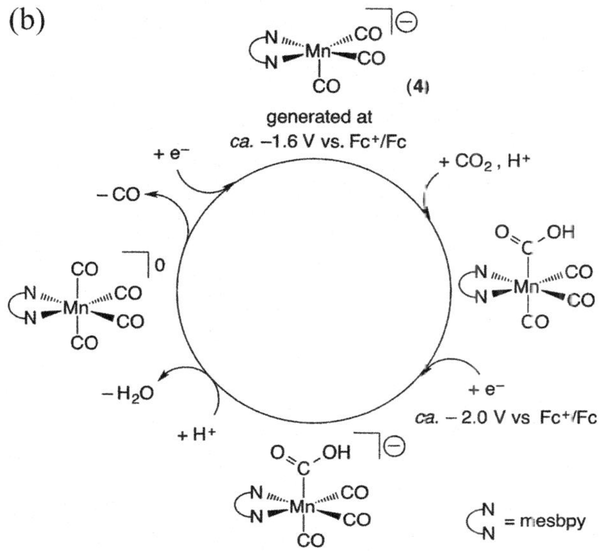

Fig. 19 (continued)

manganese complexes after the reduction would introduce difficulties during the mechanistic studies [229]. Therefore, Kubiak and coworkers utilized a bulky diimine ligand, 6,6′-dimesityl-2,2′-bipyridine, to prepare the manganese complex, which showed a single and reversible two-electron oxidation without dimerization [230]. A turnover frequency of 5,000 s^{-1} (1.4 M 2,2,2-trifluoroethanol) and a Faradaic efficiency of 98% (0.3 M 2,2,2-trifluoroethanol) for reduction of CO_2 to CO were achieved, with no observable production of hydrogen [230]. A proposed catalytic mechanism is shown in Fig. 19b.

Apart from conversion to carbon monoxide, carbon dioxide could also be reduced to formic acid. Ishitani and coworkers reported a ruthenium(II)-based photocatalytic system [231]. A trinuclear complex consisting of two photosensitizer units and one catalyst unit showed the highest quantum yield of 0.061 and TON of 671 [231]. They further loaded the ruthenium-based photosensitizer–catalyst unit onto the Ag-loaded TaON for methanol oxidation [232]. The resulting system successfully converted light energy to chemical energy in the form of formic acid and formaldehyde ($\Delta G^o = +83.0$ kJ/mol), and a Z-scheme for the photocatalytic process was proposed (Fig. 20a) [232]. Besides, iridium complexes could also function as carbon dioxide hydrogenation catalysts [233–235]. A series of high-performance iridium pincer complexes was reported by Nozaki and coworkers with turnover frequency and TON of 150,000 h^{-1} and 3,500,000, respectively [233]. Later, a proton-responsive iridium complex was synthesized by Hull, Himeda, Fujita and coworkers [234]. The catalyst operated at near-ambient condition and could reversibly absorb and release hydrogen and carbon dioxide gases, by storing in the form of formic acid. The proposed mechanism is depicted in Fig. 20b. The highest turnover rate and TON of 228,000 h^{-1} (90°C) and 308,000 (80°C) were observed, respectively [234]. A series of related mononuclear iridium complexes

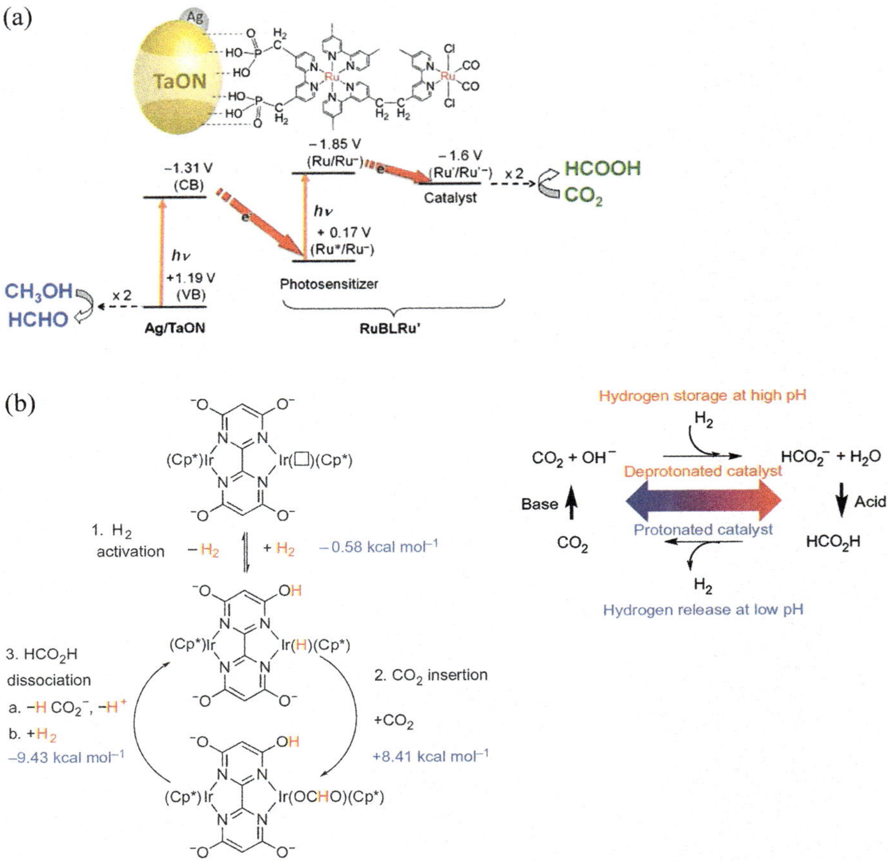

Fig. 20 (**a**) Artificial Z-scheme for photocatalytic CO_2 reduction. Reprinted with permission from Ishitani and coworkers [232]. Copyright 2013 American Chemical Society, *ACS Editors' Choice*. (**b**) Proposed mechanism of the controlled absorption and release of H_2 and CO_2. Reprinted with permission from Macmillan Publishers Ltd: Nature Chemistry [234], copyright 2012

was further investigated by Fujita, Himeda and coworkers [235]. It was found that rate enhancement of ~0.6 orders of magnitude could be achieved by simply moving the 4,4′-substituents to the 6,6′-positions. They attributed this to the presence of the pendant base in the second coordination sphere, which significantly lowered the transition state energy [235].

4 Recent Advances in Photo-switchable Transition Metal Complexes and Molecular Machines

According to the Moore's law, the number of components per integrated circuit is expected to be doubled every 2 years during the technological advancement [238]. Nevertheless, the development of a top-down silicon-based fabrication technique is getting increasingly challenging as the size of the semiconductor device is approaching that of the molecular level and the physical microfabrication limit. This motivates the development of molecular functional materials for the fabrication of devices using the bottom-up approach. In particular, photo-responsive functional materials represent an important class of potential candidates for device application [239] and have received tremendous attention since the past few decades [239–256]. These systems possess distinct and addressable states that are interconvertible by using different external stimuli. Therefore, they are potentially useful for applications in optical memories and optical switching devices. Examples of photo-switchable functional motifs include the *cis–trans* isomerization of the azo group [239–248] and alkenes [247–249], 1,2-*cis*-diarylethenes [239, 244–247, 250–252], spirooxazines [244, 247, 253, 254], spiropyrans [239, 244–247, 253–255], and others [244, 256]. The photochemical rearrangements result in structural and conformational changes, which could strongly influence their optical, electronic, and optoelectronic properties and intermolecular interactions.

Unlike pure organic compounds, transition metal complexes possess rich photophysical and electrochemical properties through involvement of metal character during the electronic transitions and redox processes. In addition, the heavy-atom effect facilitates the intersystem crossing to access the spin-forbidden triplet excited state which is more long-lived. More importantly, the synthesis of transition metal complexes offers an advantage of easy functionalization by judicious choice and stepwise incorporation of custom-made ancillary ligands. This opens up unlimited opportunities for the functionalization of the photochromic and photo-switchable molecules with ligating sites and for the coordination to the transition metal centers to afford photo-responsive behavior. For instance, electronic properties [242, 246, 247, 254, 255, 257–379], binding modes [357–360], redox behavior [361–366], magnetic properties [367–375], and NLO properties [376–379] could be controlled by external photo-stimulation. By stepwise incorporation of multicomponents into transition metal complexes, molecular machines with molecular motions occurring in a controlled manner are achievable [380–429]. External stimuli that have been employed to control molecular motions include photo-excitation, electrochemical oxidation or reduction, pH variation, and metal exchange [380–429]. In this section, recent developments, starting from 2007, on the photo-switchable transition metal complexes bearing various photochromic moieties will be introduced. The influence on their structural, chemical, optical, electronic, and optoelectronic properties will also be discussed. Molecular machines that are responsive toward photo-excitations will also be highlighted.

4.1 Photochromic Ligands and Their Transition Metal Complexes

For common organic photochromic systems including *cis–trans* isomerization of alkenes and azo compounds, diarylethenes, spirooxazines, and spiropyrans, the photochemical process often proceeds via the singlet excited state. Functionalization with ligating sites and coordination to transition metal centers would lead to the possible opening up of the triplet photochemical pathway or the photosensitization of the photochromic reaction by extending the excitation wavelength to visible light. The advantages of these kinds of photofunctional transition metal complexes over their organic counterparts include the ready access to desirable optical, electronic, and optoelectronic properties by systematic design and judicious choice of ligands, which enables the use of designated and less energetic visible-light wavelengths as excitation source instead of the need for high-energy UV irradiation commonly used in organic photochromic systems. More importantly, the photochromic reactions would change the electronic properties of the entire system, resulting in the possible alteration of energy or electron transfer directions.

With the continuing efforts in the development of photo-switchable moieties through photosensitization reactions by various research groups [242, 246, 247, 254, 257–266], the exploration of different photo-responsive metal–ligand chromophores represents an important area of research. Nishihara and coworkers investigated the photosensitization of the azobenzene-containing dithiolato-platinum(II) bipyridine complexes [267]. This class of complexes displays absorptions due to azobenzene π–π* transition, ILCT [π(ditholate)–π*(azobenzene)] transition and LLCT [π(ditholate)–π*(bipyridine-azobenzene)] transition. It has been observed that photo-excitation into the low-energy LLCT band could induce *cis*-to-*trans* isomerization [267]. By proper selection of excitation wavelength, three different conformational isomers are achievable, as shown in Fig. 21a. The same research group later reported the solid-state *cis*-to-*trans* isomerization of bis (dipyrazolylstyrylpyridine)iron(II) complex sensitized by the MLCT excited state (Fig. 21b) [268]. It is found that the *trans*-isomer displayed spin-crossover phenomenon, while the *cis*-isomer only existed in the high-spin state [268]. Recently, a highly efficient tetranuclear rhenium(I) tricarbonyl diimine complex with a stilbene-type bridging ligand has been developed by Sun, Lees and coworkers, showing an MLCT [dπ(Re)–π*(bridging ligand)]-sensitized *trans*-to-*cis* isomerization (Fig. 21b) [269].

The impact of the incorporation of transition metal centers on diimine ligands bearing two photo-active olefin bonds and their respective copper(I) and platinum(II) complexes (Fig. 22a) has also been studied by Wand and coworkers [270]. The free ligand and its copper(I) complex displayed *trans*-to-*cis* isomerization, while the platinum(II) analogue showed no isomerization upon photo-excitation. A similar observation has also been found by Katan, Malval, Fillaut and coworkers on their azobenzene-containing ruthenium(II) tris(bipyridine) complexes (Fig. 22b) [271]. It is believed that the presence of lower-lying ^{3}MLCT states would serve as a deactivation pathway to quench the photochemical reaction (Fig. 22c). Other

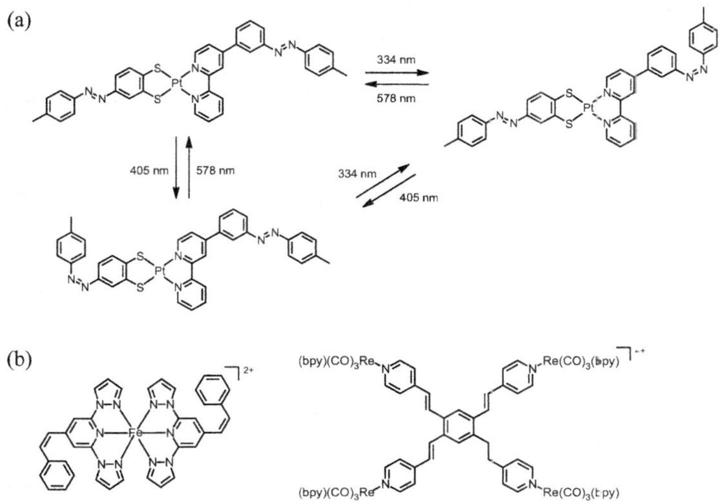

Fig. 21 (**a**) Schematic diagram showing the photo-controlled tri-stable dithiolatoplatinum (II) bipyridine complexes. Reproduced from Nishihara and coworkers [267] with permission from Wiley-VCH. (**b**) Chemical structures of photo-responsive transition metal complexes [268, 269]

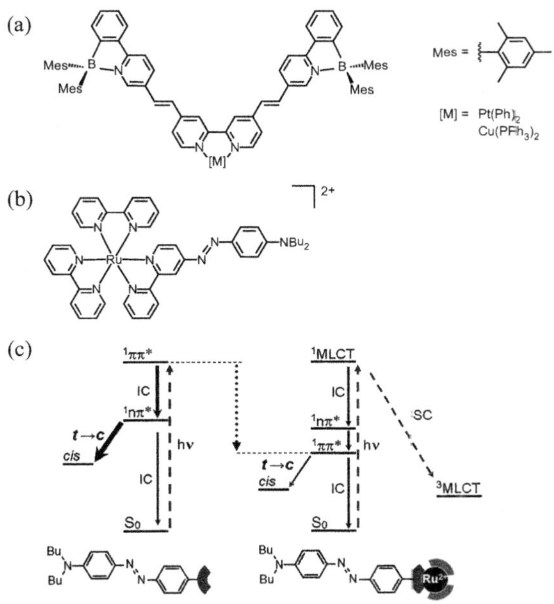

Fig. 22 Molecular structures of the photofunctional transition metal complexes synthesized by (**a**) Wand and coworker [270] and (**b**) Katan, Malval, Fillaut and coworkers [271]. (**c**) Simplified schematic diagram showing the photophysical process. Reproduced from Katan, Malval, Fillaut and coworkers [271] with permission from Wiley-VCH

transition metal systems with *cis–trans* photofunctional groups have also been developed [271, 272].

Pericyclic photocyclization of DAE was initially observed by Irie on the photoisomerization of *cis*-stilbene [273]. With the replacement of the benzene ring by thiophene ring [274], together with the introduction of a methyl group at the adjacent position [273], this system shows promising photo-responsive behaviors, such as good reversibility, fatigue resistance, fast response time, good tunability, and photo-addressable states. A lot of seminal works on the organic photochromic diarylethene have been reported by Irie [250, 252, 275–280], Feringa [281–285], Lehn [286–288], Branda [251, 289–292], Yam [293–303], Tian [255, 304–307], Kawai [278, 308–310], and others [244, 311–316]. Recently, crystalline-state photochromism with fast and reversible shape interconversion has been demonstrated by Irie and coworkers [280]. In spite of the rapid developments, their transition complex analogues have been relatively less explored. The earliest work on the photophysical and photochromic studies of the photochromic diarylethene-containing pyridine transition metal complexes was reported by Lehn and coworkers [317]. The structure–property relationship of different transition metal complexes appended with photochromic diarylethene has been studied by Lehn [318], Irie [252, 319–321], Yam [247, 322–330], Branda [331–333], Tian [304, 334–336], and others [337, 338]. In 2004, the photosensitization of diarylethene photocyclization was reported by two independent research groups at almost the same time, including the phenanthroline tricarbonylrhenium(I) complex (Fig. 23a) by Yam and coworkers [264] and dinuclear ruthenium(II)

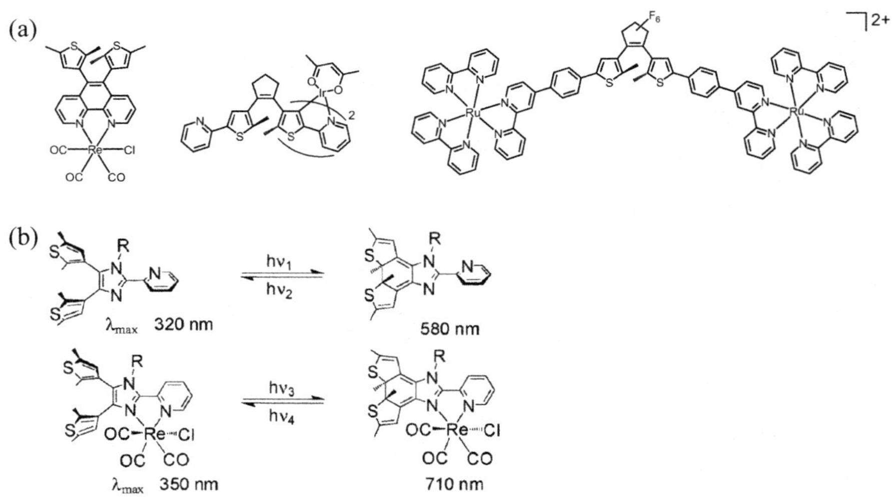

Fig. 23 (**a**) Chemical structures of diarylethene-containing photochromic transition metal complexes [264, 265]. (**b**) Schematic diagram showing the photochemical reactions of the ligands and their metal complexes. Reprinted with permission from Yam and coworkers [323]. Copyright 2007 American Chemical Society

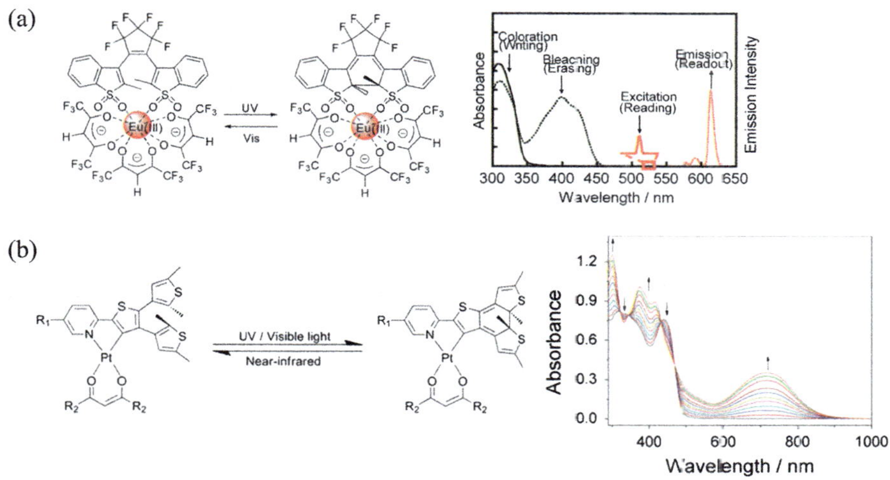

Fig. 24 (**a**) Photochromism and the spectroscopic properties of the europium(III) complex. Reproduced from Hasegawa, Kawai and coworkers [342] with permission from the Royal Society of Chemistry. Photochromism of (**b**) the platinum(II) cyclometalated complexes and their representative UV–vis spectral changes, reprinted with permission from Yam and coworkers [329]. Copyright 2011 American Chemical Society. (**c**) Platinum(II) alkynyl bridged bis (dithienylethene), reprinted with permission from Branda, Wolf and coworkers [343]. Copyright 2009 American Chemical Society. (**d**) Selected examples of diarylethene-containing metal complexes [325, 333, 344, 345]

tris(bipyridine) complex (Fig. 23a) by De Cola and coworkers [265]. Unlike the usual approach of introducing the ligation site as a pendant of the diarylethene moiety, Yam and coworkers incorporated the diarylethene group into the ligand backbone in order to provide greater perturbation on the photophysical behavior upon photochemical cyclization [247, 322–330]. The triplet sensitization pathway was then confirmed by ultrafast spectroscopic studies by the same group [266] as well as later by Indelli, Ravaglia, Scandola and coworkers on the diarylethene-containing ruthenium(II) polypyridine system [339]. Later on in 2007, Yam and coworkers extended the absorption of the closed form to the NIR region [323]. This was accomplished by using diarylethene-containing 2-(2-pyridyl)imidazole as the *N,N*-donor ligand in the tricarbonylrhenium(I) system (Fig. 23b) [323]. The metal-assisted planarization of the two heterocyclic rings is believed to enhance the π-conjugation, resulting in a drastic red shift in the absorption band. A similar NIR photochromism and luminescence switching behavior has also been subsequently reported by Tian and coworkers on the bis-cyclometalated iridium(III) acetylacetonate complex (Fig. 23a) [340, 341].

Manipulation of photophysical properties of the photochromic ligands could also be achieved by incorporation of lanthanide metal centers. Hasegawa, Kawai and coworkers studied the luminescence behavior of a photochromic europium(III) complex (Fig. 24a) [342]. The utilization of the discrete and photo-responsive absorption and emission properties of europium(III) chromophore and the ligand

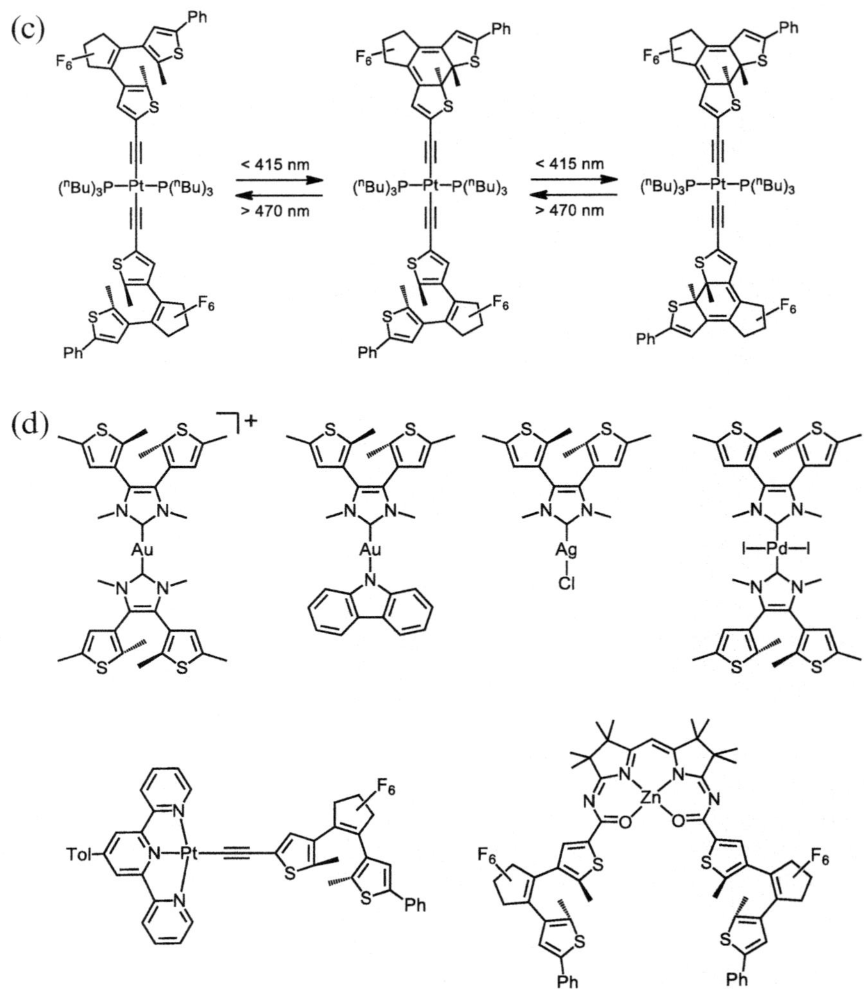

Fig. 24 (continued)

has opened up the nondestructive readout capability via the monitoring of the luminescence intensity. Wavelengths for writing, erasing, and reading processes have been identified, as shown in Fig. 24a. Photochromic diarylethene moiety has also been incorporated into other ligand frameworks by Yam and coworkers [325, 326, 328, 329]. In 2009, they reported a series of photochromic NHCs and their silver(I), gold(I), and palladium(II) complexes (Fig. 24d) [325], which were potentially applicable in photochromic catalysis. Additional ligating sites, such as pyridine, pyrimidine, and other nitrogen-containing heterocycles, were functionalized with photochromic carbenes as bidentate ligands to incorporate into tetracyanoruthenate(II) [326] and η^6-mesitylene ruthenium(II) systems [328]. Later, Yam and coworkers reported a series of diarylethene-containing cyclometalated platinum(II) complexes (Fig. 24b) [329]. Systematic variation of the substituents on both the cyclometalating ligand and β-diketonate ligand not only altered the photophysical properties of the open form but also the photocyclized closed form, demonstrating a crucial advantage of employing transition metal complex systems to fine-tune the photophysical and photochromic properties of the entire system. Unlike previous organic counterparts, a bis(dithienylethene) bridged by diphosphinoplatinum(II) bis-alkynyl complexes reported by Branda, Wolf and coworkers displayed a dual ring-closing photochromism without the observation of energy transfer quenching after the first photocyclization step (Fig. 24c) [343]. They attributed this to the localized triplet exciton and the energy barrier at the Pt linkage that slowed down the energy transfer process after the first photocyclization, showing the importance of the incorporation of heavy elements in this multi-DAE system [343]. Branda, Wolf and coworkers also investigated the linker-dependent photosensitization process in the photochromic alkynylplatinum(II) terpyridine system through ground state spectroscopic [344] and time-resolved spectroscopic [345] measurements. Reduction of communication between the metal-based and intraligand states by lengthening the spacer or introducing nonconjugated spacer resulted in decrease of photosensitization efficiency, showing how structural design could manipulate the deactivation pathway of the photo-generated excited state.

Apart from the previously mentioned photofunctional transition metal complexes, various ligands were also functionalized with diarylethene and incorporated into transition metal complexes. Their switchable photophysical properties have been investigated [322, 324, 327, 333, 346–351]. These include diarylethene-containing alkynyls [327, 346, 347], terpyridines [348, 349], bipyridines [350, 351], 1,10-phenenthrolines [322, 324], pyridylimidazoles [330], and bis (acylamidines) (Fig. 24d) [333].

Under photo-irradiation, spirooxazine undergoes ring-opening reaction to give π-extended MC via C–O bond cleavage. Yam and coworkers functionalized this type of photo-responsive group with bipyridine and incorporated it into the zinc (II) thiolate [352] and platinum(II) bis-alkynyl complex systems [353]. Upon UV excitation, the color of the platinum(II) complex in dichloromethane solution was found to change from yellow to blue due to the photo-induced ring-opening reaction (Fig. 25a). This could also be achieved by irradiating into the MLCT/LLCT transition band in the visible region, showing the successful intramolecular photosensitization reaction [353]. In the presence of the trialkoxyl chain, gelation and liquid crystal properties have also been observed [353]. The photo-induced LF modulation was also studied by Frank and coworkers on the spirooxazine-containing tetracarbonyl molybdenum complex (Fig. 25b) by spectroscopic and computational methods [354]. The LUMO+1 (phenanthroline-based) was found to be 0.2–0.3 eV lower in energy in the ring-closed (spirooxazine) form than the ring-opened (MC) form, suggesting that the spirooxazine form is a better π-acceptor.

Recently, photochromic boron-containing diarylethenes have also been investigated by Yam and coworkers [295, 297, 298, 300, 303]. NIR photochromic behavior by boron β-diketonates [295, 300], fluoride-gated photochromism by triarylboranes [297, 303], and NIR and tunable photochromism by thienylpyridine

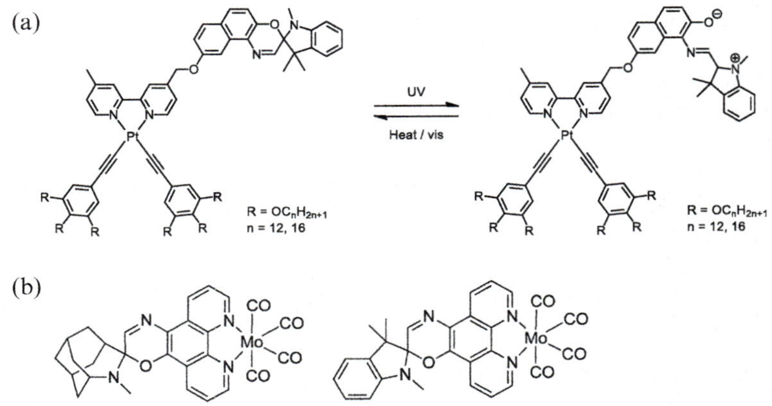

Fig. 25 (**a**) Photochromism of the spirooxazine-containing platinum(II) bipyridine bis-alkynyl complexes. Reproduced from Yam and coworkers [353] with permission from Wiley-VCH. (**b**) Chemical structures of the molybdenum complex reported by Frank and coworkers [354]

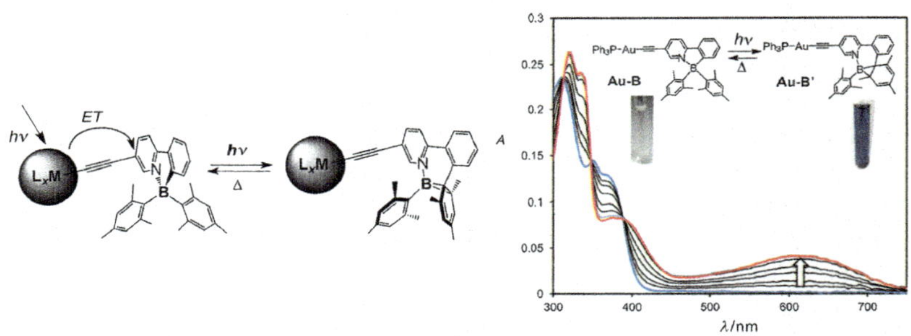

Fig. 26 Photochromism and UV–vis spectral changes of the metal-containing N^C-chelated organoboron complexes. Reproduced from Wang and coworkers [356] with permission from Wiley-VCH

bis(alkynyl)boranes [298] have been achieved. Recently, another type of photochromism based on N^C-chelated organoboron compounds [355] has been reported by Wang and coworkers, and the impact of metal coordination has also been investigated [356]. The alkynyl ligation site was functionalized onto the photoactive N^C-chelated organoboron and was incorporated into the tricarbonylrhenium(I) alkynyl, phosphinogold(I) alkynyl, and bis(phosphino)platinum(II)-bis-alkynyl systems (Fig. 26) [356]. Photosensitization has been observed in the gold(I) system via the triplet pathway, whereas the platinum(II) system showed reduced photochromic quantum efficiency, and only a single photochromic reaction was observed due to intramolecular energy transfer quenching after the first photochemical reaction. However, the rhenium(I) complex did not show any sign of photochromism, probably due to the rapid energy transfer to the low-lying triplet ILCT/MLCT excited state.

In addition to the perturbation on the photophysical properties, the photochemical process also induces conformational changes, affecting the coordination mode of the ligands. Ceroni and coworkers reported a dendrimer bearing two cyclam units bridged by the photoactive azobenzene moiety [357]. The *trans*-to-*cis* photoisomerization of the azo linkage resulted in closer proximity of the two cyclam units, and the binding stoichiometry with zinc(II) ion was changed from 1:2 to 1:1 (Fig. 27a). On the other hand, the coordination of copper(II) was found to quench the photochemical process through efficient energy transfer, showing the

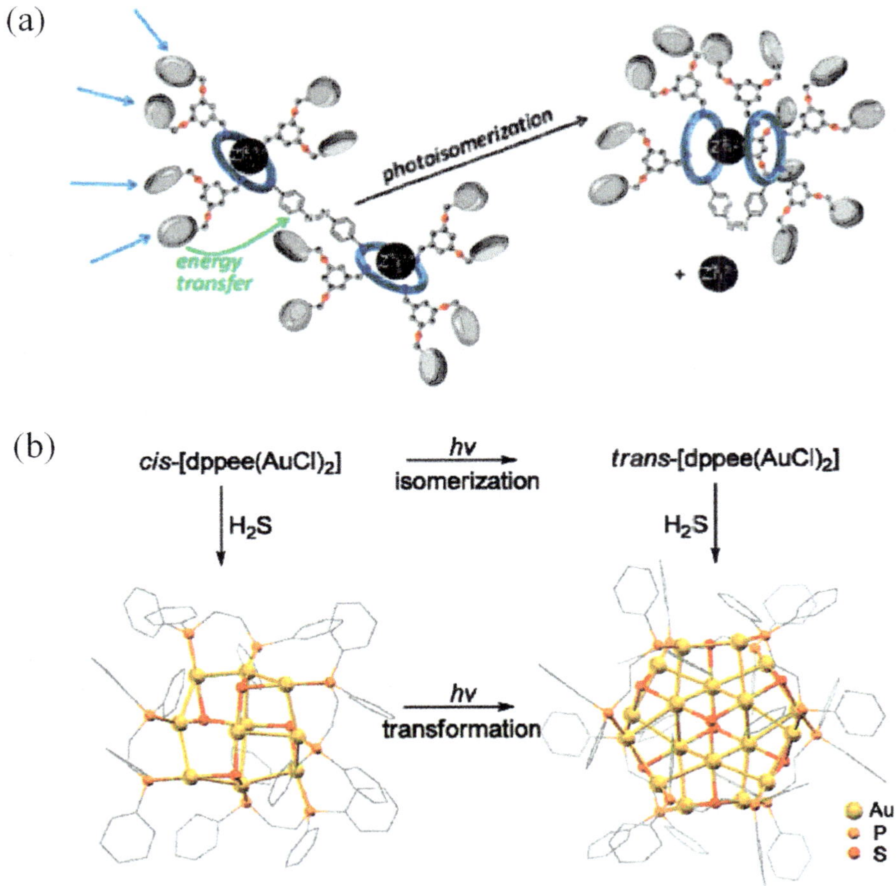

Fig. 27 (**a**) Photochromism and binding mode of the photo-responsive dendritic tweezer. Reprinted with permission from Ceroni and coworkers [357]. Copyright 2012 American Chemical Society. (**b**) Self-assembly process of gold(I) clusters and photo-induced isomerization driven transformation. Reprinted with permission from Yam and coworkers [358]. Copyright 2015 American Chemical Society

absence of *cis–trans* isomerization. Yam and coworkers constructed two gold (I) sulfido clusters by using the photoactive dppee ligand [358]. Two different clusters, $[Au_{10}(\mu\text{-}cis\text{–dppee})_4(\mu_3\text{-S})_4]^{2+}$ and $[Au_{18}(\mu\text{-}trans\text{–dppee})_6(\mu_3\text{-S})_8]^{2+}$, were successfully prepared from their corresponding *cis*–dppee and *trans*–dppee, respectively. Upon photo-irradiation, *cis*-to-*trans* isomerization and cluster-to-cluster transformation were found to occur (Fig. 27b). It has been attributed to the difference in bite angle of the bidentate *cis*–dppee and *trans*–dppee ligand, resulting in a change of the coordination mode.

Andréasson and coworkers reported a system consisting of a pyridine-appended diarylethene and a Zn–porphyrin dimer (Fig. 28a) [359]. Before photo-irradiation, the diarylethene backbone was flexible, and the two pyridine rings could coordinate with zinc. Upon photocyclization, the diarylethene moiety was rigidified and was no longer able to interact with both Zn–porphyrin chromophores. More importantly, this large change in coordination mode has resulted in a significant influence on their electronic absorption properties. A nondestructive readout has been achieved by monitoring the low-energy absorption band at 748 nm or the fluorescence intensity by 748-nm photo-excitation as this could not trigger the photocycloreversion. The same group later on utilized the same supramolecular system to investigate the photo-induced release of analytes by making use of the difference in the binding site between two conformations (Fig. 28b)

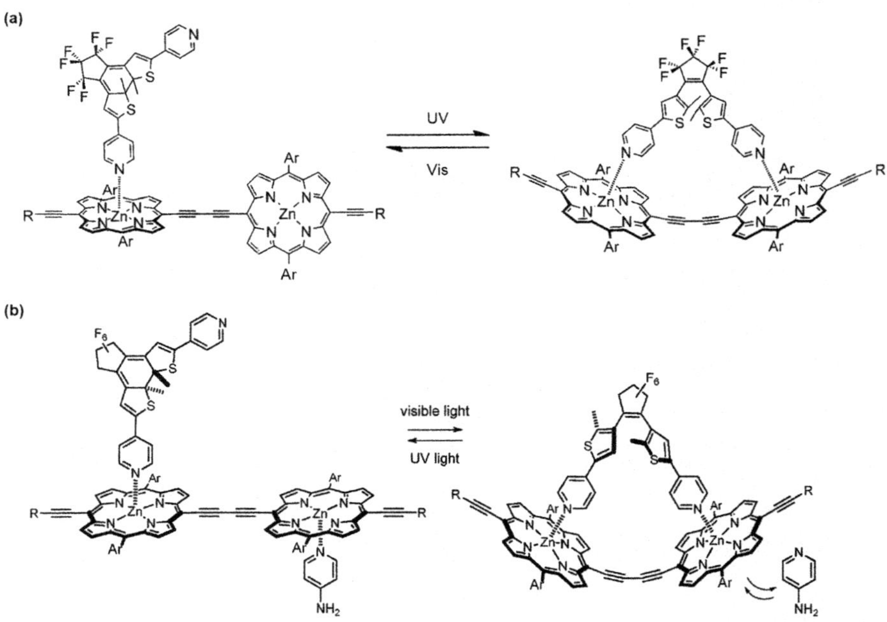

Fig. 28 (**a**) Photochromism of the pyridine-appended diarylethene-Zn-porphyrin dimer. Reproduced from Andréasson and coworkers [359] with permission from Wiley-VCH. (**b**) Schematic diagram showing the photo-induced release of analyte. Reproduced from Andréasson and coworkers [360] with permission from the Royal Society of Chemistry

[360]. A fluorescence-based nondestructive readout and a photo-controlled release of compounds have been achieved.

4.2 Multi-responsive Photofunctional Transition Metal Complexes

Recent developments on photofunctional transition metal complexes involve the introduction of additional functionalities, giving rise to interesting and fascinating multi-stimuli-responsive transition metal complexes. Nishihara and coworkers reported a redox-active ferrocene-appended azobenzene as a monolayer on the ITO electrode [361]. This class of compounds exhibits photochromism by single source of green light, assisted by the electrochemical control from the ferrocene moiety [361]. Akita and coworkers also introduced diarylethene into redox-responsive iron complexes [362, 363]. The photochromic reaction was found to perturb the communication between the two iron chromophores. Efficient molecular wirelike properties with one-electron oxidation could only be achieved in the closed form with high switching factor, giving rise to a photo-active molecular wire. The same research group also reported a similar system based on ruthenium and iron complexes, which showed electrocyclization behavior, as shown in Fig. 29a [364]. Based on the earlier example on the utilization of 9-mesityl-10-methylacridinium as photoredox catalyst reported by You, Fukuzumi, Nam and coworkers [365], the photocycloreversion quantum efficiency was enhanced by one order of magnitude using the cyclometalated iridium(III) photoredox catalyst with long-lived charge-transfer excited state [366]. By using ultrafast spectroscopy, the mechanism has been believed to involve PET, followed by electrocatalytic ring-opening and reductive termination (Fig. 29b). Detailed photophysical and electrochemical studies revealed the strong negative relation between back electron transfer rate and the Gibbs free energy changes, indicative of the occurrence of back electron transfer in the Marcus inverted region. The investigation has provided insights into the future development of photoredox catalysts for the improvement of photocycloreversion quantum yields.

Apart from manipulating the redox properties, photo-controlled magnetic properties have also received tremendous attention. In 2009, Frank and coworkers reported a photochromic spirooxazine-containing cobalt bis(dioxolene) complex (Fig. 30a) [367]. Valence tautomerism of cobalt(II) bis(dioxolene) with conversion between low-spin and high-spin states could be observed by modulating the physical environment of the complexes, for example, by changing the temperature. Attempts on investigating the photo-perturbation of the magnetic properties have been limited due to the small photochromic conversion. Later, this was realized by Khusniyarov and coworkers by replacing the spirooxazine-containing phenanthroline ligand with two 4-styrylpyridine ligands in cobalt(II) bis(dioxolene) complex (Fig. 30b) [368]. The result showed that the spin-state equilibrium by

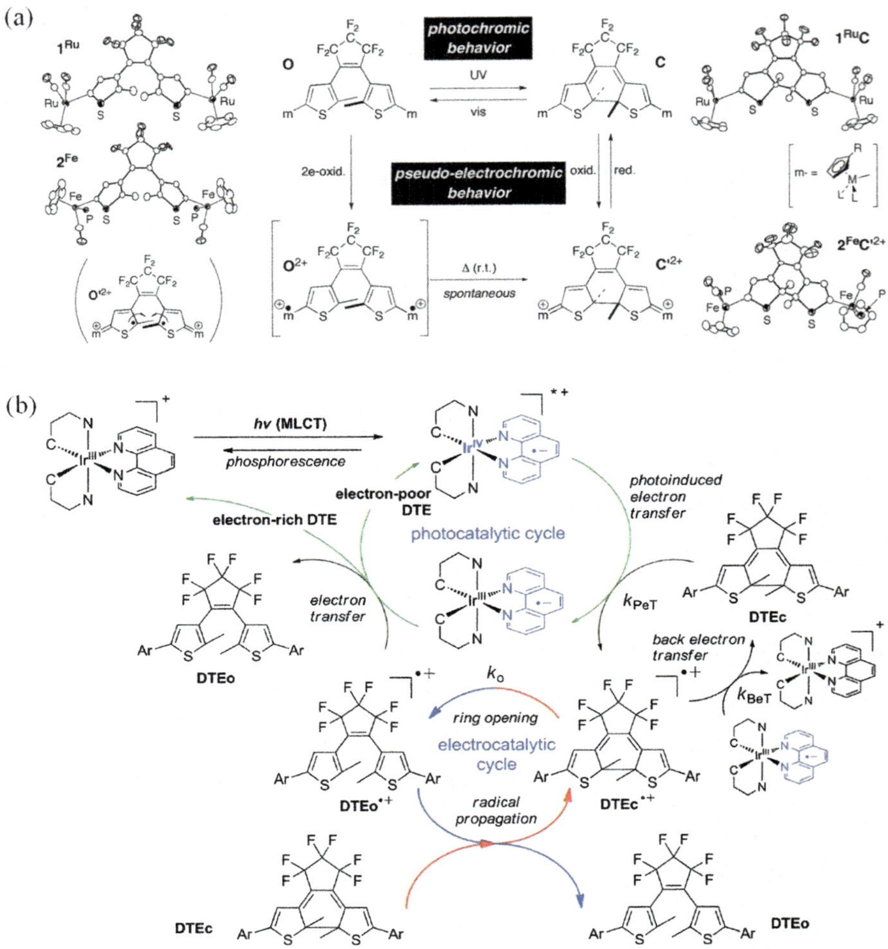

Fig. 29 (**a**) Photochromism and electrochromism of the ruthenium and iron complexes. Reproduced from Akita and coworkers [364] with permission from the Royal Society of Chemistry. (**b**) Proposed mechanism for the photocatalytic cycloreversion. Reproduced from You, Fukuzumi, Nam and coworkers [366] with permission from the Royal Society of Chemistry

valence tautomerism could be altered by photo-induced *trans*-to-*cis* isomerization at room temperature.

In 2009, Miyasaka, Irie and coworkers synthesized a diarylethene-appended mixed-valence tetranuclear $[Mn^{II}_2Mn^{III}_2]$ complex [369]. X-ray crystal structure revealed a polymeric one-dimensional linear structure of $[Mn_4]$ SMMs linked by the bis-carboxylate functionalized diarylethene ligands. Two forms of SMMs were prepared using their corresponding open and closed form of diarylethene bis-carboxylate. However, a significant difference in magnetic property via through-space interunit-interaction was only observed upon photocycloreversion

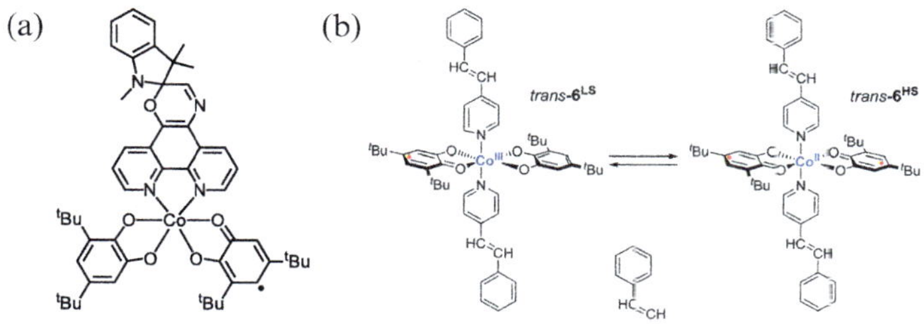

Fig. 30 (a) Chemical structure and (b) schematic diagram of the photofunctional transition metal complexes with switchable magnetic property. Reproduced from Khusniyarov and coworkers [368] with permission from the Royal Society of Chemistry

of the closed form of SMMs. Thus, this was attributed to the difference in molecular packing in the crystals. Other diarylethene-bridged SMMs have also been investigated by Yamashita and coworkers [370, 371]. A similar approach to modulate the magnetic property was also done by Oshio and coworkers [372]. They incorporated the diarylethene-containing 1,10-phenanthroline reported previously by Yam and coworkers [264] into the iron(II) metal center. Spin-state conversion of 20% was observed under UV irradiation at 173 K [372]. The same iron(II) complex has also been investigated by Khusniyarov and coworkers and revealed that the photo-induced spin crossover could occur at room temperature with 40% conversion (Fig. 31a) [373]. In 2011, Jana, Herges and coworkers reported an interesting finding on photo-induced spin-crossover phenomenon at room temperature [374]. Nickel(II) is known to possess different spin states according to the coordination mode, with 4-coordinated nickel(II) being low spin and 6-coordinated nickel(II) being high spin, while 5-coordinated nickel(II) could be either high or low spin. A photo-active azo-containing nickel-porphyrin, as shown in Fig. 31b, was synthesized [374]. In the *trans*-configuration, the square-pyramidal coordination geometry about the nickel(II) center is not favorable, whereas in the *cis*-configuration, coordination to the nickel center has become possible. In other words, this system has provided an elegant example of bringing about photo-induced switching of the coordination number and coordination geometry of the metal center. More importantly, this has resulted in the formation of a paramagnetic 5-coordinated nickel(II) complex and the reversible photo-switching of magnetic properties. Later, the same group studied the substituent effect on the photochromic group, which was also investigated using DFT calculations, and the switching efficiency was improved to 72% paramagnetic Ni-porphyrin after photo-irradiation [375].

Diarylethene photochromism is well known to exhibit photo-switchable electronic communication between two pendant groups, which represents an ideal photofunctional moiety for NLO properties. In 2008, Le Bozec and coworkers reported diarylethene-containing zinc(II) diimine diacetate complexes (Fig. 32a) [376]. Their open forms showed weak NLO response, while the closed form, after photo-excitation at 365 nm, exhibited strong NLO behavior, attributed to the improved communication between the electron-donating dimethylamino group

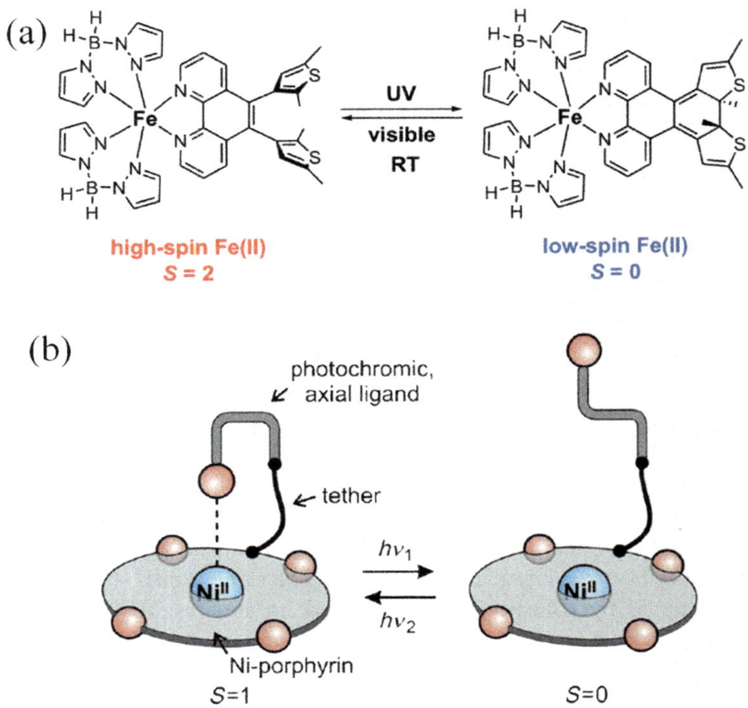

Fig. 31 (**a**) Schematic diagram of photo-induced spin crossover of diarylethene-containing iron (II) complex. Reprinted with permission from Khusniyarov and coworkers [373]. Copyright 2013 American Chemical Society. (**b**) Photo-induced switching of coordination mode of the nickel (II) porphyrin complex. Reproduced from Herges and coworkers [375] with permission from Wiley-VCH

and the electron-accepting perfluorocyclopentene moiety. Later, the investigation was extended to the tetrahedral copper(I) bis(bipyridine) complexes using a similar ligand (Fig. 32a) [377]. The copper complexes showed photochromism in the NIR region with band maximum at 701 nm and strong enhancement of the NLO response upon photocyclization. Similar to the molecular design based on transition metal-diimine-DAE donors reported by Le Bozec, Jacquemin and coworkers showed an enhancement of hyperpolarizability of the iron(II) complexes upon photocyclization (Fig. 32a) [378]. In 2014, Jacquemin, Boucekkine, Roberto and coworkers reported the photo-switchable second-order NLO properties of a series of cyclometalated platinum(II) complexes (Fig. 32b), leading to the best $\mu\beta_{\mathrm{EFISH}}$ reported to date for the platinum(II) complexes [379]. In addition, this represents the first demonstration of the photo-switchable NLO behavior of transition metal complexes in PMMA thin film, opening up a new area for further development.

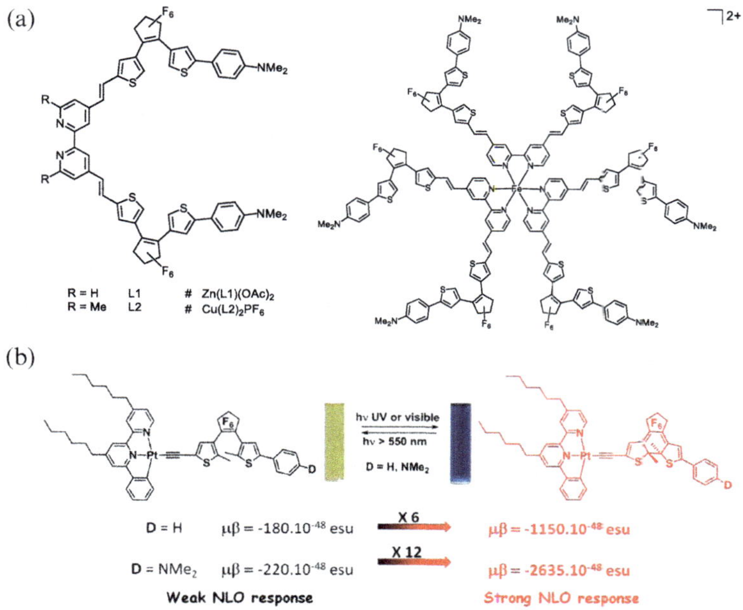

Fig. 32 (**a**) Chemical structures of photofunctional transition metal complexes with switchable NLO behavior [376–378]. (**b**) Photo-induced NLO switching of cyclometalated platinum (II) complexes. Reprinted with permission from Jacquemin, Boucekkine, Roberto and coworkers [379]. Copyright 2014 American Chemical Society

4.3 Photo-controlled Molecular Machines

Over the past years, inspiring molecular machines with molecular motions triggered by external stimuli, for example, change in pH, metal coordination, and redox environment, using rotaxanes, catenanes, or others have been reported by Lehn [380–382], Sauvage [383–390], Stoddart [391–393], Schmittel [394, 395], Balzani [396–398], Feringa [399–402], Collin [388–390, 403], Aida [404–407], Rebek [408], and others [409–417]. In particular, molecular movements operated by light irradiation possess unique advantages of easily controllable excitation wavelengths and high site selectivity. By incorporating transition metal centers, higher structural versatility and richer photophysical and electrochemical properties were demonstrated. Sauvage, Stoddart and coworkers reported the photo-responsive molecular interlocked system using a ruthenium(II) tris(2,2′-bipyridine)-containing bistable [2]rotaxane (Fig. 33a) [418]. The PET phenomena from $[Ru(bpy)_3]^{2+}$ to BIBP unit and the strong interaction between reduced BIBP cationic radical and dicationic CBPQT diradical were used to construct the light-stimulated interlocking system [418]. Without an external stimulus, CBPQT would prefer to rest on the π-electron-rich 1,5-dialkoxynaphthalene driven by donor–acceptor interactions. In the presence of a sacrificial electron donor, such as TEOA, all the BIBP and

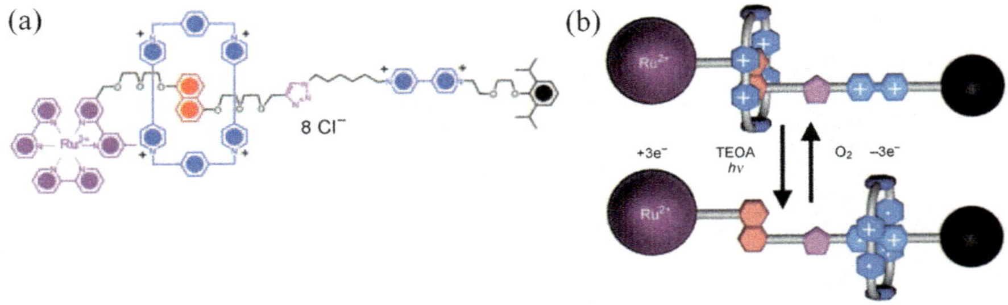

Fig. 33 (**a**) Chemical structure of the molecular switch. (**b**) Schematic diagram showing the photo-induced movement. Reproduced from Sauvage, Stoddart and coworkers [418] with permission from Wiley-VCH

CBPQT units were reduced to their corresponding cationic radicals by PET. Molecular movement was subsequently induced, stabilized by radical pairing interactions between CBPQT and BIBP cationic radicals. Upon contact with oxygen, the radicals were oxidized, and CBPQT would return to the original position by regaining the donor–acceptor interaction (Fig. 33b). The use of ruthenium chromophore in place of TTF used in most organic systems has resulted in much improved redox reversibility in aqueous solution, providing insights on the development of light-stimulated molecular switches for nanobiomechanical systems and molecular prosthetics.

The photo-triggered molecular motor is another class of molecular machines that has received much attention for multicomponent supramolecular devices [382, 383, 400, 419–424]. Schmittel and coworkers reported a nanomotor based on four-component self-assembly of two different zinc(II) porphyrins, DABCO and copper(I) ion [419]. Control of rotation speed between 97,000 s^{-1} and ~80,000 s^{-1} has been achieved by controlling the number of equivalents of copper(I) ion [419]. Further work has been done on the speed control by using light/heat as external stimulus. In the presence of photochromic 2,2′-diazastilbene, which reversibly converts between monodentate and bidentate coordination mode toward a copper(I) metal center, the speed of the molecular rotor could be reversibly altered between 81,300 s^{-1} and 38,200 s^{-1} (Fig. 34a) [420]. Alternatively, Feringa and coworkers reported a photochromic molecular rotor based on an overcrowded alkene, with the rotation speed controlled by structural design [421]. The molecular rotor was functionalized with different ligation sites and incorporated into palladium tetraphenylporphyrin [422] and ruthenium bipyridine [423] systems (Fig. 34b). The presence of metal–ligand chromophore has opened up the possibility of visible-light excitation to trigger the molecular rotor, which was believed to improve the robustness of the entire system. For the ruthenium(II) system developed by Wezenberg, Feringa and coworkers, the metal coordination to the 4,5-diazafluorenyl motif has served to reduce the steric hindrance of the overcrowded alkene and resulted in a 50-fold increase of rotation speed upon photo-irradiation when compared to the metal-free counterpart [423]. The interesting photochromic overcrowded alkene has also been utilized in the construction of

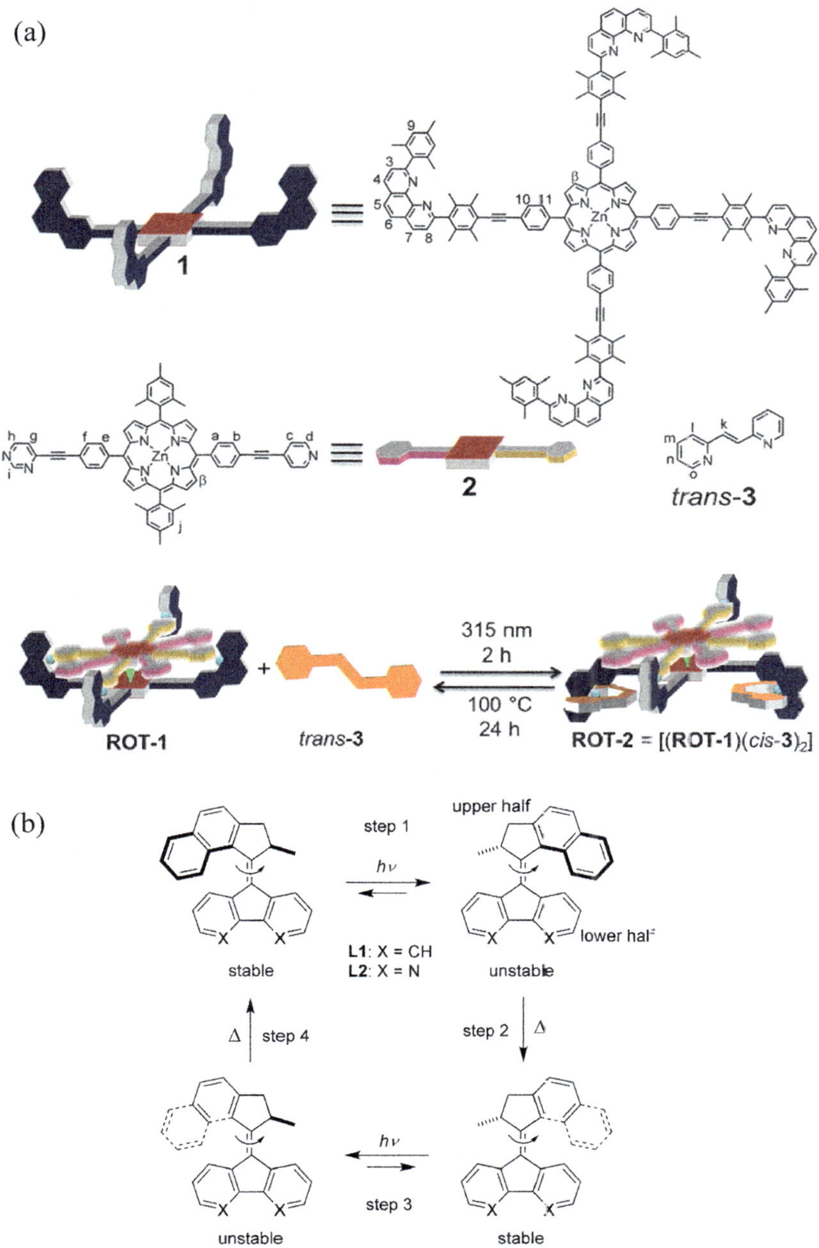

Fig. 34 (**a**) Schematic diagram showing the reversible conversion between high-speed ROT-1 and low-speed ROT-2. Reproduced from Schmittel and coworkers [420] with permission from the Royal Society of Chemistry. (**b**) Isomerization behavior of the molecular rotor. Reproduced from Feringa and coworkers [423] with permission from Wiley-VCH. (**c**) Schematic diagram showing the interconversion of the three-state photo-responsive phosphine ligand. Reprinted with permission from Macmillan Publishers Ltd: Nature Communications [424], copyright 2015

(c)

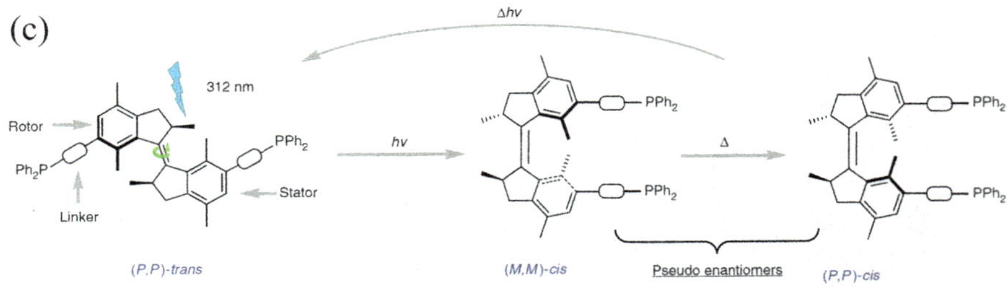

Fig. 34 (continued)

phosphine ligands to control their chirality for enantioselective catalysis [424]. The helicity of the diphosphine ligand and the spatial distance between the phosphorus centers have been carefully modulated (Fig. 34c), showing impressive results in catalysis from nearly racemic mixture for (*P*,*P*)-*trans*-state to 93/7 e.r. for (*M*,*M*)-*cis*-state and 6/94 e.r. for (*P*,*P*)-*cis*-state.

In 2007, a molecular lock was developed by Fujita and coworkers based on the photo-induced ligand dissociation process of the platinum(II)–pyridine moiety [425]. A catenated dimer was formed when the complex was irradiated in DMSO–H_2O (1:2 v/v) solvent mixture, whereas unlocking could be achieved by UV irradiation in MeCN–MeOH (1:1 v/v) solvent mixture, representing an interesting example of light-controlled supramolecular architecture (Fig. 35a) [425]. On the other hand, molecular capsules that are responsive toward light are particularly important for the development of systems for the controlled release of guest molecules. Rebek and coworkers reported the use of photochromic azobenzene as guest molecule [426] for complexing with a molecular capsule [427]. In the presence of *n*-tridecane in mesitylene solution, *trans*-4-methyl-4′-hexylazobenzene remained encapsulated due to stronger non-covalent interactions. However, upon photoisomerization, the large structural change to the *cis*-isomer resulted in its release from the capsule, allowing the labile *n*-tridecane to move in [426]. The investigation of a photo-induced guest exchange reaction in the encapsulation complex was further extended to wavelength-dependent and sequential light-triggered guest exchange by using an additional photo-active guest, HTI, which could be readily isomerized under irradiation of longer wavelengths at 410–430 nm [428]. As a result, in the presence of cavitand, glycoluril, HTI, azobenzene, 4,4′-dibromobenzil, and *p*-cymene, two types of assemblies were initially observed, encapsulating the two photo-active species with both *p*-cymene and 4,4′-dibromobenzil free in solution, as shown in Fig. 35b. Based on the previous experience in guest exchange processes [426], sequential guest exchange or parallel guest exchange reactions could be well controlled by using different excitation wavelengths. Photo-controlled guest uptake and release could also be accomplished with a photochromic cage. Clever and coworkers utilized pyridine-appended diarylethene and connected it via palladium coordination [429]. The reversible transformation between flexible open form and rigid closed form has resulted in

(a)

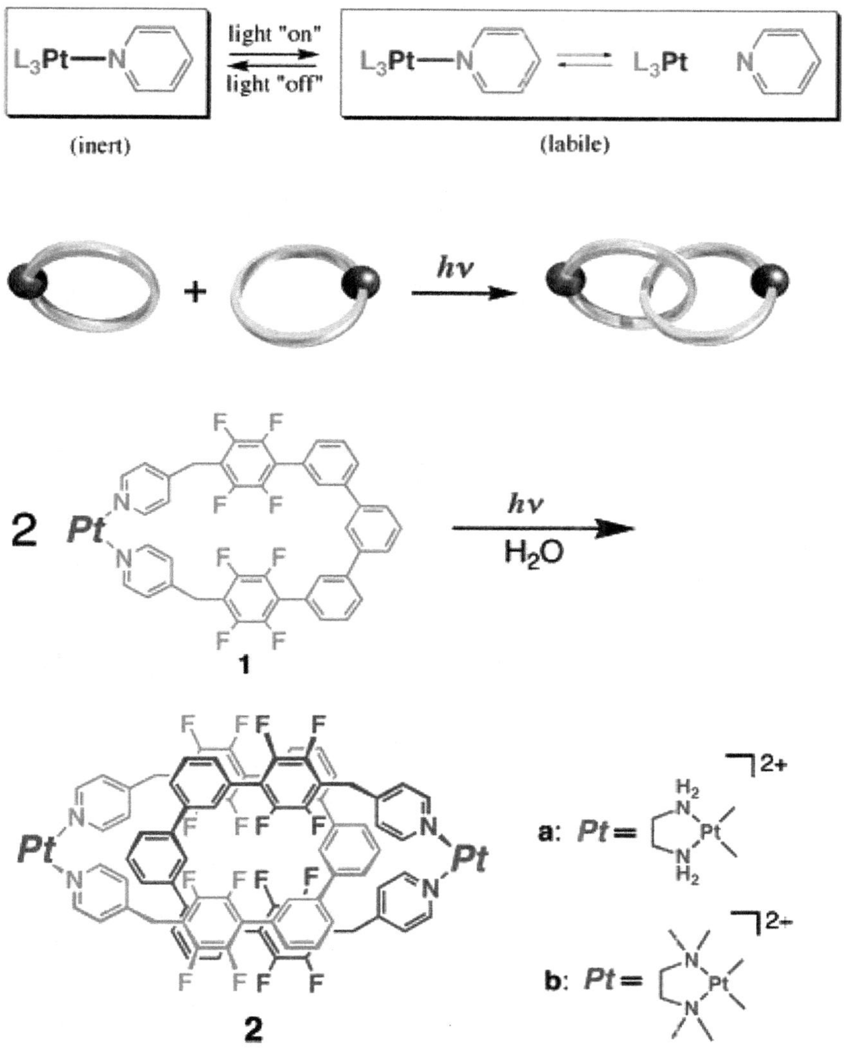

Fig. 35 (**a**) Schematic diagram showing the photo-induced molecular lock. Reprinted with permission from Fujita and coworkers [425]. Copyright 2007 American Chemical Society. (**b**) Sequential guest exchange reaction by irradiation of different wavelengths. Reproduced from Rebek and coworkers [428] with permission from Wiley-VCH. (**c**) Schematic diagram showing the photochromic coordination cage and its binding of guest. Reproduced from Clever and coworkers [429] with permission from Wiley-VCH

variation in affinity toward $B_{12}F_{12}^{2-}$ anion (Fig. 35c). These examples on the reversible molecular encapsulation should provide insights toward further applications and developments of supramolecular catalysis and drug delivery.

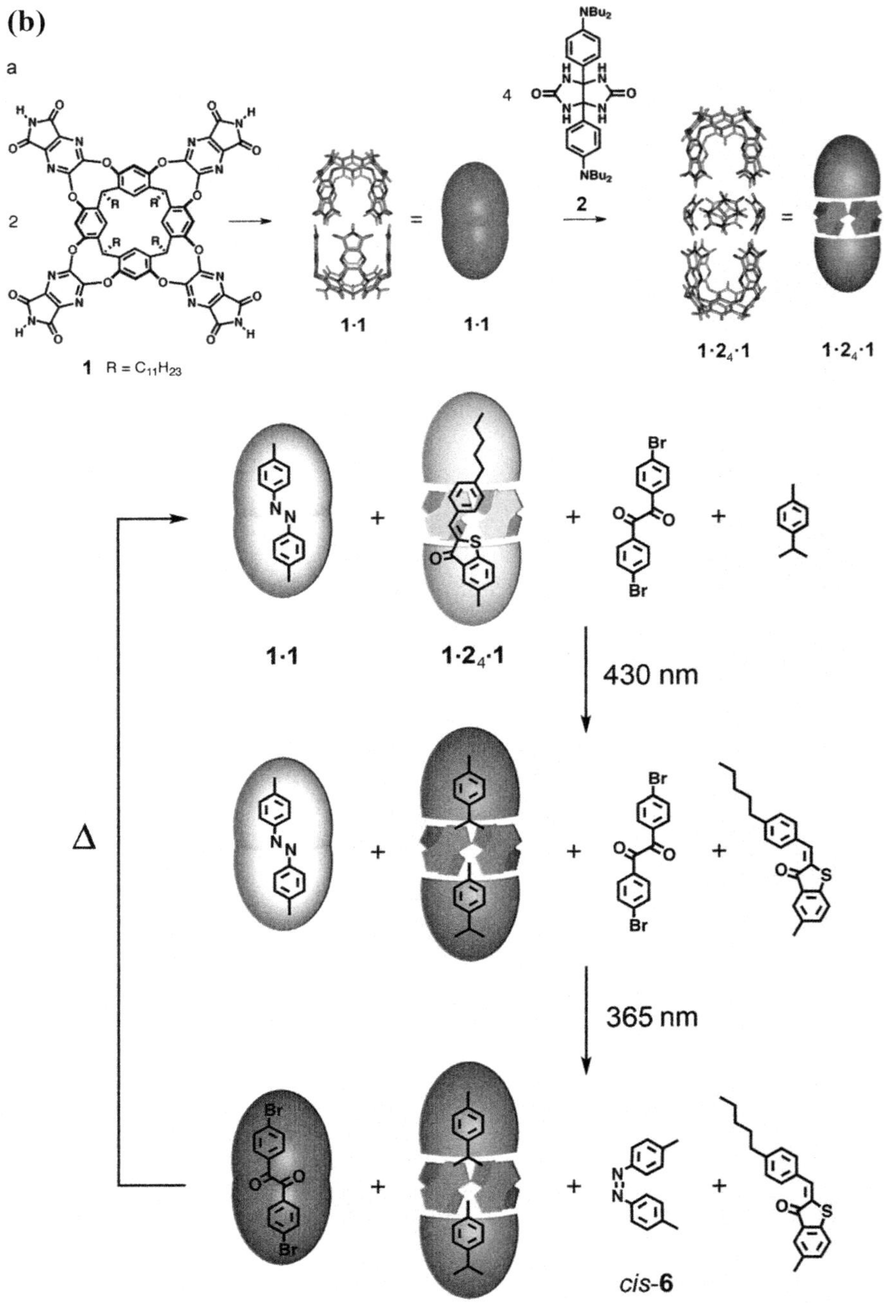

Fig. 35 (continued)

(c)

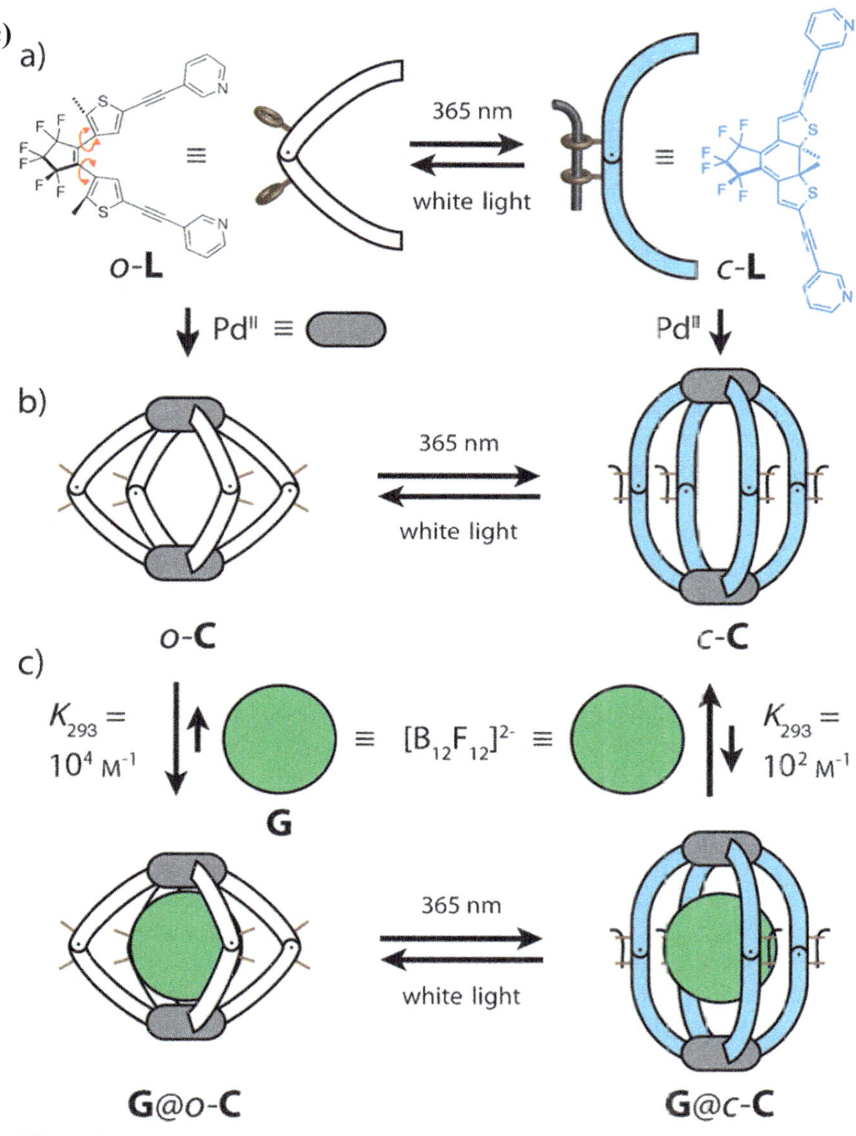

Fig. 35 (continued)

5 Advancement of Photofunctional Transition Metal Complexes in Biological Applications

Construction of in vitro biological probes based on the functionalization of transition metal complexes with substrate pendants specifically recognized by the target host molecules or with hydrophobic moieties enabling nonspecific hydrophobic interactions between the complexes and target biomolecules have been well exploited [430, 431]. Changes in local environment upon binding with targets would give rise to the alteration of the photophysical properties of transition metal complexes, which has constituted the fundamentals of the detection strategy [430, 431]. A new research dimension based on the utilization of metal–metal interactions of square-planar d^8 platinum(II) complexes upon their formation of supramolecular assembly for the detection of biomolecules has also been established. Advantaged by the incorporation of water-soluble functionalities as well as the "switching on" of ^{3}MMLCT emission in the NIR region upon self-assembly, the exploration of in vitro biological sensing capability of alkynylplatinum(II) complexes in aqueous environment has emerged. Polyanionic single-stranded DNA with a certain critical length has been demonstrated by Yam and coworkers to induce self-assembly of an achiral cationic alkynylplatinum (II) terpyridine complex via electrostatic interactions with the involvement of metal–metal and π–π stacking interactions (Fig. 36a) [432]. The resultant NIR ^{3}MMLCT emission and induced circular dichroism signal enhancement have been found to exhibit strong correlation with the primary and secondary structures of these oligonucleotides, serving as an effective way for the differentiation of these oligonucleotides. The same complex has also been utilized to probe the formation of G-quadruplex from several short G-rich oligonucleotides in the presence of K^+ ion, along with the turning on of the NIR ^{3}MMLCT emission due to a substantial increase in the local density of negative charges of the G-quadruplex conformation [433]. Owing to the close spectral overlap between the emission energy of PPE-SO_3^- and the absorption energy of aggregated dicationic alkynylplatinum (II) terpyridine complex as shown in Fig. 36b, the formation of polymer–metal complexes aggregates has led to an efficient FRET from the PPE-SO_3^- donor to the aggregated platinum(II) complex [434]. Addition of human telomere resulted in the modulation of the extent of the aggregation behavior of platinum(II) complex and

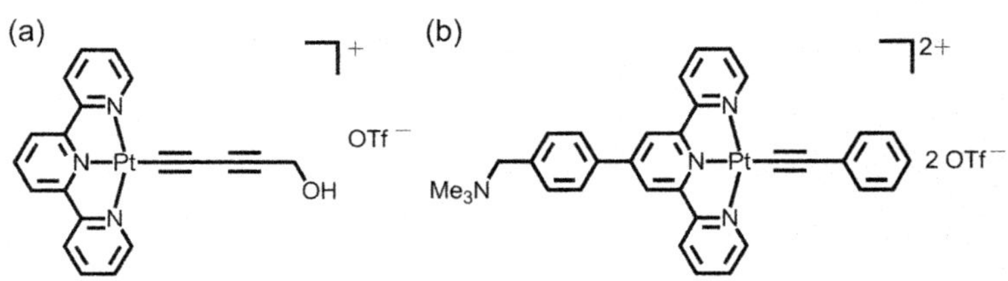

Fig. 36 Chemical structures of alkynylplatinum(II) terpyridine complexes, which are utilized for in vitro (**a**) differentiation of oligonucleotides of various sequences and probing the formation of G-quadruplex structure [432, 433] and (**b**) sensing of human telomere [434]

hence the FRET efficiency, giving rise to a ratiometric detection of human telomere through the monitoring of emission spectral changes over the visible-NIR region. Such label-free detection strategy has also been demonstrated to distinguish G-quadruplex structure formed by human telomere from other G-quadruplex-forming sequences with high selectivity.

Increasing research interest has also been directed to the exploitation of the potential of transition metal complexes to function as sensing probes for cellular detection. Dinuclear ruthenium(II) polypyridine complex prepared by the group of Battaglia and Thomas has been found to differentiate DNA duplex and quadruplex structures in living cells based on the turning on of two separate emissions (Fig. 37a) [435]. Incorporation of the phenanthridine moiety to the ruthenium (II) complex has been demonstrated by Turro and coworkers and shown to give rise to the duplex RNA probing capability along with significant emission enhancement (Fig. 37b) [436]. Williams and coworkers reported a 2-pyridyl-benzimidazole-containing iridium(III) complex capable of revealing the changes in intracellular pH with the exhibition of different emission color (Fig. 37c) [437]. The protonation of the benzimidazole ring at decreasing pH which resulted in the changes in LF strength and hence the excited-state properties of the complex could account for its pH-sensing capability. Ye, Yuan and coworkers have reported a weakly emissive ruthenium(II) complex which is applicable for real-time cellular nitric oxide detection (Fig. 37d) [438]. Such weakly emissive nature has been due to the presence of the electron-rich *o*-diaminophenyl moiety which quenched the luminescence of the metal complex via PET. Incubation of the metal complex-deposited plant or animal cells with a nitric oxide donor has been shown to result in intense red intracellular luminescence, attributed to the conversion of the *o*-diaminophenyl moiety to benzotriazole functionality. A non-emissive cyclometalated iridium(III) complex prepared by Li and coworkers has been shown to function as a reaction-based phosphorescence switch-on probe for the detection of histidine/histidine-rich proteins in the nuclei of living cells (Fig. 37e) [439]. This was ascribed to the reaction of the metal complex with the histidine/histidine-containing proteins in nuclei which led to the formation of the highly emissive adduct. An iridium(III) complex reported by Tobita and coworkers was functionalized with a tetraproline-linked coumarin 343 and exhibited dual emission of both fluorescence from coumarin and red phosphorescence from the iridium(III) phosphor moiety (Fig. 37f) [440]. Increasing cellular oxygen level in HeLa cell has resulted in the quenching of the red phosphoresce of metal complex, while its fluorescence intensity remained nearly unchanged. Such a ratiometric monitoring of the fluorescence and phosphorescence intensity changes has enabled this complex to function as a potential cellular probe for hypoxia to provide quantitative mapping of local oxygen levels in living cells. A zinc(II) salen complex with thioether functionalities reported by Zhang and coworkers has been found to respond to H_2O_2 selectively in acidic environment with the presence of myeloperoxidase with significant emission intensity enhancement than that of other biologically relevant reactive oxygen species (Fig. 37g) [441]. This has been ascribed to the specific enzymatic conversion of H_2O_2 and Cl^- to ClO^-. This complex has also been shown to predominantly localize in lysosomes and to be

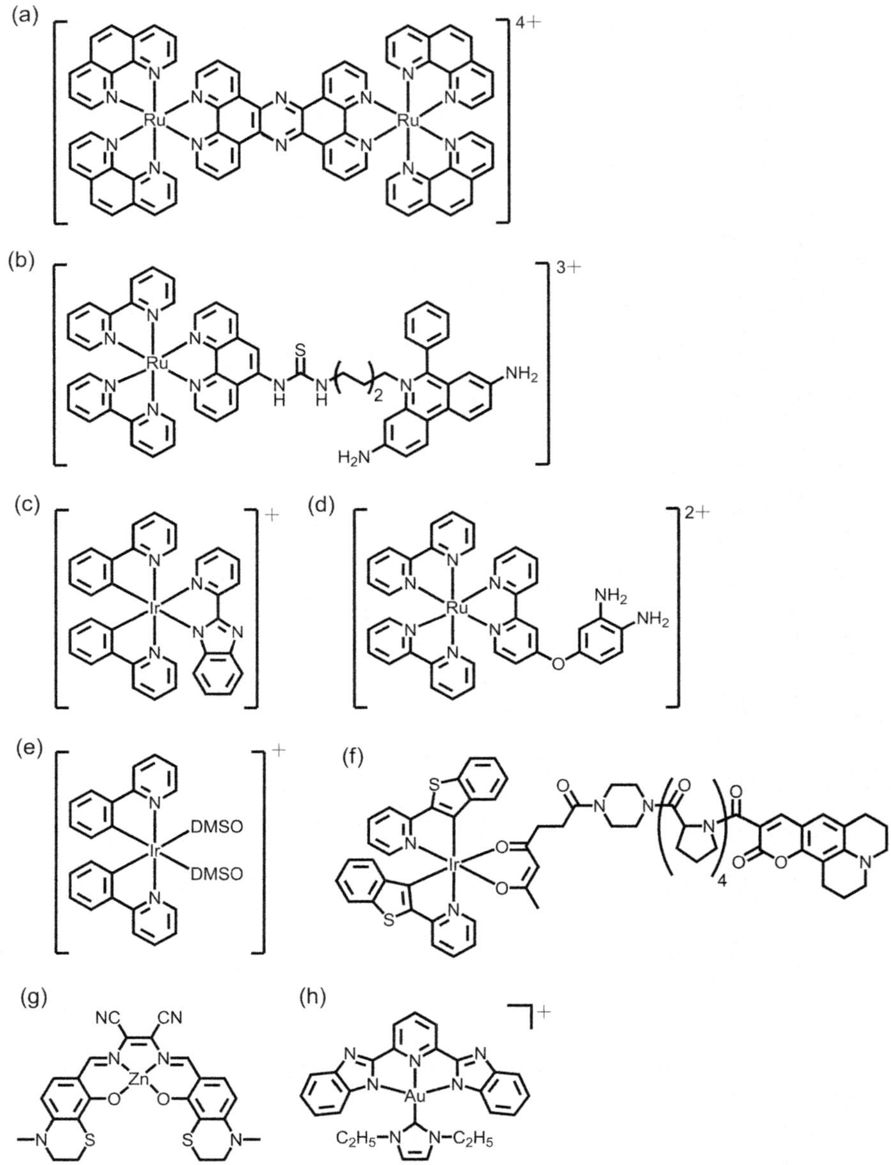

Fig. 37 Chemical structures of ruthenium(II), iridium(III), zinc(II) and gold(III) complexes, which are utilized for cellular sensing of (**a**) duplex DNA [435], (**b**) duplex RNA [436], (**c**) pH [437], (**d**) nitric oxide [438], (**e**) histidine/histidine-rich proteins [439], (**f**) oxygen level [440], (**g**) H_2O_2 [441], and (**h**) glutathione [442]

capable of detecting both exogenous and endogenous lysosomal H_2O_2 level in living cells. A non-emissive gold(III)-NHC complex reported by Che and coworkers has been found to signal the cellular glutathione content by the release

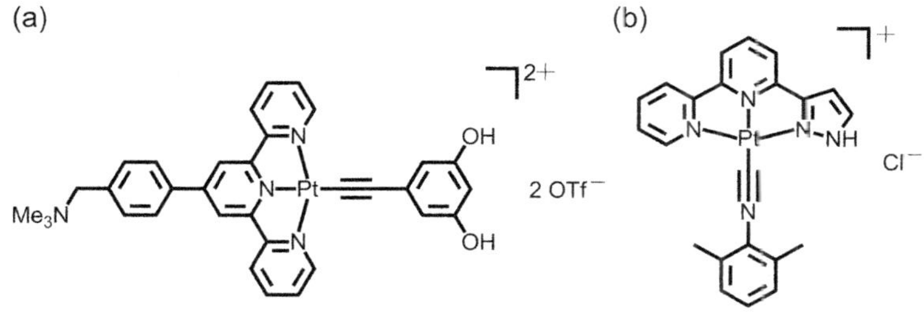

Fig. 38 Chemical structures of platinum(II) complexes reported by the groups of (**a**) Yam [443] and (**b**) Che [444], which are utilized for cellular pH sensing based on their self-assembly properties

of fluorescent bzimpy upon the reduction of gold(III)-NHC to gold(I)-NHC by glutathione, reminiscent of the activation process of platinum(IV) prodrugs by intracellular reduction to give platinum(II) complexes (Fig. 37h) [442].

The self-assembly properties of transition metal complexes have also been employed in the cellular probing strategy. Alkynylplatinum(II) complexes reported by Yam and coworkers have been demonstrated to function as a cellular pH sensor with significant NIR emission intensity enhancement at acidic pH due to the aggregation of the metal complex molecules via metal–metal and π–π stacking interactions (Fig. 38a) [443]. Significant diminution of emission intensity was observed at high pH. This has been attributed to the combination effect of enhanced hydrophilicity of the complex molecules upon deprotonation of the phenolic functionalities which gives rise to the deaggregation of the complex molecules, as well as PET quenching upon an increase in the electron richness of the alkynyl ligands. Similarly, cyclometalated platinum(II) complex reported by Che and coworkers has been shown to exhibit pH-dependent emission color in various cellular compartments (Fig. 38b) [444]. Orange ^{3}MMLCT emission would be turned on in acidic compartments attributed to the formation of the intracellular aggregates of the complex molecules associated with metal–metal and π–π stacking interactions, while green emission would be observed upon its disassembly in increasing pH environment. In addition, this complex has been shown to be capable of forming hydrogel via intermolecular hydrogen bonding interaction between the pyrazole unit, capable of functioning as a delivery vehicle for therapeutic drugs to cell.

Understanding the structural effect of transition metal complexes on cellular uptake properties will be helpful in the design of cellular probes and imaging agents. Systematic studies on the effect of charge, lipophilicity, and size of transition metal complexes on cellular uptake as well as subcellular internalizing properties have been reported [431, 445–451]. A series of ruthenium (II) dipyridophenazine complexes with various ancillary ligands of different hydrophobicity has been synthesized by Barton and coworkers, with their cellular uptake efficiency in HeLa cell investigated (Fig. 39a) [445]. HeLa cells incubated with the DIP-containing ruthenium(II) complex has been found to show more intense luminescence with respect to the case of bpy and phen as ancillary ligands, suggesting the facilitation brought about by the lipophilicity of the DIP ligand for

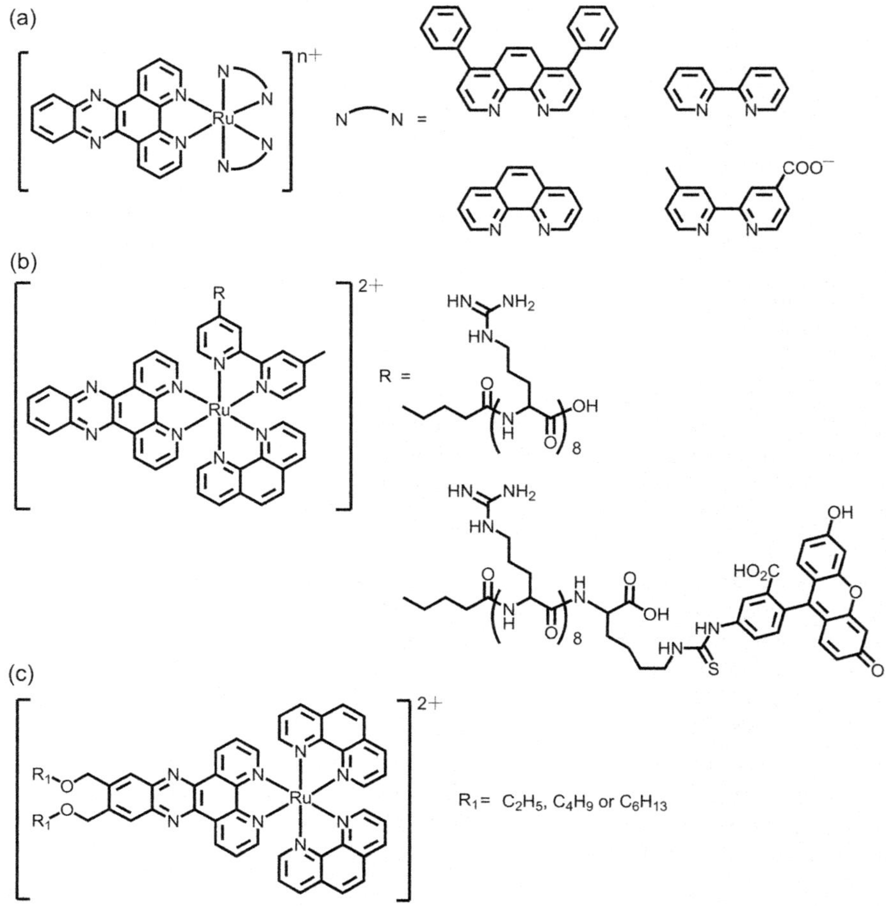

Fig. 39 Chemical structures of ruthenium(II) dipyridophenazine complexes with (**a**) ancillary ligands of different hydrophobicity [445], (**b**) cell-penetrating peptide [446], and (**c**) alkyl ether chains of various lengths [447], which are utilized as imaging agents

cellular uptake despite the larger size of the complex molecule. In addition, the importance of positive charge of the metal complexes on improving cellular uptake has been confirmed with only weak luminescence obtained for the HeLa cell incubated with the carboxylate-containing neutral ruthenium(II) complex. Conjugation of the ruthenium(II) complexes with cell-penetrating peptide has also been shown by the same group to assist cellular uptake as the octaarginine could provide extra-positive charges for cellular entry across the membrane potential (Fig. 39b) [446]. Tagging the fluorescein moiety to cell-penetrating peptide could even direct the localization of the resultant ruthenium(II) complex to the nucleus of HeLa cells, suggesting that the enhanced lipophilicity given by the fluorescein tag greatly contributes to the nucleus-staining properties of this complex. Modulation of the cellular uptake efficiency by the introduction of alkyl ether chains of various lengths to the dipyridophenazine moiety of the ruthenium(II) complexes has been demonstrated by Svensson and coworkers (Fig. 39c) [447]. Increase in alkyl ether chain length has been shown to give rise to an enhancement of the hydrophobicity of the metal complexes as well as their preferential localization in the hydrophobic membrane structures in cytoplasm rather than staining the nuclei. Detailed investigations on the effect of chemical structures of the classes of iridium(III), rhodium (III), and rhenium(I) complexes on cellular uptake have also been performed by the research groups of Lo, Li, Coogan, and Sheldrick [431, 448–451]. Other than ruthenium(II) complexes, rhenium(I), iridium(III), rhodium(III), platinum(II), and gold(I) complexes have also been demonstrated with specific staining capability for different subcellular compartments, such as cytoplasm, nucleus, nucleoli, mitochondria, endoplasmic reticulum, Golgi apparatus, lysosomes, vacuoles, and membranes [431, 448–450], enabling the visualization of various cellular compartments under microscopy.

Stimulated by the successful clinical trials involving the utilization of transition metal complexes for the treatment of various diseases, considerable efforts have been continuously devoted to the search for other potential transition metal-based therapeutics. Long-lived excited states of phosphorescent transition metal complexes have enabled their potential capability to be employed as photosensitizers for photodynamic therapy as the excited states of which would react with molecular oxygen to generate different reactive oxygen species for cell killing [452]. Integration of the target-internalizing properties and photo-induced cytotoxicity in the design of transition metal-based therapeutics would increase the effectiveness of the photodynamic treatment. Cyclometalated iridium(III) complex with poly(ethylene glycol) pendant has been reported by Lo and coworkers and demonstrated to internalize in mitochondria with negligible cytotoxicity in the dark (Fig. 40a) [453]. However, enhanced photo-induced cytotoxic activity has been observed upon photo-excitation due to the generation of singlet oxygen. Protonation of the three amino functionalities of an iridium(III) complex reported by Aiko and coworkers has been shown to give rise to the intense green emission for the imaging of acidic organelles, such as lysosomes (Fig. 40b) [454]. Interestingly, photoexcitation of this complex has led to efficient singlet oxygen generation only under acidic condition. Such pH-dependent photosensitizer has resulted in the photo-induced cell death of HeLa-S3 cells via necrosis owing to the massive breakdown of lysosomes by the reactive oxygen species. An iridium(III) complex reported by Meggers and coworkers has been shown to result in considerable enhancement in cytotoxicity upon irradiation with visible light and hence apoptotic cell death (Fig. 40c) [455]. Such photocytotoxicity and efficient cellular apoptosis have been correlated with the facile substitution of the selenocyanate by chloride ligand upon visible-light activation. Apart from that, the inhibitory effect of transition metal-based therapeutics for amyloid fibrillogenesis has also been explored. A solvato iridium(III) complex reported by Ma and coworkers has been found to probe the presence of $A\beta_{1-40}$ fibrils with emission switched on (Fig. 40d) [456]. This has been ascribed to the displacement of the labile co-ligands by the

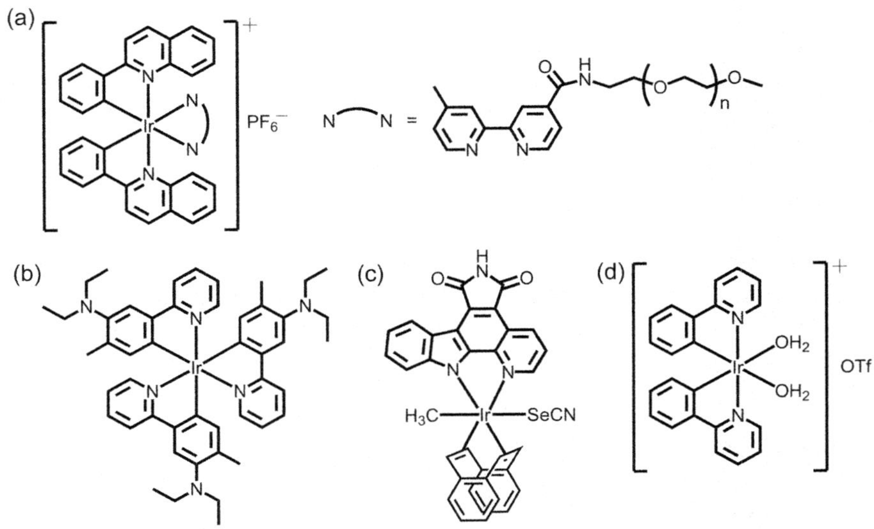

Fig. 40 Chemical structures of iridium(III)-based therapeutics reported by the groups of (**a**) Lo [453], (**b**) Aiko [454], (**c**) Meggers [455], and (**d**) Ma [456]

imidazole N-donor ligand of histidine residues of $A\beta_{1-40}$ fibrils, providing a hydrophobic environment of reduced exposure to solvent molecules which disfavors oxygen quenching. Importantly, this complex has been shown to inhibit $A\beta_{1-40}$ peptide aggregation with fibrils of significantly reduced lengths obtained, in which the interaction of phenylpyridine of the metal complex with the aromatic side chains around the *N*-terminal domain of $A\beta_{1-40}$ peptide has accounted for its inhibitory properties [456].

6 Advances in Frontiers of Supramolecular Chemistry

Continuous efforts have been made to search for new molecular-based photofunctional materials of improved properties through rational design of the molecular structures. Introduction of the concept of supramolecular chemistry to the design of photofunctional materials beyond molecular levels has opened up a new research dimension, attributed to the unlimited possibilities arising from the unique features of these assemblies of hierarchical complexities which are absent in the constituting discrete molecules. Transition metal complexes of d^8 and d^{10} electronic configurations with the respective square-planar and linear coordination geometry, which favor the axial overlap of the frontier orbitals, have been demonstrated to exhibit strong propensity to self-aggregate into supramolecular assemblies [13–16, 457–461]. The involvement of the non-covalent metal–metal and/or π–π interactions within the supramolecular architectures would give rise to the intriguing photophysical properties of these classes of complexes, rendering their potentials to function as supramolecular-based photofunctional materials. In addition to the aforementioned applications in WOLED fabrications and biological sensing, a recent review by Yam and coworkers has also described the diverse photofunctional applications of versatile d^8 and d^{10} transition metal complexes based on their supramolecular assembly features [462].

Owing to the combination of relativistic and correlation effects [457, 458], aurophilic interactions have been observed in numerous examples of low-dimensional gold(I) complexes as well as multinuclear supramolecular architectures of gold(I) clusters, with intriguing spectroscopic properties associated with the intra- and/or intermolecular gold(I)–gold(I) interactions [462–464]. The introduction of external stimuli, such as solvent vapor and mechanical stress, on the modulation of the extent of such aurophilicity of this class of complexes in the solid state, and hence their alteration of structural properties and spectroscopic changes, has been well illustrated [462–464]. In this section, selected examples of the recently reported low-dimensional gold(I) complexes and gold(I) clusters of high nuclearities have been highlighted as the illustrative models for the realization of the versatile potential applications stemming from the modulation of supramolecular interactions and, in this particular case, the aurophilicity or the aurophilic interactions.

In view of the sensitivity of aurophilic interactions in the supramolecular assemblies toward the environment, incorporation of ion receptor functionalities to the gold(I) systems would lead to the modulation of the assembly behaviors upon ion complexation and hence the changes in the luminescence properties associated with the alteration of the aurophilic interactions. Functionalization of isocyanobenzo[15]crown-5 or isocyanobenzo[18]crown-6 to the bis(alkynyl)-calix[4]arene gold(I) complexes has been demonstrated by Yam and coworkers to give rise to the K^+ and Cs^+ ion-sensing capability of the gold(I) complexes as shown in Fig. 41a [465]. The sandwiching of these ions in the benzocrown units has been shown to result in the formation of intramolecular gold(I)–gold(I) interaction with the switching on of low-energy emission. A similar strategy has also been demonstrated in the 2,6-dimethyphenyl isocyanide-containing analogue to signal the detection of Al^{3+} ion upon its coordination to the carbonyl oxygen on the amide functionalities, which would facilitate the two gold(I) moieties to come into close vicinity (Fig. 41a) [466]. The three intramolecular oligoether pendants of a tripodal alkynylgold(I) complex reported by the same group enabled the encapsulation of Mg^{2+} ion (Fig. 41b) [467]. This would lead to the close proximity of the three rigid and linear alkynylgold(I) units for the formation of gold(I)–gold(I) interaction and hence the switching on of a red emission. A tripodal alkynylgold(I) complex containing three alkynyl units has been reported by Chao and coworkers to serve as the coordination site for Cu^+ ion, which on binding caused the switching on of a lower-energy emission with the involvement of intramolecular gold(I)–gold (I) interaction (Fig. 41c) [468]. A trinuclear gold(I) pyrazolate metallacycle has been utilized by the group of Lintang and Aida to prepare a spin-coated film of organometallic-silica nanocomposite with hexagonal nanochannels by sol–gel

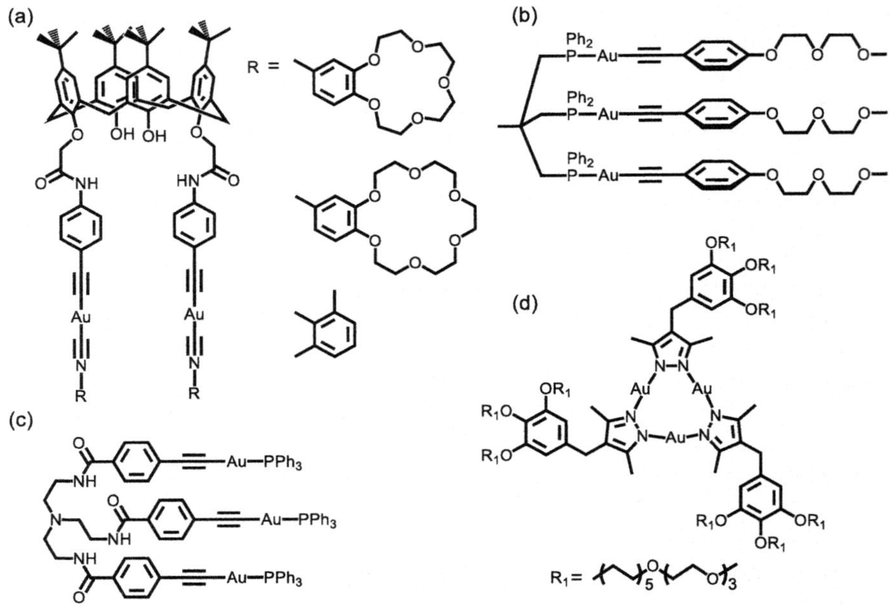

Fig. 41 Chemical structures of gold(I) complexes, which are utilized as sensing probes for (**a**) K^+, Cs^+, and Al^{3+} [465, 466], (**b**) Mg^{2+} [467], (**c**) Cu^+ [468], and (**d**) Ag^+ [469, 470] ions

synthesis (Fig. 41d) [469, 470], in which one-dimensional molecular assemblies of the trinuclear gold(I) metallacycle stabilized by gold(I)–gold(I) interactions have been found within the robust hexagonal frameworks of the mesoporous silica, giving rise to the red phosphorescence of the thin film. Exposure of this film to a Ag^+ ion solution resulted in the diminution of red phosphorescence but the emergence of green emission, suggestive of the conversion of gold(I)–gold(I) interactions to gold(I)–silver(I) metallophilic interactions upon the permeation of Ag^+ ion into the nanochannels [470]. Such finding has realized the potential of this nanocomposite to function as ion-sensing nanomaterials. An octadecanuclear gold(I) μ_3-sulfido cluster reported by Yam and coworkers has been shown to adopt thiacrown-like structure and function as a macrocyclic host for selective Ag^+ ion binding with an emergence of red emission owing to the formation of gold(I)–silver(I) adduct (Fig. 42) [471]. Addition of acetonitrile to this adduct has been shown to reverse the binding event and spectroscopic changes without the disruption of the metallosupramolecular architecture. A series of gold(I)-containing metallosupramolecular cages prepared by Yu, Yam and coworkers has been found to adopt quadruple-stranded conformation with supramolecular chirality upon the self-assembly of two achiral deep-cavitand calix[4]arene-based dithiocarbamate and eight gold(I) metal centers (Fig. 43) [472]. Variation in the peripheral substituents of dithiocarbamate ligands led to the formation of two-dimensional molecular arrays of cages of different packing modes due to the involvement of various extents of aurophilic interactions. Interestingly, these metallosupramolecular cages have been shown to exhibit selective sensing behavior toward Ag^+ ion with

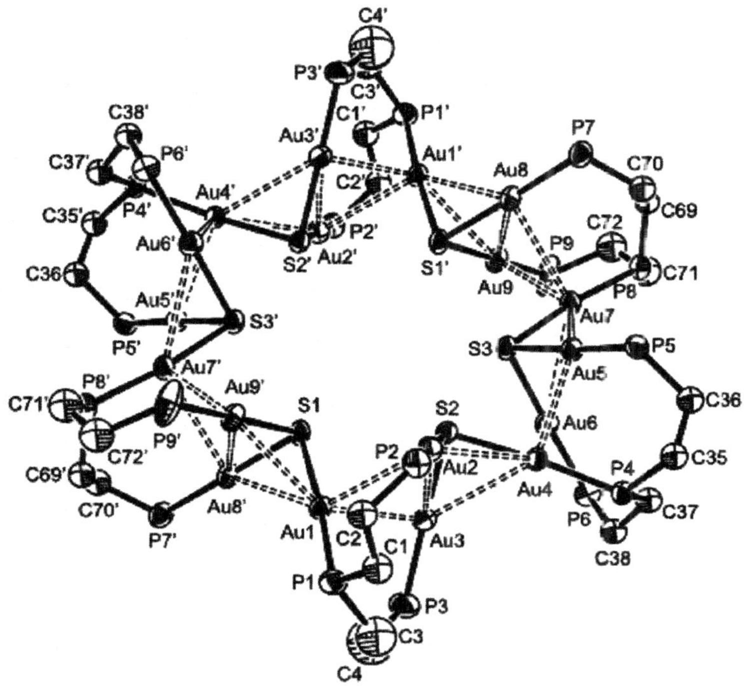

Fig. 42 Crystal structure of octadecanuclear gold(I) μ_3-sulfido cluster. Reprinted with permission from Yam and coworkers [471]. Copyright 2010 American Chemical Society

the switching on of red emission [472]. This is suggestive of the extension of gold (I)–gold(I) interactions via the formation of gold(I)–silver(I) metallophilic interaction upon Ag^+ ion inclusion. Addition of tetraethylammonium iodide would result in the disappearance of such red emission owing to the removal of Ag^+ ion upon precipitation of AgI.

In light of the fact that changes in the microenvironment would result in the alteration of the aurophilic interaction of the gold(I) systems as well as their supramolecular assembly features, apart from the investigation on the uses as ion-sensing probes, the exploration on the utilization of gold(I) complexes or clusters as luminescence-based sensors for the detection of microenvironmental changes has also been made. A dinuclear gold(I) complex with bridging diphosphine has been shown by Lee and coworkers to be strongly emissive in the solid state, but not in the solution state, owing to the substantial involvement of intermolecular gold(I)–gold(I) interactions between the adjacent molecules in the solid state as confirmed by X-ray crystallography (Fig. 44) [473]. Exposure of this complex to trifluoroacetic acid vapor has been found to result in remarkable diminution in its luminescence intensity, while the emission can be resumed upon the exposure to triethylamine vapor. Interestingly, vapor of hydrochloric acid would not lead to the observation as the case of trifluoroacetic acid. These findings are indicative of the importance of the bulkiness of trifluoroacetic acid on the

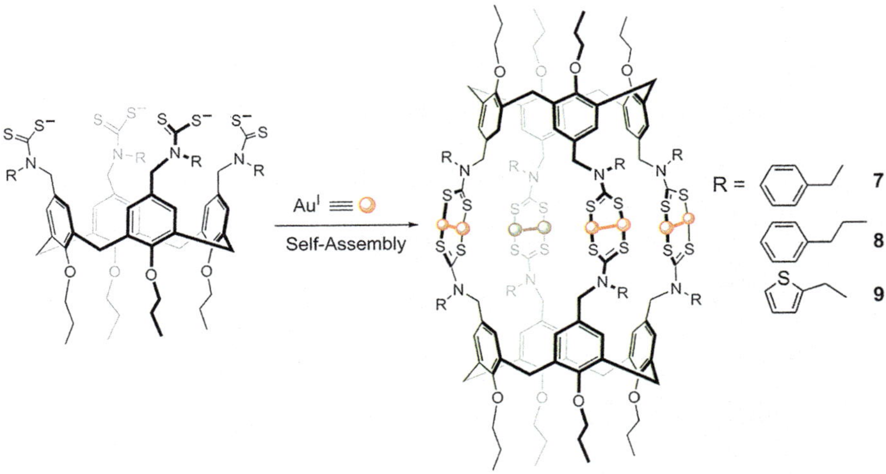

Fig. 43 Formation of gold(I)-containing metallosupramolecular cages via self-assembly. Reprinted with permission from Yu, Yam and coworkers [472]. Copyright 2014 American Chemical Society

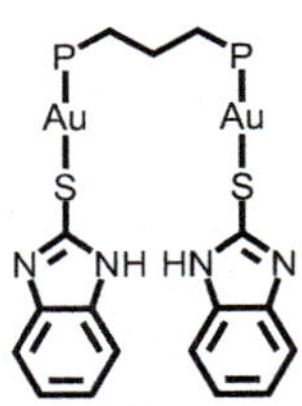

Fig. 44 Chemical structure of dinuclear gold(I) complex, which is utilized as an acid vapor-sensing probe [473]

dissociation of the intermolecular aurophilic interactions, which accounts for the trifluoroacetic acid vapor-probing capability of this dinuclear gold(I) complex. An octanuclear alkynylgold(I) diphosphine cluster prepared by the group of Koshevoy and Chou has been demonstrated to exhibit thermochromic luminescence response in the solid state, with significant red shifts and intensity enhancement upon decreasing temperature (Fig. 45) [474]. Such thermochromic behavior was lost upon grinding and was found to be absent in the solution state because of its dissociation into different species as confirmed by NMR and mass spectroscopic studies. These observations are suggestive of the temperature-dependent luminescence behavior arising from the crystal lattice rearrangement, giving rise to the potential capability of this cluster to function as thermochromic luminescence sensor.

Upon the introduction of photo-reactive or photo-responsive functionalities to the gold(I) systems, the investigation on the effect of aurophilicity on the photochemical reactivity and the modulation of the supramolecular assembly properties

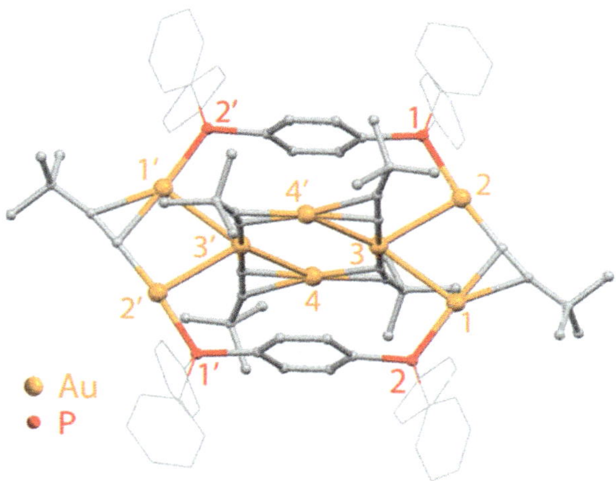

Fig. 45 Crystal structure of octanuclear gold(I) cluster, which is utilized as a thermochromic luminescence sensor. Reproduced from Koshevoy, Chou and coworkers [474] with permission from the Royal Society of Chemistry

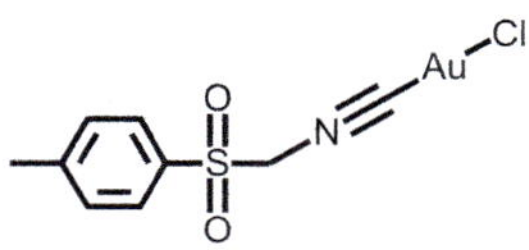

Fig. 46 Chemical structure of gold(I) isocyanide complex, which undergoes oligomerization in dichloromethane solution [475]

of this class of complexes upon photo-irradiation have been made possible. A dichloromethane solution of a gold(I) isocyanide complex has been shown by the group of Omary to exhibit oligomerization via intermolecular gold(I)–gold (I) interactions, as supported by the deviations of the lower-energy absorption from Beer's law (Fig. 46) [475]. Irradiation of this solution has been found to result in photodissociation of the isocyanide ligand, followed by disproportionation to form gold(III) and gold(0) photoproducts based on the emergence of the new lower-energy absorption bands and HPLC measurements. Continuous irradiation has led to the formation of purple suspension of colloidal gold nanoparticles from the metallic gold, as confirmed by the observation of the plasmon resonance absorption. Increasing the concentration of the metal complex solution has been demonstrated to give rise to the enhanced photoreactivity, owing to the red shifts of the absorption band assignable to the involvement of aurophilic interaction within the intermolecular aggregates which was responsible for the absorption of additional photons for the photoreaction. This work has also successfully demonstrated a method for the preparation of gold nanoparticles without the use of reducing agents. The aforementioned example on the photoisomerization of *cis*–dppee moiety in decagold(I) cluster prepared by Yam and coworkers has been shown to result in the

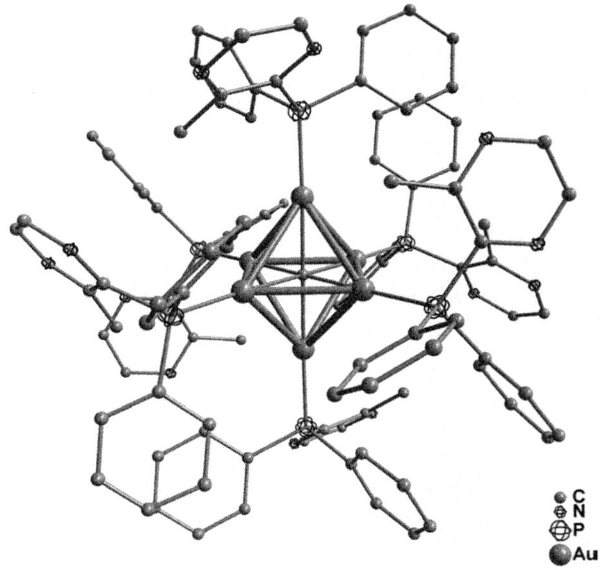

Fig. 47 Crystal structure of hexanuclear gold(I) cluster, which is utilized as a cluster linker for the construction of luminescent porous MOF. Reproduced from Wang and coworkers [476] with permission from Wiley-VCH

reconfiguration of the cluster architecture via transforming to an octadecagold (I) cluster with *trans* form of the organic diphosphine ligand (Fig. 27b) [358]. Significant red shifts and intensity enhancement have been observed in the electronic absorption and emission spectra of the decagold(I) cluster containing the *cis* form of diphosphine ligand in methanol upon photo-excitation. Such observations are attributed to the reduced HOMO–LUMO energy gap upon an increase of nuclearity in the transformation. This finding is indicative of the potential of these clusters of different symmetries to function as promising photochromic materials.

Owing to the high number of coordination sites as well as the luminescence properties associated with the aurophilic interactions which are highly responsive to the microenvironmental changes, the exploration on the use of gold(I) clusters as alternative linkers for the construction of multifunctional superstructures has emerged, of which MOFs of larger porosity with respect to those constructed from organic linkers, integrated with the intrinsic luminescence properties and high sensitivity toward changes in microenvironment that originated from the constituting metal clusters, have been anticipated. Wang and coworkers have utilized a luminescent 2-(3-methylpyrazinyl)-diphenylphosphine-containing hexanuclear gold(I) cluster as a cluster linker for the construction of luminescent porous MOF upon the addition of Ag^{+} ion (Fig. 47) [476]. The resultant cluster-based MOF, featuring with NbO topology and one-dimensional hexagonal channels, has been shown to exhibit intense luminescence in the solid state because of the enhanced rigidity upon the formation of the three-dimensional architectures through the coordination with Ag^{+} ion. Such cluster-based MOF has also been demonstrated to exhibit solvatochromism, of which its yellowish-green emission has been turned to orange upon its immersion in diethyl ether, while the emission color could be resumed upon subsequent immersion in methanol. The solvatochromic response has been ascribed to the interactions between the solvent molecules and Ag^+ ion inside the hexagonal channels, illustrating the potential use of this MOF for solvent molecule exchanges along with signals of various luminescence colors.

7 Summary and Perspectives

Unlimited combination of ligand functionalities and diverse bonding modes upon transition metal complexation has resulted in transition metal–ligand chromophoric and luminophoric complexes which constitute an important class of building blocks. Their intrinsic electronic absorption, luminescence, and excited-state redox properties not only allow for their fundamental photophysical and photochemical studies but also expand the scope to the demonstration of their potential applications in the areas of energy, catalysis, materials, biology, and diagnostics which are indispensible to our daily lives. In addition to the molecular properties, integration of the supramolecular assembly properties of the class of transition metal complexes via metal–metal interactions as well as the hierarchical complexity to the design of supramolecular-based functional materials has opened up a research dimension of unlimited possibilities because of the uniqueness arising from the supramolecular architectures. Importantly, the ease of stepwise incorporation of custom-made ancillary ligands to the transition metal centers has enabled the systematic structural modification of this class of complexes, leading to the fine-tuning of their photophysical, photoredox, as well as supramolecular assembly properties, demonstrating the importance of their structure–property relationships for the construction of photofunctional materials. The valuable information obtained from the structure–property relationships of this class of complexes is envisaged to provide an effective strategy for the improvement and optimization of their functional properties. Continuous efforts are also anticipated in the exploitation and development of new molecular- and supramolecular-based photofunctional materials in the future.

Acknowledgments We acknowledge support from The University of Hong Kong under the URC Strategic Research Theme on New Materials. Financial supports from the University Grants Committee Areas of Excellence Scheme (AoE/P-03/08) and General Research Fund (GRF) (HKU 7060/12P, HKU 7051/13P, HKU 17305614, and HKU 17302414) from the Research Grants Council of Hong Kong Special Administrative Region, China, and The University of Hong Kong are gratefully acknowledged.

References

1. Yersin H (2004) Triplet emitters for OLED applications. Mechanisms of exciton trapping and control of emission properties. Top Curr Chem 241:1–26

2. Fan C, Yang C (2014) Yellow/orange emissive heavy-metal complexes as phosphors in monochromatic and white organic light-emitting devices. Chem Soc Rev 43:6439–6469
3. Hofbeck T, Yersin H (2010) The triplet state of *fac*-Ir(ppy)$_3$. Inorg Chem 49:9290–9299
4. Djurovich PI, Thompson ME (2008) Cyclometallated organoiridium complexes as emitters in electrophosphorescent devices. In: Yersin H (ed) Highly efficient OLEDs with phosphorescent materials. Wiley, Weinheim
5. Chi Y, Chou PT (2010) Transition-metal phosphors with cyclometalating ligands: fundamentals and applications. Chem Soc Rev 39:638–655
6. Zhou G, Wong WY, Yang X (2011) New design tactics in OLEDs using functionalized 2-phenypyridine-type cyclometalates of iridium(III) and platinum(II). Chem Asian J 6:1706–1727
7. Ho CL, Wong WY (2013) Small-molecular blue phosphorescent dyes for organic light-emitting devices. New J Chem 37:1665–1683
8. Rausch AF, Homeier HHH, Yersin H (2010) Organometallic Pt(II) and Ir(III) triplet emitters for OLED applications and the role of spin-orbit coupling: a study based on high-resolution optical spectroscopy. Top Organomet Chem 29:193–235
9. Yang X, Wang Z, Madakuni S, Li J, Jabbour GE (2008) Efficient blue- and white-emitting electrophosphorescent devices based on platinum(II) [1,3-difluoro-4,6-di(2-pyridinyl)benzene] chloride. Adv Mater 20:2405–2409
10. Fleetham T, Li G, Wen L, Li J (2014) Efficient "pure" blue OLEDs employing tetradentate Pt complexes with a narrow spectral bandwidth. Adv Mater 26:7116–7121
11. Liao KY, Hsu CW, Chi Y, Hsu MK, Wu SW, Chang CH, Liu SH, Lee GH, Chou PT, Hu Y, Robertson N (2015) Pt(II) metal complexes tailored with a newly designed spiro-arranged tetradentate ligand; harnessing of charge-transfer phosphorescence and fabrication of sky blue and white OLEDs. Inorg Chem 54:4029–4038
12. D'Andrade BW, Brooks J, Adamovich V, Thompson ME, Forrest SR (2002) White light emission using triplet excimers in electrophosphorescent organic light-emitting devices. Adv Mater 14:1032–1036
13. Houlding VH, Miskowski VM (1991) The effect of linear chain structure on the electronic structure of Pt(II) diimine complexes. Coord Chem Rev 111:145–152
14. Bailey JA, Hill MG, Marsh RE, Miskowski VM, Schaefer WP, Gray HB (1995) Electronic spectroscopy of chloro(terpyridine)platinum(II). Inorg Chem 34:4591–4599
15. Jennette KW, Gill JT, Sadownick JA, Lippard SJ (1976) Metallointercalation reagents. Synthesis, characterization, and structural properties of thiolato(2,2′,2″-terpyridine)platinum(II) complexes. J Am Chem Soc 98:6159–6168
16. Büchner R, Field JS, Haines RJ, Cunningham CT, McMillin DR (1997) Luminescence properties of salts of the $[Pt(trpy)Cl]^+$ and $[Pt(trpy)(MeCN)]^{2+}$ chromophores: crystal structure of $[Pt(trpy)(MeCN)](SbF_6)_2$. Inorg Chem 36:3952–3956
17. Adamovich V, Brooks J, Tamayo A, Alexander AM, Djurovich PI, D'Andrade BW, Adachi C, Forrest SR, Thompson ME (2002) High efficiency single dopant white electrophosphorescent light emitting diodes. New J Chem 26:1171–1178
18. Zhou G, Wang Q, Ho CL, Wong WY, Ma D, Wang L (2009) Duplicating "sunlight" from simple WOLEDs for lighting applications. Chem Commun 3574–3576
19. Murphy L, Brulatti P, Fattori V, Cocchi M, Williams JAG (2012) Blue-shifting the monomer and excimer phosphorescence of tridentate cyclometallated platinum(II) complexes for optimal white-light OLEDs. Chem Commun 48:5817–5819
20. Kui SCF, Chow PK, Tong GSM, Lai SL, Cheng G, Kwok CC, Low KH, Ko MY, Che CM (2013) Robust phosphorescent platinum(II) complexes containing tetradentate O^N^C^N ligands: excimeric excited state and application in organic white-light-emitting diodes. Chem Eur J 19:69–73
21. Cheng G, Chow PK, Kui SCF, Kwok CC, Che CM (2013) High-efficiency polymer light-emitting devices with robust phosphorescent platinum(II) emitters containing tetradentate dianionic O^N^C^N ligands. Adv Mater 25:6765–6770

22. Fleetham T, Ecton J, Wang Z, Bakken N, Li J (2013) Single-doped white organic light-emitting device with an external quantum efficiency over 20%. Adv Mater 25:2573–2576
23. Wong KMC, Zhu X, Hung LL, Zhu N, Yam VWW, Kwok HS (2005) A novel class of phosphorescent gold(III) alkynyl-based organic light-emitting devices with tunable colour. Chem Commun 2906–2908
24. Au VKM, Wong KMC, Tsang DPK, Chan MY, Zhu N, Yam VWW (2010) High-efficiency green organic light-emitting devices utilizing phosphorescent bis-cyclometalated alkynylgold(III) complexes. J Am Chem Soc 132:14273–14278
25. Tang MC, Tsang DPK, Chan MMY, Wong KMC, Yam VWW (2013) Dendritic luminescent gold(III) complexes for highly efficient solution-processable organic light-emitting devices. Angew Chem Int Ed 52:446–449
26. Cheng G, Chan KT, To WP, Che CM (2014) Color tunable organic light-emitting devices with external quantum efficiency over 20% based on strongly luminescent gold(III) complexes having long-lived emissive excited states. Adv Mater 26:2540–2546
27. Chow PK, Ma C, To WP, Tong GSM, Lai SL, Kui SCF, Kwok WM, Che CM (2013) Strongly phosphorescent palladium(II) complexes of tetradentate ligands with mixed oxygen, carbon, and nitrogen donor atoms: photophysics, photochemistry, and applications. Angew Chem Int Ed 52:11775–11779
28. Zhu ZQ, Fleetham T, Turner E, Li J (2015) Harvesting all electrogenerated excitons through metal assisted delayed fluorescent materials. Adv Mater 27:2533–2537
29. Yersin H, Rausch AF, Czerwieniec R, Hofbeck T, Fischer T (2011) The triplet state of organo-transition metal compounds. Triplet harvesting and singlet harvesting for efficient OLEDs. Coord Chem Rev 255:2622–2652
30. Czerwieniec R, Yu J, Yersin H (2011) Blue-light emission of Cu(I) complexes and singlet harvesting. Inorg Chem 50:8293–8301
31. Deaton JC, Switalski SC, Kondakov DY, Young RH, Pawlik TD, Giesen DJ, Harkins SB, Miller AJM, Mickenberg SF, Peters JC (2010) E-Type delayed fluorescence of a phosphine-supported $Cu_2(\mu\text{-}NAr_2)_2$ diamond core: harvesting singlet and triplet excitons in OLEDs. J Am Chem Soc 132:9499–9508
32. Hashimoto M, Igawa S, Yashima M, Kawata I, Hoshino M, Osawa M (2011) Highly efficient green organic light-emitting diodes containing luminescent three-coordinate copper (I) complexes. J Am Chem Soc 133:10348–10351
33. Sakai Y, Sagara Y, Nomura H, Nakamura N, Suzuki Y, Miyazaki H, Adachi C (2015) Zinc complexes exhibiting highly efficient thermally activated delayed fluorescence and their application to organic light-emitting diodes. Chem Commun 51:3181–3184
34. Hofbeck T, Monkowius U, Yersin H (2015) Highly efficient luminescence of Cu (I) compounds: thermally activated delayed fluorescence combined with short-lived phosphorescence. J Am Chem Soc 137:399–404
35. Kalyanasundaram K, Grätzel M (1998) Applications of functionalized transition metal complexes in photonic and optoelectronic devices. Coord Chem Rev 77:347–414
36. Campbell WM, Burrell AK, Officer DL, Jolley KW (2004) Porphyrins as light harvesters in the dye-sensitised TiO_2 solar. Coord Chem Rev 248:1363–1379
37. Mishra A, Fischer MKR, Bäuerle P (2009) Metal-free organic dyes for dye-sensitized solar cells: from structure–property relationships to design rules. Angew Chem Int Ed 48:2474–2499
38. Grätzel M (2009) Recent advances in sensitized mesoscopic solar cells. Acc Chem Res 42:1788–1798
39. O'Regan BC, Durrant JR (2009) Kinetic and energetic paradigms for dye-sensitized solar cells: moving from the ideal to the real. Acc Chem Res 42:1799–1808
40. Imahori H, Umeyama T, Ito S (2009) Large π-aromatic molecules as potential sensitizers for highly efficient dye-sensitized solar cells. Acc Chem Res 42:1809–1818
41. Hagfeldt A, Boschloo G, Sun L, Kloo L, Pettersson H (2010) Dye-sensitized solar cells. Chem Rev 110:6595–6663

42. Li C, Liu M, Pschirer NG, Baumgarten M, Müllen K (2010) Polyphenylene-based materials for organic photovoltaics. Chem Rev 110:6817–6855
43. Clifford JN, Martínez-Ferrero E, Viterisi A, Palomares E (2011) Sensitizer molecular structure-device efficiency relationship in dye sensitized solar cells. Chem Soc Rev 40:1635–1646
44. Yin JF, Velayudham M, Bhattacharya D, Lin HC, Lu KL (2012) Structure optimization of ruthenium photosensitizers for efficient dye-sensitized solar cells – a goal toward a "bright" future. Coord Chem Rev 256:3008–3035
45. Bignozzi CA, Argazzi R, Boaretto R, Busatto E, Carli S, Ronconi F, Caramori S (2013) The role of transition metal complexes in dye sensitized solar devices. Coord Chem Rev 257:1472–1492
46. Kapilashrami M, Zhang Y, Liu YS, Hagfeldt A, Guo J (2014) Probing the optical property and electronic structure of TiO_2 nanomaterials for renewable energy applications. Chem Rev 114:9662–9707
47. Urbani M, Grätzel M, Nazeeruddin MK, Torres T (2014) Meso-substituted porphyrins for dye-sensitized solar cells. Chem Rev 114:12330–12396
48. Lee CP, Lin RYY, Lin LY, Li CT, Chu TC, Sun SS, Lin JT, Ho KC (2015) Recent progress in organic sensitizers for dye-sensitized solar cells. RSC Adv 5:23810–23825
49. O'Regan B, Grätzel M (1991) A low-cost, high-efficiency solar cell based on dye-sensitized colloidal TiO_2 films. Nature 353:737–740
50. Nazeeruddin MK, Kay A, Rodicio I, Humphry-Baker R, Müller E, Liska P, Vlachopoulos N, Grätzel M (1993) Conversion of light to electricity by *cis*-X_2Bis(2,2′-bipyridyl-4,4′-dicarboxylate)ruthenium(II) charge-transfer sensitizers ($X = Cl^-$, Br^-, I^-, CN^-, and SCN^-) on nanocrystalline TiO_2 electrodes. J Am Chem Soc 115:6382–6390
51. Kuang D, Klein C, Ito S, Moser JE, Humphry-Baker R, Evans N, Duriaux F, Grätzel C, Zakeeruddin SM, Grätzel M (2007) High-efficiency and stable mesoscopic dye-sensitized solar cells based on a high molar extinction coefficient ruthenium sensitizer and nonvolatile electrolyte. Adv Mater 19:1133–1137
52. Haque SA, Handa S, Peter K, Palomares E, Thelakkat M, Durrant JR (2005) Supermolecular control of charge transfer in dye-sensitized nanocrystalline TiO_2 films: towards a quantitative structure–function relationship. Angew Chem Int Ed 44:5740–5744
53. Jang SR, Lee C, Choi H, Ko JJ, Lee J, Vittal R, Kim KJ (2006) Oligophenylenevinylene-functionalized Ru(II)-bipyridine sensitizers for efficient dye-sensitized nanocrystalline TiO_2 solar cells. Chem Mater 18:5604–5608
54. Johansson PG, Kopecky A, Galoppini E, Meyer GJ (2013) Distance dependent electron transfer at TiO_2 interfaces sensitized with phenylene ethynylene bridged Ru^{II}–isothiocyanate compounds. J Am Chem Soc 135:8331–8341
55. Chen CY, Wu SJ, Li JY, Wu CG, Chen JG, Ho KC (2007) A new route to enhance the light-harvesting capability of ruthenium complexes for dye-sensitized solar cells. Adv Mater 19:3888–3891
56. Chen CY, Lu HC, Wu CG, Chen JG, Ho KC (2007) New ruthenium complexes containing oligoalkylthiophene-substituted 1,10-phenanthroline for nanocrystalline dye-sensitized solar cells. Adv Funct Mater 17:29–36
57. Chen CY, Chen JG, Wu SJ, Li JY, Wu CG, Ho KC (2008) Multifunctionalized ruthenium-based supersensitizers for highly efficient dye-sensitized solar cells. Angew Chem Int Ed 47:7342–7345
58. Gao F, Wang Y, Shi D, Zhang J, Wang M, Jing X, Humphy-Baker R, Wang P, Zakeeruddin SM, Grätzel M (2008) Enhance the optical absorptivity of nanocrystalline TiO_2 film with high molar extinction coefficient ruthenium sensitizers for high performance dye-sensitized solar cells. J Am Chem Soc 130:10720–10728
59. Kim JJ, Choi H, Kim C, Kang MS, Kang HS, Ko J (2009) Novel amphiphilic ruthenium sensitizer with hydrophobic thiophene or thieno(3,2-*b*)thiophene-substituted 2,2′-dipyridylamine ligands for effective nanocrystalline dye sensitized solar cells. Chem Mater 21:5719–5726

60. Yin JF, Chen JG, Lu ZZ, Ho KC, Lin HC, Lu KL (2010) Toward optimization of oligothiophene antennas: new ruthenium sensitizers with excellent performance for dye-sensitized solar cells. Chem Mater 22:4392–4399
61. Mishra A, Pootrakulchote N, Wang M, Moon SJ, Zakeeruddin SM, Grätzel M, Bäuerle P (2011) A thiophene-based anchoring ligand and its heteroleptic Ru(II)-complex for efficient thin-film dye-sensitized solar cells. Adv Funct Mater 21:963–970
62. García-Iglesias M, Pellejà L, Yun JH, González-Rodríguez D, Nazeeruddin MK, Grätzel M, Clifford JN, Palomares E, Vázquez P, Torres T (2012) Effect of bulky groups in ruthenium heteroleptic sensitizers on dye sensitized solar cell performance. Chem Sci 3:1177–1184
63. Chang WC, Chen HS, Li TY, Hsu NM, Tingare YS, Li CY, Liu YC, Su C, Li WR (2010) Highly efficient *N*-heterocyclic carbene/pyridine-based ruthenium sensitizers: complexes for dye-sensitized solar cells. Angew Chem Int Ed 49:8161–8164
64. Stengel I, Pootrakulchote N, Dykeman RR, Mishra A, Zakeeruddin SM, Dyson PJ, Grätzel M, Bäuerle P (2012) Click-functionalized Ru(II) complexes for dye-sensitized solar cells. Adv Energy Mater 2:1004–1012
65. Yen YS, Chen YC, Hsu YC, Chou HH, Lin JT, Yin DJ (2011) Heteroleptic ruthenium sensitizers that contain an ancillary bipyridine ligand tethered with hydrocarbon chains for efficient dye-sensitized solar cells. Chem Eur J 17:6781–6788
66. Nazeeruddin MK, Péchy P, Renouard T, Zakeeruddin SM, Humphry-Baker R, Comte P, Liska P, Cevey L, Costa E, Shklover V, Spiccia L, Deacon GB, Bignozzi CA, Grätzel M (2001) Engineering of efficient panchromatic sensitizers for nanocrystalline TiO_2-based solar cells. J Am Chem Soc 123:1613–1624
67. Numata Y, Singh SP, Islam A, Lwamura M, Imai A, Nozaki K, Han L (2013) Enhanced light-harvesting capability of a panchromatic Ru(II) sensitizer based on π-extended terpyridine with a 4-methylstyryl group for dye-sensitized solar cell. Adv Funct Mater 23:1817–1823
68. Chen BS, Chen K, Hong YH, Liu WH, Li TS, Lai CH, Chou PT, Chi Y, Lee GH (2009) Neutral, panchromatic Ru(II) terpyridine sensitizers bearing pyridine pyrazolate chelates with superior DSSC performance. Chem Commun 5844–5846
69. Chou CC, Wu KL, Chi Y, Hu WP, Yu SJ, Lee GH, Lin CL, Chou PT (2011) Ruthenium (II) sensitizers with heteroleptic tridentate chelates for dye-sensitized solar cells. Angew Chem Int Ed 50:2054–2058
70. Bomben PG, Robson KCD, Koivisto BD, Berlinguette CP (2012) Cyclometalated ruthenium chromophores for the dye-sensitized solar cell. Coord Chem Rev 256:1438–1450
71. Robson KCD, Bomben PG, Berlinguette CP (2012) Cycloruthenated sensitizers: improving the dye-sensitized solar cell with classical inorganic chemistry principles. Dalton Trans 41:7814–7829
72. Bessho T, Yoneda E, Yum JH, Guglielmi M, Tavernelli I, Imai H, Rothlisberger U, Nazeeruddin MK, Grätzel M (2009) New paradigm in molecular engineering of sensitizers for solar cell applications. J Am Chem Soc 131:5930–5934
73. Bomben PG, Gordon TJ, Schott E, Berlinguette CP (2011) A trisheteroleptic cyclometalated Ru^{II} sensitizer that enables high power output in a dye-sensitized solar cell. Angew Chem Int Ed 50:10682–10685
74. Pogozhev DV, Bezdek MJ, Schauer PA, Berlinguette CP (2013) Ruthenium(II) complexes bearing a naphthalimide fragment: a modular dye platform for the dye-sensitized solar cell. Inorg Chem 52:3001–3006
75. Ji Z, Natu G, Wu Y (2013) Cyclometalated ruthenium sensitizers bearing a triphenylamino group for p-type NiO dye-sensitized solar cells. ACS Appl Mater Interfaces 5:8641–8648
76. Wadman SH, Kroon JM, Bakker K, Lutz M, Spek AL, van Klink GPM, van Koten G (2007) Cyclometalated ruthenium complexes for sensitizing nanocrystalline TiO_2 solar cells. Chem Commun 1907–1909
77. Robson KCD, Koivisto BD, Yella A, Sporinova B, Nazeeruddin MK, Baumgartner T, Grätzel M, Berlinguette CL (2011) Design and development of functionalized

cyclometalated ruthenium chromophores for light-harvesting applications. Inorg Chem 50:5494–5508
78. Nazeeruddin MK, Humphry-Baker R, Officer DL, Campbell WM, Burrell AK, Grätzel M (2004) Application of metalloporphyrins in nanocrystalline dye-sensitized solar cells for conversion of sunlight into electricity. Langmuir 20:6514–6517
79. Ishida M, Park SW, Hwang D, Koo YB, Sessler JL, Kim DY, Kim D (2011) Donor-substituted β-functionalized porphyrin dyes on hierarchically structured mesoporous TiO_2 spheres. Highly efficient dye-sensitized solar cells. J Phys Chem C 115:19343–19354
80. Davis NKS, Pawlicki M, Anderson HL (2008) Expanding the porphyrin π-system by fusion with anthracene. Org Lett 10:3945–3947
81. Mozer AJ, Griffith MJ, Tsekouras G, Wagner P, Wallace GG, Mori S, Sunahara K, Miyashita M, Earles JC, Gordon KC, Du L, Katoh R, Furube A, Officer DL (2009) Zn–Zn porphyrin dimer-sensitized solar cells: toward 3-D light harvesting. J Am Chem Soc 131:15621–15623
82. Mai CL, Huang WK, Lu HP, Lee CW, Chiu CL, Liang YR, Diau EWG, Yeh CY (2010) Synthesis and characterization of diporphyrin sensitizers for dye-sensitized solar cells. Chem Commun 46:809–811
83. Davis KS, Thompson AL, Anderson HL (2010) Bis-anthracene fused porphyrins: synthesis, crystal structure, and near-IR absorption. Org Lett 12:2124–2127
84. Davis NKS, Thompson AL, Anderson HL (2011) A porphyrin fused to four anthracenes. J Am Chem Soc 133:30–31
85. Jiao C, Zu N, Huang KW, Wang P, Wu J (2011) Perylene anhydride fused porphyrins as near-infrared sensitizers for dye-sensitized solar cells. Org Lett 13:3652–3655
86. Liu Y, Lin H, Dy JT, Tamaki K, Nakazaki J, Nakayama D, Uchida S, Kubo T, Segawa H (2011) *N*-Fused carbazole–zinc porphyrin–free-base porphyrin triad for efficient near-IR dye-sensitized solar cells. Chem Commun 47:4010–4012
87. Lewtak JP, Gryko DT (2012) Synthesis of π-extended porphyrins *via* intramolecular oxidative coupling. Chem Commun 48:10069–10086
88. Yella A, Lee HW, Tsao HN, Yi C, Chandiran AK, Nazeeruddin MK, Diau EWG, Yeh CY, Zakeeruddin SM, Grätzel M (2011) Porphyrin-sensitized solar cells with cobalt (II/III)–based redox electrolyte exceed 12 percent efficiency. Science 334:629–634
89. Nusbaumer H, Moser JE, Zakeeruddin SM, Nazeeruddin MK, Grätzel M (2001) $Co^{II}(dbbip)_2{}^{2+}$ complex rivals tri-iodide/iodide redox mediator in dye-sensitized photovoltaic cells. J Phys Chem B 105:10461–10464
90. Sapp SA, Elliott CM, Contado C, Caramori S, Bignozzi CA (2002) Substituted polypyridine complexes of cobalt(II/III) as efficient electron-transfer mediators in dye-sensitized solar cells. J Am Chem Soc 124:11215–11222
91. Liu Y, Jennings JR, Huang Y, Wang Q, Zakeeruddin SM, Grätzel M (2011) Cobalt redox mediators for ruthenium-based dye-sensitized solar cells: a combined impedance spectroscopy and near-IR transmittance study. J Phys Chem C 115:18847–18855
92. Hamann TW (2012) The end of iodide? Cobalt complex redox shuttles in DSSCs. Dalton Trans 41:3111–3115
93. Mathew S, Yella A, Gao P, Humphry-Baker R, Curchod BFE, Ashari-Astani N, Tavernelli I, Rothlisberger U, Nazeeruddin MK, Grätzel M (2014) Dye-sensitized solar cells with 13% efficiency achieved through the molecular engineering of porphyrin sensitizers. Nat Chem 6:242–247
94. Yella A, Mai CL, Zakeeruddin SM, Chang SN, Hsish CH, Yeh CY, Grätzel M (2014) Molecular engineering of push–pull porphyrin dyes for highly efficient dye-sensitized solar cells: the role of benzene spacers. Angew Chem Int Ed 53:2973–2977
95. Wang Y, Chen B, Wu W, Li X, Zhu W, Tian H, Xie Y (2014) Efficient solar cells sensitized by porphyrins with an extended conjugation framework and a carbazole donor: from molecular design to cosensitization. Angew Chem Int Ed 53:10779–10783

96. Balzani V, Moggi L, Manfrin MF, Bolletta F, Gleria M (1975) Solar energy conversion by water photodissociation. Science 189:852–856
97. Bolton JR (1978) Solar fuels. Science 202:705–710
98. Grätzel M (1981) Artificial photosynthesis: water cleavage into hydrogen and oxygen by visible light. Acc Chem Res 14:376–384
99. Lewis NS, Nocera D (2006) Powering the planet: chemical challenges in solar energy utilization. Proc Natl Acad Sci U S A 103:15729–15735
100. Meyer TJ, Huynh MHV, Thorp HH (2007) The possible role of proton-coupled electron transfer (PCET) in water oxidation by photosystem II. Angew Chem Int Ed 46:5284–5304
101. Concepcion JJ, Jurss JW, Brennaman MK, Hoertz PG, Patrocinio AOT, Iha NYM, Templeton JL, Meyer TJ (2009) Making oxygen with ruthenium complexes. Acc Chem Res 42:1954–1965
102. Puntoriero F, Sartorel A, Orlandi M, La Ganga G, Serroni S, Bonchio M, Scandola F, Campagna S (2011) Photoinduced water oxidation using dendrimeric Ru(II) complexes as photosensitizers. Coord Chem Rev 255:2594–2601
103. Nocera DG (2012) The artificial leaf. Acc Chem Res 45:767–776
104. Concepcion JJ, House RL, Papanikolas JM, Meyer TJ (2012) Chemical approaches to artificial photosynthesis. Proc Natl Acad Sci U S A 109:15560–15564
105. Cox N, Pantazis DA, Neese F, Lubitz W (2013) Biological water oxidation. Acc Chem Res 46:1588–1596
106. Han Z, Eisenberg R (2014) Fuel from water: the photochemical generation of hydrogen from water. Acc Chem Res 47:2537–2544
107. Kok B, Forbush B, McGloin M (1970) Cooperation of charges in photosynthetic O_2 evolution-I. A linear four step mechanism. Photochem Photobiol 11:457–475
108. Warren JJ, Tronic TA, Mayer JM (2010) Thermochemistry of proton-coupled electron transfer reagents and its implications. Chem Rev 110:6961–7001
109. Mayer JM (2011) Understanding hydrogen atom transfer: from bond strengths to Marcus theory. Acc Chem Res 44:36–46
110. Costentin C, Robert M, Saveant JM (2010) Concerted proton-electron transfers: electrochemical and related approaches. Acc Chem Res 43:1019–1029
111. Saveant JM (2012) Electrochemical approach to proton-coupled electron transfers: recent advances. Energy Environ Sci 5:7718–7731
112. Artero V, Chavarot-Kerlidou M, Fontecave M (2011) Splitting water with cobalt. Angew Chem Int Ed 50:7238–7266
113. Tran PD, Artero V, Fontecave M (2010) Water electrolysis and photoelectrolysis on electrodes engineered using biological and bio-inspired molecular systems. Energy Environ Sci 3:727–747
114. Andreiadis ES, Chavarot-Kerlidou M, Fontecave M, Artero V (2011) Artificial photosynthesis: from molecular catalysts for light-driven water splitting to photoelectrochemical cells. Photochem Photobiol 87:946–964
115. Walter MG, Warren EL, McKone JR, Boettcher SW, Mi Q, Santori EA, Lewis NS (2010) Solar water splitting cells. Chem Rev 110:6446–6473
116. McKone JR, Lewis NS, Gray HB (2014) Will solar-driven water-splitting devices see the light of day? Chem Mater 26:407–414
117. Haussener S, Xiang C, Spurgeon JM, Ardo S, Lewis NS, Weber AZ (2012) Modeling, simulation, and design criteria for photoelectrochemical water-splitting systems. Energy Environ Sci 5:9922–9935
118. Lewis NS (2007) Toward cost-effective solar energy use. Science 315:798–801
119. Mi Q, Zhanaidarova A, Brunschwig BS, Gray HB, Lewis NS (2012) A quantitative assessment of the competition between water and anion oxidation at WO_3 photoanodes in acidic aqueous electrolytes. Energy Environ Sci 5:5694–5700

120. Seabold JA, Choi KS (2012) Efficient and stable photo-oxidation of water by a bismuth vanadate photoanode coupled with an iron oxyhydroxide oxygen evolution catalyst. J Am Chem Soc 134:2186–2192
121. Park Y, McDonald KJ, Choi KS (2013) Progress in bismuth vanadate photoanodes for use in solar water oxidation. Chem Soc Rev 42:2321–2337
122. Heller A (1986) Optically transparent metallic catalysts on semiconductors. Pure Appl Chem 58:1189–1192
123. McKone JR, Pieterick AP, Gray HB, Lewis NS (2013) Hydrogen evolution from Pt/Ru-coated p-type WSe_2 photocathodes. J Am Chem Soc 135:223–231
124. Jaramillo TF, Jørgensen KP, Bonde J, Nielsen JH, Horch S, Chorkendorff I (2007) Identification of active edge sites for electrochemical H_2 evolution from MoS_2 nanocatalysts. Science 317:100–102
125. Popczun EJ, McKone JR, Read CG, Biacchi AJ, Wiltrout AM, Lewis NS, Schaak RE (2013) Nanostructured nickel phosphide as an electrocatalyst for the hydrogen evolution reaction. J Am Chem Soc 135:9267–9270
126. Shao Z, Halle SM (2004) A high-performance cathode for the next generation of solid-oxide fuel cells. Nature 431:170–173
127. Suntivich J, May KJ, Gasteiger HA, Goodenough JB, Shao-Horn Y (2011) A perovskite oxide optimized for oxygen evolution catalysis from molecular orbital principles. Science 334:1383–1385
128. Kanan MW, Nocera DG (2008) *In situ* formation of an oxygen-evolving catalyst in neutral water containing phosphate and Co^{2+}. Science 321:1072–1075
129. Geletii YV, Botar B, Kögerler P, Hillesheim DA, Musaev DG, Hill CL (2008) An all-inorganic, stable, and highly active tetraruthenium homogeneous catalyst for water oxidation. Angew Chem Int Ed 47:3896–3899
130. Sartorel A, Carraro M, Scorrano G, De Zorzi R, Geremia S, McDaniel ND, Bernhard S, Bonchio M (2008) Polyoxometalate embedding of a tetraruthenium(IV)-oxo-core by template-directed metalation of $[\gamma\text{-}SiW_{10}O_{36}]^{8-}$: a totally inorganic oxygen-evolving catalyst. J Am Chem Soc 130:5006–5007
131. Dincă M, Surendranath Y, Nocera DG (2010) Nickel-borate oxygen-evolving catalyst that functions under benign conditions. Proc Natl Acad Sci U S A 107:10337–10341
132. Yin Q, Tan JM, Besson C, Geletii YV, Musaev DG, Kuznetsov AE, Luo Z, Hardcastle KI, Hill CL (2010) A fast soluble carbon-free molecular water oxidation catalyst based on abundant metals. Science 328:342–345
133. Boettcher SW, Warren EL, Putnam MC, Santori EA, Turner-Evans D, Kelzenberg MD, Walter MG, McKone JR, Brunschwig BS, Atwater HA, Lewis NS (2011) Photoelectrochemical hydrogen evolution using Si microwire arrays. J Am Chem Soc 133:1216–1219
134. Kong D, Wang H, Cha JJ, Pasta M, Koski KJ, Yao J, Cui Y (2013) Synthesis of MoS_2 and $MoSe_2$ films with vertically aligned layers. Nano Lett 12:1341–1347
135. Trotochaud L, Ranney JK, Williams KN, Boettcher SW (2012) Solution-cast metal oxide thin film electrocatalysts for oxygen evolution. J Am Chem Soc 134:17253–17261
136. Mullins CS, Pecoraro VL (2007) Reflections on small molecule manganese models that seek to mimic photosynthetic water oxidation chemistry. Coord Chem Rev 252:416–443
137. Dismukes GC, Brimblecombe R, Felton GAN, Pryadun RS, Sheats JE, Spiccia L, Swiegers GF (2009) Development of bioinspired Mn_4O_4–cubane water oxidation catalysts: lessons from photosynthesis. Acc Chem Res 42:1935–1943
138. Kanady JS, Tsui EY, Day MW, Agapie T (2011) A synthetic model of the Mn_3Ca subsite of the oxygen-evolving complex in photosystem II. Science 333:733–736
139. Tsui EY, Agapie T (2013) Reduction potentials of heterometallic manganese–oxido cubane complexes modulated by redox-inactive metals. Proc Natl Acad Sci U S A 110:10084–10088
140. Kanady JS, Lin PH, Carsch KM, Nielsen RJ, Takase MK, Goddard WA III, Agapie T (2014) Toward models for the full oxygen-evolving complex of photosystem II by ligand

coordination to lower the symmetry of the Mn_3CaO_4 cubane: demonstration that electronic effects facilitate binding of a fifth metal. J Am Chem Soc 136:14373–14376
141. Zhang C, Chen C, Dong H, Shen JR, Dau H, Zhao J (2015) A synthetic Mn_4Ca-cluster mimicking the oxygen-evolving center of photosynthesis. Science 348:690–693
142. Youngblood WJ, Lee SHA, Kobayashi Y, Hernandez-Pagan EA, Hoertz PG, Moore TA, Moore AL, Gust D, Mallouk TE (2009) Photoassisted overall water splitting in visible light-absorbing dye-sensitized photoelectrochemical cell. J Am Chem Soc 131 926–927
143. Brimblecombw R, Koo A, Dismukes GC, Swiegers GF, Spiccia L (2010) Solar driven water oxidation by a bioinspired manganese molecular catalyst. J Am Chem Soc 132:2892–2894
144. Al-Oweini R, Sarotrel A, Bassil BS, Natali M, Berardi S, Scandola F, Kortz U, Bonchio M (2014) Photocatalytic water oxidation by a mixed-valent $Mn^{III}{}_3Mn^{IV}O_3$ manganese oxo core that mimics the natural oxygen-evolving center. Angew Chem Int Ed 53:11182–11185
145. Santoni MP, La Ganga G, Nardo VM, Natali M, Puntoriero F, Scandola F, Campagna S (2014) The use of a vanadium species as a catalyst in photoinduced water oxidation. J Am Chem Soc 136:8189–8192
146. Tanaka S, Annaka M, Sakai K (2012) Visible light-induced water oxidation catalyzed by molybdenum-based polyoxometalates with mono- and dicobalt(III) cores as oxygen-evolving centers. Chem Commun 48:1653–1655
147. Gersten SW, Sammuels GJ, Meyer TJ (1982) Catalytic oxidation of water by an oxo-bridged ruthenium dimer. J Am Chem Soc 104:4029–4030
148. Liu F, Concepcion JJ, Hurss JW, Cardolaccia T, Templeton JL, Meyer TJ (2008) Mechanism of water oxidation from the blue dimer to photosystem II. Inorg Chem 47:1727–1752
149. Concepcion JJ, Jurss JW, Templeton JL, Meyer TJ (2008) Mediator-assisted water oxidation by the ruthenium "blue dimer" *cis,cis*-$[(bpy)_2(H_2O)RuORu(OH_2)(bpy)_2]^{4+}$. Proc Natl Acad Sci U S A 105:17632–17635
150. Yang X, Baik MH (2006) *cis,cis*-$[(bpy)_2Ru^VO]_2O^{4+}$ Catalyzes water oxidation formally *via in situ* generation of radicaloid Ru^{IV}–O•. J Am Chem Soc 128:7476–7485
151. Masaoka S, Sakai K (2009) Clear evidence showing the robustness of a highly active oxygen-evolving mononuclear ruthenium complex with an aqua ligand. Chem Lett 38:182–183
152. Nakazono T, Parent AR, Sakai K (2013) Cobalt porphyrins as homogeneous catalysts for water oxidation. Chem Commun 49:6325–6327
153. Li L, Duan L, Xu Y, Gorlov M, Hagfeldt A, Sun L (2010) A photoelectrochemical device for visible light driven water splitting by a molecular ruthenium catalyst assembled on dye-sensitized nanostructured TiO_2. Chem Commun 46:7307–7309
154. Duan L, Araujo CM, Ahlquist MSG, Sun L (2012) Highly efficient and robust molecular ruthenium catalysts for water oxidation. Proc Natl Acad Sci U S A 109:15584–15588
155. Li F, Jiang Y, Zhang B, Huang F, Gao Y, Sun L (2012) Towards a solar fuel device: light-driven water oxidation catalyzed by a supramolecular assembly. Angew Chem Int Ed 51:2417–2420
156. Ashford DL, Sherman BD, Binstead RA, Templeton JL, Meyer TJ (2015) Electro-assembly of a chromophore-catalyst bilayer for water oxidation and photocatalytic water splitting. Angew Chem Int Ed 54:4778–4781
157. Li H, Li F, Zhang B, Zhou X, Yu F, Sun L (2015) Visible light-driven water oxidation promoted by host-guest interaction between photosensitizer and catalyst with a high quantum efficiency. J Am Chem Soc 137:4332–4335
158. McDaniel ND, Coughlin FJ, Tinker LL, Bernhard S (2008) Cyclometalated iridium(III) aquo complexes: efficient and tunable catalysts for the homogeneous oxidation of water. J Am Chem Soc 130:210–217
159. Hull JF, Balcells D, Blakemore JD, Incarvito CD, Eisenstein O, Brudvig GW, Crabtree RH (2009) Highly active and robust Cp* iridium complexes for catalytic water oxidation. J Am Chem Soc 131:8730–8731

160. Blakemore JD, Schley ND, Balcells D, Hull JF, Olack GW, Incarvito CD, Eisenstein O, Brudvig GW, Crabtree RH (2010) Half-sandwich iridium complexes for homogeneous water-oxidation catalysis. J Am Chem Soc 132:16017–16029
161. Schley ND, Blakemore JD, Subbaiyan NK, Incarvito CD, D'Souza F, Crabtree RH, Brudvig GW (2011) Distinguishing homogeneous from heterogeneous catalysis in electrode-driven water oxidation with molecular iridium complexes. J Am Chem Soc 133:10473–10481
162. Wang C, Wang JL, Lin W (2012) Elucidating molecular iridium water oxidation catalysts using metal−organic frameworks: a comprehensive structural, catalytic, spectroscopic, and kinetic study. J Am Chem Soc 134:19895–19908
163. Lalrempuia R, McHaniel ND, Müller-Bunz H, Bernhard S, Albrecht M (2010) Water oxidation catalyzed by strong carbene-type donor-ligand complexes of iridium. Angew Chem Int Ed 49:9765–9768
164. Moore GF, Blakemore JD, Milot RL, Hull JF, Song H, Cai L, Schmuttenmaer CA, Crabtree RH, Brudvig GW (2011) A visible light water-splitting cell with a photoanode formed by codeposition of a high-potential porphyrin and an iridium water-oxidation catalyst. Energy Environ Sci 4:2389–2392
165. Lewis EA, Tolman WB (2004) Reactivity of dioxygen-copper systems. Chem Rev 104:1047–1076
166. Iton S, Fukuzuki S (2007) Monooxygenase activity of type 3 copper proteins. Acc Chem Res 40:592–600
167. Que L Jr, Tolman WB (2008) Biologically inspired oxidation catalysis. Nature 455:333–340
168. Gagnon N, Tolman WB (2015) $[CuO]^+$ and $[CuOH]^{2+}$ complexes: intermediates in oxidation catalysis? Acc Chem Res 48:2126–2131
169. Halfen JA, Mahapatra S, Wilkinson EC, Kaderli SM, Young VG Jr, Que L Jr, Zuberbühler AD, Tolman WB (1996) Reversible cleavage and formation of the dioxygen O−O bond within a dicopper complex. Science 271:1397–1400
170. Donoghue PJ, Tehranchi J, Cramer CJ, Sarangi R, Solomon EI, Tolman WB (2011) Rapid C−H bond activation by a monocopper(III)−hydroxide complex. J Am Chem Soc 133:17602–17605
171. Dhar D, Tolman WB (2015) Hydrogen atom abstraction from hydrocarbons by a copper (III)−hydroxide complex. J Am Chem Soc 137:1322–1329
172. Conde A, Vilella L, Balcells D, Díaz-Requejo MM, Lledós A, Pérez PJ (2013) Introducing copper as catalyst for oxidative alkane dehydrogenation. J Am Chem Soc 135:3887–3896
173. Dietl N, van der Linde C, Schlangen M, Beyer MK, Schqarz H (2011) Diatomic $[CuO]^+$ and its role in the spin-selective hydrogen- and oxygen-atom transfers in the thermal activation of methane. Angew Chem Int Ed 50:4966–4969
174. Connolly P, Espenson JH (1986) Cobalt-catalyzed evolution of molecular hydrogen. Inorg Chem 25:2684–2688
175. Dempsey JL, Brunschwig BS, Winkler JR, Gray HB (2009) Hydrogen evolution catalyzed by cobaloximes. Acc Chem Res 42:1995–2004
176. Hu X, Cossairt BM, Brunschwig BS, Lewis NS, Peters JC (2005) Electrocatalytic hydrogen evolution by cobalt difluoroboryl-diglyoximate complexes. Chem Commun 4723–4725
177. Berben LA, Peters JC (2010) Hydrogen evolution by cobalt tetraimine catalysts adsorbed on electrode surfaces. Chem Commun 46:398–400
178. Hu X, Brunschwig BS, Peters JC (2007) Electrocatalytic hydrogen evolution at low overpotentials by cobalt macrocyclic glyoxime and tetraimine complexes. J Am Chem Soc 129:8988–8998
179. Dempsey JL, Winkler JR, Gray HB (2010) Mechanism of H_2 evolution from a photogenerated hydridocobaloxime. J Am Chem Soc 132:16774–16776
180. Stubbert BD, Peters JC, Gray HB (2011) Rapid water reduction to H_2 catalyzed by a cobalt bis(iminopyridine) complex. J Am Chem Soc 133:18070–18073

181. Valdez CN, Dempsey JL, Brunschwig BS, Winkler JR, Gray HB (2012) Catalytic hydrogen evolution from a covalently linked dicobaloxime. Proc Natl Acad Sci U S A 109:15589–15593
182. Hawecker J, Lehn JM, Ziessel R (1983) Efficient homogeneous photochemical hydrogen generation and water reduction mediated by cobaloxime or macrocyclic cobalt complexes. Nouv J Chim 7:271–277
183. Fihri A, Artero V, Razavet M, Baffert C, Leibl W, Fontecave M (2008) Cobaloxime-based photocatalytic devices for hydrogen production. Angew Chem Int Ed 47:564–567
184. McNamara WR, Han Z, Alperin PJ, Brennessel WW, Holland PL, Eisenberg R (2011) A cobalt-dithiolene complex for the photocatalytic and electrocatalytic reduction of protons. J Am Chem Soc 133:15368–15371
185. McNamara WR, Han Z, Yin CJ, Brennessel WW, Holland PL, Eisenberg R (2012) Cobalt-dithiolene complexes for the photocatalytic and electrocatalytic reduction of protons in aqueous solutions. Proc Natl Acad Sci U S A 109:15594–15599
186. Sakai K, Ozawa H (2007) Homogeneous catalysis of platinum(II) complexes in photochemical hydrogen production from water. Coord Chem Rev 251:2753–2766
187. Ozawa H, Haga M, Sakai K (2006) A photo-hydrogen-evolving molecular device driving visible-light-induced EDTA-reduction of water into molecular hydrogen. J Am Chem Soc 128:4926–4927
188. Kobayashi M, Masaoka S, Sakai K (2012) Photoinduced hydrogen evolution from water by a simple platinum(II) terpyridine derivative: a Z-scheme photosynthesis. Angew Chem Int Ed 51:7431–7434
189. Du P, Schneider J, Jarosz P, Eisenberg R (2006) Photocatalytic generation of hydrogen from water using a platinum(II) terpyridyl acetylide chromophore. J Am Chem Soc 128:7726–7727
190. Du P, Knowles K, Eisenberg R (2008) A homogeneous system for the photogeneration of hydrogen from water based on a platinum(II) terpyridyl acetylide chromophore and a molecular cobalt catalyst. J Am Chem Soc 130:12576–12577
191. Du P, Schneiderr J, Luo G, Brennessel WW, Eisenberg R (2009) Visible light-driven hydrogen production from aqueous protons catalyzed by molecular cobaloxime catalysts. Inorg Chem 48:4952–4962
192. Wang X, Goeb S, Ji Z, Pogulaichenko NA, Castellano FN (2011) Homogeneous photocatalytic hydrogen production using π-conjugated platinum(II) arylacetylide sensitizers. Inorg Chem 50:705–707
193. Lowry MS, Goldsmith JI, Slinker JD, Rohl R, Pascal RA Jr, Malliaras GG, Bernhard S (2005) Single-layer electroluminescent devices and photoinduced hydrogen production from an ionic iridium(III) complex. Chem Mater 17:5712–5719
194. Goldsmith JI, Hudson WR, Lowry MS, Anderson TH, Bernhard S (2005) Discovery and high-throughput screening of heteroleptic iridium complexes for photoinduced hydrogen production. J Am Chem Soc 127:7502–7510
195. Whang DR, Sakai K, Park SY (2013) Highly efficient photocatalytic water reduction with robust iridium(III) photosensitizers containing arylsilyl substituents. Angew Chem Int Ed 52:11612–11615
196. Wilson AD, Newell RH, McNevin MJ, Muckerman JT, DuBois MR, DuBios DL (2006) Hydrogen oxidation and production using nickel-based molecular catalysts with positioned proton relays. J Am Chem Soc 128:358–366
197. Wilson AD, Shoemaker RK, Miedaner A, Muckerman JT, DuBois DL, DuBois MR (2007) Nature of hydrogen interactions with Ni(II) complexes containing cyclic phosphine ligands with pendant nitrogen bases. Proc Natl Acad Sci U S A 104:6951–6956
198. Kilgore UJ, Roberts JAS, Pool DH, Appel AM, Stewart MP, DuBois MR, Dougherty WG, Kassel WS, Bullock RM, DuBois DL (2011) $[Ni(P^{Ph}{}_2N^{C6H4X}{}_2)_2]^{2+}$ complexes as electrocatalysts for H_2 production: effect of substituents, acids, and water on catalytic rates. J Am Chem Soc 133:5861–5872

199. Helm ML, Stewart MP, Bullock RM, DuBois MR, DuBois DL (2011) A synthetic nickel electrocatalyst with a turnover frequency above 100,000 s^{-1} for H_2 production. Science 333:863–866
200. Matsuika S, Yamamoto K, Ogata T, Kusaba M, Nakashima N, Fujita E, Yanagida S (1993) Efficient and selective electron mediation of cobalt complexes with cyclam and related macrocycles in the *p*-terphenyl-catalyzed photoreduction of CO_2. J Am Chem Soc 115:601–609
201. Ogata T, Yanagida S, Brunschwig BS, Fujita E (1995) Mechanistic and kinetic studies of cobalt macrocycles in a photochemical CO_2 reduction system: evidence of Co-CO_2 adducts as intermediates. J Am Chem Soc 117:6708–6716
202. Fujita E (1999) Photochemical carbon dioxide reduction with metal complexes. Coord Chem Rev 185–186:373–384
203. Simón-Manso E, Kubiak CP (2005) Dinuclear nickel complexes as catalysts for electrochemical reduction of carbon dioxide. Organometallics 24:96–102
204. Morris AJ, Meyer GJ, Fujita E (2009) Molecular approaches to the photocatalytic reduction of carbon dioxide for solar fuels. Acc Chem Res 42:1983–1994
205. Benson EE, Kubiak CP, Sathrum AJ, Smieja JM (2009) Electrocatalytic and homogeneous approaches to conversion of CO_2 to liquid fuels. Chem Soc Rev 38:89–99
206. DuBois MR, DuBois DL (2009) Development of molecular electrocatalysts for CO_2 reduction and H_2 production/oxidation. Acc Chem Res 42:1974–1982
207. Takeda H, Ishitani O (2010) Development of efficient photocatalytic systems for CO_2 reduction using mononuclear and multinuclear metal complexes based on mechanistic studies. Coord Chem Rev 254:346–354
208. Appel AM, Bercaw JE, Bocarsly AB, Dobbek H, DuBois DL, Dupuis M, Ferry JG, Fujita E, Hille R, Kenis PJA, Kerfeld CA, Morris RH, Peden CHF, Portis AR, Ragsdale SW, Rauchfuss TB, Reek JNH, Seefeldt LC, Thauer RK, Waldrop GL (2013) Frontiers, opportunities, and challenges in biochemical and chemical catalysis of CO_2 fixation. Chem Rev 113:6621–6658
209. Jeoung JH, Dobbek H (2007) Carbon dioxide activation at the Ni, Fe-cluster of anaerobic carbon monoxide dehydrogenase. Science 318:1461–1464
210. Meshitsuka S, Ichikawa M, Tamaru K (1974) Electrocatalysis by metal phthalocyanines in the reduction of carbon dioxide. J Chem Soc Chem Commun 158–159
211. Fisher BJ, Eisenberg R (1980) Electrocatalytic reduction of carbon dioxide by using macrocycles of nickel and cobalt. J Am Chem Soc 102:7361–7363
212. Beley M, Collin JP, Ruppert R, Sauvage JP (1986) Electrocatalytic reduction of carbon dioxide by nickel cyclam^{2+} in water: study of the factors affecting the efficiency and the selectivity of the process. J Am Chem Soc 108:7461–7467
213. Bhugun I, Lexa D, Savéant JM (1996) Catalysis of the electrochemical reduction of carbon dioxide by iron(0) porphyrins: synergistic effect of weak Broensted acids. J Am Chem Soc 118:1769–1776
214. Hawecker J, Lehn JM, Ziessel R (1984) Electrocatalytic reduction of carbon dioxide mediated by $Re(bipy)(CO)_3Cl$ (bipy = 2,2′-bipyridine). J Chem Soc Chem Commun 328–330
215. Ishida H, Tanaka K, Tanaka T (1987) Electrochemical CO_2 reduction catalyzed by ruthenium complexes $[Ru(bpy)_2(CO)_2]^{2+}$ and $[Ru(bpy)_2(CO)Cl]^{+}$. Effect of pH on the formation of CO and $HCOO^{-}$. Organometallics 6:181–186
216. Bolinger CM, Story N, Sullivan BP, Meyer TJ (1988) Electrocatalytic reduction of carbon dioxide by 2,2′-bipyridine complexes of rhodium and iridium. Inorg Chem 27:4582–4587
217. Slater S, Wagenknecht JH (1984) Electrochemical reduction of carbon dioxide catalyzed by $Rh(diphos)_2Cl$. J Am Chem Soc 106:5367–5368
218. Fujita E, Szalda DJ, Creutz C, Sutin N (1988) Carbon dioxide activation: thermodynamics of carbon dioxide binding and the involvement of two cobalt centers in the reduction of carbon dioxide by a cobalt(I) macrocycle. J Am Chem Soc 110:4870–4871

219. DuBois DL, Miedaner A, Haltiwanger RC (1991) Electrochemical reduction of carbon dioxide catalyzed by [Pd(triphosphine)(solvent)]$(BF_4)_2$ complexes: synthetic and mechanistic studies. J Am Chem Soc 113:8753–8764
220. DeLaet DL, Del Rosario R, Fanwick PE, Kubiak CP (1987) Carbon dioxide chemistry and electrochemistry of a binuclear cradle complex of nickel(0), $Ni_2(\mu$-CNMe) $(CNMe)_2(PPh_2CH_2PPh_2)_2$. J Am Chem Soc 109:754–758
221. Morgenstern DA, Ferrence GM, Washington J, Henderson JI, Rosenhein L, Heise JD, Fanwick PE, Kubiak CP (1996) A class of halide-supported trinuclear nickel clusters $[Ni_3(\mu_3\text{-L})(\mu_3\text{-X})(\mu_2\text{-dppm})_3]^{n+}$ (L = I^-, Br^-, CO, CNR; X = I^-, Br^-; n = 0, 1; dppm = $Ph_2PCH_2PPh_2$): novel physical properties and the Fermi resonance of symmetric μ_3-η_1 bound isocyanide ligands. J Am Chem Soc 118:2198–2207
222. Grodkowski J, Neta P, Fujita E, Mahammed A, Simkhovich L, Gross Z (2002) Reduction of cobalt and iron corroles and catalyzed reduction of CO_2. J Phys Chem A 106:4772–4778
223. Schneider J, Jia H, Kobiro K, Cabelli DE, Muckerman JT, Fujita E (2012) Nickel (II) macrocycles: highly efficient electrocatalysts for the selective reduction of CO_2 to CO. Energy Environ Sci 5:9502–9510
224. Takeda H, Koike K, Inoue H, Ishitani O (2008) Development of an efficient photocatalytic system for CO_2 reduction using rhenium(I) complexes based on mechanistic studies. J Am Chem Soc 130:2023–2031
225. Benson EE, Kubiak CP (2012) Structural investigations into the deactivation pathway of the CO_2 reduction electrocatalyst Re(bpy)$(CO)_3$Cl. Chem Commun 48:7374–7376
226. Smieja JM, Benson EE, Kumar B, Grice KA, Seu CS, Miller AJM, Mayer JM, Kubiak CP (2012) Kinetic and structural studies, origins of selectivity, and interfacial charge transfer in the artificial photosynthesis of CO. Proc Natl Acad Sci U S A 109:15646–15650
227. Benson EE, Sampson MD, Grice KA, Smieja JM, Froehlich JD, Friebel D, Keith JA, Carter EA, Nilsson A, Kubiak CP (2013) The electronic states of rhenium bipyridyl electrocatalysts for CO_2 reduction as revealed by X-ray absorption spectroscopy and computational quantum chemistry. Angew Chem Int Ed 52:4841–4844
228. Keith JA, Grice KA, Kubiak CP, Carter EA (2013) Elucidation of the selectivity of proton-dependent electrocatalytic CO_2 reduction by *fac*-Re(bpy)$(CO)_3$Cl. J Am Chem Soc 135:15823–15829
229. Bourrez M, Molton F, Chardon-Noblat S, Deronzier A (2011) [Mn(bipyridyl)$(CO)_3$Br]: an abundant metal carbonyl complex as efficient electrocatalyst for CO_2 reduction. Angew Chem Int Ed 50:9903–9906
230. Sampson MD, Nguyen AD, Grice KA, Moore CE, Rheingold AL, Kubiak CP (2014) Manganese catalysts with bulky bipyridine ligands for the electrocatalytic reduction of carbon dioxide: eliminating dimerization and altering catalysis. J Am Chem Soc 136:5460–5471
231. Tamaki Y, Morimoto T, Koike K, Ishitani O (2012) Photocatalytic CO_2 reduction with high turnover frequency and selectivity of formic acid formation using Ru(II) multinuclear complexes. Proc Natl Acad Sci 109:15673–15678
232. Sekizawa K, Maeda K, Domen K, Koike K, Ishitani O (2013) Artificial Z-scheme constructed with a supramolecular metal complex and semiconductor for the photocatalytic reduction of CO_2. J Am Chem Soc 135:4596–4599
233. Tanaka R, Yamashita M, Nozaki K (2009) Catalytic hydrogenation of carbon dioxide using Ir (III)-pincer complexes. J Am Chem Soc 131:14168–14169
234. Hull JF, Himeda Y, Wang WH, Hashiguchi B, Periana R, Szalda DJ, Muckerman JT, Fujita E (2012) Reversible hydrogen storage using CO_2 and a proton-switchable iridium catalyst in aqueous media under mild temperatures and pressures. Nat Chem 4:383–388
235. Wang WH, Hull JF, Muckerman JT, Fujita E, Himeda Y (2012) Second-coordination-sphere and electronic effects enhance iridium(III)-catalyzed homogeneous hydrogenation of carbon dioxide in water near ambient temperature and pressure. Energy Environ Sci 5:7923–7926

236. Yam VWW, Tang RPL, Wong KMC, Cheung KK (2001) Synthesis, luminescence, electrochemistry, and ion-binding studies of platinum(II) terpyridyl acetylide complexes. Organometallics 20:4476–4482
237. Yang QZ, Wu LZ, Wu ZX, Zhang LP, Tung CH (2002) Long-lived emission from platinum (II) terpyridyl acetylide complexes. Inorg Chem 41:5653–5655
238. Moore GE (1975) Progress in digital integrated electronics. In: 1975 international electron devices meeting IEEE text speech. pp 11–13
239. Orgiu E, Samorì P (2014) 25th Anniversary article: organic electronics marries photochromism: generation of multifunctional interfaces, materials, and devices. Adv Mater 26:1827–1845
240. Hartley GS (1937) The *cis*-form of azobenzene. Nature 140:281
241. Griffiths J (1972) II. Photochemistry of azobenzene and its derivatives. Chem Soc Rev 1:481–493
242. Nishihara H (2005) Combination of redox- and photochemistry of azo-conjugated metal complexes. Coord Chem Rev 249:1468–1475
243. Bandara HMD, Burdette SC (2012) Photoisomerization in different classes of azobenzene. Chem Soc Rev 41:1809–1825
244. Fihey A, Perrier A, Browne WR, Jacquemin D (2015) Multiphotochromic molecular systems. Chem Soc Rev 44:3719–3759
245. Kawata S, Kawata Y (2000) Three-dimensional optical data storage using photochromic materials. Chem Rev 100:1777–1788
246. Hasegawa Y, Nakagawa T, Kawai T (2010) Recent progress of luminescent metal complexes with photochromic units. Coord Chem Rev 254:2643–2651
247. Ko CC, Yam VWW (2010) Transition metal complexes with photochromic ligands – photosensitization and photoswitchable properties. J Mater Chem 20:2063–2070
248. Dugave C, Demange L (2003) *Cis-trans* isomerization of organic molecules and biomolecules: implications and applications. Chem Rev 103:2475–2532
249. Polo AS, Itokazu MK, Frin KM, Patrocínio AOT, Iha NYM (2006) Light driven *trans*-to-*cis* isomerization of stilbene-like ligands in *fac*-$[Re(CO)_3(NN)(\textit{trans}\text{-}L)]^+$ and luminescence of their photoproducts. Coord Chem Rev 250:1669–1680
250. Irie M (2000) Diarylethenes for memories and switches. Chem Rev 100:1685–1716
251. Myles AJ, Branda NR (2002) 1,2-Dithienylethene photochromes and non-destructive erasable memory. Adv Funct Mater 12:167–173
252. Irie M, Fukaminato T, Matsuda K, Kobatake S (2014) Photochromism of diarylethene molecules and crystals: memories, switches, and actuators. Chem Rev 114:12174–12277
253. Berkovic G, Krongauz V, Weiss V (2000) Spiropyrans and spirooxazines for memories and switches. Chem Rev 100:1741–1754
254. Guerchais V, Ordronneau L, Le Bozec H (2010) Recent developments in the field of metal complexes containing photochromic ligands: modulation of linear and nonlinear optical properties. Coord Chem Rev 254:2533–2545
255. Zhang J, Zou Q, Tian H (2013) Photochromic materials: more than meets the eye. Adv Mater 25:378–399
256. Yokoyama Y (2000) Fulgides for memories and switches. Chem Rev 100:1717–1740
257. Whitten DG, Zamegar PP (1971) Photochemistry of ruthenium complexes. Ligand isomerization *via* orbitally different excited states. J Am Chem Soc 93:3776–3777
258. Zamegar PP, Bock CR, Whitten DG (1973) Photoreactions of transition metal complexes. Ligand reactivity as a probe for excited-state characterization. J Am Chem Soc 95:4367–4372
259. Wrighton M, Morse DL (1974) Nature of the lowest excited state in tricarbonylchloro-1,10-phenanthrolinerhenium(I) and related complexes. J Am Chem Soc 96:998–1003
260. Wrighton MS, Morse DL, Pdungsap L (1975) Intraligand lowest excited states in tricarbonylhalobis(styrylpyridine)rhenium(I) complexes. J Am Chem Soc 97:2073–2079

261. Yam VWW, Lau VCY, Cheung KK (1995) Synthesis, photophysics and photochemistry of novel luminescent rhenium(I) photoswitchable materials. J Chem Soc Chem Commun 259–261
262. Yam VWW, Ko CC, Wu LX, Wong KMC, Cheung KK (2000) Syntheses, crystal structure, and photochromic properties of rhenium(I) complexes containing the spironaphthoxazine moiety. Organometallics 19:1820–1822
263. Ko CC, Wu LX, Wong KMC, Zhu N, Yam VWW (2004) Synthesis, characterization and photochromic studies of spirooxazine-containing 2,2′-bipyridine ligands and their rhenium (I) tricarbonyl complexes. Chem Eur J 10:766–776
264. Yam VWW, Ko CC, Zhu N (2004) Photochromic and luminescence switching properties of a versatile diarylethene-containing 1,10-phenanthroline ligand and its rhenium(I) complex. J Am Chem Soc 126:12734–12735
265. Jukes RTF, Adamo V, Hartl F, Belser P, De Cola L (2004) Photochromic dithienylethene derivatives containing Ru(II) or Os(II) metal units. Sensitized photocyclization from triplet state. Inorg Chem 43:2779–2792
266. Ko CC, Kwok WM, Yam VWW, Phillips DL (2006) Triplet MLCT photosensitization of the ring-closing reaction of diarylethenes by design and synthesis of a photochromic rhenium (I) complex of a diarylethene-containing 1,10-phenanthroline ligand. Chem Eur J 12:5840–5848
267. Sakamoto R, Kume S, Sugimoto M, Nishihara H (2009) *Trans–cis* photoisomerization of azobenzene-conjugated dithiolato-bipyridine platinum(II) complexes: extension of photoresponse to longer wavelengths and photocontrollable tristability. Chem Eur J 15:1429–1439
268. Hasegawa Y, Takahashi K, Kume S, Nishihara H (2011) Complete solid state photoisomerization of bis(dipyrazolylstyrylpyridine)iron(II) to change magnetic properties. Chem Commun 47:6846–6848
269. Lin JL, Chen CW, Sun SS, Lees AJ (2011) Photoswitching tetranuclear rhenium (I) tricarbonyl diimine complexes with a stilbene-like bridging ligand. Chem Commun 47:6030–6032
270. Baik C, Wand S (2011) Inhibiting olefin *cis*, *trans*-photoisomerization and enhancing electron-accepting ability of a diboryl compound by metal chelation Chem Commun 47:9432–9434
271. Amar A, Savel P, Akdas-Kilig H, Katan C, Meghezzi H, Boucekkine A, Malval JP, Fillaut JL (2015) Photoisomerisation in aminoazobenzene-substituted ruthenium(II) tris(bipyridine) complexes: influence of the conjugation pathway. Chem Eur J 21:8262–8270
272. Park JS, Lifschitz AM, Young RM, Mendez-Arroyo JM, Wasielewski MR, Stern CL, Mirkin CA (2013) Modulation of electronics and thermal stabilities of photochromic phosphino–aminoazobenzene derivatives in weak-link approach coordination complexes. J Am Chem Soc 135:16988–16996
273. Irie M, Masaaki M (1988) Thermally irreversible photochromic systems. Reversible photocyclization of diarylethene derivatives. J Org Chem 53:803–808
274. Kellogg RM, Groen MB, Wynberg H (1967) Photochemically induced cyclization of some furyl- and thienylethenes. J Org Chem 32:3093–3100
275. Irie M, Sakemura K, Okinaka M, Uchida K (1995) Photochromism of dithienylethenes with electron-donating substituents. J Org Chem 60:8305–8309
276. Matsuda K, Irie M (2000) A diarylethene with two nitronyl nitroxides: photoswitching of intramolecular magnetic interaction. J Am Chem Soc 122:7195–7201
277. Irie M, Kobatake S, Horichi M (2001) Reversible surface morphology changes of a photochromic diarylethene single crystal by photoirradiation. Science 291:1769–1772
278. Irie M, Fukaminato T, Sasaki T, Tamai N, Kawai T (2002) Organic chemistry: a digital fluorescent molecular photoswitch. Nature 420:759–760
279. Higashiguchi K, Matsuda K, Irie M (2003) Photochromic reaction of a fused dithienylethene: multicolor photochromism. Angew Chem Int Ed 42:3537–3540

280. Kobatake S, Takami S, Muto H, Ishikawa T, Irie M (2007) Rapid and reversible shape changes of molecular crystals on photoirradiation. Nature 446:778–781
281. Lucas LN, van Esch J, Kellogg RM, Feringa BL (1998) A new class of photochromic 1,2-diarylethenes; synthesis and switching properties of bis(3-thienyl)cyclopentenes. Chem Commun 21:2313–2314
282. De Jong JJD, Lucas LN, Kellogg RM, van Esch JH, Feringa BL (2004) Reversible optical transcription of supramolecular chirality into molecular chirality. Science 304:278–281
283. Areephong J, Kudernac T, De Jong JJD, Carroll GT, Pantorott D, Hjelm J, Browne WR, Feringa BL (2008) On/off photoswitching of the electropolymerizability of terthiophenes. J Am Chem Soc 130:12850–12851
284. Uchida K, Sukata SI, Matsuzawa Y, Akazawa M, De Jong JJD, Katsonis N, Kojima Y, Nakamura S, Areephong J, Meetsma A, Feringa BL (2008) Photoresponsive rolling and bending of thin crystals of chiral diarylethenes. Chem Commun 326–328
285. van Dijken DJ, Beierle JM, Stuart MCA, Szymański W, Browne WR, Feringa BL (2014) Autoamplification of molecular chirality through the induction of supramolecular chirality. Angew Chem Int Ed 53:5073–5077
286. Tsivgoulis GM, Lehn JM (1995) Photonic molecular devices: reversibly photoswitchable fluorophores for nondestructive readout for optical memory. Angew Chem Int Ed Engl 34:1119–1122
287. Gilat SL, Kawai SH, Lehn JM (1995) Light-triggered molecular devices: photochemical switching of optical and electrochemical properties in molecular wire type diarylethene species. Chem Eur J 1:275–284
288. Kawai SH, Gilat SL, Ponsinet R, Lehn JM (1995) A dual-mode molecular switching device: bisphenolic diarylethenes with integrated photochromic and electrochromic properties. Chem Eur J 1:285–293
289. Norsten TB, Branda NR (2001) Photoregulation of fluorescence in a porphyrinic dithienylethene photochrome. J Am Chem Soc 123:1784–1785
290. Peters A, Branda NR (2003) Electrochromism in photochromic dithienylcyclopentenes. J Am Chem Soc 125:3404–3405
291. Lemieux V, Branda NR (2005) Reactivity-gated photochromism of 1,2-dithienylethenes for potential use in dosimetry applications. Org Lett 7:2969–2972
292. Boyer JC, Carling CJ, Gates BD, Branda NR (2010) Two-way photoswitching using one type of near-infrared light, upconverting nanoparticles, and changing only the light intensity. J Am Chem Soc 132:15766–15772
293. Ko CC, Lam WH, Yam VWW (2008) Photochromic oligothienoacene derivatives with photo-switchable luminescene properties and computational studies. Chem Commun 5203–5205
294. Wong HL, Ko CC, Lam WH, Zhu N, Yam VWW (2009) Design and synthesis of a new class of photochromic diarylethene-containing dithieno[3,2-b:2′,3′- d]pyrroles and their switchable luminescence properties. Chem Eur J 15:10005–10009
295. Poon CT, Lam WH, Wong HL, Yam VWW (2010) A versatile photochromic dithienylethene-containing β-diketonate ligand: near-infrared photochromic behavior and photoswitchable luminescence properties upon incorporation of a boron(III) center. J Am Chem Soc 132:13992–13993
296. Duan G, Zhu N, Yam VWW (2010) Syntheses and photochromic studies of dithienylethene-containing imidazolium derivatives and their reactivity towards nucleophiles. Chem Eur J 16:13199–13209
297. Poon CT, Lam WH, Yam VWW (2011) Gated photochromism in triarylborane-containing dithienylethenes: a new approach to a "lock–unlock" system. J Am Chem Soc 133:19622–19625
298. Wong HL, Wong WT, Yam VWW (2012) Photochromic thienylpyridine bis(alkynyl)borane complexes: toward readily tunable fluorescence dyes and photoswitchable materials. Org Lett 14:1862–1865

299. Chan JCH, Lam WH, Wong HL, Wong WT, Yam VWW (2013) Tunable photochromism in air-stable, robust dithienylethene-containing phospholes through modifications at the phosphorus center. Angew Chem Int Ed 52:11504–11508
300. Poon CT, Lam WH, Yam VWW (2013) Synthesis, photochromic, and computational studies of dithienylethene-containing β-diketonate derivatives and their near-infrared photochromic behavior upon coordination of a boron(III) center. Chem Eur J 19:3467–3476
301. Chan JCH, Lam WH, Yam VWW (2014) A highly efficient silole-containing dithienylethene with excellent thermal stability and fatigue resistance: a promising candidate for optical memory storage materials. J Am Chem Soc 136:16994–16997
302. Chan JCH, Wong HL, Wong WT, Yam VWW (2015) Tunable photochromism in the robust dithienylethene-containing phospholes: design, synthesis, characterization, electrochemistry, photophysics, and photochromic studies. Chem Eur J 21:6936–6948
303. Poon CT, Lam WH, Wong HL, Yam VWW (2015) Photochromic dithienylethene-containing triarylborane derivatives: facile approach to modulate photochromic properties with multi-addressable functions. Chem Eur J 21:2182–2192
304. Tian H, Qin B, Yao R, Zhao X, Yang S (2003) A single photochromic molecular switch with four optical outputs probing four inputs. Adv Mater 15:2104–2107
305. Jiang G, Wang S, Yuan W, Jiang L, Song Y, Tian H, Zhu D (2006) Highly fluorescent contrast for rewritable optical storage based on photochromic bisthienylethene-bridged naphthalimide dimer. Chem Mater 18:235–237
306. Li W, Jiao C, Li X, Xie Y, Nakatani K, Tian H, Zhu W (2014) Separation of photoactive conformers based on hindered diarylethenes: efficient modulation in photocyclization quantum yields. Angew Chem Int Ed 53:4603–4607
307. Li W, Li X, Xie Y, Wu Y, Li M, Wu XY, Zhu WH, Tian H (2015) Enantiospecific photoresponse of sterically hindered diarylethenes for chiroptical switches and photomemories. Sci Rep 5:9186
308. Kawai S, Nakashima T, Atsumi K, Sakai T, Harigai M, Imamoto Y, Kamikubo H, Kataoka M, Kawai T (2007) Novel photochromic molecules based on 4,5-dithienyl thiazole with fast thermal bleaching rate. Chem Mater 19:3479–3483
309. Nakashima T, Goto M, Kawai S, Kawai T (2008) Photomodulation of ionic interaction and reactivity: reversible photoconversion between imidazolium and imidazolinium. J Am Chem Soc 130:14570–14575
310. Fukumoto S, Nakashima T, Kawai T (2011) Photon-quantitative reaction of a dithiazolylarylene in solution. Angew Chem Int Ed 50:1565–1568
311. Frigoli M, Mehl GH (2005) Multiple addressing in a hybrid biphotochromic system. Angew Chem Int Ed 44:5048–5052
312. Corredor CC, Huang ZL, Belfield KD (2006) Two-photon 3D optical data storage *via* fluorescence modulation of an efficient fluorene dye by a photochromic diarylethene. Adv Mater 18:2910–2914
313. Göstl R, Hecht S (2014) Controlling covalent connection and disconnection with light. Angew Chem Int Ed 53:8784–8787
314. Yagai S, Iwai K, Yamauchi M, Karatsu T, Kitamura A, Uemura S, Morimoto M, Wang H, Würthner F (2014) Photocontrol over self-assembled nanostructures of π–π stacked dyes supported by the parallel conformer of diarylethene. Angew Chem Int Ed 53:2602–2606
315. Li C, Yan H, Zhao LX, Zhang GF, Hu Z, Huang ZL, Zhu MQ (2014) A trident dithienylethene-perylenemonoimide dyad with super fluorescence switching speed and ratio. Nat Commun 5:5709
316. Gemayel ME, Börjesson K, Herder M, Duong DT, Hutchison JA, Ruzié C, Schweicher G, Salleo A, Geerts Y, Hecht S, Orgiu E, Samori P (2015) Optically switchable transistors by simple incorporation of photochromic systems into small-molecule semiconducting matrices. Nat Commun 6:6330
317. Fernández-Acebes A, Lehn JM (1998) Optical switching and fluorescence modulation in photochromic metal complexes. Adv Mater 10:1519–1522

318. Fernandez-Acebes A, Lehn JM (1999) Optical switching and fluorescence modulation properties of photochromic metal complexes derived from dithienylethene ligands. Chem Eur J 5:3285–3292
319. Takayama K, Matsuda K, Irie M (2003) Photoswitching of the magnetic interaction between a copper(II) ion and a nitroxide radical by using a photochromic spin coupler. Chem Eur J 9:5605–5609
320. Matsuda K, Shinkai Y, Irie M (2004) Photochromism of metal complexes composed of diarylethene ligands and $ZnCl_2$. Inorg Chem 43:3774–3776
321. Matsuda K, Takayama K, Irie M (2004) Photochromism of metal complexes composed of diarylethene ligands and Zn(II), Mn(II), and Cu(II) hexafluoroacetylacetonates. Inorg Chem 43:482–489
322. Lee JKW, Ko CC, Wong KMC, Zhu N, Yam VWW (2007) A photochromic platinum(II) bis-(alkynyl) complex containing a versatile 5,6-dithienyl-1,10-phenanthroline. Organometallics 26:12–15
323. Lee PHM, Ko CC, Zhu N, Yam VWW (2007) Metal coordination-assisted near-infrared photochromic behavior: a large perturbation on absorption wavelength properties of *N,N*-donor ligands containing diarylethene derivatives by coordination to the rhenium(I) metal center. J Am Chem Soc 129:6058–6059
324. Ngan TW, Ko CC, Zhu N, Yam VWW (2007) Syntheses, luminescence switching, and electrochemical studies of photochromic dithienyl-1,10-phenanthroline zinc(II) bis(thiolate) complexes. Inorg Chem 46:1144–1152
325. Yam VWW, Lee JKW, Ko CC, Zhu N (2009) Photochromic diarylethene-containing ionic liquids and *N*-heterocyclic carbene. J Am Chem Soc 131:912–913
326. Duan G, Yam VWW (2010) Syntheses and photophysical properties of *N*-pyridylimidazol-2-ylidene tetracyanoruthenates(II) and photochromic studies of their dithienylethene-containing derivatives. Chem Eur J 16:12642–12649
327. Wong HL, Tao CH, Zhu N, Yam VWW (2011) Photochromic alkynes as versatile building blocks for metal alkynyl systems: design, synthesis, and photochromic studies of diarylethene-containing platinum(II) phosphine alkynyl complexes. Inorg Chem 50:471–481
328. Duan G, Wong WT, Yam VWW (2011) Synthesis and photochromic studies of η^6-mesitylene ruthenium(II) complexes bearing *N*-heterocyclic carbene ligands with the dithienylethene moiety. New J Chem 35:2267–2278
329. Chan JCH, Lam WH, Wong HL, Zhu N, Wong WT, Yam VWW (2011) Diarylethene-containing cyclometalated platinum(II) complexes: tunable photochromism *via* metal coordination and rational ligand design. J Am Chem Soc 133:12690–12705
330. Wong HL, Zhu N, Yam VWW (2014) Photochromic alkynylplatinum(II) diimine complexes containing a versatile dithienylethene-functionalized 2-(2′-pyridyl)imidazole ligand. J Organomet Chem 751:430–437
331. Murguly E, Norsten TB, Branda NR (2001) Nondestructive data processing based on chiroptical 1,2-dithienylethene photochromes. Angew Chem Int Ed 40:1752–1755
332. Norsten TB, Branda NR (2001) Axially coordinated porphyrinic photochromes for non-destructive information processing. Adv Mater 13:347–349
333. Zhao H, Al-Atar U, Pace TCS, Bohne C, Branda NR (2008) High-contrast fluorescence switching using a photoresponsive dithienylethene coordination compound. J Photochem Photobiol A 200:74–82
334. Tian H, Chen B, Tu H, Müllen K (2002) Novel bisthienylethene-based photochromic tetraazaporphyrin with photoregulating luminescence. Adv Mater 14:918–923
335. Luo Q, Chen B, Wang M, Tian H (2003) Mono-bisthienylethene ring-fused versus multi-bisthienylethene ring-fused photochromic hybrids. Adv Funct Mater 13:233–239
336. Chen S, Chen LJ, Yang HB, Tian H, Zhu W (2012) Light-triggered reversible supramolecular transformations of multi-bisthienylethene hexagons. J Am Chem Soc 134:13596–13599

337. Munakata M, Wu LP, Kuroda-Sowa T, Maekawa M, Suenaga Y, Furuichi K (1996) Reversible photochromism of a crystalline dithienylethene copper(I) polymer. J Am Chem Soc 118:3305–3306
338. Kim HJ, Jang JH, Choi H, Lee T, Ko J, Yoon M, Kim HJ (2008) Photoregulated fluorescence switching in axially coordinated tin(IV) porphyrinic dithienylethene. Inorg Chem 47:2411–2415
339. Indelli MT, Carli S, Ghirotti M, Chiorboli C, Ravaglia M, Garavelli M, Scandola F (2008) Triplet pathways in diarylethene photochromism: photophysical and computational study of dyads containing ruthenium(II) polypyridine and 1,2-bis(2-methylbenzothiophene-3-yl)-maleimide units. J Am Chem Soc 130:7286–7299
340. Tan W, Zhang Q, Zhang J, Tian H (2009) Near-infrared photochromic diarylethene iridium (III) complex. Org Lett 11:161–164
341. Monaco S, Semeraro M, Tan W, Tian H, Ceroni P, Credi A (2012) Multifunctional switching of a photo- and electro-chemiluminescent iridium–dithienylethene complex. Chem Commun 48:8652–8654
342. Nakagawa T, Hasegawa Y, Kawai T (2009) Nondestructive luminescence intensity readout of a photochromic lanthanide(III) complex. Chem Commun 5630–5632
343. Roberts MN, Carling CJ, Nagle JK, Branda NR, Wolf MO (2009) Successful bifunctional photoswitching and electronic communication of two platinum(II) acetylide bridged dithienylethenes. J Am Chem Soc 131:16644–16645
344. Roberts MN, Nagle JK, Finden JG, Branda NR, Wolf MO (2009) Linker-dependent metal-sensitized photoswitching of dithienylethenes. Inorg Chem 48:19–21
345. Roberts MN, Nagle JK, Majewski MB, Finden JG, Branda NR, Wolf MO (2011) Charge transfer and intraligand excited state interactions in platinum-sensitized dithienylethenes. Inorg Chem 50:4956–4966
346. Lin Y, Yin J, Yuan J, Hu M, Li Z, Yu GA, Liu SH (2010) Synthesis, characterization, and properties of binuclear gold(I) phosphine alkynyl complexes. Organometallics 29:2808–2814
347. Harvey EC, Areephong J, Cafolla AA, Long C, Brown WR, Feringa BL, Pryce MT (2014) Incorporating cobalt carbonyl moieties onto ethynylthiophene-based dithienylcyclopentene switches. 1. Photochemistry. Organometallics 33:447–456
348. Zhong YW, Vilà N, Henderson JC, Flores-Torres S, Abruña HD (2007) Dinuclear transition-metal terpyridine complexes with a dithienylcyclopentene bridge directed toward molecular electronic applications. Inorg Chem 46:10470–10472
349. Zhong YW, Vilà N, Henderson JC, Abruña HD (2009) Dithienylcyclopentenes-containing transition metal bisterpyridine complexes directed toward molecular electronic applications. Inorg Chem 48:991–999
350. Zhong YW, Vilà N, Henderson JC, Abruña HD (2009) Transition-metal tris-bipyridines containing three dithienylcyclopentenes: synthesis, photochromic, and electrochromic properties. Inorg Chem 48:7080–7085
351. Aubert V, Ordronneau L, Escadeillas M, Williams JAG, Boucekkine A, Coulaud E, Dragonetti C, Righetto S, Roberto D, Ugo R, Valore A, Singh A, Zyss J, Ledoux-Rak I, Le Bozec H, Guerchais V (2011) Linear and nonlinear optical properties of cationic bipyridyl iridium(III) complexes: tunable and photoswitchable? Inorg Chem 50:5027–5038
352. Bao Z, Ng KY, Yam VWW, Ko CC, Zhu N, Wu L (2008) Syntheses, characterization, and photochromic studies of spirooxazine-containing 2,2′-bipyridine ligands and their zinc (II) thiolate complexes. Inorg Chem 47:8912–8920
353. Li Y, Tam AYY, Wong KMC, Li W, Wu L, Yam VWW (2011) Synthesis, characterization, and the photochromic, luminescence, metallogelation and liquid-crystalline properties of multifunctional platinum(II) bipyridine complexes. Chem Eur J 17:8048–8059
354. Paquette MM, Patrick BO, Frank NL (2011) Determining the magnitude and direction of photoinduced ligand field switching in photochromic metal–organic complexes: molybdenum–tetracarbonyl spirooxazine complexes. J Am Chem Soc 133:10081–10093

355. Rao YL, Amame H, Wang S (2012) Photochromic four-coordinate N,C-chelate boron compounds. Coord Chem Rev 256:759–770
356. Wang N, Ko SB, Lu JS, Chen LD, Wang S (2013) Tuning the photoisomerization of a N^C-chelate organoboron compound with a metal acetylide unit. Chem Eur J 19:5314–5323
357. Marchi E, Baroncini M, Bergamini G, van Heyst J, Vögtle F, Ceroni P (2012) Photoswitchable metal coordinating tweezers operated by light-harvesting dendrimers. J Am Chem Soc 134:15277–15280
358. Yao LY, Yam VWW (2015) Photoinduced isomerization-driven structural transformation between decanuclear and octadecanuclear gold(I) sulfido clusters. J Am Chem Soc 137:3506–3509
359. Kärnbratt J, Hammarson M, Li S, Anderson HL, Albinsson B, Andréasson J (2010) Photochromic supramolecular memory with nondestructive readout. Angew Chem Int Ed 49:1854–1857
360. Nilsson JR, O'Sullivan MC, Li S, Anderson HL, Andréasson J (2015) A photoswitchable supramolecular complex with release-and-report capabilities. Chem Commun 51:847–850
361. Namiki K, Sakamoto A, Murata M, Kume S, Nishihara H (2007) Reversible photochromism of a ferrocenylazobenzene monolayer controllable by a single green light source. Chem Commun 4650–4652
362. Tanaka Y, Inagaki A, Akita M (2007) A photoswitchable molecular wire with the dithienylethene (DTE) linker, (dppe)(μ^5-C_5Me_5)Fe–C≡C–DTE–C≡C–Fe(μ^5-C_5Me_5)(dppe). Chem Commun 1169–1171
363. Tanaka Y, Ishisaka T, Inagaki A, Koike T, Lapinte C, Akita M (2010) Photochromic organometallics with a dithienylethene (DTE) bridge, [Y−C≡C−DTE−C≡C−Y] (Y={MCp*(dppe)}): photoswitchable molecular wire (M=Fe) versus dual photo- and electrochromism (M=Ru). Chem Eur J 16:4762–4776
364. Motoyama K, Koike T, Akita M (2008) Remarkable switching behavior of bimodally stimuli-responsive photochromic dithienylethenes with redox-active organometallic attachments. Chem Commun 5812–5814
365. Lee S, You Y, Ohkubo K, Fukuzumi S, Nam W (2012) Photoelectrocatalysis to improve cycloreversion quantum yields of photochromic dithienylethene compounds. Angew Chem Int Ed 51:13154–13158
366. Lee S, You Y, Ohkubo K, Fukuzumi S, Nam W (2014) Highly efficient cycloreversion of photochromic dithienylethene compounds using visible light-driven photoredox catalysis. Chem Sci 5:1463–1474
367. Paquette MM, Kopelman RA, Beitler E, Frank NL (2009) Incorporating optical bistability into a magnetically bistable system: a photochromic redox isomeric complex. Chem Commun 5424–5426
368. Witt A, Heinemann FW, Khusniyarov MM (2015) Bidirectional photoswitching of magnetic properties at room temperature: ligand-driven light-induced valence tautomerism. Chem Sci 6:4599–4609
369. Morimoto M, Miyasaka H, Yamashita M, Irie M (2009) Coordination assemblies of [Mn_4] single-molecule magnets linked by photochromic ligands: photochemical control of the magnetic properties. J Am Chem Soc 131:9823–9835
370. Pinkowicz D, Ren M, Zheng LM, Sato S, Hasegawa M, Morimoto M, Irie M, Breedlove BK, Cosquer G, Katoh K, Yamashita M (2014) Control of the single-molecule magnet behavior of lanthanide-diarylethene photochromic assemblies by irradiation with light. Chem Eur J 20:12502–12513
371. Cosquer G, Morimoto M, Irie M, Fetoh A, Breedlove BK, Yamashita M (2015) Photo-control of the magnetic properties of Dy(III) and Ho(III) homometal coordination polymers bridged by a diarylethene ligand. Dalton Trans 44:5996–6002
372. Nihei M, Suzuki Y, Kimura N, Kera Y, Oshio H (2013) Bidirectional photomagnetic conversions in a spin-crossover complex with a diarylethene moiety. Chem Eur J 19:6946–6949

373. Milek M, Heinemann FW, Khusniyarov MM (2013) Spin crossover meets diarylethenes: efficient photoswitching of magnetic properties in solution at room temperature. Inorg Chem 52:11585–11592
374. Venkataramani S, Jana U, Dommaschk M, Sönnichsen FD, Tuczek F, Herges R (2011) Magnetic bistability of molecules in homogeneous solution at room temperature. Science 331:445–448
375. Thies S, Sell H, Bornholdt C, Schütt C, Köhler F, Tuczek F, Herges R (2012) Light-driven coordination-induced spin-state switching: rational design of photodissociable ligands. Chem Eur J 18:16358–16368
376. Aubert V, Guerchais V, Ishow E, Hoang-Thi K, Ledoux I, Nakatani K, Le Bozec H (2008) Efficient photoswitching of the nonlinear optical properties of dipolar photochromic zinc (II) complexes. Angew Chem Int Ed 47:577–580
377. Nitadori H, Ordronneau L, Boixel J, Jacquemin D, Boucekkine A, Singh A, Akita M, Ledoux I, Guerchais V, Le Bozec H (2012) Photoswitching of the second-order nonlinearity of a tetrahedral octupolar multi DTE-based copper(I) complex. Chem Commun 48:10395–10397
378. Ordronneau L, Aubert V, Guerchais V, Boucekkine A, Le Bozec H, Singh A, Ledoux I, Jacquemin D (2013) The first hexadithienylethene-substituted tris(bipyridine)metal complexes as quadratic NLO photoswitches: combined experimental and DFT studies. Chem Eur J 19:5845–5849
379. Boixel J, Guerchais V, Le Bozec H, Jacquemin D, Amar A, Boucekkine A, Colombo A, Dragonetti C, Marinotto D, Roberto D, Righetto S, De Angelis R (2014) Second-order NLO switches from molecules to polymer films based on photochromic cyclometalated platinum (II) complexes. J Am Chem Soc 136:5367–5375
380. Petitjean A, Khoury RG, Kyritsakas N, Lehn JM (2004) Dynamic devices. Shape switching and substrate binding in ion-controlled nanomechanical molecular tweezers. J Am Chem Soc 126:6637–6647
381. Lehn JM (2007) From supramolecular chemistry towards constitutional dynamic chemistry and adaptive chemistry. Chem Soc Rev 36:151–160
382. Greb L, Lehn JM (2014) Light-driven molecular motors: imine as four-step or two-step unidirectional rotors. J Am Chem Soc 136:13114–13117
383. Sauvage JP (1998) Transition metal-containing rotaxanes and catenanes in motion: toward molecular machines and motors. Acc Chem Res 31:611–619
384. Mobian P, Kern JM, Sauvage JP (2004) Light-driven machine prototypes based on dissociative excited states: photoinduced decoordination and thermal recoordination of a ring in a ruthenium(II)-containing [2]catenane. Angew Chem Int Ed 43:2392–2395
385. Champin B, Mobian P, Sauvage JP (2007) Transition metal complexes as molecular machine prototypes. Chem Soc Rev 36:358–366
386. Durola F, Sauvage JP (2007) Fast electrochemically induced translation of the ring in a copper-complexed [2]rotaxane: the biisoquinoline effect. Angew Chem Int Ed 46:3537–3540
387. Durot S, Reviriego F, Sauvage JP (2010) Copper-complexed catenanes and rotaxanes in motion: 15 years of molecular machines. Dalton Trans 39:10557–10570
388. Collin JP, Dietrich-Buchecker CD, Gaviña P, Jimenez-Molero MC, Sauvage JP (2001) Shuttles and muscles: linear molecular machines based on transition metals. Acc Chem Res 34:477–487
389. Collin JP, Durola F, Lux J, Sauvage JP (2009) A rapidly shuttling copper-complexed [2]-rotaxane with three different chelating groups in its axis. Angew Chem Int Ed 48:8532–8535
390. Joosten A, Trolez Y, Collin JP, Heitz V, Sauvage JP (2012) Copper(I)-assembled [3]rotaxane whose two rings act as flapping wings. J Am Chem Soc 134:1802–1809
391. Saha S, Stoddart JF (2007) Photo-driven molecular devices. Chem Soc Rev 36:77–93
392. Coskun A, Banaszak M, Astumian RD, Stoddart JF, Grzybowski BA (2012) Great expectations: can artificial molecular machines deliver on their promise? Chem Soc Rev 41:19–30

393. Bruns CJ, Stoddart JF (2014) Rotaxane-based molecular muscles. Acc Chem Res 47:2186–2199
394. Schmittel M, De S, Pramanik S (2012) Reversible on/off nanoswitch for organocatalysis: mimicking the locking and unlocking operation of CaMKII. Angew Chem Int Ed 51:3832–3836
395. Pramanik S, De S, Schmittel M (2014) Bidirectional chemical communication between nanomechanical switches. Angew Chem Int Ed 53:4709–4713
396. Ballardini R, Balzani V, Credi A, Gandolfi MT, Venturi M (2001) Artificial molecular-level machines: which energy to make them work? Acc Chem Res 34:445–455
397. Balzani V, Credi A, Silvi S, Venturi M (2006) Artificial nanomachines based on interlocked molecular species: recent advances. Chem Soc Rev 35:1135–1149
398. Balzani V, Bergamini G, Ceroni P (2008) From the photochemistry of coordination compounds to light-powered nanoscale devices and machines. Coord Chem Rev 252:2456–2469
399. Koumura N, Zijistra RWJ, van Delden RA, Harada N, Feringa BL (1999) Light-driven monodirectional molecular rotor. Nature 401:152–155
400. Feringa BL (2001) In control of motion: from molecular switches to molecular motors. Acc Chem Res 34:504–513
401. Fletcher SP, Dumur F, Pollard MM, Feringa BL (2005) A reversible, unidirectional molecular rotary motor driven by chemical energy. Science 310:80–82
402. Vicario J, Katsonis N, Ramon BS, Bastiaansen CWM, Broer DJ, Feringa BL (2006) Nanomotor rotates microscale objects. Nature 440:163
403. Bonnet S, Collin JP (2008) Ruthenium-based light-driven molecular machine prototypes: synthesis and properties. Chem Soc Rev 37:1207–1217
404. Muraoka T, Kinbara K, Kobayashi Y, Aida T (2003) Light-driven open–close motion of chiral molecular scissors. J Am Chem Soc 125:5612–5613
405. Kinbara K, Aida T (2005) Toward intelligent molecular machines: directed motions of biological and artificial molecules and assemblies. Chem Rev 105:1377–1400
406. Muraoka T, Kinbara K, Aida T (2006) Mechanical twisting of a guest by a photoresponsive host. Nature 440:512–515
407. Muraoka T, Kinbara K, Aida T (2006) A self-locking molecule operative with a photoresponsive key. J Am Chem Soc 128:11600–11605
408. Durola F, Rebek J (2010) The ouroborand: a cavitand with a coordination-driven switching device. Angew Chem Int Ed 49:3189–3191
409. Brouwer AM, Frochot C, Gatti FG, Leigh DA, Mottier L, Paolucci F, Roffia S, Wurpel GWH (2001) Photoinduction of fast, reversible translational motion in a hydrogen-bonded molecular shuttle. Science 291:2124–2128
410. Amendola V, Fabbrizzi L, Mangano C, Pallavicini P (2001) Molecular machines based on metal ion translocation. Acc Chem Res 34:488–493
411. Hernández JV, Kay ER, Leigh DA (2004) A reversible synthetic rotary molecular motor. Science 306:1532–1537
412. Serreli V, Lee CF, Kay ER, Leigh DA (2007) A molecular information ratchet. Nature 445:523–527
413. Kume S, Nomoto K, Kusamoto T, Nishihara H (2009) Intramolecular electron arrangement with a rotative trigger. J Am Chem Soc 131:14198–14199
414. Canary JW, Mortezaei S, Liang J (2010) Transition metal-based chiroptical switches for nanoscale electronics and sensors. Coord Chem Rev 254:2249–2266
415. Nishikawa M, Nomoto K, Kume S, Nishihara H (2012) Reversible copper(II)/(I) electrochemical potential switching driven by visible light-induced coordinated ring rotation. J Am Chem Soc 134:10543–10553
416. van Dongen SFM, Cantekin S, Elemans JAAW, Rowan AE, Nolte RJM (2014) Functional interlocked systems. Chem Soc Rev 43:99–122
417. McConnell AJ, Wood CS, Neelakandan PP, Nitschke JR (2015) Stimuli-responsive metal–ligand assemblies. Chem Rev 115:7729–7793

418. Li H, Fahrenbach AC, Coskun A, Zhu Z, Barin G, Zhao YL, Botros YY, Sauvage JP, Stoddart JF (2011) A light-stimulated molecular switch driven by radical–radical interactions in water. Angew Chem Int Ed 50:6782–6788
419. Samanta SK, Schmittel M (2013) Four-component supramolecular nanorotors. J Am Chem Soc 135:18794–18797
420. Samanta SK, Bats JW, Schmittel M (2014) A five-component nanorotor with speed regulation. Chem Commun 50:2364–2366
421. Vicario J, Walko M, Meetsma A, Feringa BL (2006) Fine tuning of the rotary motion by structural modification in light-driven unidirectional molecular motors. J Am Chem Soc 128:5127–5135
422. Cnossen A, Hou L, Pollard MM, Wesenhagen V, Browne WR, Feringa BL (2012) Driving unidirectional molecular rotary motors with visible light by intra- and intermolecular energy transfer from palladium porphyrin. J Am Chem Soc 134:17613–17619
423. Wezenberg SJ, Cen KY, Feringa BL (2015) Visible-light-driven photoisomerization and increased rotation speed of a molecular motor acting as a ligand in a ruthenium (II) complex. Angew Chem Int Ed 54:11457–11461
424. Zhao D, Neubauer TM, Feringa BL (2015) Dynamic control of chirality in phosphine ligands for enantioselective catalysis. Nat Commun 6:6652
425. Yamashita KI, Kawano M, Fujjita M (2007) Photoswitchable molecular lock. One-way catenation of a Pt(II)-linked coordination ring *via* the photolabilization of a Pt(II)–pyridine bond. J Am Chem Soc 129:1850–1851
426. Dube H, Ajami D, Rebek J Jr (2010) Photochemical control of reversible encapsulation. Angew Chem Int Ed 49:3192–3195
427. Heinz T, Rudkevich DM, Rebek J Jr (1998) Pairwise selection of guests in a cylindrical molecular capsule of nanometre dimensions. Nature 394:764–766
428. Dube H, Rebek J Jr (2012) Selective guest exchange in encapsulation complexes using light of different wavelengths. Angew Chem Int Ed 51:3207–3210
429. Han M, Michel R, He B, Chen YS, Stalke D, John M, Clever GH (2013) Light-triggered guest uptake and release by a photochromic coordination cage. Angew Chem Int Ed 52:1319–1323
430. Lo KKW (2007) Luminescent transition metal complexes as biological labels and probes. Struct Bond 123:205–245
431. Lo KKW, Choi AWT, Law WHT (2012) Applications of luminescent inorganic and organometallic transition metal complexes as biomolecular and cellular probes. Dalton Trans 41:6021–6047
432. Yu C, Chan KHY, Wong KMC, Yam VWW (2006) Single-stranded nucleic acid-induced helical self-assembly of alkynylplatinum(II) terpyridyl complexes. Proc Natl Acad Sci U S A 103:19652–19657
433. Yu C, Chan KHY, Wong KMC, Yam VWW (2009) Nucleic acid-induced self-assembly of a platinum(II) terpyridyl complex: detection of G-quadruplex formation and nuclease activity. Chem Commun 3756–3758
434. Chung CYS, Yam VWW (2013) Selective label-free detection of G-quadruplex structure of human telomere by emission spectral changes in visible-and-NIR region under physiological condition through the FRET of a two-component PPE-SO_3^-–Pt(II) complex ensemble with Pt···Pt, electrostatic and π–π interactions. Chem Sci 4:377–387
435. Gill MR, Garcia-Lara J, Foster SJ, Smythe C, Battaglia G, Thomas JA (2009) A ruthenium (II) polypyridyl complex for direct imaging of DNA structure in living cells. Nat Chem 1:662–667
436. O'Connor NA, Stevens N, Samaroo D, Solomon MR, Martí AA, Dyer J, Vishwasrao H, Akins DL, Kandel ER, Turro NJ (2009) A covalently linked phenanthridine-ruthenium(II) complex as a RNA probe. Chem Commun 2640–2642
437. Murphy L, Congreve A, Pålsson LO, Williams JAG (2010) The time domain in co-stained cell imaging: time-resolved emission imaging microscopy using a protonatable luminescent iridium complex. Chem Commun 46:8743–8745

438. Zhang R, Ye Z, Wang G, Zhang W, Yuan J (2010) Development of a ruthenium(II) complex based luminescent probe for imaging nitric oxide production in living cells. Chem Eur J 16:6884–6891
439. Li C, Yu M, Sun Y, Wu Y, Huang C, Li F (2011) A nonemissive iridium(III) complex that specifically lights-up the nuclei of living cells. J Am Chem Soc 133:11231–11239
440. Yoshihara T, Yamaguchi Y, Hosaka M, Takeuchi T, Tobita S (2012) Ratiometric molecular sensor for monitoring oxygen levels in living cells. Angew Chem Int Ed 51:4148–4151
441. Jing J, Zhang JL (2013) Combining myeloperoxidase (MPO) with fluorogenic ZnSalen to detect lysosomal hydrogen peroxide in live cells. Chem Sci 4:2947–2952
442. Zou T, Lum CT, Chui SSY, Che CM (2013) Gold(III) complexes containing N-heterocyclic carbene ligands: thiol "switch-on" fluorescent probes and anti-cancer agents. Angew Chem Int Ed 52:2930–2933
443. Chung CYS, Li SPY, Louie MW, Lo KKW, Yam VWW (2013) Induced self-assembly and disassembly of water-soluble alkynylplatinum(II) terpyridyl complexes with "switchable" near-infrared (NIR) emission modulated by metal–metal interactions over physiological pH: demonstration of pH-responsive NIR luminescent probes in cell-imaging studies. Chem Sci 4:2453–2462
444. Tsai JLL, Zou T, Liu J, Chen T, Chan AOY, Yang C, Lok CN, Che CM (2015) Luminescent platinum(II) complexes with self-assembly and anti-cancer properties: hydrogel, pH dependent emission color and sustained-release properties under physiological conditions. Chem Sci 4:3823–3830
445. Puckett CA, Barton JK (2007) Methods to explore cellular uptake of ruthenium complexes. J Am Chem Soc 129:46–47
446. Puckett CA, Barton JK (2009) Fluorescein redirects a ruthenium-octaarginine conjugate to the nucleus. J Am Chem Soc 131:8738–8739
447. Svensson FR, Matson M, Li M, Lincoln P (2010) Lipophilic ruthenium complexes with tuned cell membrane affinity and photoactivated uptake. Biophys Chem 149:102–106
448. Zhao Q, Huang C, Li F (2011) Phosphorescent heavy-metal complexes for bioimaging. Chem Soc Rev 40:2508–2524
449. Fernández-Moreira V, Thorp-Greenwood FL, Coogan MP (2010) Application of d^6 transition metal complexes in fluorescence cell imaging. Chem Commun 46:186–202
450. Thorp-Greenwood FL, Balasingham RG, Coogan MP (2012) Organometallic complexes of transition metals in luminescent cell imaging applications. J Organomet Chem 714:12–21
451. Geldmacher Y, Oleszak M, Sheldrick WS (2012) Rhodium(III) and iridium(III) complexes as anticancer agents. Inorg Chim Acta 393:84–102
452. Stacey OJ, Pope SJA (2013) New avenues in the design and potential application of metal complexes for photodynamic therapy. RSC Adv 3:25550–25564
453. Li SPY, Lau CTS, Louie MW, Lam YW, Cheng SH, Lo KKW (2013) Mitochondria-targeting cyclometalated iridium(III)–PEG complexes with tunable photodynamic activity. Biomaterials 34:7519–7532
454. Moromizato S, Hisamatsu Y, Suzuki T, Matsuo Y, Abe R, Aoki S (2012) Design and synthesis of a luminescent cyclometalated iridium(III) complex having *N*, *N*-diethylamino group that stains acidic intracellular organelles and induces cell death by photoirradiation. Inorg Chem 51:12697–12706
455. Kastl A, Wilbuer A, Merkel AL, Feng L, Fazio PD, Ocker M, Meggers E (2012) Dual anticancer activity in a single compound: visible-light-induced apoptosis by an antiangiogenic iridium complex. Chem Commun 48:1863–1865
456. Man BYW, Chan HM, Leung CH, Chan DSH, Bai LP, Jiang ZH, Li HW, Ma DL (2011) Group 9 metal-based inhibitors of β-amyloid (1–40) fibrillation as potential therapeutic agents for Alzheimer's disease. Chem Sci 2:917–921
457. Schmidbaur H (1995) Ludwig Mond lecture. High-cart gold compounds. Chem Soc Rev 24:391–400

458. Pyykkö P (1997) Strong closed-shell interactions in inorganic chemistry. Chem Rev 97:597–636
459. Vogler A, Kunkely H (1988) Absorption and emission spectra of tetrameric gold (I) complexes. Chem Phys Lett 150:135–137
460. Puddephatt RJ (2001) Coordination polymers: polymers, rings and oligomers containing gold (I) centres. Coord Chem Rev 216–217:313–332
461. Yam VWW, Cheng ECC (2008) Highlights on the recent advances in gold chemistry – a photophysical perspective. Chem Soc Rev 37:1806–1813
462. Yam VWW, Au VKM, Leung SYL (2015) Light-emitting self-assembled materials on d^8 and d^{10} transition metal complexes. Chem Rev 115:7589–7728
463. Balch AL (2007) Remarkable luminescence behaviors and structural variations of two-coordinate gold(I) complexes. Struct Bond 123:1–40
464. Schmidbaur H, Schier A (2012) Aurophilic interactions as a subject of current research: an up-date. Chem Soc Rev 41:370–412
465. He X, Lam WH, Zhu N, Yam VWW (2009) Design and synthesis of calixarene-based bis-alkynyl-bridged dinuclear Au^I isonitrile complexes as luminescent ion probes by the modulation of Au···Au interactions. Chem Eur J 15:8842–8851
466. Hau FKW, He X, Lam WH, Yam VWW (2011) Highly selective ion probe for Al^{3+} based on Au(I)···Au(I) interactions in a bis-alkynyl calix[4]arene Au(I) isocyanide scaffold. Chem Commun 47:8778–8780
467. He X, Cheng ECC, Zhu N, Yam VWW (2009) Selective ion probe for Mg^{2+} based on Au(I)··· Au(I) interactions in a tripodal alkynylgold(I) complex with oligoether pendants. Chem Commun 4016–4018
468. Zhou YP, Liu EB, Wang J, Chao HY (2013) Highly Ag^+ selective tripodal gold(I) acetylide-based "off–on" luminescence chemosensors based on $^3(\pi\pi^*)$ emission switching. Inorg Chem 52:8629–8637
469. Lintang HO, Kinbara K, Tanaka K, Yamashita T, Aida T (2010) Self-repair of a one-dimensional molecular assembly in mesoporous silica by a nanoscopic template effect. Angew Chem Int Ed 49:4241–4245
470. Lintang HO, Kinbara K, Yamashita T, Aida T (2012) Metal-ion permeation in congested nanochannels: the exposure effect of Ag^+ ions on the phosphorescent properties of a gold(I)-pyrazolate complex that is confined in the nanoscopic channels of mesoporous silica. Chem Asian J 7:2068–2072
471. Lee TKM, Zhu N, Yam VWW (2010) An unprecedented luminescent polynuclear gold(I) μ_3-sulfido cluster with a thiacrown-like architecture. J Am Chem Soc 132:17646–17648
472. Jiang XF, Hau FKW, Sun QF, Yu SY, Yam VWW (2014) From {Au^I···Au^I}-coupled cages to the cage-built 2-D {Au^I···Au^I} arrays: Au^I···Au^I bonding interaction driven self-assembly and their Ag^I sensing and photo-switchable behavior. J Am Chem Soc 136:10921–10929
473. Han S, Yoon YY, Jung OS, Lee YA (2011) Luminescence on–off switching *via* reversible interconversion between inter- and intramolecular aurophilic interactions. Chem Commun 47:10689–10691
474. Koshevoy IO, Lin CL, Karttunen AJ, Haukka M, Shih CW, Chou PT, Tunik SP, Pakkanen TA (2011) Octanuclear gold(I) alkynyl-phosphine clusters showing thermochromic luminescence. Chem Commun 47:5533–5535
475. Elbjeirami O, Omary MA (2007) Photochemistry of neutral isonitrile gold(I) complexes: modulation of photoreactivity by aurophilicity and π-acceptance ability. J Am Chem Soc 129:11384–11393
476. Lei Z, Pei XL, Jiang ZG, Wang QM (2014) Cluster linker approach: preparation of a luminescent porous framework with NbO topology by linking silver ions with gold (I) cluster. Angew Chem Int Ed 53:12771–12775

Struct Bond (2016) 172: 291–312
DOI: 10.1007/430_2015_195

Published online: 14 February 2016

Spectroscopy and Chemical Bonding in Transition Metal Complexes

Andreas Hauser and Christian Reber

Abstract Optical spectroscopy of transition metal complexes plays an important role in establishing excited-state electronic and nuclear structures and thus in the elucidation of the multitude of photophysical and photochemical relaxation processes. The most important advances in this area of research over the past decade are due to the development of new experimental techniques such as ultrafast spectroscopy as well as structure determination in conjunction with other methods such as high-pressure and variable temperature techniques. In this contribution, several paradigmatic systems, namely, of complexes of chromium(III), iron(II), ruthenium(II), nickel(II), platinum(II) and palladium(II), are discussed with regard to their excited electronic and nuclear structures and photophysical relaxation processes.

Keywords High pressure • Intersystem crossing • Photophysics and photochemistry • Spectroscopy • Spin crossover • Transition metal complexes • Ultrafast methods

Contents

A. Hauser (✉)
Département de chimie physique, Université de Genève, 1211 Genève, Switzerland
e-mail: andreas.hauser@unige.ch

C. Reber (✉)
Département de chimie, Université de Montréal, Montréal, QC H3C 3J7, Canada
e-mail: christian.reber@umontreal.ch

1 Introduction

Volumes 106 and 107 of *Structure and Bonding* entitled "Optical Spectra and Chemical Bonding in Inorganic Compounds" [1] edited by Th. Schönherr were dedicated to the memory of Christian Klixbüll Jørgensen, author of the very first article in Vol. 1 of *Structure and Bonding*, and his contributions to electronic structure theory in compounds containing transition metal ions and lanthanides [2–5]. Optical spectroscopy is the tool of choice to investigate both the complex electronic and nuclear structures and thus chemical bonding in excited states of these compounds. This is essential for understanding their photophysical properties, which find applications, for instance, in dye-sensitised solar cells [6, 7], lighting in lamps based on mercury discharge excitation [8] or in OLEDS [9, 10], in solid-state lasers [11] and in biomedical research as fluorescent markers and for phototherapy [12, 13]. The field of photophysics and photochemistry of transition metal complexes and compounds is vast and still very active as borne out by important international conferences [14] and special issues of peer-reviewed journals [15] and monographs dedicated to this topic [16–18].

The most important advances over the past 10 years in the field came with the development of ultrafast spectroscopic [19, 20] and structure determining methods [21, 22] for the investigation of photophysical processes down to the femtosecond timescale. Together with the advancement of mostly density functional theory (DFT)-based computational approaches for open-shell systems [23–25], this has resulted in a giant step forward with regard to the understanding of the sequence of events and the dynamics of elementary photophysical steps following the initial absorption of a photon, from intramolecular vibrational relaxation and vibrational cooling [26] to internal conversion and intersystem crossing [27, 28] and from light-induced excitation energy ([29] and references therein) and electron transfer [16–18] to proton-coupled electron transfer [30, 31] and photochemical reactions [32–34] involving transition metal complexes.

It is of course impossible to cover all of the above aspects in the comparatively restricted space available for the topic of optical spectroscopy and chemical bonding allocated in this anniversary volume of *Structure and Bonding*. In the following we shall show how the understanding of the excited-state electronic and nuclear structures and the photophysical properties of transition metal compounds has evolved over the past decade as a result of the abovementioned new experimental techniques. We will begin with the historically important and comparatively simple d^3 chromium(III) systems and go on to discuss the more complex and fascinating iron(II) complexes, with their many low-lying ligand-field states leading to temperature-, pressure- and light-induced spin crossover. The latter will naturally lead to the other d^6 transition metal ion, namely, ruthenium(II) and a discussion of the luminescence quenching by a low-lying ligand-field state in the family of polypyridyl complexes. Finally, square-planar d^8 nickel(II), palladium(II) and platinum(II) complexes are very susceptible to the application of external pressure, and corresponding experiments can teach us a lot on ground- and excited-state geometries and bonding in such complexes.

2 Paradigmatic Case Studies

2.1 *Chromium(III)*

Chromium(III) holds a special place in the development of electronic structure theory of transition metal ions, going back to the historic experiments of Becquerel [35] on the determination of the lifetime of the sharp line luminescence in ruby, that is, sapphire doped with Cr^{3+}, Al_2O_3:Cr^{3+}. As nicely summarised by Imbusch and Yen [36], ruby has been instrumental in the development of modern ligand-field theory by Tanabe and Sugano [37] and in establishing basic principles governing the photophysical properties of transition metal complexes in particular with respect to geometries of excited states as, for instance, the Jahn–Teller distortion in the $^4T_2(t_{2g}{}^2e_g{}^1)$ state [38, 39]. Likewise, the sharp emission doublet at 693 nm could be attributed to the zero-field split origins of the $^2E{\rightarrow}^4A_2$ spin-flip transition, both states belonging to the same $t_{2g}{}^3$ electronic configuration and therefore having the same bonding characteristics and equilibrium geometries. Ruby was used as active medium in the first laser [40] and henceforth served to demonstrate at the time novel optical phenomena such as fluorescence line narrowing [41], transient photophysical hole burning [42], photon echo [43], ODMR [44] or more recently the creation of slow light [45] to name but a few. Only over the past two decades did some of these methods find their way toward the investigation of a number of interesting photophysical phenomena in coordination compounds of chromium (III). For instance, Riesen et al. [46, 47] discovered a very efficient mechanism for persistent photophysical spectral hole burning in the electronic origins of the $^4A_2{\rightarrow}^2E$ transition based on flipping partially deuterated water molecules in the 2D oxalate network $NaMgAl(ox)_3{\cdot}9H_2O$ doped with Cr^{3+}, and Hauser et al. [48] evidenced very efficient resonant energy migration within this state in the 3D oxalate network $[NaCr(ox)_3][Rh(bpy)_3]ClO_4$, ox = $C_2O_4{}^{2-}$, bpy = 2,2′-bipyridine, at 1.3 K. The latter contrasts with the more common phonon-assisted energy migration usually found for this process in more concentrated systems [36, 49, 50]. Chromium(III) systems are also model systems for vibronic coupling between excited states. For ligand-field strengths for which the ^2E and the $^4T_2$ states are almost equienergetic, this leads to Fano resonances [51], indirectly illustrating fs-time-domain dynamics as described theoretically by Neuhauser et al. [52] and

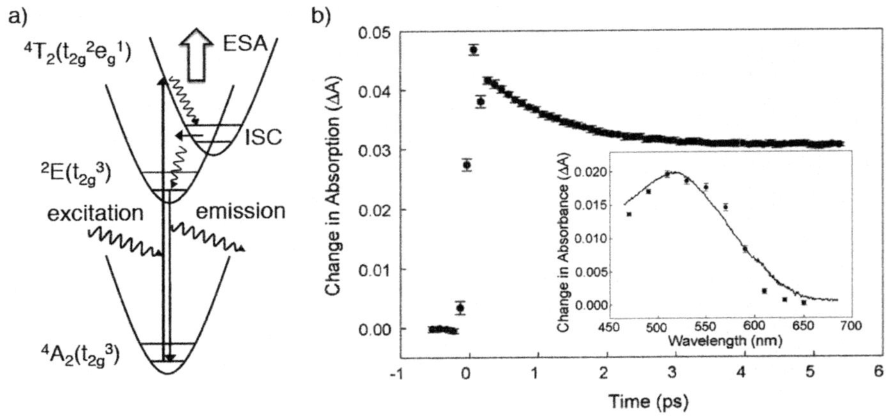

Fig. 1 (**a**) Potential energy diagram of the lowest-energy ligand-field states for Cr^{3+}, ISC (intersystem crossing) and ESA (excited-state absorption); (**b**) transient absorption at 480 nm measured in $Cr(acac)_3$ following pulsed excitation at 650 nm, that is, into the spin-allowed ligand-field transition. *Inset*: excitation spectrum at a delay of 5 ps (●) overlaid with the ground-state absorption spectrum (—) in the region of the $^4A_2 \rightarrow {}^4T_2$ transition (Adapted from [56])

applied to the antiresonance observed for the coupled transition for a Cr^{3+}-doped zirconium oxide glass [53]

Thus, one very important process, for instance, for the operation in the three-level laser and also for the photophysical and photochemical properties of Cr^{3+} complexes in general [54, 55], is the intersystem crossing process from the initially excited $^4T_2$ state to the ^2E (see Fig. 1). Indirect estimates of this process showed it to be very fast [57] for a number of photochemically interesting metal–organic complexes, but a direct observation proved only possible with the advent of ultrafast laser systems and sufficiently sensitive pump–probe techniques to monitor the small population changes and transient absorbance changes achievable when working with parity-forbidden ligand-field transitions. Juban et al. [56, 58, 59] thus investigated the $Cr(acac)_3$ complex (acac = acetylacetonate) in solution using direct excitation into the $^4A_2 \rightarrow {}^4T_2$ absorption band. The ultrafast evolution of the excited-state absorption (ESA) clearly showed that the ^2E state is populated within less than 100 fs following the excitation, which the authors interpreted as the so-called prompt process occurring directly from vibrationally excited states of the $^4T_2$ manifold. Thermalisation in the ^2E state subsequently occurred with $\tau = 1.1$ ps. As will become evident in the other examples discussed below, such ultrafast intersystem crossing in transition metal complexes seems to be the rule rather than the exception.

2.2 *Iron(II)*

Iron(II) coordination compounds were more famous for their magnetic properties, in particular the spin-crossover phenomenon [60–62], and not so much for their photophysical properties up to 1984, when the phenomenon of light-induced spin-state trapping (LIESST) was discovered [63]. In short, in octahedral iron(II) spin-crossover compounds, the ligand-field strength is such that for the $^1A_1(t_{2g}^6)$ low-spin (LS) state having a shorter metal–ligand bond length than the $^5T_2(t_{2g}^4e_g^2)$ high-spin state (HS) with two of the electrons in the antibonding e_g orbitals, the splitting of the d orbitals is larger than the spin-pairing energy, whereas for the latter state, it is smaller [64]. As a result, the zero-point energy difference between the two states is small enough such that at low temperature only the LS state is populated, but that at elevated temperature an almost quantitative, entropy-driven population of the HS state can occur. Typical iron(II) spin-crossover compounds have $[FeN_6]$ coordination sphere with at least some of the donating N atoms belonging to aromatic pyridine, triazole or tetrazole moieties or derivatives thereof. At low temperatures, typically below 50 K, the HS state can be trapped as metastable state with lifetimes of up to many days via irradiation into either the ligand-field or, depending on their energies, into the metal–ligand charge transfer (MLCT) absorption bands [65–68]. On irradiation of the trapped species in the near infrared, that is, into the spin-allowed ligand-field transition of the HS state, the LS state can be partially recovered via reverse LIESST [69]. Figure 2 depicts schematically the proposed mechanisms for the light-induced processes leading from the initially excited state to the final state via double intersystem crossing. However, as discussed below, it took almost 30 years to arrive at a more detailed understanding of the mechanisms of these phenomena.

Also for spin-crossover complexes in solution, the spin equilibrium can be perturbed via pulsed irradiation, but at higher temperatures the return to equilibrium typically occurs within a few μs [70–74]. This can conveniently be monitored by the ground-state bleaching (GBS) of the intense ^{1}MLCT band of the LS species. Likewise, pure LS complexes such as the prototypical $[Fe(bpy)_3]^{2+}$ complex can be converted to a transient, nonluminescent HS state via irradiation into the ^{1}MLCT band with quantum efficiencies approaching unity [75], but due to the larger driving force for the nonradiative relaxation back to the LS state, with much shorter lifetimes both at low temperatures and at elevated temperature. Thus, for the abovementioned complex at room temperature, the lifetime of the light-induced HS state is around 0.5 ns and increases to several μs depending on the surrounding medium at low temperatures [76]. That the transient state in this LS complex is indeed the $^5T_2$ state was established by the comparison of the lifetime measured in the same matrix optically with the one determined via time-resolved Mössbauer spectroscopy [77]. Picosecond transient X-ray absorption (Fig. 3) and emission spectroscopy with $[Fe(bpy)_3]^{2+}$ [78, 79] and other LS complexes [80–83] in solution at room temperature furthermore established that the bond length difference between the light-induced HS state and the LS ground state is indeed equal to the

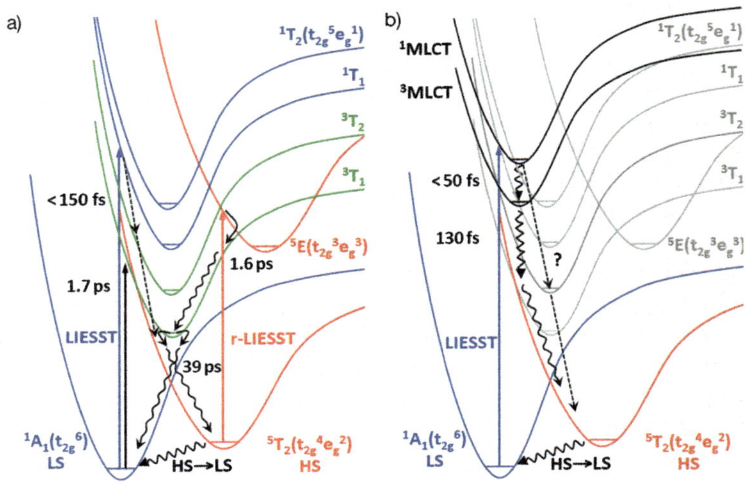

Fig. 2 Ground- and excited-state potential energy curves along the metal–ligand bond lengths for iron(II) complexes: (**a**) the ligand-field states and the mechanisms for LIESST (*broken arrows*) and r-LIESST (*full arrows*) for a system with no low-energy MLCT states; (**b**) the controversy concerning the mechanism for LIESST for a system with low-energy MLCT states: *curly arrows* via the ^{3}MLCT state or *broken arrows* via the $^3T_2$ state

typical 0.2 Å also found in iron(II) spin-crossover complexes via temperature-dependent single-crystal X-ray diffraction [84, 85] and LIESST experiments [86–88], as was theoretically predicted by DFT calculations [89–91].

For the determination of the HS→LS relaxation, conventional nanosecond transient absorption spectroscopy was for the most part sufficient [64, 70–74], but to actually pin down the exact mechanism of LIESST and reverse LIESST required faster and more sensitive methods. McCusker et al. [92] were the first to investigate the fast relaxation from the initially excited ^{1}MLCT state to the HS state using picosecond pump–probe spectroscopy with ps time resolution. As they could not detect any intermediate state in their experiments, they surmised that upon irradiation into the ^{1}MLCT band, the system must convert extremely rapidly directly to the HS state, thereby bypassing the low-lying ligand-field states (Fig. 2b). In the following a discussion started as to whether this is really the case, and what the difference to systems with only comparatively high-energy MLCT states was, for which the ligand-field states undoubtedly play an important role (see below). With the advent of femtosecond systems and an enormous increase in sensitivity, McCusker et al. [93, 94] and Chergui et al. [95–97] first showed that in solution at room temperature, the passage from the initially excited ^{1}MLCT state to the $^5T_2$ takes in fact only around 150 fs followed by vibrational cooling within a few picoseconds. The latter concluded that the first step in the relaxation process must be an intersystem crossing process from the ^{1}MLCT to the ^{3}MLCT state within less than 50 fs because of the almost identical evolution of the transient spectrum and some hot emission from the ^{1}MLCT state on this timescale in $[Ru(bpy)_3]^{2+}$ (see

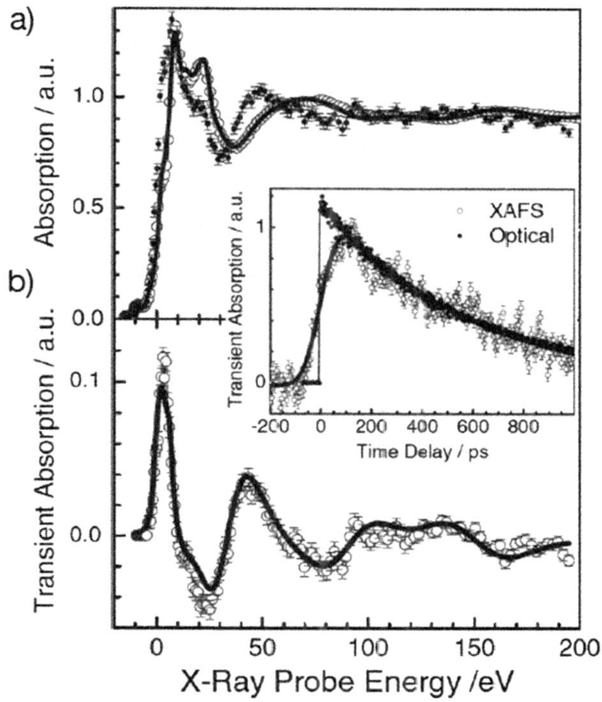

Fig. 3 (**a**) K-edge X-ray absorption spectra of $[Fe(bpy)_3]^{2+}$ in aqueous solution: (○) experimental LS ground-state spectrum, (—) theoretical, (●) extrapolated for the HS state; (**b**) transient difference spectrum 50 ps after the excitation laser pulse: (○) experimental, (—) theoretical fit; *inset*: comparison of the decay of the light-induced HS state measured by XAFS and optical spectroscopy (From [78])

below). In their seminal paper, they then demonstrated vibrational coherence being transferred to the final state, that is, the HS state, by monitoring a feature corresponding to a transient absorption of this species. They regarded this as experimental confirmation of the postulated direct relaxation to the HS state from the vibrationally hot ^{3}MLCT manifold (Fig. 4). A further experiment using subpicosecond XAS [98, 99] seemed to endorse this interpretation (Fig. 5a). This experiment was based on a single energy transient feature, which essentially monitored the arrival of the system in the HS state. Subsequently, Zhang et al. [100] recorded femtosecond time-resolved X-ray fluorescence spectra (Fig. 5b), which confirmed the population of the HS state within less than 150 fs. In their full spectra, they observed some indications of a transient state with an estimated lifetime of 70 fs, which, in contrast to Chergui et al. [95–97], they assigned to either the $^3T_1$ or the $^3T_2$ ligand-field state. Computational work seems to favour the higher-energy $^3T_2$ state as transient state based on the larger spin–orbit coupling matrix element to the $^5T_2$ state [101]. As shown below, this cannot be entirely ruled out, but as pointed out by McCusker [102], it is questionable to talk about a transient state, which has a lifetime of only a fraction of the time it takes for one vibrational period along the reaction coordinate.

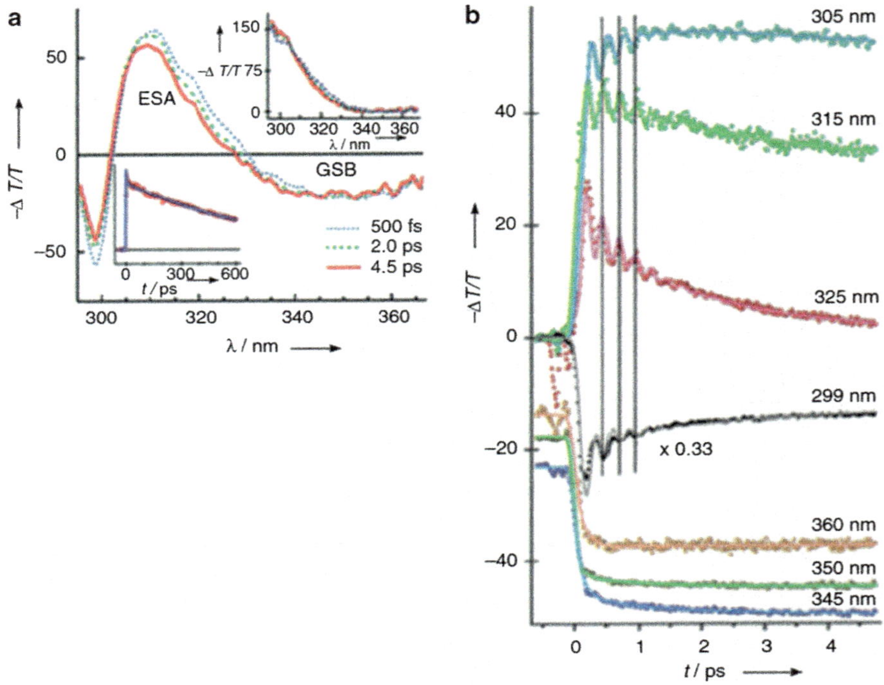

Fig. 4 (**a**) Transient absorption spectra in the region of ESA of the light-induced HS species of $[Fe(bpy)_3]^{2+}$ on irradiation into the ^{1}MLCT band at 530 nm; (**b**) time profiles for different wavelengths across the ESA of the HS species with damped oscillations indicating vibrational coherence in the final state (From [95])

The controversy with regard to the role of the excited singlet, triplet and even quintet ligand-field states needs further discussion. That they play an important role is beyond question. LIESST works perfectly well for spin-crossover complexes with no low-lying MLCT states such as $[Fe(ptz)_6](BF_4)_2$ (ptz = 1-propyltetrazole) for irradiation directly into the spin-allowed $^1A_1 \rightarrow ^1T_1$ and $^1A_1 \rightarrow ^1T_2$ as well as into the spin-forbidden $^1A_1 \rightarrow ^3T_1$ and $^1A_1 \rightarrow ^3T_2$ ligand-field transitions, and reverse LIESST via irradiation into the spin-allowed $^5T_2 \rightarrow ^5E$ ligand-field transition also involves exclusively ligand-field states as depicted in Fig. 2a [64]. However, up to recently, pump–probe transient absorption spectroscopy was not sensitive enough to allow pumping and probing with the parity-forbidden and therefore rather weak ligand-field transition. Marino et al. [103] resolved this problem for both LIESST and reverse LIESST in $[Fe(ptz)_6](BF_4)_2$ or rather in the mixed crystal $[Zn_{1-x}Fe_x(ptz)_6](BF_4)_2$, $x = 0.1$, by pumping into the spin-allowed ligand-field $^1A_1 \rightarrow ^1T_1$ transition at 532 nm for the former and into the $^5T_2 \rightarrow ^5E$ transition at 830 nm for the latter, but monitoring the transient absorption at 300 nm (Fig. 6).

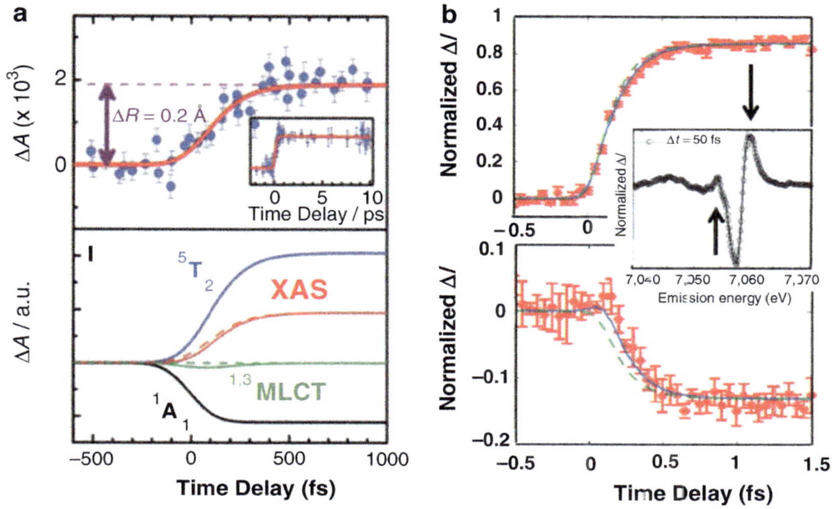

Fig. 5 (**a**) Transient X-ray absorption measured at 7.12 keV for $[Fe(bpy)_3]^{2+}$ in aqueous solution: (●) experimental, (—) theoretical based on sequence $^1A_1 \rightarrow {}^1MLCT \rightarrow {}^3MLCT \rightarrow {}^5T_2$ (*top*) and corresponding evolution of the species (*bottom*) from [98]; (**b**) transient X-ray emission profile for $[Fe(bpy)_3]^{2+}$ in aqueous solution at two different energies: (●) experimental, (—) theoretical based on a sequence $^1A_1 \rightarrow {}^1MLCT \rightarrow {}^3T \rightarrow {}^5T_2$. *Inset*: transient X-ray emission spectrum with a delay of 50 fs after the excitation pulse (Adapted from [100])

This is in the region of the very intense MLCT transitions. They performed the experiment at 125 K. At this temperature, within the thermal transition curve, the HS fraction γ_{HS} is equal to 0.85 and the LS⇔ HS equilibration time is 0.3 ms. The high absorption cross section for the ^{1}MLCT band at 300 K and the comparatively high concentration of the iron(II) complex in the crystal assured the necessary sensitivity for the detection of transient species. The equilibration time of 0.3 ms at 125 K allowed the use of kHz repetition rate for the experiment, and the equilibrium value of $\gamma_{HS} = 0.85$ allowed to perturb the equilibrium in both directions at the same temperature. The key result of this work is that for reverse LIESST (Fig. 6a), there is a thermalised intermediated state in the passage from the initially excited ^{5}E state to the final $^1A_1$ state with a lifetime of 39 ps at 125 K [103]. This manifests itself in a minimum in the ESA following the ultrafast initial decay immediately after the excitation followed by a rise to the ESA of the final state. The intermediate state can be assigned to the lowest-energy triplet ligand-field state, the $^3T_1$ state, in line with the previously estimated quantum efficiencies of around 10% for reverse LIESST [64]. Thus, reverse LIESST can indeed be described as a sequential double intersystem crossing process with the $^3T_1$ state as a well-defined intermediate state decaying with a branching ratio of around 1:4 to either the LS or the HS state according to non-adiabatic multiphonon relaxation based on Fermi's golden

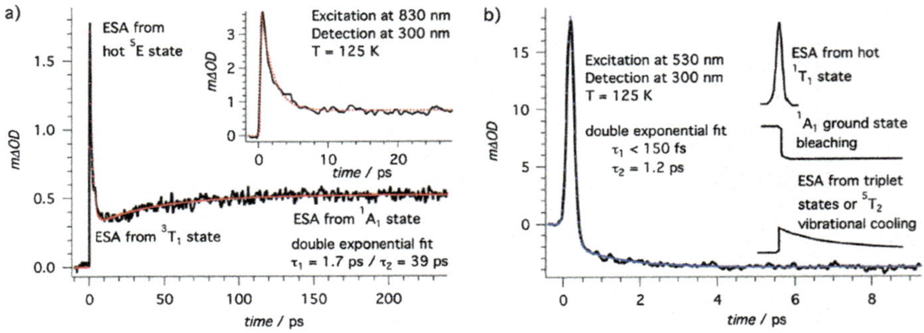

Fig. 6 Transient absorption on $[Zn_{1-x}Fe_x(ptz)_6](BF_4)_2$, $x = 0.1$, at 125 K monitored at 300 nm, that is, at the wavelength of the strong ^{1}MLCT transition for (**a**) irradiation into the $^5T_2 \rightarrow ^5E$ ligand-field transition at 830 nm (reverse LIESST); (**b**) irradiation into the $^1A_1 \rightarrow ^1T_1$ ligand-field transition at 532 nm (LIESST) according to [103]

rule and the crude Born–Oppenheimer approximation [104]. Things look very different for LIESST (Fig. 6b) using the $^1A_1 \rightarrow ^1T_1$ absorption band. The light-induced transition to the HS state is as fast as for irradiation into the ^{1}MLCT bands of systems with low-lying MLCT states, namely, the excited-state absorption signal of the $^1T_1$ at 300 nm decays within less than 150 fs and is replaced by bleaching characteristic of the ^{1}MLCT transition, which settles down within 1.5 ps. Obviously, for LIESST there is no clear evidence for thermalisation in an intermediate state as this would inevitably lead to the population of the $^3T_1$ state which would have to show up with a transient signal having the same associated lifetime of the abovementioned 39 ps at 125 K. Thus, for LIESST the triplet states are not real intermediate states. Of course they serve as what would probably be better termed mediator states for the ultrafast transition from the vibrationally hot $^1T_1$ state to very high vibrationally excited $^5T_2$ levels. This process is beyond a description by sequential relaxation processes. Rather it should be described as a coherent evolution of the vibronic wave function of the initially created vibrational packet in the $^1T_1$ state on coupled and complex hypersurfaces involving states of singlet, triplet and quintet character. The theoretical concepts for such a description still need to be developed. The same of course holds for LIESST on irradiation into the ^{1}MLCT bands of systems with low-lying MLCT states. Thus, the apparent controversy on the exact pathway and possible intermediate states for LIESST in the latter case is not really all that meaningful.

2.3 *Ruthenium(II)*

The literature on ruthenium(II) polypyridyl complexes is vast, totalling several thousand publications in peer-reviewed journals alone with regard to their photophysical properties ([105] and references therein). This is due to fundamental studies on light-induced excitation energy transfer and electron transfer [32–34, 106] processes, leading to their application, for instance, as sensitisers in dye-sensitised solar cells [6, 7] or in cancer phototherapy [107–112]. In essence, ruthenium(II) complexes are isoelectronic with iron(II) complexes, but for 4d metal ions the ligand-field strength for a given ligand is around 50% higher than for 3d metal ions [113], such that for polypyridyl complexes the first excited ligand-field state, the $^3T_1(t_{2g}{}^5e_g{}^1)$ state, is of comparable energy as the famous luminescent ^{3}MLCT state, first assigned by Demas et al. [114, 115] in 1971 for $[Ru(bpy)_3]^{2+}$. Subsequently, van Houten and Watts [116] attributed the quenching of the ^{3}MLCT luminescence at higher temperatures to the thermal population of the $^3T_1$ ligand-field state, which in this complex is around 3,000 cm^{-1} higher in energy than the ^{3}MLCT state and which decays non-radiatively to the $^1A_1(t_{2g}{}^6)$ ground state. The relative energies of the two states can be modulated physically [76, 117] or by chemical substitution on the ligands [118]. Indeed for ligands with lower ligand-field strengths, the $^3T_1$ state drops to below the ^{3}MLCT state thus quenching the luminescence down to low temperatures, as schematically shown in Fig. 7.

In addition to the quenching of the luminescence by the ligand-field state, the time it takes for the intersystem crossing from the initially excited ^{1}MLCT to the ^{3}MLCT state is also of key importance because of the potential for hot electron injection into the conduction band of a semiconductor in dye-sensitised solar cells,

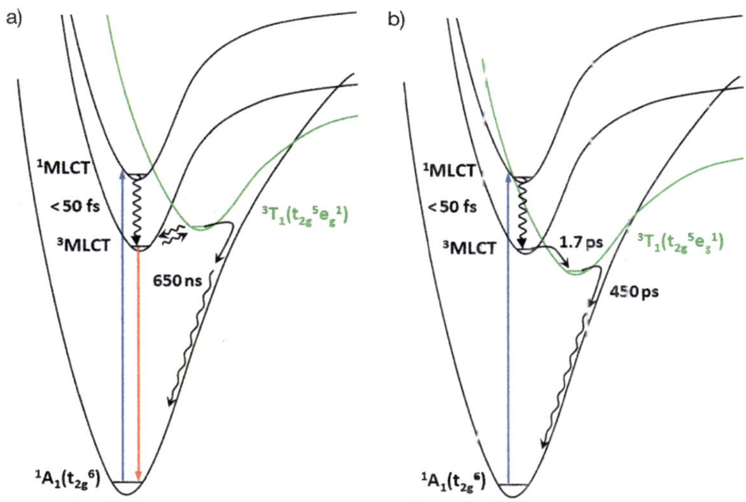

Fig. 7 Potential energy curves along the metal–ligand bond length for typical ruthenium (II) polypyridyl complexes: (**a**) with the $^3T_1$ ligand field state above the ^{3}MLCT state. The thermal population of this state quenches the luminescence only at higher temperatures as, for instance, for $[Ru(bpy)_3]^{2+}$; (**b**) with the $^3T_1$ state below the ^{3}MLCT state, which quenches the luminescence from the ^{3}MLCT state down to low temperatures as, for instance, for $[Ru(mbpy)_3]^{2+}$. The internal conversion to the $^3T_1$ state is in the Marcus normal region, while the intersystem crossing from the $^3T_1$ state back to the ground state is in the inverted region

provided the lifetime of the ^{1}MLCT state is long enough [119, 120]. Furthermore, an ongoing point of discussion concerned the localisation of the electron on a single ligand in the ^{3}MLCT state and possibly its hopping rate from one ligand to another. Ultrafast pump–probe spectroscopy served to find some more definite answers to the above questions. McCusker et al. [121, 122] showed that in $[Ru(bpy)_3]^{2+}$ the intersystem crossing from the ^{1}MLCT to the ^{3}MLCT state takes only 50 fs. This was confirmed by Chergui et al. [123, 124], who, based on time-resolved luminescence up-conversion, identified it as occurring from hot vibrational states of the ^{1}MLCT state within 30 fs and with extremely fast energy dissipation manifesting itself with a quasi-instantaneous Stokes shift. Finally, Yeh et al. [125] and Hammarstrom et al. [126] showed experimentally that in polar solvents solvent-driven localisation of the electron on one ligand in conjunction with vibrational cooling took on the order of picoseconds at room temperature, which could be rationalised using a DFT-based theoretical approach to molecular modelling [127].

The above still left the experimental characterisation of the $^3T_1$ state as elusive as before. In systems, for which the luminescence at room temperature is partially quenched via thermal activation, the concentration of the $^3T_1$ state is always very low because its lifetime is shorter than the process feeding it. In systems for which the luminescence is fully quenched, that is, when the $^3T_1$ state lies at substantially lower energy than the ^{3}MLCT state, all processes are very fast, and even if it has an appreciable transient concentration, it is difficult to pick up as it is not expected to have a strong spectroscopic signature. That is, in theory a spin-allowed MLCT transition from the $^3T_1$ state is possible, but it is expected to be much weaker than the ^{1}MLCT transition from the ground state, because of the substantially longer metal–ligand bond length due to the population of the antibonding e_g orbital. However, a judicious choice of the ligand in the form of 6-methyl-2,2′-bipyridine (mbpy) finally allowed tracking down the $^3T_1$ state. In $[Ru(mbpy)_3]^{2+}$ the methyl groups in the 6 position force slightly longer Ru–N bond lengths, which lower the energy of the $^3T_1$ state just enough for it to be almost equienergetic with the ^{3}MLCT state [128, 129]. At room temperature in solution, the ^{3}MLCT luminescence of this complex is completely quenched. Whereas in transient absorption spectra of $[Ru(bpy)_3]^{2+}$ in deoxygenated acetonitrile solution at room temperature the decay of the excited-state absorption at 380 nm characteristic for the ^{3}MLCT state [130], the recovery of the ground-state bleaching of the ^{1}MLCT transition at 458 nm, and the ^{3}MLCT luminescence decay all are single exponential with the same lifetime of 900 ns, transient absorption spectroscopy reveals that the relaxation of $[Ru(mbpy)_3]^{2+}$ from the ^{3}MLCT state is a two-step process occurring on two very different timescales (see Fig. 8). Very importantly, the characteristic excited-state absorption of the ^{3}MLCT state at 380 disappears within 1.6 ps, whereas ground-state recovery only occurs with a lifetime of 450 ps together with the decay of weak excited-state absorption between 600 and 850 nm. It stands to reason that the nonluminescent intermediate state with a lifetime of 450 ps can be attributed to

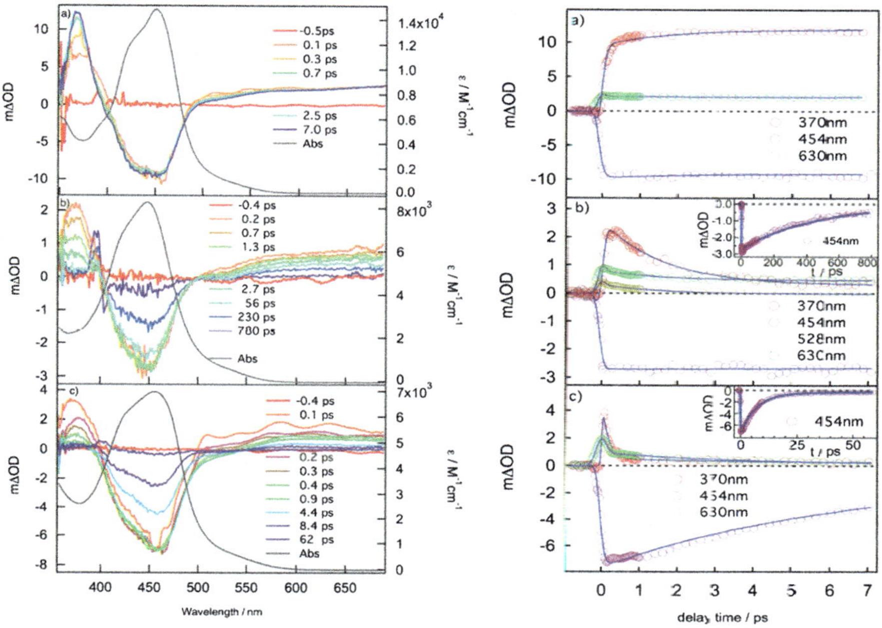

Fig. 8 Transient absorption spectra (*left*) and time profiles at selected wavelengths (*right*) for (*a*) $[Ru(bpy)_3]^{2+}$, (*b*) $[Ru(mbpy)_3]^{2+}$ and (*c*) $[Ru(tmbpy)_3]^{2+}$ following excitation at 400 nm in acetonitrile at room temperature (From [128, 129]). For $[Ru(bpy)_3]^{2+}$ all transient signals decay with the luminescence lifetime of 650 ns. For $[Ru(mbpy)_3]^{2+}$, the marker band for the ^{3}MLCT state at 380 nm decays within 1.6 ps, while ground-state recovery occurs with 450 ps. For $[Ru(tmbpy)_3]^{2+}$ the corresponding time constants are <150 fs and 4.8 ps (Adapted from [128])

the lowest component of the $^3T_1$ manifold. Interestingly, for $[Ru(tmbpy)_3]^{2+}$, tmbpy = 4,4′,6,6′-tetramethyl-2,2′-bipyridine, for which the sterically hindered methyl groups push the $^3T_1$ state to even lower energy with respect to the ^{3}MLCT state, both processes are faster. This can be explained by the fact that the internal conversion from the ^{3}MLCT to the $^3T_1$ state is in the Marcus normal region, whereas the intersystem crossing from the latter state to the ground state is in the inverted region.

The attribution of the intermediate state to the triplet ligand-field state can be further tested by the application of external pressure, which switches on the luminescence in $[Ru(mbpy)_3]^{2+}$ already for the comparatively modest pressure of 0.5 GPa [128, 129] with only a small shift of the actual luminescence maximum with increasing pressure. This indicates that external pressure destabilises the quencher state, and thus it must have a much larger molecular volume but a very similar electronic structure, namely, a d^6 configuration, compared to the ground state. This is in line with the assignment of this state to the $^3T_1(t_{2g}{}^5e_g{}^1)$ state with 1D electron in the antibonding e_g orbitals.

2.4 Nickel(II), Platinum(II) and Palladium(II)

Square-planar complexes with a d^8 electron configuration are molecular systems with a rich variety of spectroscopic properties and high symmetry [131–133]. Of particular interest are the nature of their lowest-energy excited states and their unsaturated coordination sphere, providing an attractive terrain for spectroscopic and photochemical studies [16–18].

Many square-planar platinum(II) and palladium(II) complexes show d–d luminescence in the red to near-infrared spectral regions [133, 134]. Broad luminescence bands are observed with vibronic structure indicating excited-state distortions along several normal modes involving the metal centre and coordinated ligand atoms [134, 135]. Some aspects of the ligand-field states of these complexes with a totally symmetric, nondegenerate electronic ground state are straightforward: the lowest-energy excited state is a triplet state, with a corresponding singlet state arising from the same electron configuration at higher energy. One electron occupying the σ-antibonding $d_{x^2-y^2}$ orbital in these excited states leads to bond weakening and broad absorption and luminescence bands. A challenging aspect of the electronic structure of such systems is the small energy differences separating the occupied d orbitals, and it has been shown with DFT calculations that the energy order expected from traditional ligand-field arguments is incorrect for many of the compounds [136]. Experimental evidence for the nature of the lowest-energy excited state is obtained from low-temperature luminescence spectra with resolved vibronic structure, and the presence of vibronic progressions involving non-totally symmetric stretching and bending modes is indicative of a degenerate excited state with a Jahn–Teller distortion [134]. Such degenerate excited states occur if the electron promoted to the $d_{x^2-y^2}$ orbital originates from the degenerate $d_{xz,yz}$ set, an experimental observation supporting the calculated energy order for simple complexes such as $PdBr_4^{2-}$ or $Pt(SCN)_4^{2-}$ [136] and leading to an emitting state with a unique excited-state structure and distinct possibilities to vary the efficiency of competing relaxation processes. Small structure variations, such as those resulting from temperature or pressure changes, can lead to very significant effects. An illustrative example is given by the luminescence spectra of $(n\text{-}Bu_4N)_2Pd(SCN)_4$ [134]. At room temperature, a weak luminescence band is observed with a maximum at 820 nm (12,200 cm^{-1}). The luminescence intensity increases significantly with increasing external pressure, as shown in Fig. 9. The luminescence lifetime also increases, from 0.3 to 53 μs between ambient pressure and 3 GPa, indicating that the intensity increase is due to less efficient nonradiative relaxation processes at higher pressure. The potential energy curves along the S–Pd–S bending normal coordinate shown in Fig. 9 qualitatively illustrate this behaviour: external pressure decreases the offset ΔQ of the emitting state potential energy minima, leading to a higher barrier for crossing to the ground-state potential energy surface and therefore to less efficient nonradiative relaxation and higher luminescence intensities. It is

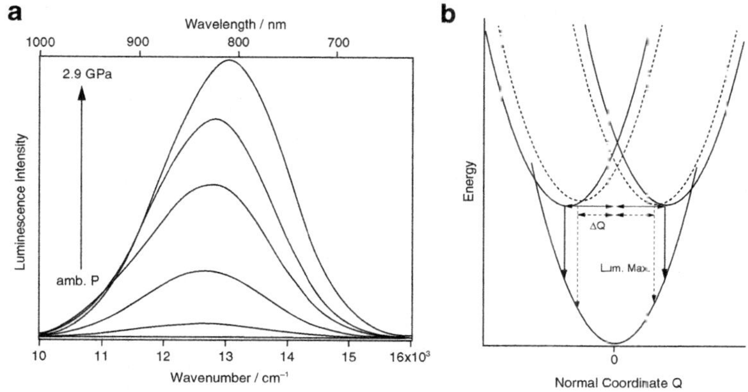

Fig. 9 (**a**) Luminescence spectra of $(n\text{-}Bu_4N)_2Pd(SCN)_4$ at variable pressure; (**b**) the *solid and dotted potential energy curves* represent ambient and high pressure, respectively. The decrease of the distortion ΔQ is indicated by the difference between *solid and dotted horizontal arrows*

worth to note that the luminescence band maximum shifts to higher energy with increasing pressure. This shift of +290 cm^{-1}/GPa increases the number of high-frequency C–N vibrational quanta needed to bridge the gap to the ground state from 6 to 6.2, an increase too small to rationalise the several orders of magnitude of intensity change shown in Fig. 9. This example thus illustrates the importance of small variations of the excited-state structure, as opposed to the often dominant variation of excited-state energies, discussed, for instance, in the preceding section for ruthenium(II) systems.

Charge-transfer excited states have been studied extensively for a wide variety of platinum(II) complexes with polypyridyl ligands [14, 132]. Charge-transfer processes in such systems have been among the coordination compounds where ultrafast transient vibrational spectroscopy has been successfully applied to characterise the effects of the charge redistribution on vibrational frequencies [20, 137]. Figure 10 shows one of the pioneering experiments in this area, illustrating the frequency shift of the CO stretching frequency of square-planar Pt $(4,4'\text{-}(CO_2Et)_2\text{-}2,2'\text{-bpy})Cl_2$ observed at 1,733 cm^{-1} in the ground-state IR spectrum, given as trace (a) in Fig. 10. Upon excitation at 400 nm, that is, near the maximum of the lowest-energy intense absorption band, the IR absorption at 1,733 cm^{-1} decreases and a lower-frequency absorption at approximately 1,710 cm^{-1} appears, indicative of the additional π-antibonding electron density resulting from a metal–ligand charge transfer process. The transient spectrum disappears within less than 50 ps, with much slower kinetics observed for platinum(II) complexes with more complex ligand systems involving multiple chromophores [20].

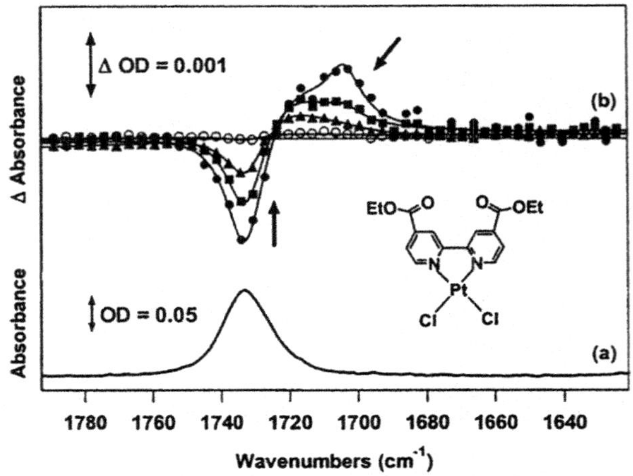

Fig. 10 (**a**) Ground-state FTIR spectrum of Pt (4,4′-$(CO_2Et)_2$-2,2′-bpy)Cl_2 in CH_2Cl_2 solution. (**b**) Time-resolved IR spectra obtained at (*solid circles*) 1.5 ps, (*squares*) 4 ps, (*triangles*) 11 ps and (*open circles*) 50 ps following 400 nm (ca. 150 fs FWHM) photolysis of this solution. *Solid lines* represent least-squares fits, and *arrows* indicate movement of the bands with increasing time delay following excitation (From [137])

An interesting feature illustrated in Fig. 10 is the shift of the transient maximum at approximately 1,710 cm^{-1} (1.5 ps delay) to higher frequencies at delays 4 and 11 ps. This shift is attributed to early relaxation processes of the photoexcited molecule, such as cooling or solvation, illustrating significant variations of fundamental molecular properties such as frequencies at the short time scales summarised here. Such effects are observed even in complexes such as where multiple excited states do not appear to play as important, but intricate vibronic dynamics still play a key role.

A characteristic aspect of the electronic structure of square-planar platinum (II) complexes is the possibility of metal–metal stacking interactions perpendicular to the molecular plane. Electronic spectra of such stacks or bimetallic molecular complexes show MMLCT transitions with energies strongly dependent on Pt(II)–Pt (II) distances and low-energy excimer luminescences [138]. Variable temperature and pressure again strongly influence the dynamics and spectroscopic signatures of such effects [138]. Interactions of the metal centre in square-planar chromophores with nearby groups other than neighbouring metals have not yet been extensively characterised by ultrafast optical spectroscopy. Recent work [139] combining variable-pressure crystallography and vibrational spectroscopy on a square-planar nickel(II) model system in order to characterise agostic metal–CH interactions points toward highly relevant future areas for the application of ultrafast spectroscopic techniques and spectroscopic studies under variable conditions in order to gain quantitative chemical insight.

3 Conclusions and Perspectives

In conclusion, the ultrafast spectroscopic methods, coupled with other state-of-the-art experimental and computational approaches, have allowed a much more detailed understanding of ground- and excited-state chemical bonding. The above examples show how the development of new experimental techniques leads to deeper insight into the dynamics of fundamental processes and a quantitative understanding. Of course other examples could have served the same purpose, and the ones chosen here are to some extent our personal preference. Nevertheless they are exemplary and allow fundamental conclusions, transferable to many other systems. In particular the ultrafast techniques showed that intersystem crossing can occur on ultrafast time scales, sometimes within much less than one vibrational period along the reaction coordinate even for overall $\Delta S = 2$ processes in iron (II) low-spin and spin-crossover systems. This means that more often than not, the processes occur from excited vibrational states and are in direct competition with intramolecular vibrational relaxation and vibrational cooling. In order to describe these processes correctly, theoretical tools going beyond the description of relaxation processes via Fermi's golden rule and the crude Born–Oppenheimer approximation need to be expanded from the current state of the art for small molecules [140–143] to the more complex open-shell systems [144]. The experimental identification of the $^3T_1$ ligand-field state in ruthenium(II) polypyridyl complexes is of practical importance for their technological applications and verifies the growing literature on DFT-based mechanistic studies of their photochemical and photophysical properties [145–151]. Charge-transfer processes in square-planar platinum(II) complexes have been extensively studied and time scales for charge separation tuned by chromophore design [19, 20]. The quantitative comparison of dynamics in isoelectronic nickel(II), palladium(II) and platinum (II) compounds, which have mostly been explored by steady-state structural and spectroscopic techniques, provides a promising perspective in this area [152].

In the future we will see more structural studies not only from time-resolved X-ray absorption spectroscopy but also from time-resolved X-ray diffraction [153–155]. Indeed first results on iron(II) complexes are already available [156–159] and allow to follow the structural evolution at early times as well as subsequent lattice effects and intermolecular dynamics. The ultimate goal of such experiments is to achieve sub-femtosecond resolution in order to also follow the redistribution of electronic density in situ. An alternative to X-ray diffraction is provided by time-resolved TEM and electron diffraction [160, 161]. Ultrafast time-resolved IR and Raman spectroscopy [162, 163] will likewise give insight into vibrational and vibronic coupling. And finally, ultrafast spectroscopic methods will also be applied to more complex systems, for instance, mixed valence systems, valence tautomeric systems or polynuclear compounds with combinations of different metal centres.

Acknowledgements We thank all our collaborators and friends who over the years have contributed in one way or another to our research and our understanding of the photophysical and photochemical properties of transition metal complexes and compounds. We acknowledge G3 travel funding.

References

1. Schönherr T (ed) (2004) Structure and bonding, vols 106, 107. Springer, Berlin
2. Jørgensen CK (1966) Struct Bond 1:3
3. Jørgensen CK (1962) Absorption spectra and chemical bonding in complexes. Pergamon, Oxford
4. Schäffer CE, Jørgensen CK (1958) J Inorg Nucl Chem 8:143
5. Jørgensen CK, Reisfeld R (1982) Top Curr Chem 100:127
6. Grätzel M (2001) Nature 414:338
7. Grätzel M (2005) Inorg Chem 44:6941
8. Jüstel T, Nikol H, Ronda C (1998) Angew Chem Int Ed 37:3084
9. Nazeeruddin MK, Grätzel M (2007) Struct Bond 123:113
10. Brütting W, Frischeisen J, Schmidt TD, Scholz BJ, Mayr C (2013) Phys Status Solidi A 210:44
11. Denker B, Shklovsky E (eds) (2013) Handbook of solid state lasers. Woodhead, Philadelphia
12. Stochel G, Brindell M, Macyk W, Stasicka Z, Szacilowski K (2009) Bioinorganic photochemistry. Wiley, Chichester
13. Abdel-Kadar MH (ed) (2014) Photodynamic therapy: from theory to application. Springer, Heidelberg
14. Lever ABP (ed) (2006) Proceedings of the 16th–20th international symposium on photophysics and photochemistry of coordination compounds. Coord Chem Rev (2006) 250:1621–1842, (2008) 252:2445–2612, (2010) 254:2447–2702, (2012) 256:1437–1786, (2015) 282–283:1–158
15. Weinstein J (ed) (2014) Themed issue on inorganic spectroscopy. Dalton Trans 43:17565–17870
16. Yam V (ed) (2007) Structure and bonding, vol 123. Springer, Heidelberg
17. Lo KKW (ed) (2015) Structure and bonding, vol 165. Springer, Heidelberg
18. Balzani V, Campagna S (eds) (2007) Topics in current chemistry, vols 280, 281. Springer, Heidelberg
19. McCusker JK (2003) Acc Chem Res 36:878
20. McCusker JK, Vlcek A Jr (eds) (2015) Ultrafast excited state processes in inorganic systems. Acc Chem Res 48:774–877, 1115–1148, 1207–1208, 1423–1449
21. Srinivasan R, Feenstra J, Park ST, Xu S, Zewail AH (2005) Science 307:558
22. Chergui M, Zewail AH (2009) ChemPhysChem 10:28
23. Daniel C (2015) Coord Chem Rev 282–283:19
24. Mingos DMP, Day P, Dahl JP (eds) (2012) Structure and bonding, vols 142, 143. Springer, Berlin
25. Daniel C (2006) Photochemistry of transition metal complexes: theory, encyclopedia of inorganic chemistry. Wiley, New York
26. Elsaesser T, Kaiser W (1991) Annu Rev Phys Chem 42:83
27. DeArmond MK (1974) Acc Chem Res 7:309
28. Demas JN (1983) J Chem Educ 60:803
29. Barbieri A, Ventura B, Ziessel R (2012) Coord Chem Rev 256:1732
30. Mayer JM (2004) Annu Rev Phys Chem 55:363
31. Wenger OS (2015) Coord Chem Rev 282–283:150
32. Maldotti A (2009) Photochemistry 37:240
33. Balzani V, Ceroni P, Juris A (2014) Photochemistry and photophysics: concepts, research and applications. Wiley, Weinheim
34. Wagenknecht PS, Ford PC (2011) Coord Chem Rev 256:591
35. Becquerel E (1887) La lumière, ses causes et ses effets. Librairie de Firmin Didot Frères, Fils et Cie, Paris
36. Imbusch GF, Yen WM (1987) Lasers, spectroscopy and new ideas, vol 54, Springer series in optical sciences. Springer, Berlin, p 258

37. Sugano S, Tanabe Y, Kamimura H (1970) Multiplets of transition metal ions in crystals, vol 33, Pure and applied physics. Academic, New York
38. Duval E, Louat R, Lacroix R (1972) Phys Status Solidi B 50:627
39. Güdel HU, Snellgrove TR (1978) Inorg Chem 17:1617
40. Maiman TH (1960) Nature 187:493
41. Szabo A (1970) Phys Rev Lett 25:924
42. Szabo A (1975) Phys Rev B 11:4512
43. Kurnit NA, Abella ID, Hartmann SR (1964) Phys Rev Lett 13:567
44. Geschwind S, Collins RJ, Schawlow AL (1959) Phys Rev Lett 3:545
45. Riesen H, Rebane A, Szabo A, Carceller I (2012) Opt Express 20:19039
46. Lewis ML, Riesen H (2001) PhysChemComm 26:1
47. Riesen H, Rae AD (2008) Dalton Trans 4717
48. Hauser A, von Arx ME, Langford VS, Oetliker U, Kairouani S, Pillonnet A (2004) Top Curr Chem 241:65
49. Selzer PM, Hamilton DS, Yen WM (1977) Phys Rev Lett 38:858
50. Henderson B, Imbusch GF (1989) Optical spectroscopy of inorganic solids. Clarendon, Oxford
51. Fano U (1961) Phys Rev 124:1866
52. Neuhauser D, Park TJ, Zink JI (2000) Phys Rev Lett 85:5304
53. Bussière G, Reber C, Walter D, Neuhauser D, Zink JI (2003) J Phys Chem A 107:1258
54. Kirk AD (1999) Chem Rev 99:1607
55. Kane-Maguire NAP (2007) Top Curr Chem 280:37
56. Juban EA, McCusker JK (2005) J Am Chem Soc 127:6857
57. Forster LS (2006) Coord Chem Rev 250:2023
58. Juban EA, Smeigh AL, Monat JE, McCusker JK (2006) Coord Chem Rev 250:1783
59. Schrauben JN, Dillmann KL, Beck WF, McCusker JK (2010) Chem Sci 1:405
60. Gütlich P, Link R, Trautwein A (1978) Mössbauer spectroscopy and transition metal chemistry, vol 3, Inorganic chemistry concepts. Springer, Heidelberg
61. Gütlich P, Goodwin HA (eds) (2004) Spin crossover in transition metal compounds I–III, vol 333–335, Topics in current chemistry. Springer, Heidelberg
62. Halcrow MA (ed) (2013) Spin-crossover materials. Wiley, Chichester
63. Decurtins S, Gütlich P, Köhler CP, Spiering H, Hauser A (1984) Chem Phys Lett 105:1
64. Hauser A (2004) Top Curr Chem 233:49
65. Gütlich P, Hauser A, Spiering H (1994) Angew Chem 106:2971
66. Hauser A (2004) Top Curr Chem 234:155
67. Hauser A (1991) J Chem Phys 94:2741
68. Gütlich P, Hauser A, Spiering H (1994) Angew Chem Int Ed 33:2024
69. Hauser A (1986) Chem Phys Lett 124:543
70. Lawthers I, McGarvey JJ (1984) J Am Chem Soc 106:4280
71. McGarvey JJ, Lawthers I (1982) J Chem Soc Chem Commun 1982:906
72. Beattie JK (1988) Adv Inorg Chem 32:1
73. Brady C, McGarvey JJ, McCusker JK, Toftlund H, Hendrickson DN (2004) Top Curr Chem 235:1
74. König E (1991) Struct Bond 76:151
75. Hauser A (1990) Chem Phys Lett 173:507
76. Hauser A, Amstutz N, Delahaye S, Schenker S, Sadki A, Sieber R, Zerara M (2003) Struct Bond 106:81
77. Hauser A, Adler P, Deisenroth S, Gütlich P, Hennen C, Spiering H, Vef A (1994) Hyperfine Interact 90:77
78. Gawelda W, Pham VT, Benfatto M, Zaushitsyn Y, Kaiser M, Grolimund D, Johnson SL, Abela R, Hauser A, Chergui M, Bressler C (2007) Phys Rev Lett 98:057401
79. Milne CJ, Penfold TJ, Chergui M (2014) Coord Chem Rev 277–278:44

80. Khalil M, Marcus MA, Smeigh AL, McCusker JK, Chong HHW, Schoenlein RW (2006) J Phys Chem A 110:38
81. Huse N, Kim TK, Jamula L, McCusker JK, de Groot FMF, Schoenlein RW (2010) J Am Chem Soc 36:876
82. Haldrup K, Vanko G, Gawelda W, Galler A, Doumy G, March AM, Kanter EP, Bordage A, Dohn A, van Driel TB, Kjær KS, Lemke HT, Canton SE, Uhlig J, Sundstrøm V, Young L, Southworth SH, Nielsen MM, Bressler C (2012) J Phys Chem A 116:9878
83. Canton SE, Zhang X, Lawson Daku LM, Smeigh AL, Zhang J, Wallentin CJ, Liu Y, Attenkofer K, Jennings G, Kurtz CA, Gosztola D, Wärnmark K, Hauser A, Sundström V (2014) J Phys Chem C 118:4536
84. Guionneau P, Marchivie M, Bravic G, Létard JF, Chasseau D (2004) Top Curr Chem 234:97
85. Kusz J, Gütlich P, Spiering H (2004) Top Curr Chem 234:129
86. Marchivie M, Guionneau P, Howard JAK, Goeta AE, Chastenet G, Létard JF, Chasseau D (2002) J Am Chem Soc 124:194
87. Kusz J, Gütlich P, Spiering H (2000) J Appl Crystallogr 33:201
88. Chakraborty P, Enachescu C, Bronisz R, Pillet S, Bendeif E, Hauser A (2013) Chem Eur J 19:11104
89. Paulsen H, Trautwein AX (2004) Top Curr Chem 235:197
90. Lawson Daku ML, Vargas A, Hauser A, Fouqueau A, Casida ME (2005) ChemPhysChem 6:1393
91. Rudavskyi A, Sousa C, de Graaf C, Havenith RWA, Broer R (2014) J Chem Phys 140:184318
92. McCusker JK, Walda KN, Dunn RC, Simon JD, Magde D, Hendrickson DN (1993) J Am Chem Soc 115:298
93. Monat JE, McCusker JK (2000) J Am Chem Soc 122:4097
94. Smeigh AL, Creeelman M, Mathies RA, McCusker JK (2008) J Am Chem Soc 130:14105
95. Cosani C, Prémont-Schwarz M, ElNahhas A, Bressler C, van Mourik F, Cannizzo A, Chergui M (2009) Angew Chem Int Ed 48:7184
96. Cannizzo A, Milne CJ, Consani C, Gawelda W, Bressler C, van Mourik F, Chergui M (2010) Coord Chem Rev 254:2677
97. Auböck G, Chergui M (2015) Nat Chem 7:629
98. Bressler C, Milne C, Pham VT, ElNahhas A, van der Veen R, Gawelda W, Johnson S, Beaud P, Grolimund D, Kaiser M, Borca CN, Ingold G, Abela R, Chergui M (2009) Science 323:489
99. Lemke HT, Bressler C, Chen LX et al (2013) J Phys Chem A 117:735
100. Zhang W, Alonso-Mori R, Bergmann U et al (2014) Nature 509:345
101. Sousa C, de Graaf C, Rudavskyi A, Broer R, Tatchen J, Etinski M, Marian CM (2013) Chem Eur J 19:17541
102. McCusker JK (2014) Nat Phys 10:476
103. Marino A, Servol M, Lorenc M, Chakraborty P, Collet E, Hauser A (2014) Angew Chem Int Ed 53:3863
104. Buhks E, Navon G, Bixon M, Jortner J (1980) J Am Chem Soc 102:2918
105. Thompson DW, Ito A, Meyer TJ (2013) Pure Appl Chem 85:1257
106. Belser P, von Zelewsky A, Frank M, Seel C, Vögtle F, DeCola L, Barigelletti F, Balzani V (1993) J Am Chem Soc 115:4076
107. Higgins SLH, Brewer KJ (2012) Angew Chem Int Ed 51:11420
108. Howerton BS, Heidary DK, Glazer EC (2012) J Am Chem Soc 134:8324
109. Sgambellone MA, David A, Garner RN, Dunbar KR, Turro C (2013) J Am Chem Soc 135:11274
110. Song H, Kaiser JT, Barton JK (2012) Nat Chem 4:615
111. Glazer EC (2013) Isr J Chem 53:391
112. Bugarcic T, Habtermariam A, Deeth RJ, Fabbiani FPA, Parsons S, Sadler PJ (2009) Inorg Chem 48:9444
113. Figgis BN, Hitchman MA (2000) Ligand field theory and its applications. Wiley, New York

114. Demas JN, Crosby GA (1971) J Am Chem Soc 93:2841
115. Harrigan RW, Crosby GA (1973) J Chem Phys 59:3468
116. Van Houten J, Watts RJ (1976) J Am Chem Soc 98:6
117. Maruszewski K, Strommen DP, Kincaid JR (1993) J Am Chem Soc 115:8345
118. Vos JG, Kelly JM (2006) Dalton Trans 4969
119. Moser JE, Bonnôte P, Grätzel M (1998) Coord Chem Rev 171:245
120. Li G, Yi C, Knappenberger KL, Meyer GJ, Gorelsky SI, Shatruk M (2013) J Phys Chem C 117:17399
121. Damrauer NH, Cerullo G, Yeh A, Boussie TR, Shank CV, McCusker JK (1997) Science 275:54
122. Damrauer NH, McCusker JK (1999) J Phys Chem A 103:8440
123. Cannizzo A, van Mourik F, Gawelda W, Zgrablic G, Bressler C, Chergui M (2006) Angew Chem 118:3246
124. Gawelda W, Johnson M, de Groot FMF, Abela R, Bressler C, Chergui M (2006) J Am Chem Soc 128:5001
125. Yeh AT, Shank CV, McCusker JK (2000) Science 289:5481
126. Wallin S, Davidsson J, Modin J, Hammarstrom L (2005) J Phys Chem A 109:4697
127. Moret ME, Tavernelli I, Chergui M, Röthlisberger U (2010) Chem Eur J 16:5889
128. Sun Q, Mosquera-Vasquez S, Lawson Daku LM, Guénée L, Goodwin HA, Vauthey E, Hauser A (2013) J Am Chem Soc 135:13660
129. Sun Q, Mosquera-Vazquez S, Suffren Y, Hankache J, Amstutz N, Lawson Daku LM, Vauthey E, Hauser A (2015) Coord Chem Rev 282–283:87
130. Thompson DW, Wishart JF, Brunschwig BS, Sutin N (2001) J Phys Chem A 105:8117
131. Gray HB, Ballhausen CJ (1963) J Am Chem Soc 85:260
132. McGuire R Jr, Clark McGuire M, McMillin D (2010) Coord Chem Rev 254:2574
133. Preston DW, Güntner W, Lechner A, Gliemann G, Zink JI (1988) J Am Chem Soc 110:5628
134. Grey JK, Butler IS, Reber C (2003) Inorg Chem 42:6503
135. Lanthier E, Reber C, Carrington T Jr (2006) Chem Phys 329:90
136. Deeth RJ (2003) Faraday Discuss 124:379
137. Weinstein JA, Grills DC, Towrie M, Matousek P, Parker AW, George MW (2002) Chem Commun 382
138. Delahaye S, Loosli C, Liu S-X, Decurtins S, Labat G, Neels A, Loosli A, Ward TR, Hauser A (2006) Adv Funct Mater 16:286
139. Scherer W, Dunbar AC, Barquera-Lozada JE, Schmitz D, Eickerling G, Kratzert D, Stalke D, Lanza A, Macchi P, Casati NPM, Ebad-Allah J, Kuntscher C (2015) Angew Chem Int Ed 54:2505
140. Casida ME, Huix-Rotllant M (2012) Annu Rev Phys Chem 63:287
141. Robb MA, Garvelli M, Olivucci M, Bernardi F (2000) Rev Comput Chem 15:87
142. Matsika S, Krause P (2011) Annu Rev Phys Chem 62:621
143. Bussiere G, Beaulac R, Cardinal-David B, Reber C (2001) Coord Chem Rev 219:509
144. Eng J, Gourlaouen C, Gindensperger E, Daniel C (2015) Acc Chem Res 48:809
145. Nazeeruddin MK, De Angelis F, Fantacci S, Selloni A, Viscardi G, Liska P, Ito S, Bessho T, Grätzel M (2005) J Am Chem Soc 127:16835
146. Lever ABP (2010) Coord Chem Rev 254:1397
147. Salassa L, Garino C, Salassa G, Nervi C, Gobetto R, Lamberti C, Gianolio D, Bizzarri R, Sadler PJ (2009) Inorg Chem 48:1469
148. Camillo MR, Cardoso CR, Carlos RM, Lever ABP (2014) Inorg Chem 53:3694
149. Thomas RA, Tsai CN, Mazumeder S, Lu IC, Lord RL, Schlegel HB, Chen YJ, Endicott JF (2015) J Phys Chem B 119:7393
150. Vlcek A, Zalis S (2007) Coord Chem Rev 251:258
151. Alary F, Broggio-Pasquera M, Heully JL, Marsden CJ, Vicendo P (2008) Inorg Chem 47:5259

152. Gupta AN, Kumar V, Singh V, Manar KK, Drew MGB, Singh N (2014) CrystEngComm 16:9299
153. Naumov P (2012) Top Curr Chem, 315:111
154. Coppens P, Vorontsov II, Graber T, Gembicky M, Kovalevsky AY (2005) Acta Crystallogr A 61:162
155. Patterson BD (2014) Crystallogr Rev 20:242
156. Lorenc M, Herbert J, Moisan N, Trzop E, Servol M, Buron-Le Cointe M, Cailleau H, Boillot ML, Pontecorvo E, Wulff M, Koshihara S, Collet E (2009) Phys Rev Lett 103:028301
157. Marino A, Buron-Le Cointe M, Lorenc M, Toupet L, Henning R, DiChiara AD, Moffat K, Bréfuel N, Collet E (2015) Faraday Discuss 177:363
158. Cailleau H, Lorenc M, Buron-le Cointe M, Servol M, Cammarata M, Collet E (2013) Eur Phys J Spec Top 222:1077
159. Collet E, Lorenc M, Cammarata M et al (2012) Chem Eur J 18:2051
160. Sciaini G, Miller RJD (2011) Rep Prog Phys 74:096101
161. van der Veen RM, Kwon OH, Tissot A, Hauser A, Zewail AH (2013) Nat Chem 5:395
162. Dattelbaum DM, Omberg KM, Schoonover JR, Martin RL, Meyer TJ (2002) Inorg Chem 41:6071
163. Fedoseeva M, Delor M, Parker SC, Sazanovich IV, Towrie M, Parker AW, Weinstein JA (2015) PCCP 17:1688

Struct Bond (2016) 172: 313–340
DOI: 10.1007/430_2015_200

Published online: 9 April 2016

Ligand Field Theory and the Fascination of Colours: Oxidic Iron(III) Solids as the Omnipresent Examples

Peter Köhler and Dirk Reinen

Abstract The treatment of the high-spin d^5-configurated iron(III) cation in 6- and 4-coordinate ligand fields is a highly complex matter. The $^6A_1$-ground state allows only spin-forbidden transitions, here of relevance to ten spin-reduced $^4A_1$-, $^4A_2$-, $^4E(2x)$-, $^4T_1(3x)$-, $^4T_2(3x)$-states, which are not easy to handle. Though the literature offers a series of carefully prepared solids with beautifully resolved ligand field spectra, the philosophy of utilising these in terms of their binding character, particularly in respect to the d-electron cloud density between cation and the anions, diverges. Accordingly, the magnitudes of reported ligand field parameters Δ and of the Racah parameters of interelectronic repulsion B and C, which parameterise the mentioned effects, differ, and comparisons become difficult. In this review we propose a well-founded and comprehensible calculational procedure, in order to, as the main matter of concern, convince the readers that the ligand field spectra also sensibly reflect finer perturbational details of local or even cooperative binding quality. The origin of the latter effects is from the chemical environment beyond the first coordination sphere of a central, say $Fe^{III}L_n$-complex ($n = 6{,}4$) in an extended solid. Already subtle disturbances of this kind will modify the shade of colour. An essential point of the discussion is the symmetry analysis, which provides rigorous constraints and sets strict conditions on what can be experimentally observed. We will first discuss manganese(II) in oxidic solids. Because a divalent cation is associated with rather ionic binding properties towards oxygen, crystal field theory is the appropriate analytical instrument. Iron(III) provides a situation, which requires more sophistication and a refinement of the theory by taking also bond

Dirk Reinen is the only surviving member of the founding editorial board of Structure and Bonding in 1966. C.K. Jorgensen, J.B. Neilands, R.S. Nyholm and R.J.P. Williams sadly are no longer with us.

P. Köhler and D. Reinen (✉)
Fachbereich Chemie and Zentrum für Materialwissenschaften, Philipps-Universität, 35043 Marburg, Germany
e-mail: reinen@chemie.uni-marburg.de

covalency into account. A symmetry-based reinterpretation of the additional absorptions in the d–d-spectra of Fe^{III} and Cr^{III} in corundum-type solids is presented. This treatment sheds new light on the finer roots of the impressive red-to-green colour change of Cr^{3+} in mixed crystals $Al_{2-x}Cr_xO_3$ with increasing x. Particular examples are discussed, where absorptions due to octahedral and tetrahedral iron(III) overlap in the ligand field spectra of spinel- and garnet-type solids, which model the hue in a predictable way. Finally, though not directly related to the primary topic, the charge-transfer properties of oxidic iron(III) are briefly examined. These absorptions often stray far into the visible region, with a very frequently significant influence on the apparent hue.

Keywords d^5 ligand field theory and spectra of Mn^{II}, Cr^{III}, Fe^{III} • Higher-sphere impacts in solids; their symmetry and energy criteria • Metal–metal interactions in corundum-type structures

Contents

1 Historical Preface

The appeal of colours frequently originates from solids, in which transition metal ions with an incomplete d-shell, embedded, for example, into oxidic host structures, play a deciding role. Specifically these kinds of pigments and the search for the scientific roots of their coloration have occupied spectroscopists and experimentalists from inorganic and physical chemistry vehemently in the decade following 1960 – initiated by the early publication of Ilse and Hartmann [1]. Many books covering this topic were published ([2–4], to give a selection of many) and also the first article in *Structure and Bonding* [5] was devoted to this phenomenon. A flow of innovative ideas established a sound basis for a semiquantitative understanding of the data provided by the d–d-spectra. Here, we particularly mention the spectrochemical and nephelauxetic series of cations and ligands [6]. Early volumes of *Structure and Bonding* documented this development and in particular emphasised the semiempirical aspects, where experiment and theory meet in a synergistic manner. We will substantiate in the following that – considering oxidic iron(III)-type solids with their sustainable gentle hues – the origin of the observed colour diversity is highly complex and needs careful investigation and interpretation. The authors think that this choice is not only reasonable for historical reasons, but also because of a current trend in inorganic chemistry, which one may call a revival of experiment-based ligand field theory. Progress in science is indeed based on the symbiosis of achievements from old sources and novel insights and findings, which originate from an extended-knowledge horizon. A novel perspective onto old and carefully worked-out spectroscopic results is always worthwhile. A concise summary of the theoretical basis of the ligand field treatment and basic informations as to how theory can be applied to the experimental landscape are given in the first two appendices.

2 The Colour Properties and the Ligand Field Spectra of Octahedrally and Tetrahedrally Coordinated Manganese(II) as Introducing Examples

A compact summary of the crystal field theory for a d^5-cation in octahedral and tetrahedral coordination, based on the spectroscopic properties of the free cation, is given in Appendix 1. The energies of the possible transitions from the (here only considered) high-spin ground state ${}^6A_{1g}$ with $S = 5/2$ to excited quartet states with $S = 3/2$ (${}^4A_1$, ${}_a{}^4E$; ${}_b{}^4E$; ${}_a{}^4T_2$; ${}_b{}^4T_2$; ${}_a{}^4T_1$; ${}_b{}^4T_1$, leaving aside the three highest-in-energy states) have been listed in dependence on the crystal field energy parameters B and Δ in Appendix 2 (Eq. 8), particularly for manganese(II). Because these transitions are spin-forbidden, the intensities are usually very low. Inspecting the rock salt-type mixed crystal series $Mg_xMn_{1-x-y}Ca_yO$, the colour varies for large to small x-values and $y = 0$ from weakly yellow via yellowish-green to finally bright-green [7]. The hue is mainly due to the deep minimum between the first and second absorption band and shifts from about 17,000 in the yellow to 19,000 cm^{-1} in the green spectral region, which may be followed via inspecting Fig. 1 and the colour diagram therein.

Crystal field theory predicts a shift of the two lowest energy bands to higher energies, if the bond length increases, due to the substitution of Mg^{2+} by the larger Mn^{2+}-cation [7]. This is so because of the negative Δ-dependence (see Appendix 2, Eq. 8) of these ${}^6A_{1g} \rightarrow {}_a{}^4T_{1g}$, ${}_a{}^4T_{2g}$ transitions. The pronounced rise in colour intensity with increasing Mn^{2+} concentration is induced by antiferromagnetic spin–spin interactions between the cations via the bridging oxygen atoms in the rock-salt structure, which makes the consideration of single-cation spins obsolete.

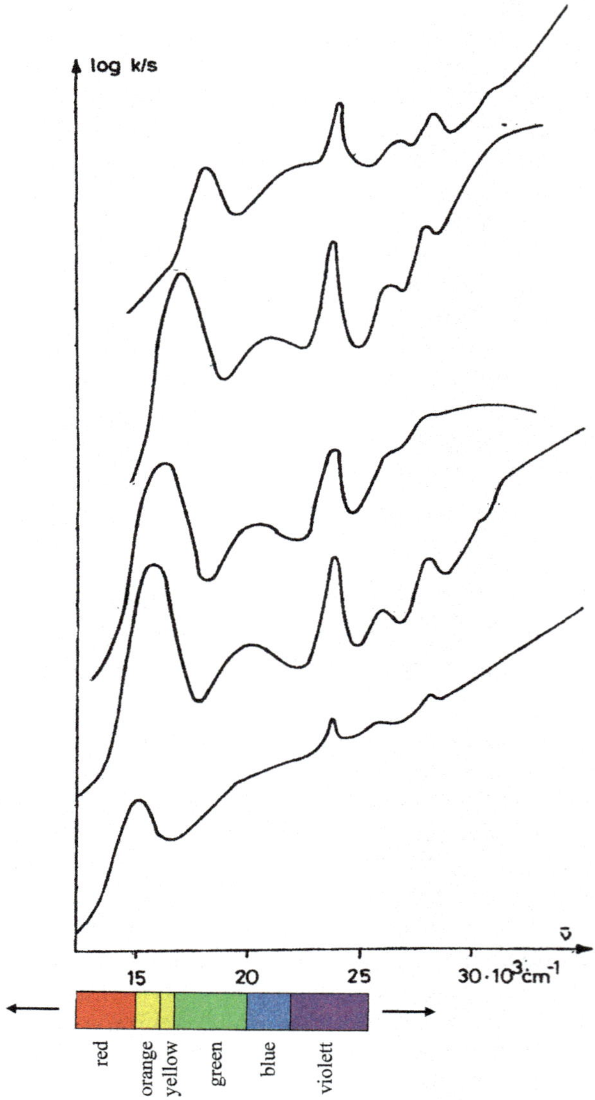

Fig. 1 Powder reflectance spectra (298 K) of mixed crystals $Mg_{0.9}Mn_{0.1}O$ (partly $Mg_{0.75}Mn_{0.25}O$, $Mg_{0.1}Mn_{0.9}O$, $Mn_{0.5}Ca_{0.5}O$ and $Mn_{0.1}Ca_{0.9}O$ (from *bottom coloured*) to *top*, adopted from [7]). The colour scaling below is thought to refer to the absorption minima, which characterise the pigments' hues (see Table 1)

If Mn^{2+} is diluted by the larger Ca^{2+} via increasing y (with $x = 0$) in the mixed crystals, an intensity decrease, and the expected shift of the two broad lowest energy bands and of the minimum between them to larger energies is observed. It generates a blue-green colour at $y = 0.5$ ($x = 0$). Enhancing the Ca^{2+} concentration further, the solids adopt pink to pale-red hues, because in these cases the deep gap on the low-energy side of the first absorption band now governs the colour (Fig. 1). Figure 2 presents the band assignment for one of the discussed mixed crystals [7] and Table 1 summarises spectral parameters for some octahedrally coordinated

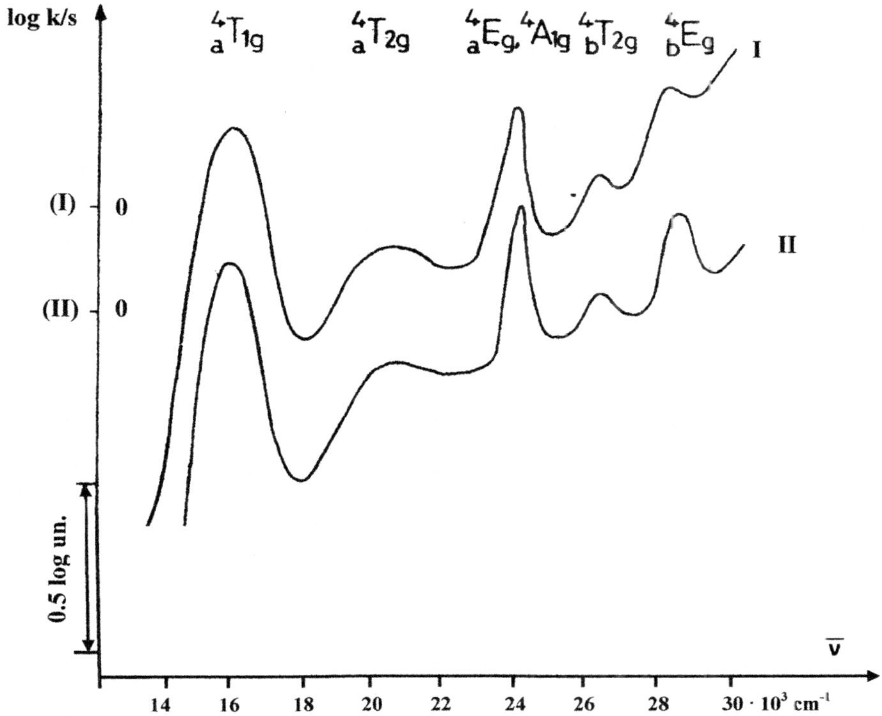

Fig. 2 Powder reflectance spectra of $Mn_{0.6}Mg_{0.4}O$ at 298 K (I) and 5 K (II), with band assignment (adapted from [7])

Table 1 Ligand field parameters (in units of 10^{-3} cm^{-1}) of octahedral Mn^{2+}-polyhedra in various solids (nephelauxetic ratio β referring to $B_0 = 750$ cm^{-1}; $C/B = 5.2$) and of Mn^{2+} in O_h- and T_d-oxo coordination in $MnAlO_4$

	Δ	*B*	*β*	Reference
MnF_2	7.4	0.70	0.93_5	[8]
$Mg_{0.5}Mn_{0.5}O$	9.8	0.65_5	0.87_5	[7]
$Mn_{0.5}Ca_{0.5}O$	8.3	0.65_5	0.87_5	[7]
$RbMnCl_3$	6.1	0.66	0.88	[9]
$MnAl_2O_4$ (O_h)	10.4	0.65	0.86_5	[7]
$MnAl_2O_4$ (T_d)	5.8	0.61_5	0.82	[7]

Mn^{2+} complexes in various solids, deduced by using the calculation procedure sketched in the appendices.

Figure 3 shows the crystal field spectrum of the spinel $MnAl_2O_4$ [7], displaying besides octahedral very distinct tetrahedral features, at 5K(II) well resolved. One should have in mind that – due to the lack of an inversion centre – electric dipole transitions in tetrahedra are parity-allowed. Accordingly, already a small

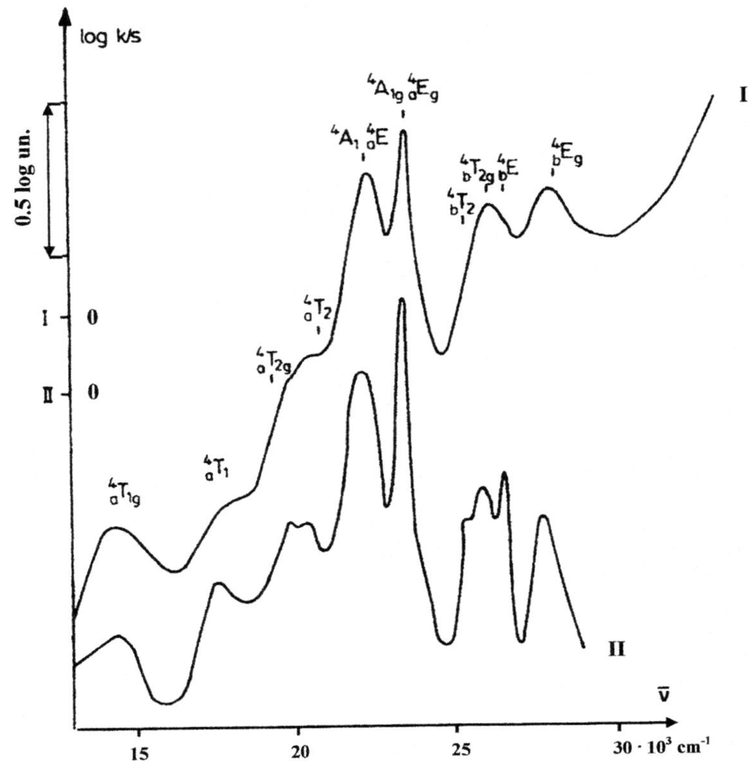

Fig. 3 Powder reflectance spectra of the spinel $MnAl_2O_4$ at 298 K (I) and 5 K (II) (adopted from [7]). The crystal field parameters derived from the indicated calculated band positions according to the given assignment are listed in Table 1 (the ${}^6A_{1(g)} \rightarrow {}^4T_{1(g)}$-transitions are not included into the energy fitting to Δ and B, see text)

percentage of Mn^{2+} in T_d-coordination should have a strong appearance in the spectrum. Leaving aside the transitions to ${}_a{}^4T_{1g}$ and ${}_a{}^4T_1$ (the one to ${}_b{}^4T_{1g}$ is expected in the region of the ${}_b{}^4E_g$-band around 28,000 cm^{-1}) – see Appendix 2 for the respective critical details – a very satisfactory agreement is achieved between experiment and theory, with the parameters summarised in Table 1. Though there is a coarse agreement with Jørgensens' wonderfully helpful spectrochemical and nephelauxetic series of ligands [6], additional matrix effects may often have a pronounced bearing on the polyhedron properties. This is the case if the considered polyhedron is embedded into an extended lattice and, specifically, if network-forming ligands, such as oxygen, are considered. The following structural implications significantly affect the energy and intensity landscape of the ligand field spectra and accordingly modify the colour:

(a) Substitutions into the octahedral sites of smaller or larger divalent cations modify the Δ parameter of Mn^{2+}, with predictable shifts of specific absorption bands. The nephelauxetic ratio β (see Appendix 2) indicates a slightly more covalent bond in T_d- than in O_h-ligand fields; this is so because the ligand-to-cation charge transfer per oxygen ligator atom is larger in the case of the smaller coordination number, which renders the Mn–O-bond in O_h to be more ionic than in T_d. The Δ_T/Δ_o ratio is, with 0.56, near to the expected value of about 0.5.

(b) Spin–spin interactions between neighboured colour centres in solids may release the respective selection rule, with the consequence that MnO has a bright-green colour, whilst Mn^{2+}-doped MgO and CaO appear as solids with only faint hues. If Mn^{2+} switches from the O_h- to a T_d-coordination, the vibronically assisted, now parity-allowed, transitions grow substantially in intensity.

3 Iron(III) as the Hue-Generating Cation

3.1 Critical Inspection of the Ligand Field Spectra and the Influence of Higher-Sphere Coordinations

As is pointed out in Appendix 2, crystal field theory – which has been shown to work fairly well for Mn^{2+} in its binding to rather ionic ligands – has to be essentially modified, if applied to the spectra of a trivalent cation, such as Fe^{3+}. Distinct covalency contributions alter the quality of the cation-to-ligand bonds significantly and have led to introduce additional semi-empirical parameters. These allow one, for example, to differentiate between non-bonding or π- and σ-anti-bonding properties, when inspecting the t_{2g}, e_g- and e, t_2-molecular orbital sets in O_h- and T_d-ligand fields, respectively. There are good reasons, however (see Appendix 2), to stick to a simple model with two unknowns, Δ and B, to parameterize ligand-field spectral results. This helps to avoid the disturbing feature to introduce additional parameters, which are equal in number to available experimental data. Though this approach frequently implies larger numerical deviations of the calculated from the experimental absorption band-energy positions, exceptional aberrations are nevertheless discernible. The interpretation may then be based on innovative perturbational approaches and/or by calling off additional evidence from suitable supplementing investigation methods. Always, the symmetry conditions are extremely helpful here and have to be strictly obeyed. If one switches from a di- to a trivalent cation, incorporated into an extended solid, the interaction between the cation and its higher-coordination sphere becomes more distinct and more significant in energy as well, with the advantage of providing a higher degree and quality of information. Some of the following considerations are thought to illustrate this perception.

The two solids with FeF_6^{3-} octahedra in Table 2 possess elpasolite-type structures. The cationic coordination of the anions is also octahedral, according to a *trans*-KFFe grouping, supplemented by four Rb^+- or K^+-cations, respectively

Table 2 Ligand field parameters (in units of 10^{-3} cm^{-1}) of octahedral (o) and tetrahedral (t) Fe^{3+} in various solids

	Δ	*B*	Δ/*B*	Colour	Ref.	Temp. (K)
Rb_2KFeF_6 ᵒ	14.0	0.74	18.9	Pale	[a]	298
K_3FeF_6 ᵒ	13.2	0.76	17.4	Pale	[10][b]	298
$Cs_2NaFeCl_6$ ᵒ	10.7	0.53	20.1	Redd. orange	[11][b]	298; 5
$CsFeCl_4$ ᵗ	≅7.0	0.53	≅13.2	Yellow green	[a]	298; 5
$NH_4FeP_2O_7$ ᵒ and	13.2	0.70	18.9	Pale	[12][b]	298
$Al(Fe)PO_4 \cdot 2H_2O$ ᵒ				Pale	[13][b,c]	298
$Al(Fe)PO_4$ ᵗ and	6.7	0.64	10.5	Pale	[13][b,c]	298
$Ga(Fe)PO_4$ ᵗ				Pale	[13][b,c]	298

[a]This work
[b]In the cited references a deviating adaptation philosophy from the one applied here was used – see Appendix 2, iron(III)
[c]Single crystal data

[14]. The slightly differing *B*-values possibly indicate a smaller iconicity of the iron (III)–fluoride bond in the Rb_2K-solid, which would match with the reduced contrapolarising force of the larger Rb cations, in comparison to K^+, acting upon the fluoride ligands. The indeed present weakening of the Fe–F bond by the "K_5"--contrapolarisation, as compared to a "KRb_4"-contrapolarisation, is mirrored by a nearly 6% decrease of Δ (Table 2). The corresponding higher-energy shifts of the $^6A_{1g} \rightarrow {}_a{}^4T_{1g}$, ${}_a{}^4T_{2g}$ spectral transitions, in which the two effects come out to be additive, are, with about 1,500 cm^{-1}, considerable. In contrast to the pale-shaded fluorides, the elpasolite $Cs_2NaFeCl_6$ possesses a distinct reddish-orange colour, as particularly the minimum around 15,000 cm^{-1} in the spectrum (Fig. 4) suggests. The more covalent chloride ligand causes the steep intensity increase towards the charge-transfer region to occur already very early. The lower-energy electron transfer transitions of Fe^{3+}-doped Cs_2NaYCl_6 in this region have been shown to involve the non-bonding t_{2u}- and the weakly as well as the strongly anti-bonding t_{1u}-molecular orbital sets (see Sect. 3.4). They are observed as nicely resolved bands between 25 and 44,000 cm^{-1} [15]. Because these transitions are not Laporte forbidden, they predominantly model the colour shading by leaving the red-to-orange spectral region open.

The comparison with the remeasured spectra of solids A^IFeCl_4 (Fig. 4) is instructive. There, a *B*-parameter equal to that of $Cs_2NaFeCl_6$ is deduced, when utilising the Δ-independent transitions. On the other hand, the lower-energy bands are shifted to distinctly higher energies in respect to those of $Cs_2NaFeCl_6$, which suggests a drastically reduced Δ-value. The latter indicates tetrahedral Fe^{3+}, though the $\Delta_{tet}/\Delta_{oct}$-ratio is, with 0.66, rather large. The possible reason is the reduced contrapolarisation by only one soft countercation (Cs^+, $N(CH_3)_4{}^+$). The considerable band splitting in the $^6A_1 \rightarrow {}_a{}^4T_1$, ${}_a{}^4T_2$ spectral regions signals markedly distorted tetrahedra (Table 2). Obviously the energy gap between bonding and anti-bonding MOs is significantly larger (see Sect. 3.4) in a T_d- than in the O_h ligand field – the intensity increase towards the charge-transfer region starts only at

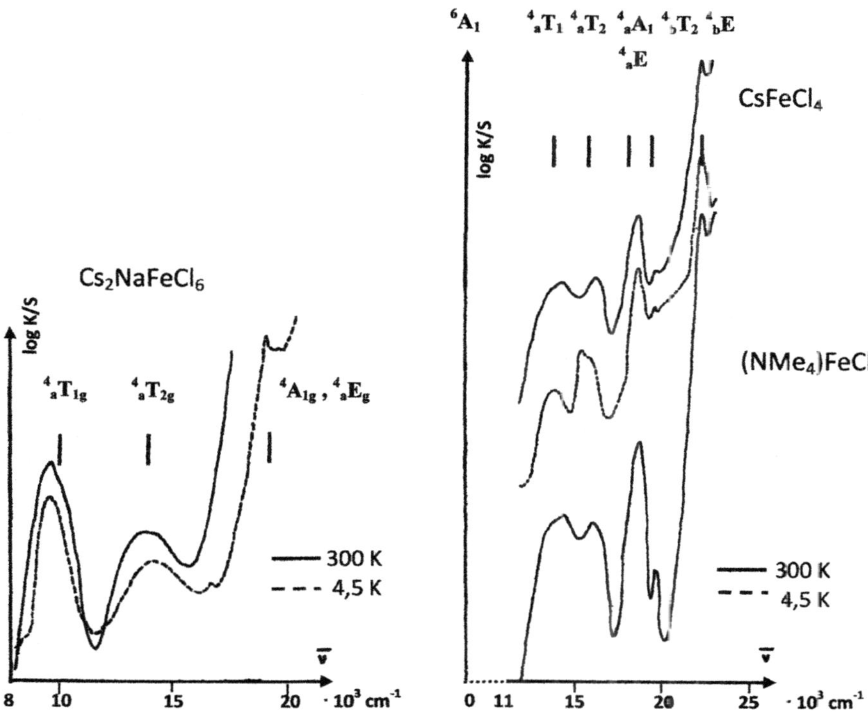

Fig. 4 The ligand field spectra of iron(III) in chloride coordination. (*Left*: octahedral; yielding, with the indicated calculated band positions, Δ and *B*, as listed in Table 2. *Right*: distorted tetrahedral; fitting with the coarse assumption of T_d-symmetry, see text, yielding the parameters in Table 2)

about 21,000 cm^{-1}. On the other hand, the tetrahedral d–d bands are not Laporte forbidden and model the colour shading as well. Both findings match with the observed yellow-green hue, according to minima around 17,000 and 20,000 cm^{-1} (Fig. 4).

In $NH_4FeP_2O_7$ and Fe^{3+}-doped $AlPO_4 \cdot 2\ H_2O$, the ligand field exerted on the $3d^n$-cation is largely governed by the strong contrapolarising effect of six P^V- and of four P^V- and two H^+-centres, respectively, on the oxygen ligands (Table 2). The accordingly highly ionic, but also weakened iron–oxygen bonds are reflected by large *B*-values and comparatively small ligand field parameters near to those of fluoride. Accordingly, charge-transfer-absorptions are only expected in the UV. Because, on the other hand, the Fe^{3+} cations lie isolated in the lattice, a noticeable coloration is not expected. In the Fe^{3+}-doped solids, $Al(Ga)PO_4$, which crystallise in a quartz-type structure with an ordered distribution of the trivalent cations and P(V) over the tetrahedral Si(IV) positions, the contrapolarisation by P (V) plays again a significant role. It modifies the Racah parameter and the ligand field strength, though, due to the reduced coordination number, to a lesser extent. The $\Delta_{tet}/\Delta_{oct}$-ratio is, with 0.51, near to what is expected.

This section is intended to demonstrate by a few examples that solid-state effects from the chemical surrounding beyond the first anionic coordination sphere frequently essentially modify the energy landscape and the colour shading. Effects of this type are already rich in information in the case of divalent 3 d^n-cations [16, 17], but are greatly enhanced, if trivalent cations, such as Cr^{3+} and Fe^{3+}, are involved. We refer here in particular to Sect. 3.3, where one learns that the colour may even reveal intercationic binding effects.

3.2 The Presence of Both Octahedral and Tetrahedral Iron (III) in Oxidic Solids

The ligand field spectra of spinel mixed crystals of the composition $ZnFe_xGa_{2-x}O_4$ (Fig. 5) [12, 18] can be understood as originating from octahedrally coordinated

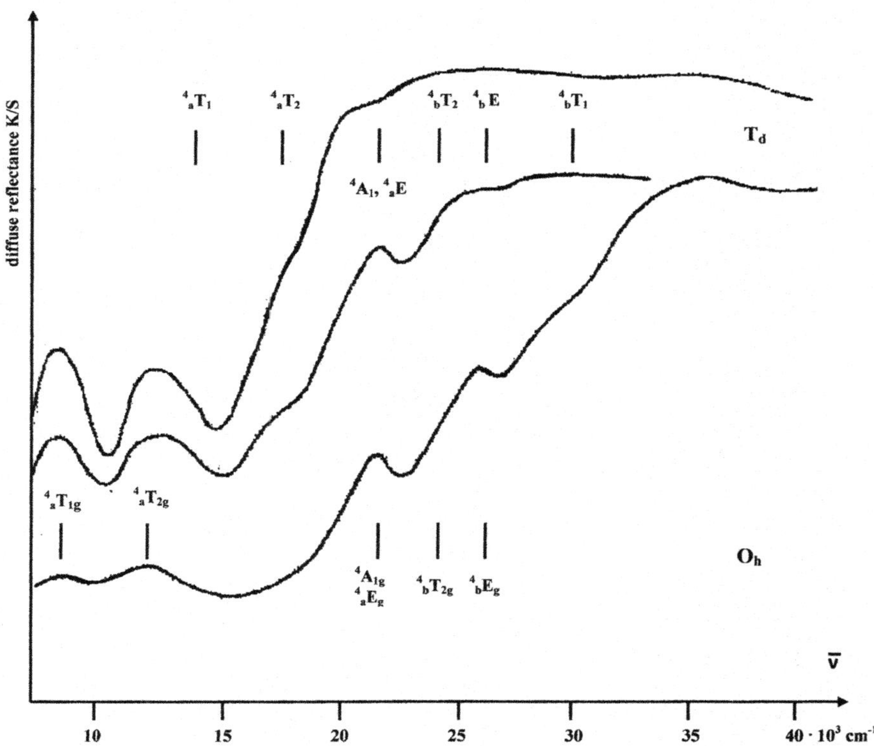

Fig. 5 The ligand field spectra of spinel mixed crystals $ZnGa_{2-x}Fe_xO_4$ (after [12]). The assignment and energy fitting of the, the intensity by far dominating, absorption bands from octahedral iron(III) and of the additional spectral features due to iron(III) in tetrahedral sites has been accomplished with the parameters, listed in Eq. (1). The Fe^{3+} concentration rises from $x = 0.1$ via 0.5 to 2.0, from below

Fe^{3+}-colour centres, with some weak tetrahedral features at about 17.5; 23.5 and 28.5×10^3 cm^{-1} superimposed [12]. The following ligand field parameters are deduced and estimated, respectively, adopting (according to experience) a slightly smaller B-value in T_d than in O_h:

$$\begin{array}{ll} \text{oct.} & \Delta = 15.7,\ B = 0.61 \times 10^3 \text{cm}^{-1};\ \Delta/B = 25.7 \\ \text{tet.} & \Delta \cong 9.2\ ,\ B \cong 0.59 \times 10^3 \text{cm}^{-1};\ \Delta/B \approx 10.7 \end{array} \quad (1)$$

The charge-transfer region has shifted considerably towards lower energies, when inspecting iron(III) in comparison to manganese(II). While at low x-values a gradual intensity rise is observed between 20 and 35×10^3 cm^{-1}, it is steep for $ZnFe_2O_4$ and occurs around 17,000 cm^{-1}, with considerable influence on the coloration (vide infra and Sect. 3.4).

$ZnGa_2O_4$ and $ZnFe_2O_4$ are both described as phases with a normal cation distribution [19]. In agreement, the weak tetrahedral absorption bands in the spectrum indicate – because they are not Laporte forbidden – an only tiny percentage of iron(III) on the T_d-site of the spinel structure, which might indeed (depending on the preparation procedure) still be present. Lenglet et al. proposed [18] that the extra-absorptions originate from electronically coupled pairs of neighboured $Fe^{III}O_6$-octahedra. Following this argument, a closer inspection of the spinel structure is necessary. The basic frame consists of edge-connected octahedra of trigonal D_{3d}-symmetry. The oxygen ligands possess a near-to-tetrahedral cationic coordination of three "octahedral" and one "tetrahedral" cations, the latter lies isolated in the structure and is corner-connected to the octahedral frame (Fig. 6). The octahedra are regular for an oxygen parameter of $u = 3/8$ and trigonally compressed for $u < 3/8$, as here, bringing the cationic oxygen coordination closer towards tetrahedral. Obviously, the four O^{2-}-electron pairs are fully

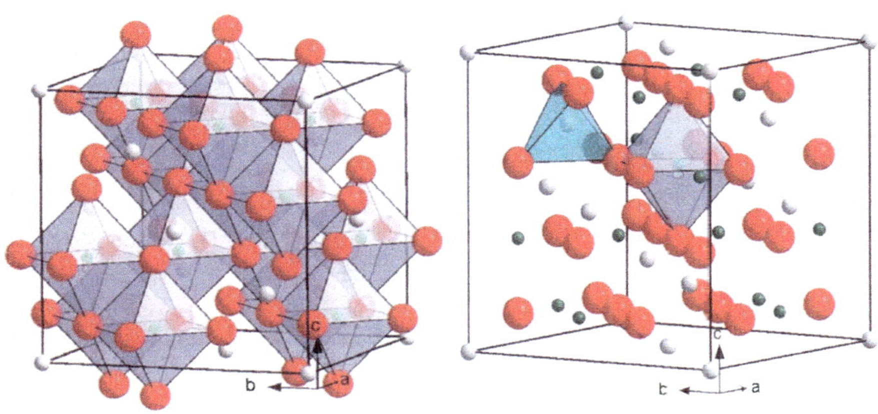

Fig. 6 The polyhedra connections in the spinel structure of $ZnFe_2O_4$ (looked upon as possessing an undisturbed normal cation distribution), with the *white*, *green* and *red spheres* representing Zn^{2+}, Fe^{3+} and oxygen, respectively. The frame of edge-connected octahedra with D_{3d}-symmetry (*left*) and their linking with the isolated tetrahedra in the structure via common corners (*right*)

occupied in σ-bonds (in fair approximation). The t_{2g}-orbitals are left as (nearly) non-bonding and oriented towards the centres of the common edges between two octahedra. They thus possess the correct symmetry for a direct metal–metal overlap (Fig. 6), with the ability to transpose T_{2g}-type electron clouds into a π-anti-bonding status. Exclusively states with this symmetry can shape this interaction in the spinel lattice. We first inspect octahedral Cr^{3+} with a $^4A_{2g}(t_{2g}{}^3)$ ground state in spinel mixed crystals $MgAl_{2-x}Cr_xO_4$ [20, 21], because, in this case, a high sensitivity in regard to phenomena involving t_{2g}-electrons can be expected.

The first absorption band $^4A_{2g} \rightarrow {}^4T_{2g}$ in the ligand field spectrum (Fig. 7, left) directly mirrors Δ, which parameter immediately reflects changes in the binding character of T_{2g}-type electron clouds. Because the phenomenon under examination is a two-centre binding effect, a splitting or striking broadening of the first absorption band is indeed expected in view of the presence of two types of chromium(III) octahedra (isolated and coupled, respectively). Though the differing Δ-values should most likely be visible at x-values around 1.0, a splitting is not detected (Fig. 7, left). This is in contrast to the corundum-type mixed crystals $Al_{2-x}Cr_xO_3$, where a distinct lower-energy shoulder of the $^4A_{2g} \rightarrow {}^4T_{2g}$ absorption develops with

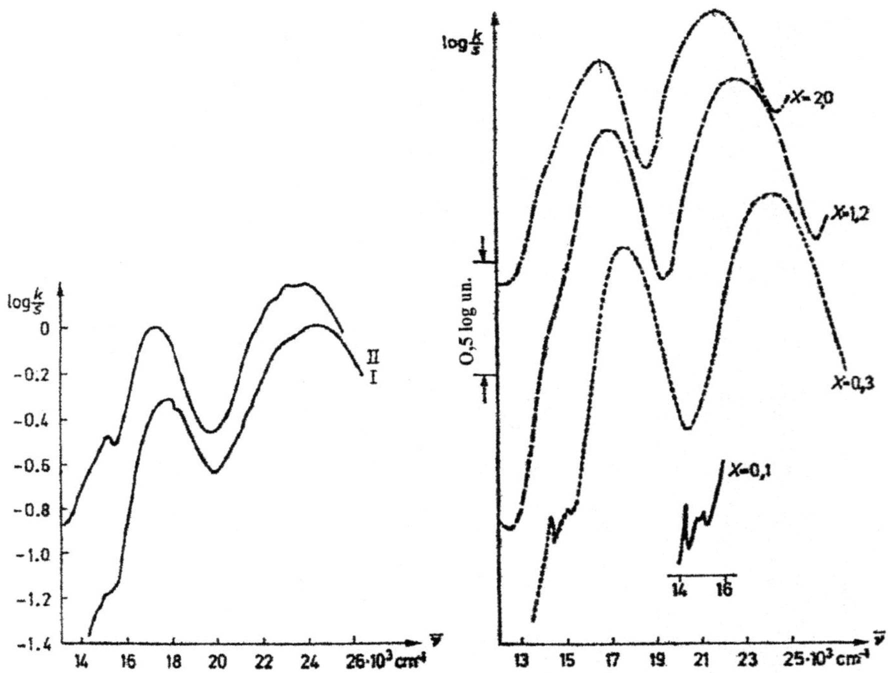

Fig. 7 The powder reflection spectra of octahedrally coordinated chromium(III) in spinel mixed crystals $MgAl_{2-x}Cr_xO_4$ (I and II: $x = 1$ and 2, respectively) (*left*) and in corundum-type phases $Al_{2-x}Cr_xO_3$ (*right*) both adapted from [21]. The two main absorption bands are the transitions $^4A_{2g} \rightarrow {}^4T_{2g}$ at lower and $\rightarrow {}_a{}^4T_{1g}$ at higher energies; the weak sharp features correspond to spin-forbidden, largely Δ-independent absorptions [21]

increasing x. It overlaps and finally dominates the region around 15,000 cm^{-1} in intensity, where also weak spin-forbidden, quartet–doublet transitions appear (Fig. 7, right). We will extend and consolidate these considerations in the following section and return to Fe^{3+} as the colour-generating cation. It is noteworthy that the additional low-intensity absorptions in the spectra of the $ZnGa_{2-x}Fe_xO_4$-mixed crystals (Fig. 5) are not the result of cooperative binding, as proposed by Lenglet et al. [18]. In contrast, distinct evidence for spectral effects of this kind is present, when inspecting iron(III) in the corundum structure. The phenomenon is symmetry restricted, solely referring to the $^6A_{1g} \rightarrow {}^4T_{2g}$ transitions and main subject of the next section.

It is challenging to correlate the above findings with their colour images. While the $MgAl_{2-x}Cr_xO_4$-mixed crystals are pale-pink at low x-values – due to the deep and open gap below the $^4A_{2g} \rightarrow {}^4T_{2g}$ transition at about 18,000 cm^{-1} – they turn to greenish at very high Cr^{3+}-concentrations [20, 21]. There, because of the ionic-size-induced lower-energy shift of the two spin-allowed absorptions, the minimum between, at about 19,500 cm^{-1}, determines the hue. The first absorption band has now widely closed the lower-energy gap (Fig. 7, left). In the case of α-Al_2O_3 as the host, ruby-red is the colour at low and intense-green at higher concentrations above 50% of chromium(III), when replacing Al^{3+} by Cr^{3+} (Fig. 7, right). One might speculate that – without the Cr(III)–Cr(III) interactions – the (bright) green colour of Cr_2O_3 (Fig. 8) would only persist above 60 to 65 mol% of chromium(III) in the mixed crystal phases, due to the then larger apparent Δ-values. Passing to iron(III), $ZnFe_2O_4$ is a light brown and $MgFe_2O_4$ a darker brown pigment (Fig. 8), which hues are coarsely understood when looking at their spectral prints. While the former solid is rather open in the red spectral region, preceding the steep increase towards the charge-transfer region (Fig. 5) [12, 18], the latter spectrum is less contoured not only in its pre-edge absorption [18]. The latter feature is certainly due to the drastically enhanced percentage of iron(III) on T_d-sites, having in mind that $MgFe_2O_4$ is reported to be an (approximately) inverse spinel phase [19].

The garnet structure also offers the possibility, to study the binding and colour properties of iron(III) in octahedral and tetrahedral coordination [22, 23]. These polyhedra constitute a corner-connected frame, in which the tetrahedra lie isolated; additional dodecahedral polyhedra complete the network. The ligands are four-coordinated by one tetrahedral, one octahedral and two dodecahedral cations. We reproduce the ligand field spectra of $Y_3FeGa_4O_{12}$, equilibrated at different temperatures, in Fig. 9, where about 20% of iron(III) is found in T_d-sites [23]. Nevertheless,

Fig. 8 The bright (*brownish*) green hue of Cr_2O_3 (*left*) and the *light* and *dark brown* hues of $ZnFe_2O_4$ and $MgFe_2O_4$, respectively

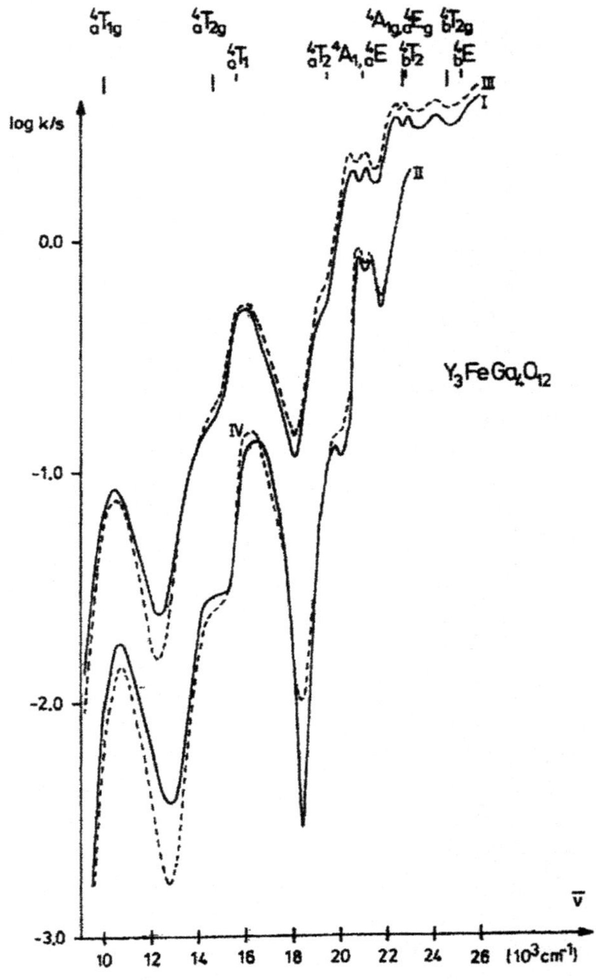

Fig. 9 The powder reflection ligand field spectra of the garnet phase $Y_3FeGa_4O_{12}$, equilibrated at 1,050° (I, II) and 1,350°C (III, IV) and measured at 298 (I, III) and 5 K (II, V), respectively (reproduced from [23]). The intensities of the 298 K curves are shifted to by 0.5 larger log *k*/*s*-values

the latter absorptions dominate the intensity distinctly, as the low-energy $^6A_{1g} \rightarrow {}_a{}^4T_{1g}$, ${}_a{}^4T_{2g}$ in comparison to the $^6A_1 \rightarrow {}_a{}^4T_1$, ${}_a{}^4T_2$-transitions clearly indicate. Though the octahedral and tetrahedral spectral features widely overlap at higher energies, a consistent assignment is possible. The, via the master equations (Eq. 9), derived parameters:

$$\begin{array}{ll} \text{oct.} & \Delta = 15.3, B = 0.65 \times 10^3\text{cm}^{-1} \\ \text{tet.} & \Delta = \ 7.2\ , B = 0.60 \times 10^3\text{cm}^{-1} \end{array} \qquad (2)$$

are consistent in yielding $B(\text{tet.}) < B(\text{oct.})$ and a Δ_T/Δ_o-relation near 0.5 (=0.47).

3.3 *Symmetry and Spectroscopy: Evidence for Metal–Metal Interactions in the Corundum Structure*

In the corundum structure of α-Al_2O_3, there exist pairs of face-connected octahedra: . . .$O_3AlO_3AlO_3$. . .. They impose a local (very near to) C_{3v} distortion on the AlO_6 octahedra, which is induced by the electrostatic repulsion between the two cations via the common face (Fig. 10). While the two asymmetric metal-to-oxygen distance triples differ considerably in length, with 1.97 and 1.86 Å, respectively, the bond length anisotropy is only small in the isostructural Cr_2O_3 (2.02 and 1.97 Å) [24]. Obviously, the σ-electron donor activity of oxygen has become more distinct, in comparison, by the participation of the binding e_g^4 subshell of chromium (III). It lowers the effective charge of the metal cations and reduces the inter-metal repulsion. Both, the charge effect and the decrease of the intercationic distance, mediate a metal–metal orbital overlap, thereby also shaping the ligand field spectrum. As pointed out above, the T_{2g}-electron density distribution is symmetry-adapted to such an inter-metal binding. It is, favourably, (near-to) non-bonding in respect to oxygen, because the four O^{2-}-electron pairs are involved in metal-ligand σ-bonds to their four cationic neighbours. If the threefold axis is chosen as the axis of quantisation (z), as the C_{3v}-symmetry demands, the t_{2g}-orbital set transforms and splits into a_1 and e. a_1, corresponding to d_{z^2}, promotes the direct metal–metal overlap, whilst e represents a non-bonding $d_{x^2-y^2}$, d_{xy}-orbital pair. The e_g-orbitals in O_h remain degenerate in C_{3v} and are involved into the σ-bonding towards the $3d^n$-cation. Figure 10 visualises how the face-connected double-octahedra are linked via common corners in the corundum structure.

From Fig. 11, the energy implications, if Cr^{3+} is substituted into α-Al_2O_3, can be deduced. The extreme situations are the ones where no Cr^{III}–Cr^{III} pairs exist in the

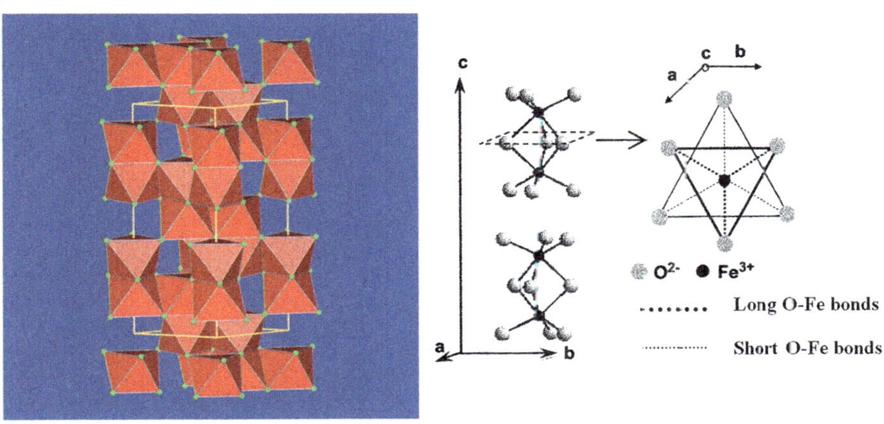

Fig. 10 The corundum structure of α-Al_2O_3, Cr_2O_3 and α-Fe_2O_3, with pairs of face-connected octahedra as the exceptional feature. The bond distance anisotropy due to the repulsion between the two cations along the C_3-axis (‖*c*) is also schematically visualised (*right*, [25])

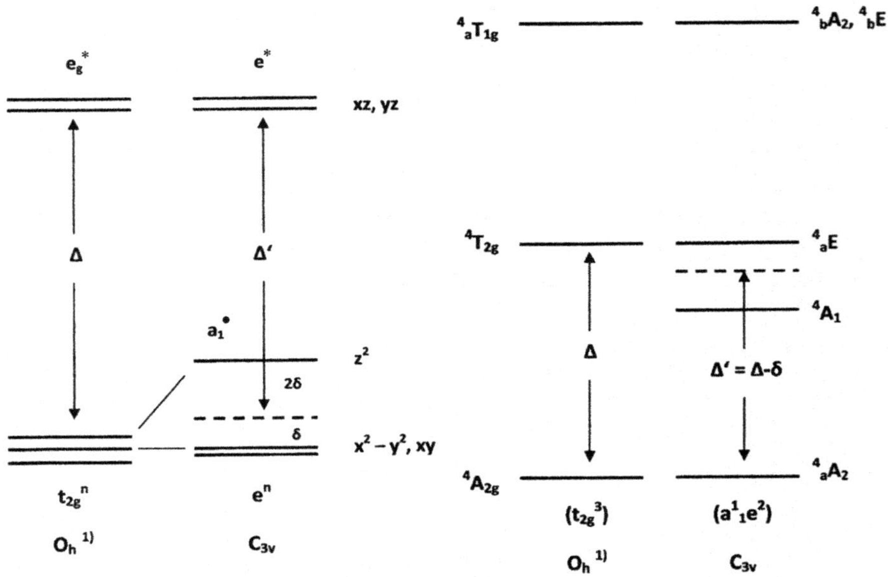

Fig. 11 The energy implications, induced by the possibility of a direct metal–metal d–d-overlap in the corundum structure along the *z*-axis in the C_{3v}-point group. The d-orbital schemes (*left*) and the d^3-state diagrams, referring to $Al_{2-x}Cr_xO_3$-mixed crystals (*right*), sketch the symmetry and energy conditions in C_{3v}. Without inter-metal overlap, only a tiny energy splitting (see: 1)) according to $t_{2g} \rightarrow a_1$, e and $T_{2g} \rightarrow A_1$, E; $T_{1g} \rightarrow A_2$, E is expected. The *superscripts* •, *n* and * stand for metal–metal σ-antibonding, non-bonding and (predominantly) metal-ligand σ-antibonding, respectively

structure (at very low dopings) and that of Cr_2O_3, respectively. However, already without the d–d interactions within the pairs – which, for symmetry reasons, concern the $^4T_{2g}$ state exclusively – band splittings are expected due to the imposed C_{3v} symmetry. They are very small, however, and neglected here, because the bond length anisotropy (1.995 ± 0.025 Å) is tiny. We are now ready to assign and to interpret the split bands of the octahedral $^4A_{2g} \rightarrow {}^4T_{2g}$ transition correctly, according to Fig. 11. The results are summarised in the survey of Eq. (3):

$$\begin{array}{llllr} Al(Ga)_{2-x}Cr_xO_3 & \Delta & B & & \\ x = 0.1 \quad (Al) & 17.8 & 0.64 \times 10^3 cm^{-1} & a & \\ x = 0.07 \quad (Ga) & 16.6 & 0.62 & b & (3) \\ x = 2.0 & 15.8^x & 0.57 & c & \\ (^x\text{split bands at } 14.1 \text{ and } 16.7 \times 10^3 cm^{-1}) & & & & \end{array}$$

The decrease of Δ from *b* to *c* by 5% and of *B* by about 10% reflects the σ-anti-bonding effect on the T_{2g}-electron density due to the metal–metal bonding. The interpretation exceeds and essentially refines the one in [21], which did not consider this interaction. A closer inspection of the symmetry and chemistry landscape, which characterises the d-electron distribution of chromium(III) in the corundum lattice, is found in Appendix 3.

Turning to the binding status of iron(III) in α-Fe_2O_3, there is a significant difference in the bond lengths in respect to α-Cr_2O_3, which widely diverge, with 1.94_5 and 2.11_5 Å (towards the common face) [25] even more than in the case of α-Al_2O_3 (vide supra). The two additional σ-anti-bonding e_g-electrons possibly explain this discrepancy, because their presence reduces the d-orbital donor-strength of oxygen, thus leaving a larger positive effective charge on iron than this was the case for chromium. One might further argue that, in compensation of this charge disbalance, the bond lengths opposite to the common face are considerably shortened. Proceeding to the utilisation of the ligand field spectra of iron(III) in α-Al_2O_3, thereby having the results of chromium(III) in the same host solid in mind, the following symmetry and experimental conditions have to be considered:

1. The theoretical analysis yields that the inclusion of the pair overlap in the C_{3v}-point group affords the redefinition of the parameters Δ and B according to Eq. (4a). Here, δ and analogously b have the meaning as anticipated in Fig. 11, left. Symmetry-underlayed considerations moreover demand a pronounced splitting of the $^4T_{2g}$ states; their A_1-split components occur at energies, characterised by parameters as in Eq. (4b):

$$\Delta' = \Delta - \delta \text{ and } B' = B - b \tag{4a}$$

$$\Delta - 3\delta(\Delta' - 2\delta) \text{ and } B - 3b(B' - 2b) \tag{4b}$$

While the latter $^4A_1(^4T_{2g})$ energy displacements mirror specifically the pair interactions, one should keep in mind that the large bond length anisotropy due to the C_{3v}-polyhedron distortion may induce additional T-state splittings, which are not considered here. Appendix 3 treats and deepens this matter in greater detail.

2. After all, the spectra of mixed crystals $Al_{2-x}Fe_2O_3$, with $x < 2$ (Fig. 12), represent a highly complex binding situation. One constituent involves iron(III) without (Δ, B) and one component with pair interactions (Δ-δ, B-b) – the latter with the additional symmetry condition in Eq. (4b). Though these species overlap with differing contributions, the ligand field spectra are in essential agreement with the above presumptions. The polarised single crystal spectra of Fe^{3+}-doped Al_2O_3 by Lehmann and Harder [26] image the described symmetry and energy conditions according to Eqs. 4a and 4b rather well (Fig. 12, I), as is pointed out subsequently.

The spectral fitting is based on the four parameters, Δ, B, δ and b, which reproduce the energy landscape with three broad, one broad-footed and two neighboured narrow transitions quite well (Table 3). The broad-footed band is resolved into four components, as reported in a later study by Ferguson and Fielding [27]. The latter authors investigated, as also Krebs and Maisch [28] did, single crystals of natural sapphires by optical spectroscopy down to 5K. Increasing the Fe^{3+} concentration shifts the

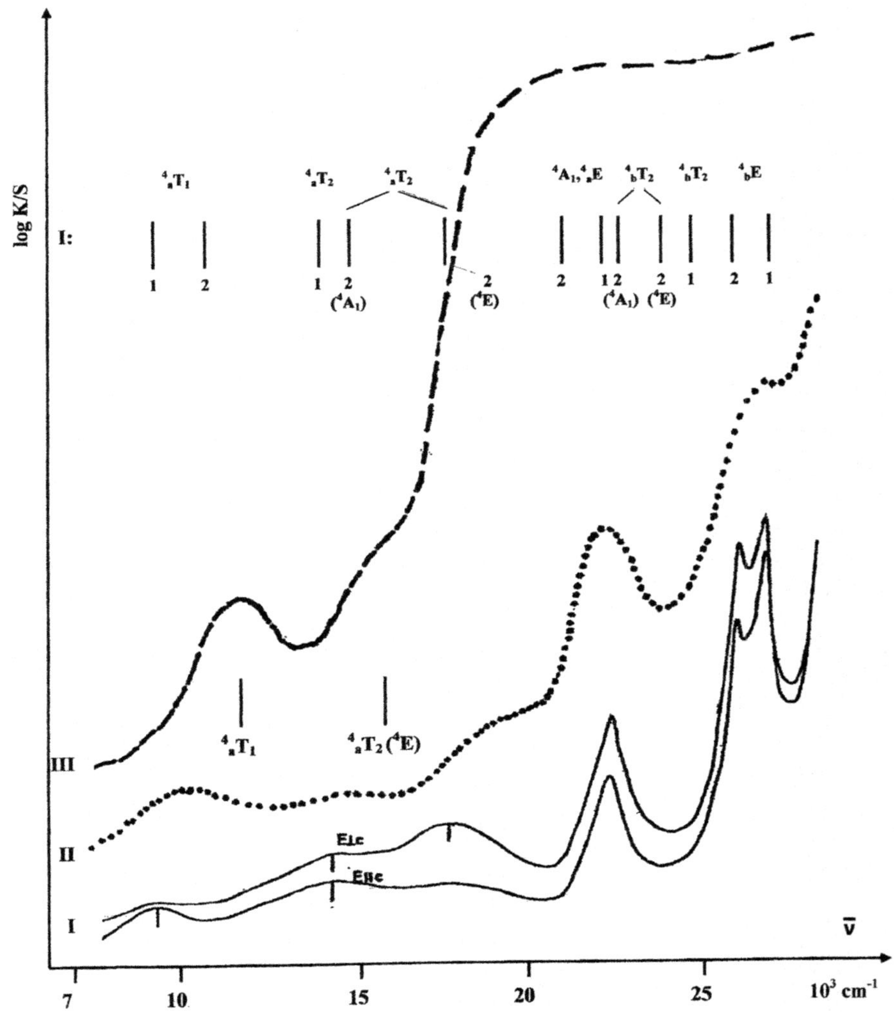

Fig. 12 The ligand field spectra of α-$Al_{2-x}Fe_xO_3$ mixed crystals with x-dopants (I), $x = 0.2$ (II) and $x = 2.0$ (III), respectively. Spectrum I is of polarised single crystal type and reproduced from [26]. Fittings 1 and 2 refer to isolated octahedra and to octahedra with pair interactions, respectively. The derived parameters are collected in Table 3, together with those which were coarsely estimated for α-Fe_2O_3. For the sake of keeping the survey, the indicated state symbols are those of the O-point group; only, where T_2-states are involved, C_{3v}-nomenclature is utilised for the split-states (in *parentheses*)

Δ-dependent transitions to higher energies, this indicating the expected bond length elongation. In the case of α-Fe_2O_3, the fitting of the low-energy $^6A_1 \rightarrow {}_a{}^4A_2$, ${}_a{}^4E$ (${}_a{}^4T_1$)- and $\rightarrow {}_b{}^4E$ (${}_a{}^4T_2$)-broad bands yields, with the tentative choice of

Table 3 The ligand field parameters of α-$Al_{2-x}Fe_xO_3$-mixed crystals at very low iron(III) concentrations, according to their spectral images in Fig. 12, I

(a)	$\Delta = 15.0$, $B = 0.63_5 \times 10^3$ cm^{-1}; $\Delta/B = 23.6$
(b)	$\Delta' = 13.2$, $B' = 0.62 \times 10^3$ cm^{-1}; $\Delta'/B' = 21.3$ with $\delta = 1{,}800$ and $b = 15$ cm^{-1}; the ligand field parameters for the $^4A_1$ ($_a{}^4T_2$)-split state are: $\Delta - 3\delta = 9.6$, $B - 3b = 0.59 \times 10^3$ cm^{-1}

The data refer to isolated octahedra (a) and dimeric octahedra (b) – see the notations 1 and 2, respectively, in Fig. 12, I and also Eqs. (4a and 4b)

$B \cong 600$ cm^{-1} (slightly smaller than the value for B' in Table 3, accounting for the Al(III)-by-Fe(III)substitution), a ligand field parameter Δ of 11,700 cm^{-1} (spectrum III in Fig. 12). It is of reasonable magnitude if compared with 13,200 cm^{-1} for the doped compound. The Δ decrease, which models the transition from isolated to dimeric colour centres and simultaneously the Al(III)-by-Fe(III) substitution, is with $\cong$3,300 cm^{-1}, distinctly larger than the one for the α-$Al_{2-x}Cr_xO_3$-mixed crystals (Eq. 3), where it is 2,000 cm^{-1}. Similar to Lenglet et al. in the spinel case [18], the above-mentioned authors [27, 28] utilised the model of pair excitations in order to understand the appearance of the extra-transition at about 17,500 cm^{-1} (Fig. 12). As has been pointed out above in length, this treatment is not symmetry conform with C_{3v}. Not only the latter excitation but also the sharp double peaks at about 27,000 cm^{-1} and the broadened absorption around 22,000 cm^{-1} find an obvious and near-to-the experiment interpretation, if the here proposed approach is chosen, yielding the ligand field parameters in Table 3. According to the Racah parameter $B = 655$ cm^{-1} and the C/B-ratio of 4.73, derived by Lehmann and Harder [26] from the $^6A_1 \rightarrow {}^4A_1$, $_a{}^4E$- and $\rightarrow {}_b{}^4E$-transition energies, the (practically Δ independent; see Eq. 9) $^6A_1 \rightarrow {}_b{}^4T_2$ absorption is calculated to appear at 24,000 cm^{-1}. This is indeed very far off the chosen assignment of the sharp peak at 25,700 cm^{-1} to this transition, which was adopted in [27, 28] as well.

The spectra of oxidic iron(III) solids are – at high concentrations of the hue-lending cation, where cooperativity within the lattice governs the intensity mechanisms – dominated by a rather steep than gradual intensity increase in the red-to-orange colour space towards the charge-transfer region. Accordingly, reddish-brown colours are observed (e.g., see the haematite samples, in Fig. 13), with various shades due to the specific spectral structure between 14,000 and about 18,000 cm^{-1}. In the next section, we will shortly discuss the origin of the charge-transfer absorption, which characterises oxidic iron(III) solids. Though not directly subject of this contribution, such phenomena essentially control the coloration in very many cases.

Fig. 13 The reddish-brown hues of α-Fe_2O_3 pigments of different particle size (~0.1; ~0.25; ~0.75 μm from *left* to *right*)

3.4 Origin of Colour Due To the Charge-Transfer Absorption of Oxide Pigments

The lowest energy charge-transfer transitions of octahedral iron(III) are those from the approximately non-bonding molecular orbitals t_{1g}, t_{1u}, t_{2u} with predominating ligand character to the π- and σ-anti-bonding t_{2g}* and e_g* MO-sets, which mainly originate from the cationic 3d-orbitals. In a pioneering study [15], based on the spectra of iron(III)-doped Cs_2NaYCl_6, it has been shown that the energy sequence of the lowest spin- and parity-allowed transitions is $t_{1u} \rightarrow t_{2g}$*, $t_{2u} \rightarrow t_{2g}$*, $t_{1u} \rightarrow e_g$* and $t_{2u} \rightarrow e_g$* between 25 and 45×10^3 cm^{-1}. The Δ-parameter of 7.100 cm^{-1}, derived from these numbers (without accounting for interelectronic repulsions via *B*, *C*), is considerably smaller than the one, reported for $Cs_2NaFeCl_6$ in Table 2 (10,700 cm^{-1}) – most probably induced by the large ionic radii difference between Y(III) (0.90) Å and Fe(III) (0.645 Å). The parity-forbidden transition $t_{1g} \rightarrow t_{2g}$* is detected as a peak of very weak intensity at about 22,400 cm^{-1} [15]. The mentioned transitions mirror Fe–Cl bond properties via the anti-bonding 3d-MOs and are a crucial probe for the extent of their participation in the cation-ligand interaction. With this background information in mind, one can define what is called the optical electronegativity χ_{opt} for the cation and the ligand, respectively [29]. These parameters are linked with the energy (wavenumber ν) of the lowest parity-allowed CT-transition via the relation:

$$\nu \cong 30\left(\chi_{\mathrm{opt}}(\mathrm{lig}) - \chi_{\mathrm{opt}}(\mathrm{cat})\right) \times 10^3 \mathrm{cm}^{-1} \tag{5a}$$

$\chi_{\mathrm{opt}}(Fe^{3+})$ is 2.5, and a refinement, when also coarsely accounting for the $d^5 \rightarrow d^6$-specific change of the interelectronic repulsion during the charge-transfer excitation, yields:

$$\nu \cong \left(30\chi_{\mathrm{opt}}(\mathrm{lig}) - 75 + 0.019B\right) \times 10^3 \mathrm{cm}^{-1} \tag{5b}$$

Inspecting Fe^{3+}-doped Cs_2NaYCl_6, one derives, with $\chi_{\mathrm{opt}}(\text{chloride}) = 3.0$ and $B = 530\ \mathrm{cm}^{-1}$ (Table 2), $\nu \cong 25{,}000\ \mathrm{cm}^{-1}$ – very well in accord with the position of the first parity-allowed transition [15]. In the case of $Cs_2NaFeCl_6$ (Fig. 4), however, the steep ascent towards the charge-transfer region occurs already at about 18,000 cm^{-1} (vide infra). If Fe^{3+} is tetrahedrally coordinated, a shift of the charge-transfer bands to higher energies is expected, because particularly the effective charge on the cation less distinctly deviates from 3 than in an octahedral ligand field. Indeed, the steep increase towards the first CT-transition of the solids A^IFeCl_4 (Fig. 4) is observed only at about 22,000 cm^{-1}. A critical evaluation of Eqs. 5a and 5b in respect to experimental and DFT-calculated charge-transfer energies – for trivalent $3d^n$-cations in octahedral chloride coordination – has been given elsewhere [11]. It is expected that χ_{opt} widely varies for oxygen as the ligand; this is so because the contrapolarising influence of its cationic neighbours significantly models the quality of the cation–oxygen bond [16], particularly if a higher-valent cation such as iron(III) is considered. Here, $\chi_{\mathrm{opt}}(\text{oxygen})$ varies from 3.7 or 3.5, if S(VI) or P(V) and Si(IV) are the contrapolarising species, via 3.0 for Ca^{2+} to 2.2 for Cs^+, which represents the softest cation in the scaling [29]. The general statement is that the ligand-to-metal charge-transfer energies decrease, the more covalent the iron(III)-oxygen bond becomes (small $\chi_{\mathrm{opt}}(\text{oxygen})$- and B-values). Applying these considerations to the garnet $Y_3Ga_4FeO_{12}$ (Fig. 9) and the spinel mixed crystals $ZnGa_{2-x}Fe_xO_4$ (Fig. 5), the first charge-transfer band is calculated (Eq. 5b) to appear at 33 and $30 \times 10^3\ \mathrm{cm}^{-1}$, respectively, if χ_{opt} is suggested to be $\cong 3.2$ and 3.1, here also following the magnitudes of B, with 650 cm^{-1} for the garnet and 610 cm^{-1} for the spinels. The reality is different, with an only vague correspondence; the intensity increase towards the first charge-transfer band is estimated to set in at $\approx$24,000 cm^{-1} in the former and at about 17,000 cm^{-1} in the latter case. One crucial reason is surely, that covalent-type cooperative interactions between the iron(III) cations in the solid matrix come in, with the tendency to constitute a broad 3d-band. Particularly the spectra of the spinel mixed crystals – this structure favours intercationic interactions via the edge-connected octahedra – display respective features. Whilst at low iron(III) concentrations the intensity increase towards the charge-transfer region extends over a wide energy range between about 22 and 33,000 cm^{-1}, it is steep in the case of $ZnFe_2O_4$ – between 16 and 20,000 cm^{-1} (Fig. 5). Striking is the extent of this shift in the spectra of the $Fe_xAl_{2-x}O_3$-mixed crystal series, from 29,000 cm^{-1} in the case of the doped solid [26] to about 16,000 cm^{-1} for α-Fe_2O_3 (Fig. 12). The phenomenon suggests a pronounced cooperativity effect with the formation of an expanded anti-bonding d-band, which is broader than that for the spinel phases. The very distinct intercationic interactions (direct and via oxygen bridging), indicated in the spectrum, are modulated by rather intense lower-energy ligand field bands. The latter give rise to a minimum in the red and a pronounced shoulder in the orange spectral region. It is mainly the transformation of the local anti-bonding 3d-states into a broad band, however, which generates the typical brown-shaded hues (Fig. 13), omnipresent in nature. There is no complete mixed crystals series $Al_{2-x}Fe_xO_3$, because the ionic radii difference of the two cations is, with 0.11 Å, rather large. We conclude by mentioning a recent study, which investigates via density functional calculations [30] the influence of a substitution of iron(III) in α-Fe_2O_3 by various 3d-cations on the band structure.

4 Concluding Remarks

Crystallographic investigations yield the basic information on the structure of a solid, which characterizes the geometric details and the symmetry conditions within an extended lattice. Probe methods, such as ligand field spectroscopy, possess the potential to refine such results by providing additional insight in respect to specifically the probe-cation within its chemical environment. One can even analyse such properties and their dependence on the concentrations, in order to further learn about intercationic interactions. This information, in the context of this contribution, supplies detailed knowledge about the participation of the 3d-electrons in the bulk iron(III)-ligand bond. One learns about the various interactions of the probe-cation with its higher-sphere environment, which may reveal subtle qualities of the chemical bonding beyond the direct anionic neighbours. Interestingly enough, this detail knowledge comes from inspecting electronic excitations within the 3d-shell. Thus, although enclosing excited energy landscapes, the Franck–Condon transitions provide valid information via the projection of excited-state properties onto the ground state. A fascinating story is that the shortly sketched abstract knowledge is perceived and translated by the human senses into the colour variety of the optical spectra. An early book publication [31] internalizes the ambiguity of colours as being both, an emotional phenomenon and an abstract scientific event. Indeed, finer changes in the symmetry and structure of, say, an oxidic iron(III) pigment find their equivalent in specific shades of hue, which are aesthetic qualities with emotional implications. The $Fe_xAl_{2-x}O_3$-mixed crystals are particularly interesting. Here, even binding peculiarities, such as the pair formation via a direct metal–metal interaction, are manifested by the presence of additional symmetry-dictated absorption bands. The very often observed brown tinges of iron(III) oxide-type solids are usually caused by spectroscopic charge-transfer phenomena, which accompany the changeover from isolated 3d states towards an anti-bonding 3d-band. In conclusion, the authors think that the theoretical and experimental background knowledge about the colour of oxide iron(III) pigments is detailed enough to allow a predictable manipulation of the hue.

Acknowledgement The technical assistance by Mrs. D. Kloss and Mrs. M. Köhler in transforming the manuscript and the figures, respectively, into a publishable shape is gratefully recognised.

Appendix 1: The Crystal Field Treatment of a d^5-Cation in a Cubic Environment

The state energies of a free d^5-cation, in relation to its high-spin ^6S-ground state, are, in terms of the Racah parameters B_0 and C_0 which parameterize the interelectronic repulsion [2]:

$$\begin{array}{ll} \rightarrow {}^4\mathrm{G}\ \ 10B_0 + 5\,C_0; & \rightarrow {}^4\mathrm{P}\ \ 7\,B_0 + 7C_0; \\ \rightarrow {}^4\mathrm{D}\ \ 17B_0 + 5C_0; & \rightarrow {}^4\mathrm{F}\ \ 22\,B_0 + 7C_0 \end{array} \tag{6}$$

These states transform and split, by symmetry, in an octahedral crystal field into the components: $^4A_{1g}$, ^4E_g, $^4T_{1g}$, $^4T_{2g}$; $^4T_{1g}$; ^4E_g, $^4T_{2g}$; $^4A_{2g}$, $^4T_{1g}$, $^4T_{2g}$ – where $^6A_{1g}$ is the (high-spin) ground state. The energies are (see also Eq. 6) [2]:

$$\begin{array}{ll} {}^4\mathrm{A}_{1g}\ \ 10B + 5C; & {}^4\mathrm{E}_g\ \ 10B + 5C \\ {}^4\mathrm{A}_{2g}\ \ 22B + 7C; & \qquad 17B + 5C \end{array} \tag{7}$$

$${}^4\mathrm{T}_{1g}\begin{vmatrix} -\Delta - E & 3B\sqrt{2} & -C \\ 3B\sqrt{2} & 9B + C - E & -3B\sqrt{2} \\ -C & -3B\sqrt{2} & \Delta - E \end{vmatrix} = 0, \quad \text{with } 10B + 6\,C \text{ to be added to the three solutions}$$

$${}^4\mathrm{T}_{2g}\begin{vmatrix} -\Delta - E & -\sqrt{6}B & -(4B + C) \\ -\sqrt{6}B & -5B - C - E & -\sqrt{6}B \\ -(4B + C) & -\sqrt{6}B & \Delta - E \end{vmatrix} = 0, \quad \text{with diagonal additions of } 18B + 6C$$

Here, the ligand field parameter Δ and the modified Racah parameters B and C account for the energy modifications due to placing the cation into an external crystal field. Accordingly, ten $S = 5/2$ – to – $S = 3/2$-transitions may be observed at the most, from which one ($^4A_{2g}$) and those with a positive Δ dependence (${}_c{}^4T_{1g}$, ${}_c{}^4T_{2g}$) are expected only at very high energies; furthermore, two ($^4A_{1g}$, ${}_a{}^4E_g$) are degenerate in energy. The ^4T_g-matrices indicate that – according to a (near-to) $(-\Delta)$- or $(+\Delta)$-dependence – the spin pairing may occur in the t_{2g}-($t_{2g}{}^4e_g{}^1$) or in the e_g-subshell($t_{2g}{}^2e_g{}^3$), respectively, while the ${}_b{}^4T_{1g}$- and ${}_b{}^4T_{2g}$-transitions are expected to depend only weakly on Δ. The specific symmetry of the two ^4T_g matrices in respect to Δ is the reason why the Eq. (7) can be used in the case of tetrahedral crystal fields as well. Though usually a change of sign due to the reverse splitting into a lower-energy e- and a higher-energy t_2-orbital set in T_d as compared to O_h has to be applied, this is needless here.

Appendix 2: A Critical Review Towards the Numerical Instruments of Ligand Field Theory in Respect to Mn^{2+} and Fe^{3+} in O_h- and T_d-Coordination

As has been discussed in detail elsewhere [7], spectral data for the free manganese (II)-cation [32], specifically the electronic transitions from ^6S to ^4G, ^4P, ^4D and ^4F, can be rather precisely fitted to the Racah parameters B_0 and C_0, with numerical values of $B_0 = 750$ cm^{-1} and $C_0/B_0 = 5.2$ – though with the exception of the $^6S \rightarrow {}^4P$-transition, which is calculated to occur at a, by about 10% too low energy. Thus – refraining from the doubtful-debatable procedure to introduce one additional parameter in order to achieve an improved fit to the experiment (Trees correction, see [7]) – one may confidently use the crystal field energies of Eq. (7) in Appendix 1 to parameterize experimental data, thereby leaving aside the $^6A_{1g} \rightarrow {}^4T_{1g}$-transitions in the numerical calculation, however, for obvious reasons. The following master energy equations (utilising the $B/C = B_0/C_0$-ratio of 5.2 and referring to Δ/B-ratios in the range of those, deduced for oxidic manganese(II)-solids in T_d(left) and O_h(right) crystal fields) are easily derived – restricting to the six lowest energy transitions [7]:

$$^6A_1 \rightarrow {}^4A_1,\ {}_a{}^4E \quad 36B \qquad {}_b{}^4E \quad 43B \tag{8}$$

and, for $\Delta/B \cong 7$ (left) and $\cong 14$ (right), respectively:

$$\begin{array}{lll} ^6A_1 \rightarrow \ {}_a{}^4T_1 & -0.84\Delta + 38.0B & -0.94\Delta + 39.1B \\ \quad\quad\ \ {}_a{}^4T_2 & -0.34\Delta + 37.3B & -0.70\Delta + 41.0B \\ \quad\quad\ \ {}_b{}^4T_2 & -0.28\Delta + 43.5B & -0.14\Delta + 42.0B \\ \quad\quad\ \ {}_b{}^4T_1 & +0.68\Delta + 42.0B & +0.48\Delta + 44.5B \end{array}$$

After all, one is in the comfortable situation, that there are frequently four experimental data ($^6A_1 \rightarrow {}^4A_1$, ${}_a{}^4E$; ${}_b{}^4E$; ${}_a{}^4T_2$; ${}_b{}^4T_2$), from which only two unknown parameters have to be deduced. The interelectronic repulsion parameter B can be derived from the Δ-independent transitions, where the minimum positions of the ground- and excited-state potential curves coincide. These minima are displaced in respect to each other, if Δ is involved. Though the Stoke shift, which is the energy difference between a transition in absorption and emission, is rather small in the case of a divalent cation such as Mn^{2+}, the derived Δ parameter is nevertheless – though modestly – semi-empirical in nature. It is hence expected that – within the discussed limits – the nephelauxetic ratio $\beta = B/B_0$ images the covalency properties near to reality and that also Δ is a parameter, which reflects changes in the binding quality in a sensitive manner, if one sticks to the outlined critical calculation procedure. Having this in mind, both parameters, B and Δ, constitute as reliable members of the nephelauxetic and spectrochemical series of ligands [6].

Iron(III) differs from Mn(II) in its binding properties, by the distinctly enhanced covalent overlap between the metal and the ligands, and demands to classify the t_{2g}-

and e_g-electrons in O_h according to their π- and σ-anti-bonding character, respectively; similarly, the e- and t_2-electrons in T_d are π-anti-bonding in the former and intermixed between σ- and π-anti-bonding in the latter MOs. Accordingly, the Racah parameters differ, depending on whether they refer to t_2- or e-electrons. For example, in the case of Cr^{3+} in the octahedral coordination of fluoride elpasolites, B varies considerably between 790, ≅685 and ≅520 cm^{-1}, for a $t_{2g}{}^3$-, $t_{2g}{}^2e_g{}^1$- and $t_{2g}{}^1e_g{}^2$-configuration, respectively [11]. Having covalence phenomena of this type in mind, a coarser approach than the one introduced for Mn^{2+} seems appropriate – and this with the clear intention to not overload the treatment with too many not sufficiently meaningful parameters. We will accordingly follow an approach, in which only a singular B-value is used and in which also the $^6A_1 \rightarrow {}^4T_1$-transitions are included into the fitting procedure, thereby utilising a C/B-ratio of 5.0. Though well defined, this approach certainly implies a larger uncertainty in the numerical fixation of the ligand field parameters. Nevertheless, "quod est demonstrandum", rather satisfactory fits to the iron(III)-ligand field spectra are possible, with consistent interpretations of the derived binding parameters Δ and B. The basic demand is, as always, the assistance of an appropriate symmetry classification. As for Mn^{2+}, we use in the following master equations for the transition energies of iron(III) in O_h- and T_d-ligand fields at (for the Δ-dependent absorption bands) Δ/B-ratios, which are particularly applicable to the oxidic iron(III) solids, considered here, with $C/B = 5.0$:

$$^6A_1 \rightarrow {}^4A_1,\ {}_a{}^4E\ :\ 35B \qquad \rightarrow {}_b{}^aE\ :\ 42B \tag{9}$$

and, for $\Delta/B \cong 12$ (left) and 20 (right), respectively:

$$\begin{array}{llll} ^6A_1 \rightarrow & {}_a{}^4T_1 & -0.93\Delta + 37.7B & -0.97\Delta + 38.3B \\ & {}_a{}^4T_2 & -0.77\Delta + 41.7B & -0.90\Delta + 43.7B \\ & {}_b{}^4T_2 & -0.02\Delta + 38.3B & -0.01\Delta + 38.1B \\ & {}_b{}^4T_1 & +0.57\Delta + 41.8B & +0.23\Delta + 47.2B \end{array}$$

Appendix 3: The Symmetry Relations and the d-Electron Density Distribution of Chromium(III) and Iron(III) in the Corundum Structure

If one marks the t_{2g}- and e_g-wave functions in O_h with ζ, η, ξ and Θ, ε, respectively, the following linear combinations result, if the axis of quantisation is now C_3 instead of C_4 (see column C_{3v} (1)):

$$
\begin{array}{llllll}
O_h & & C_{3v}(1) & & C_{3v}(2) & \\
t_{2g} & \zeta & 1/\sqrt{3}(\zeta+\eta+\xi) & a_1 & d_z{}^2 & \\
 & \eta & 1/\sqrt{6}(2\zeta-\eta-\xi) & e & d_{x^2-y^2} & \\
 & \xi & 1/\sqrt{2}(\eta-\xi) & & d_{xy} & \\
e_g & \Theta & \Theta & e & d_{xz} & \\
 & \varepsilon & \varepsilon & & d_{yz} &
\end{array}
\tag{10}
$$

Adopting the C_3- as the new molecular *z*-axis, the d-wave functions transform as listed under C_{3v} (2); the latter are the ones used in the main text and in Fig. 11. According to the presence of the threefold axis, any cyclic permutation of the wave functions under C_{3v} (1) is also a valid choice. The $d_z{}^2$-wave function models the metal pair interaction in the corundum lattice via a_1. The two e-doublets come out to be non-bonding (from t_{2g}) and σ-anti-bonding (from e_g), respectively, if otherwise the octahedral symmetry is retained – as this is nearly so in Cr_2O_3. In difference, the structural C_{3v} distortion is very significant in α-Fe_2O_3, and the two e-orbital sets are expected to intermix. The distance between the two, displaced, octahedral cations within the $O_3MO_3MO_3$-binuclear pairs (Fig. 10) along the C_3-axis is about 2.12, 2.17 and 2.27 Å for M being Al(III), Cr(III) and Fe(III), respectively. These values coarsely follow the sequence of the octahedral ionic radii, which are reported to be [33, 34] 0.53_5, 0.61_5 and 0.64_5 Å, only in the case of the two 3d-cations. For Al(III) this spacing is about 0.1 Å larger, which becomes apparent, if one corrects the internuclear distances in respect to the ionic radii differences. This observation indeed matches with the lack of intercationic binding in that case. Focussing on Cr (III) and Fe(III) accordingly, one further notes that the bond lengths towards the oxygen ligands in the boundary faces of the binuclear pairs (1.97 and 1.94_5 Å, respectively) indicate a by nearly 3% (again radii-corrected) reduced effective value in the latter case. Recalling the spectral results in addition, one has to conclude as follows: The distinct C_{3v}- structural component, considering α-Fe_2O_3, has introduced, via the $e(t_{2g}) - e(e_g)$-interaction (Fig. 11, left), particularly a more pronounced t_{2g}-splitting; this energy effect is nicely imaged via the by a factor of two enhanced parameter δ (see Eq. (3) and Table 3 in Sect. 3.3), and attended by the conversion of the two symmetry-equivalent orbital doublets into linear combinations (see Eqs (10)). Also the derived ligand field parameters (Δ, in 10^3 cm^{-1}) for the here and in [21] discussed spinel-, garnet- and corundum-type phases indicate the exceptional position of α-Fe_2O_3:

Compound	Δ	B
$ZnCr_2O_4$	17.4	0.64
$ZnFe_2O_4$	15.7	0.65
$Y_3CrGa_4O_{12}$	16.1_5	0.63
$Y_3FeGa_4O_{12}$	15.3	0.65
α-Cr_2O_3	15.8	0.57
α-Fe_2O_3	$\cong 11.7$	$\cong 0.60$

(11)

It is, induced by the dramatic rise of the intercationic overlap parameter δ, the pronounced decrease of Δ, which mirrors the binding landscape specifically in the α-$Fe_xAl_{2-x}O_3$-mixed crystals (see Table 3). Δ becomes smaller by at least 20%, when replacing chromium(III) by iron(III) in the corundum structure, in comparison to 10% or even less in spinel- or garnet-type phases. Considering the above outlined model arguments, this is an expected experimental result. Already discussed is the reduction of B (in 10^3 cm^{-1}, see Eq. 11) by about 10%, which is observed, if the within-pair overlap comes up in addition (Table 3). We think that – though the iron(III) bonding properties in the corundum lattice are highly complex – the proposed model is an acceptable symmetry-adjusted approach to what one may call the truth.

References

1. Ilse FE, Hartmann H (1951) Z Phys Chem 197:239
2. Griffith JS (1971) The theory of transition metal ions. Cambridge University Press, London
3. Figgis BN, Hitchman MA (2000) Ligand field theory and its applications. Wiley-VCH, New York
4. Lever ABP (1984) Inorganic electronic spectroscopy. Elsevier, Amsterdam
5. Jorgensen CK (1966) Struct Bond 1:1
6. Jorgensen CK (1968) Oxidation numbers and oxidation states. Springer, New York
7. Reinen D, Schwab G, Günzler V (1984) Z Anorg Allg Chem 515:140
8. Stout JW (1959) J Chem Phys 31:709
9. Srivastava JP, Mehra A (1972) J Chem Phys 57:1587
10. Allen GC, Warren KD (1971) Struct Bond 9:49
11. Reinen D, Atanasov M, Köhler P, Babel D (2010) Coord Chem Rev 254:2703
12. Reinen D (2014) Z Anorg Allg Chem 640:2677 (The author apologises for misprints in Fig 2: $_b{}^4E_g$, $_b{}^4T_{1g}$ should read as $_b{}^4T_{2g}$, $_b{}^4E_g$, respectively)
13. Lehmann G (1970) Z Phys Chem 72:279
14. Babel D (1967) Struct Bond 3:1
15. Neuenschwander K, Güdel HU, Collingwood JC, Schatz PN (1983) Inorg Chem 22:1712
16. Reinen D, Atanasov M, Lee SL (1998) Coord Chem Rev 175:91
17. Atanasov M, Reinen D (2003) Comprehensive coordination chemistry, vol I fundamental. Elsevier, London, p 669
18. Lenglet M, Bizi M, Jorgensen CK (1990) J Solid State Chem 86:82
19. Weiss A, Witte H (1983) Kristallstruktur und Chemische Bindung. Verlag Chemie, Weinheim
20. Schmitz-DuMont O, Reinen D (1960) Z Elektrochem Ber Bunsenges Phys Chem 64:330
21. Reinen D (1969) Struct Bond 6:30
22. Köhler P, Amthauer G (1979) J Solid State Chem 28:329

23. Amthauer G, Günzler V, Hafner SS, Reinen D (1982) Z Kristallogr 161:167
24. Moss SC, Newnham RE (1964) Z Kristallogr 120:359
25. Pailhé N, Wattiaux A, Gaudon M, Demourgues A (2010) J Solid State Chem. doi:10.1016/j.jssc.2010.04.043
26. Lehmann G, Harder H (1970) Am Mineral 55:98
27. Ferguson J, Fielding PE (1972) Aust J Chem 25:1371
28. Krebs JJ, Maisch WG (1971) Phys Rev B 4:757
29. Duffy JA (1990) Bonding, energy levels and bands in inorganic solids. Longman, London
30. Huda MN, Walsh A, Yanta Y, Wei SH, Al-Jassim MM (2010) J Appl Phys 107:123712
31. Nassau K (1983) The Physics and Chemistry of Color. Wiley (see particularly part III)
32. Moore CE (1952) Atomic energy levels, Vol II Nat Bur Stand Circ 467
33. Shannon RD, Prewitt CT (1969) Acta Crystallogr B 25:925
34. Shannon RD (1976) Acta Crystallogr A 32:751

Struct Bond (2016) 172: 341–342
DOI: 10.1007/430_2016_213

Published online: 10 August 2016

Erratum to: Synthetic Bioinorganic Chemistry: Scorpionates Turn 50

Daniel Rabinovich

Erratum to: Struct Bond
DOI: 10.1007/430_2015_212

Figure 10 of this chapter was inadvertently displayed incorrect. The correct form of the figure should be displayed as follows:

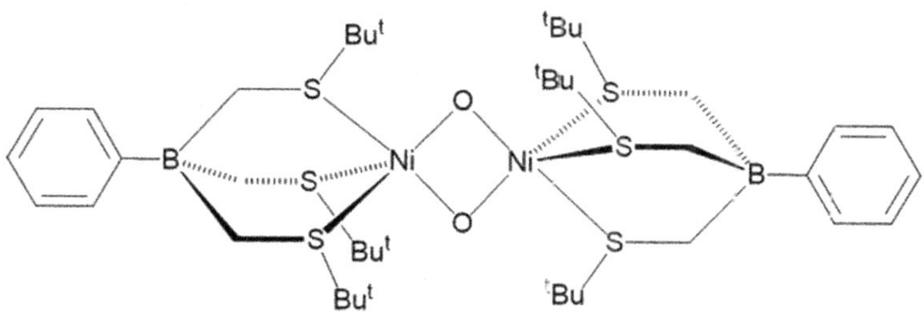

Fig. 10 Dinuclear Ni(III) complex with a $Ni_2(\mu\text{-}O)_2$ core

---

The updated online version of the original chapter can be found under DOI 10.1007/430_2015_212

D. Rabinovich (✉)
Department of Chemistry, The University of North Carolina at Charlotte, 9201 University City Boulevard, Charlotte, NC 28223, USA
e-mail: drabinov@uncc.edu

Index

CPSIA information can be obtained
at www.ICGtesting.com
Printed in the USA
LVOW02*2205211016
509756LV00003B/57/P